Statistics, Data Mining, and Machine Learning in Astronomy

PRINCETON SERIES IN MODERN OBSERVATIONAL ASTRONOMY

David N. Spergel, SERIES EDITOR

Written by some of the world's leading astronomers, the Princeton Series in Modern Observational Astronomy addresses the needs and interests of current and future professional astronomers. International in scope, the series includes cutting-edge monographs and textbooks on topics generally falling under the categories of wavelength, observational techniques and instrumentation, and observational objects from a multiwavelength perspective.

Statistics, Data Mining, and Machine Learning in Astronomy

A PRACTICAL PYTHON GUIDE FOR THE ANALYSIS OF SURVEY DATA

Željko Ivezić, Andrew J. Connolly,
Jacob T. VanderPlas, and Alexander Gray

PRINCETON UNIVERSITY PRESS • PRINCETON AND OXFORD

Published by Princeton University Press, 41 William Street,
Princeton, New Jersey 08540

In the United Kingdom: Princeton University Press, 6 Oxford Street,
Woodstock, Oxfordshire OX20 1TW

press.princeton.edu

ISBN 978-0-691-15168-7

Library of Congress Control Number: 2013951369

British Library Cataloging-in-Publication Data is available

This book has been composed in Minion Pro w/ Universe light condensed for display
Printed on acid-free paper ∞

Typeset by S R Nova Pvt Ltd, Bangalore, India
Printed in the United States of America

10 9 8 7 6 5 4 3 2 1

Contents

Preface

Astronomy and astrophysics are witnessing dramatic increases in data volume as detectors, telescopes, and computers become ever more powerful. During the last decade, sky surveys across the electromagnetic spectrum have collected hundreds of terabytes of astronomical data for hundreds of millions of sources. Over the next decade, data volumes will enter the petabyte domain, and provide accurate measurements for billions of sources. Astronomy and physics students are not traditionally trained to handle such voluminous and complex data sets. Furthermore, standard analysis methods employed in astronomy often lag far behind the rapid progress in statistics and computer science. The main purpose of this book is to help minimize the time it takes a student to become an effective researcher.

This book provides the interface between astronomical data analysis problems and modern statistical methods. It is aimed at physical and data-centric scientists who have an understanding of the science drivers for analyzing large data sets but may not be aware of developments in statistical techniques over the last decade. The book targets researchers who want to use existing methods for the analysis of large data sets, rather than those interested in the development of new methods. Theoretical discussions are limited to the minimum required to understand the algorithms. Nevertheless, extensive and detailed references to relevant specialist literature are interspersed throughout the book.

We present an example-driven compendium of modern statistical and data mining methods, together with carefully chosen examples based on real modern data sets, and of current astronomical applications that will illustrate each method introduced in the book. The book is loosely organized by practical analysis problems, and offers a comparative analysis of different techniques, including discussions of the advantages and shortcomings of each method, and their scaling with the sample size. The exposition of the material is supported by appropriate publicly available Python code (available from the book website, rather than fully printed here) and data to enable a reader to reproduce all the figures and examples, evaluate the techniques, and adapt them to their own field of interest. To some extent, this book is an analog of the well-known *Numerical Recipes* book, but aimed at the analysis of massive astronomical data sets, with more emphasis on modern tools for data mining and machine learning, and with freely available code.

From the start, we desired to create a book which, in the spirit of reproducible research, would allow readers to easily replicate the analysis behind every example and figure. We believe this feature will make the book uniquely valuable as a practical guide. We chose to implement this using Python, a powerful and flexible programming language that is quickly becoming a standard in astronomy (a number of next-generation large astronomical surveys and projects use Python, e.g., JVLA, ALMA, LSST). The Python code base associated with this book, called AstroML, is maintained as a live web repository (GitHub), and is intended to be a growing collection of well-documented and well-tested tools for astronomical research. Any astronomical researcher who is currently developing software for the analysis of massive survey data is invited and encouraged to contribute their own tools to the code.

The target audience for this text includes senior undergraduate and graduate students in physics and astronomy, as well as researchers using large data sets in a scientific context. Familiarity with calculus and other basic mathematical techniques is assumed, but no extensive prior knowledge in statistics is required (e.g., we assume that readers have heard of the Gaussian distribution, but not necessarily of the Lorentzian distribution). Though the examples in this book are aimed at researchers in the fields of astronomy and astrophysics, the organization of the book allows for easy mapping of relevant algorithms to problems from other fields. After the first introductory chapter, data organization and some aspects of fast computation are discussed in chapter 2, statistical foundations are reviewed in chapters 3–5 (statistical distributions, maximum likelihood and other classical statistics, and Bayesian methodology), exploratory data analysis is described in chapters 6 and 7 ("Searching for Structure in Point Data"; "Dimensionality and its Reduction"), and data-based prediction methods are described in chapters 8–10 ("Regression and Model Fitting"; "Classification"; "Time Series Analysis").

Finally, we are indebted to a number of colleagues whose careful reading and resulting comments significantly improved this book. A summer study group consisting of Bob Abel, Yusra AlSayyad, Lauren Anderson, Vaishali Bhardwaj, James Davenport, Alexander Fry, Bryce Kalmbach, and David Westman identified many rough edges in the manuscript and tested the AstroML code. We thank Alan Weinstein for help and advice with LIGO data, and Carl Carter-Schwendler for motivational and expert discussions about Bayesian statistics. In addition, Tim Axelrod, Andy Becker, Joshua Bloom, Tamás Budavári, David Hogg, Robert Lupton, Chelsea MacLeod, Lovro Palaversa, Fernando Perez, Maria Süveges, Przemek Woźniak, and two anonymous reviewers provided extensive expert comments. Any remaining errors are entirely our own.

We dedicate this book to Cristin, Hieu, Ian, Nancy, Pamela, Tom, and Vedrana for their support, encouragement, and understanding during the periods of intensive work and absent-mindedness along the way to this finished text.

Authors, Seattle and Atlanta, 2012

PART I
Introduction

1 About the Book and Supporting Material

"Even the longest journey starts with the first step." (Lao-tzu paraphrased)

This chapter introduces terminology and nomenclature, reviews a few relevant contemporary books, briefly describes the Python programming language and the Git code management tool, and provides details about the data sets used in examples throughout the book.

1.1. What Do Data Mining, Machine Learning, and Knowledge Discovery Mean?

Data mining, *machine learning*, and *knowledge discovery* refer to research areas which can all be thought of as outgrowths of multivariate statistics. Their common themes are analysis and interpretation of data, often involving large quantities of data, and even more often resorting to numerical methods. The rapid development of these fields over the last few decades was led by computer scientists, often in collaboration with statisticians. To an outsider, data mining, machine learning, and knowledge discovery compared to statistics are akin to engineering compared to fundamental physics and chemistry: applied fields that "make things work." The techniques in all of these areas are well studied, and rest upon the same firm statistical foundations. In this book we will consider those techniques which are most often applied in the analysis of astronomical data.

While there are many varying definitions in the literature and on the web, we adopt and are happy with the following:

- **Data mining** is a set of techniques for analyzing and describing structured data, for example, finding patterns in large data sets. Common methods include density estimation, unsupervised classification, clustering, principal component analysis, locally linear embedding, and projection pursuit. Often, the term "knowledge discovery" is used interchangeably with data mining. Although there are many books written with "knowledge discovery" in their title, we shall uniformly adopt "data mining" in this book. The data mining techniques result in the understanding of data set properties, such as "My measurements of the size and temperature of stars form a well-defined sequence in the size–temperature diagram, though I find some stars in three clusters far away from this sequence." From the data mining point of view, it is not important to immediately contrast these data with a model (of stellar structure in this case), but rather to quantitatively describe the "sequence," as well as the behavior of measurements falling "far away" from it. In short, data mining is about what the data themselves are telling us. Chapters 6 and 7 in this book primarily discuss data mining techniques.

- **Machine learning** is an umbrella term for a set of techniques for interpreting data by comparing them to models for data behavior (including the so-called nonparametric models), such as various regression methods, supervised classification methods, maximum likelihood estimators, and the Bayesian method. They are often called inference techniques, data-based statistical inferences, or just plain old "fitting." Following the above example, a physical stellar structure model can predict the position and shape of the so-called main sequence in the size–temperature diagram for stars, and when combined with galaxy formation and evolution models, the model can even predict the distribution of stars away from that sequence. Then, there could be more than one competing model and the data might tell us whether (at least) one of them can be rejected. Chapters 8–10 in this book primarily discuss machine learning techniques.

Historically, the emphasis in data mining and knowledge discovery has been on what statisticians call *exploratory data analysis*: that is, learning qualitative features of the data that were not previously known. Much of this is captured under the heading of "unsupervised learning" techniques. The emphasis in machine learning has been on *prediction* of one variable based on the other variables—much of this is captured under the heading of "supervised learning." For further discussion of data mining and machine learning in astronomy, see recent informative reviews [3, 7, 8, 10].

Here are a few concrete examples of astronomical problems that can be solved with data mining and machine learning techniques, and which provide an illustration of the scope and aim of this book. For each example, we list the most relevant chapter(s) in this book:

- Given a set of luminosity measurements for a sample of sources, quantify their luminosity distribution (the number of sources per unit volume and luminosity interval). *Chapter 3*
- Determine the luminosity distribution if the sample selection function is controlled by another measured variable (e.g., sources are detected only if brighter than some flux sensitivity limit). *Chapters 3 and 4*
- Determine whether a luminosity distribution determined from data is statistically consistent with a model-based luminosity distribution. *Chapters 3–5*
- Given a signal in the presence of background, determine its strength. *Chapter 5*
- For a set of brightness measurements with suspected outliers, estimate the best value of the intrinsic brightness. *Chapter 5*
- Given measurements of sky coordinates and redshifts for a sample of galaxies, find clusters of galaxies. *Chapter 6*
- Given several brightness measurements per object for a large number of objects, identify and quantitatively describe clusters of sources in the multi-dimensional color space. Given color measurements for an additional set of

sources, assign to each source the probabilities that it belongs to each of the clusters, making use of both measurements and errors. *Chapters 6 and 9*
- Given a large number of spectra, find self-similar classes. *Chapter 7*
- Given several color and other measurements (e.g., brightness) for a galaxy, determine its most probable redshift using (i) a set of galaxies with both these measurements and their redshift known, or (ii) a set of models predicting color distribution (and the distribution of other relevant parameters). *Chapters 6–8*
- Given a training sample of stars with both photometric (color) measurements and spectroscopic temperature measurements, develop a method for estimating temperature using only photometric measurements (including their errors). *Chapter 8*
- Given a set of redshift and brightness measurements for a cosmological supernova sample, estimate the cosmological parameters and their uncertainties. *Chapter 8*
- Given a set of position (astrometric) measurements as a function of time, determine the best-fit parameters for a model including proper motion and parallax motion. *Chapters 8 and 10*
- Given colors for a sample of spectroscopically confirmed quasars, use analogous color measurements to separate quasars from stars in a larger sample. *Chapter 9*
- Given light curves for a large number of sources, find variable objects, identify periodic light curves, and classify sources into self-similar classes. *Chapter 10*
- Given unevenly sampled low signal-to-noise time series, estimate the underlying power spectrum. *Chapter 10*
- Given detection times for individual photons, estimate model parameters for a suspected exponentially decaying burst. *Chapter 10*

1.2. What Is This Book About?

This book is about extracting knowledge from data, where "knowledge" means a quantitative summary of data behavior, and "data" essentially means the results of measurements. Let us start with the simple case of a scalar quantity, x, that is measured N times, and use the notation x_i for a single measurement, with $i = 1, \ldots, N$. We will use $\{x_i\}$ to refer to the set of all N measurements. In statistics, the data x are viewed as realizations of a random variable X (random variables are functions on the sample space, or the set of all outcomes of an experiment). In most cases, x is a real number (e.g., stellar brightness measurement) but it can also take discrete values (e.g., stellar spectral type); missing data (often indicated by the special IEEE floating-point value NaN—Not a Number) can sometimes be found in real-life data sets.

Possibly the most important single problem in data mining is how to estimate the distribution $h(x)$ from which values of x are drawn (or which "generates" x). The function $h(x)$ quantifies the probability that a value lies between x and $x + dx$, equal to $h(x)\,dx$, and is called a *probability density function* (pdf). Astronomers sometimes use the terms "differential distribution function" or simply "probability distribution." When x is discrete, statisticians use the term "probability mass function" (note that "density" and "mass" are already reserved words in physical sciences, but the confusion should be minimal due to contextual information). The integral of the pdf,

$$H(x) = \int_{-\infty}^{x} h(x') \, dx', \tag{1.1}$$

is called the "cumulative distribution function" (cdf). The inverse of the cumulative distribution function is called the "quantile function."

To distinguish the true pdf $h(x)$ (called the *population pdf*) from a data-derived estimate (called the *empirical pdf*), we shall call the latter $f(x)$ (and its cumulative counterpart $F(x)$).[1] Hereafter, we will assume for convenience that both $h(x)$ and $f(x)$ are properly normalized probability density functions (though this is not a necessary assumption), that is,

$$H(\infty) = \int_{-\infty}^{+\infty} h(x') \, dx' = 1 \tag{1.2}$$

and analogously for $F(\infty)$. Given that data sets are never infinitely large, $f(x)$ can never be exactly equal to $h(x)$. Furthermore, we shall also consider cases when measurement errors for x are not negligible and thus $f(x)$ will not tend to $h(x)$ even for an infinitely large sample (in this case $f(x)$ will be a "broadened" or "blurred" version of $h(x)$).

$f(x)$ is a *model* of the true distribution $h(x)$. Only samples from $h(x)$ are observed (i.e., data points); the functional form of $h(x)$, used to constrain the model $f(x)$, must be guessed. Such forms can range from relatively simple *parametric* models, such as a single Gaussian, to much more complicated and flexible *nonparametric* models, such as the superposition of many small Gaussians. Once the functional form of the model is chosen, the best-fitting member of that model family, corresponding to the best setting of the model's parameters (such as the Gaussian's mean and standard deviation) must be chosen.

A model can be as simple as an analytic function (e.g., a straight line), or it can be the result of complex simulations and other computations. Irrespective of the model's origin, it is important to remember that we can never prove that a model is correct; we can only test it against the data, and sometimes reject it. Furthermore, within the Bayesian logical framework, we cannot even reject a model if it is the only one we have at our disposal—we can only compare models against each other and rank them by their success.

These analysis steps are often not trivial and can be quite complex. The simplest nonparametric method to determine $f(x)$ is to use a histogram; bin the x data and count how many measurements fall in each bin. Very quickly several complications arise: First, what is the optimal choice of bin size? Does it depend on the sample size, or other measurement properties? How does one determine the count error in each bin, and can we treat them as Gaussian errors?

---

[1] Note that in this book we depart from a common notation in the statistical literature in which the true distribution is called $f(x)$ (here we use $h(x)$), and the data-derived estimate of the distribution is called $\widehat{f}(x)$ (here we use $f(x)$).

An additional frequent complication is that the quantity x is measured with some uncertainty or error distribution, $e(x)$, defined as the probability of measuring value x if the true value is μ,

$$e(x) = p(x|\mu, I), \tag{1.3}$$

where I stands for all other information that specifies the details of the error distribution, and "|" is read as "given." Eq. 1.3 should be interpreted as giving a probability $e(x)\,dx$ that the measurement will be between x and $x + dx$.

For the commonly used Gaussian (or normal) error distribution, the probability is given by

$$p(x|\mu, \sigma) = \frac{1}{\sigma\sqrt{2\pi}} \exp\left(\frac{-(x-\mu)^2}{2\sigma^2}\right), \tag{1.4}$$

where in this case I is simply σ, the standard deviation (it is related to the uncertainty estimate popularly known as the "error bar"; for further discussion of distribution functions, see §3.3). The error distribution function could also include a bias b, and $(x - \mu)$ in the above expression would become $(x - b - \mu)$. That is, the bias b is a *systematic* offset of all measurements from the true value μ, and σ controls their "scatter" (bias is introduced formally in §3.2.2). How exactly the measurements are "scattered around" is described by the *shape* of $e(x)$. In astronomy, error distributions are often non-Gaussian or, even when they are Gaussian, σ might not be the same for all measurements, and often depends on the signal strength (i.e., on x; each measured x_i is accompanied by a different σ_i). These types of errors are called *heteroscedastic*, as opposed to *homoscedastic* errors in which the error distribution is the same for each point.

Quantities described by $f(x)$ (e.g., astronomical measurements) can have different meanings in practice. A special case often encountered in practice is when the "intrinsic" or "true" (population pdf) $h(x)$ is a delta function, $\delta(x)$; that is, we are measuring some specific single-valued quantity (e.g., the length of a rod; let us ignore quantum effects here and postulate that there is no uncertainty associated with its true value) and the "observed" (empirical pdf) $f(x)$, sampled by our measurements x_i, simply reflects their error distribution $e(x)$. Another special case involves measurements with negligible measurement errors, but the underlying intrinsic or true pdf $h(x)$ has a finite width (as opposed to a delta function). Hence, in addition to the obvious effects of finite sample size, the difference between $f(x)$ and $h(x)$ can have *two very different origins* and this distinction is often not sufficiently emphasized in the literature: at one extreme it can reflect our measurement error distribution (we measure the same rod over and over again to improve our knowledge of its length), and at the other extreme it can represent measurements of a number of different rods (or the same rod at different times, if we suspect its length may vary with time) with measurement errors *much smaller* than the expected and/or observed length variation. Despite being extremes, these two limiting cases are often found in practice, and may sometimes be treated with the same techniques because of their mathematical similarity (e.g., when fitting a Gaussian to $f(x)$, we do not distinguish the case where its width is due to measurement errors from the case when we measure a population property using a finite sample).

The next level of complication when analyzing $f(x)$ comes from the sample size and dimensionality. There can be a large number of different scalar quantities, such as x, that we measure for each object, and each of these quantities can have a different error distribution (and sometimes even different selection function). In addition, some of these quantities may not be statistically independent. When there is more than one dimension, analysis can get complicated and is prone to pitfalls; when there are many dimensions, analysis is always complicated. If the sample size is measured in hundreds of millions, even the most battle-tested algorithms and tools can choke and become too slow.

Classification of a set of measurements is another important data analysis task. We can often "tag" each x measurement by some "class descriptor" (such quantities are called "categorical" in the statistics literature). For example, we could be comparing the velocity of stars, x, around the Galaxy center with subsamples of stars classified by other means as "halo" and "disk" stars (the latter information could be assigned codes H and D, or 0/1, or any other discrete attribute). In such cases, we would determine two independent distributions $f(x)$—one for each of these two subsamples. Any new measurement of x could then be classified as a "halo" or "disk" star. This simple example can become nontrivial when x is heteroscedastic or multidimensional, and also raises the question of completeness vs. purity trade-offs (e.g., do we care more about never ever misclassifying a halo star, or do we want to minimize the total number of misclassifications for both disk and halo stars?). Even in the case of discrete variables, such as "halo" or "disk" stars, or "star" vs. "galaxy" in astronomical images (which should be called more precisely "unresolved" and "resolved" objects when referring to morphological separation), we can assign them a continuous variable, which often is interpreted as the probability of belonging to a class. At first it may be confusing to talk about the probability that an object is a star vs. being a galaxy because it cannot be both at the same time. However, in this context we are talking about *our current state of knowledge about a given object* and its classification, which can be elegantly expressed using the framework of probability.

In summary, this book is mostly about how to estimate the empirical pdf $f(x)$ from data (including multidimensional cases), how to statistically describe the resulting estimate and its uncertainty, how to compare it to models specified via $h(x)$ (including estimates of model parameters that describe $h(x)$), and how to use this knowledge to interpret additional and/or new measurements (including best-fit model reassessment and classification).

1.3. An Incomplete Survey of the Relevant Literature

The applications of data mining and machine learning techniques are not limited to the sciences. A large number of books discuss applications such as data mining for marketing, music data mining, and machine learning for the purposes of counter-terrorism and law enforcement. We shall limit our survey to books that cover topics similar to those from this book but from a different point of view, and can thus be used as supplemental literature. In many cases, we reference specific sections in the following books.

Numerical Recipes: The Art of Scientific Computing by Press, Teukolsky, Vetterling, and Flannery [27] is famous for its engaging text and concise mathematical and algorithmic explanations (its Fortran version has been cited over 8000 times at the time of writing this book, according to the SAO/NASA Astrophysics Data System). While the whole book is of great value for the topics covered here, several of its 22 chapters are particularly relevant ("Random Numbers," "Sorting and Selection," "Fourier and Spectral Applications," "Statistical Description of Data," "Modeling of Data," "Classification and Inference"). The book includes commented full listings of more than 400 numerical routines in several computer languages that can be purchased in machine-readable form. The supplemental code support for the material covered in the book served as a model for our book. We refer to this book as NumRec.

The Elements of Statistical Learning: Data Mining, Inference, and Prediction by Hastie, Tibshirani, and Friedman [16] is a classic book on these topics, and highly recommended for further reading. With 18 chapters and about 700 pages, it is more comprehensive than this book, and many methods are discussed in greater detail. The writing style is not heavy on theorems and the book should be easily comprehensible to astronomers and other physical scientists. It comes without computer code. We refer to this book as HTF09.

Two books by Wasserman, *All of Nonparametric Statistics* [39] and *All of Statistics: A Concise Course in Statistical Inference* [40] are closer to the statistician's heart, and do not shy away from theorems and advanced statistics. Although "*All*" may imply very long books, together they are under 700 pages. They are good books to look into for deeper and more formal expositions of statistical foundations for data mining and machine learning techniques. We refer to these books as Wass10.

Statistics in Theory and Practice by Lupton [23] is a concise (under 200 pages) summary of the most important concepts in statistics written for practicing scientists, and with close to 100 excellent exercises (with answers). For those who took statistics in college, but need to refresh and extend their knowledge, this book is a great choice. We refer to this book as Lup93.

Practical Statistics for Astronomers by Wall and Jenkins [38] is a fairly concise (under 300 pages) summary of the most relevant contemporary statistical and probabilistic technology in observational astronomy. This excellent book covers classical parametric and nonparametric methods with a strong emphasis on Bayesian solutions. We refer to this book as WJ03.

Statistics: A Guide to the Use of Statistical Methods in the Physical Sciences by Barlow [4] is an excellent introductory text written by a physicist (200 pages). We highly recommend it as a starting point if you feel that the books by Lupton, and Wall and Jenkins are too advanced. We refer to this book as Bar89.

Data Analysis: A Bayesian Tutorial by Sivia [31] is an excellent short book (under 200 pages) to quickly learn about basic Bayesian ideas and methods. Its examples are illuminating and the style is easy to read and does not presume any prior knowledge of statistics. We highly recommend it! We refer to this book as Siv06.

Bayesian Logical Data Analysis for the Physical Sciences by Gregory [13] is more comprehensive (over 400 pages) than Sivia's book, and covers many topics discussed here. It is a good book to look into for deeper understanding and implementation details for most frequently used Bayesian methods. It also provides code support (for Mathematica). We refer to this book as Greg05.

Probability Theory: The Logic of Science by Jaynes [20], an early and strong proponent of Bayesian methods, describes probability theory as extended logic. This monumental treatise compares Bayesian analysis with other techniques, including a large number of examples from the physical sciences. The book is aimed at readers with a knowledge of mathematics at a graduate or an advanced undergraduate level. We refer to this book as Jay03.

Bayesian Methods in Cosmology provides an introduction to the use of Bayesian methods in cosmological studies [17]. Contributions from 24 cosmologists and statisticians (edited by M. P. Hobson, A. H. Jaffe, A. R. Liddle, P. Mukherjee, and D. Parkinson) range from the basic foundations to detailed descriptions of state-of-the-art techniques. The book is aimed at graduate students and researchers in cosmology, astrophysics, and applied statistics. We refer to this book as BayesCosmo.

Advances in Machine Learning and Data Mining for Astronomy is a recent book by over 20 coauthors from mostly astronomical backgrounds (edited by M. J. Way, J. D. Scargle, K. Ali, and A. N. Srivastava) [41]. This book provides a comprehensive overview (700 pages) of various data mining tools and techniques that are increasingly being used by astronomers, and discusses how current problems could lead to the development of entirely new algorithms. We refer to this book as WSAS.

Modern Statistical Methods for Astronomy With R Applications by Feigelson and Babu [9] is very akin in spirit to this book. It provides a comprehensive (just under 500 pages) coverage of similar topics, and provides examples written in the R statistical software environment. Its first chapter includes a very informative summary of the history of statistics in astronomy, and the number of references to statistics literature is larger than here. We refer to this book as FB2012.

Although not referenced further in this book, we highly recommend the following books as supplemental resources.

Pattern Recognition and Machine Learning by Bishop [6] provides a comprehensive introduction to the fields of pattern recognition and machine learning, and is aimed at advanced undergraduates and graduate students, as well as researchers and practitioners. The book is supported by a great deal of additional material, including lecture slides as well as the complete set of figures used in the book. It is of particular interest to those interested in Bayesian versions of standard machine learning methods.

Information Theory, Inference, and Learning Algorithms by MacKay [25] is an excellent and comprehensive book (over 600 pages) that unites information theory and statistical inference. In addition to including a large fraction of the material covered in this book, it also discusses other relevant topics, such as arithmetic coding for data compression and sparse-graph codes for error correction. Throughout, it addresses a wide range of topics—from evolution to sex to crossword puzzles—from the viewpoint of information theory. The book level and style should be easily comprehensible to astronomers and other physical scientists.

In addition to books, several other excellent resources are readily available.

The R language is familiar to statisticians and is widely used for statistical software development and data analysis. R is available as a free software environment[2] for statistical computing and graphics, and compiles and runs on a wide variety of UNIX

[2]http://www.R-project.org/

platforms, Windows and Mac OS. The capabilities of R are extended through user-created packages, which allow specialized statistical techniques, graphical devices, import/export capabilities, reporting tools, etc.

The Auton Lab, part of Carnegie Mellon University's School of Computer Science, researches new approaches to statistical data mining. The Lab is "interested in the underlying computer science, mathematics, statistics and AI of detection and exploitation of patterns in data." A large collection of software, papers, and other resources are available from the Lab's homepage.[3]

The IVOA (International Virtual Observatory Alliance) Knowledge Discovery in Databases group[4] provides support to the IVOA by developing and testing scalable data mining algorithms and the accompanying new standards for VO interfaces and protocols. Their web pages contain tutorials and other materials to support the VO users (e.g., "A user guide for Data Mining in Astronomy").

The Center for Astrostatistics at Penn State University organizes annual summer schools in statistics designed for graduate students and researchers in astronomy. The school is an intensive week covering basic statistical inference, applied statistics, and the R computing environment. The courses are taught by a team of statistics and astronomy professors with opportunity for discussion of methodological issues. For more details, please see their website.[5]

The burgeoning of work in what has been called "astrostatistics" or "astroinformatics," along with the slow but steady recognition of its importance within astronomy, has given rise to recent activity to define and organize more cohesive communities around these topics, as reflected in manifestos by Loredo et al. [22] and Borne et al. [8]. Recent community organizations include the American Astronomical Society Working Group in Astroinformatics and Astrostatistics, the International Astronomical Union Working Group in Astrostatistics and Astroinformatics, and the International Astrostatistics Association (affiliated with the International Statistical Institute). These organizations promote the use of known advanced statistical and computational methods for astronomical research, encourage the development of new procedures and algorithms, organize multidisciplinary meetings, and provide educational and professional resources to the wider community. Information about these organizations can be found at the Astrostatistics and Astroinformatics Portal.[6]

With all these excellent references and resources already available, it is fair to ask why we should add yet another book to the mix. There are two main reasons that motivate this book: First, it is convenient to have the basic statistical, data mining, and machine learning techniques collected and described in a single book, and at a level of mathematical detail aimed at researchers entering into astronomy and physical sciences. This book grew out of materials developed for several graduate classes. These classes had to rely on a large number of textbooks with strongly varying styles and difficulty levels, which often caused practical problems. Second, when bringing a new student up to speed, one difficulty with the current array of texts on data mining and machine learning is that the implementation of the discussed methods is typically

---

[3]http://www.autonlab.org/

[4]http://www.ivoa.net/cgi-bin/twiki/bin/view/IVOA/IvoaKDD

[5]http://astrostatistics.psu.edu/

[6]http://asaip.psu.edu

left up to the reader (with some exceptions noted above). The lack of ready-to-use tools has led to a situation where many groups have independently implemented desired methods and techniques in a variety of languages. This reinventing of the wheel not only takes up valuable time, but the diverse approaches make it difficult to share data and compare results between different groups. With this book and the associated online resources, we hope to encourage, and to contribute to a common implementation of the basic statistical tools.

1.4. Introduction to the Python Language and the Git Code Management Tool

The material in this book is supported by publicly available code, available from http://www.astroML.org. The site includes the Python code to repeat all the examples in this text, as well as to reproduce all the figures from the book. We do not refer by name to code used to produce the figures because the code listing on the website is enumerated by the figure number in the book and thus is easy to locate. We believe and hope that these code examples, with minor modifications, will provide useful templates for your own projects. In this section, we first introduce the Python programming language and then briefly describe the code management tool Git.

1.4.1. Python

Python is an open-source, object-oriented interpreted language with a well-developed set of libraries, packages, and tools for scientific computation. In appendix A, we offer a short introduction to the key features of the language and its use in scientific computing. In this section, we will briefly list some of the scientific computing packages and tools available in the language, as well as the requirements for running the examples in this text.

The examples and figures in this text were created with the Python package AstroML, which was designed as a community resource for fast, well-tested statistical, data mining, and machine learning tools implemented in Python (see appendix B). Rather than reimplementing common algorithms, AstroML draws from the wide range of efficient open-source computational tools available in Python. We briefly list these here; for more detailed discussion see appendix A.

The core packages for scientific computing in Python are NumPy,[7] SciPy,[8] and Matplotlib.[9] Together, these three packages allow users to efficiently read, store, manipulate, and visualize scientific data. Many of the examples and figures in this text require only these three dependencies, and they are discussed at length in appendix A.

There are a large number of other packages built upon this foundation, and AstroML makes use of several of them. An important one is Scikit-learn,[10] a large and very well-documented collection of machine learning algorithms in Python.

---

[7] Numerical Python; http://www.numpy.org

[8] Scientific Python; http://www.scipy.org

[9] http://matplotlib.org

[10] http://scikit-learn.org

Scikit-learn is used extensively in the examples and figures within this text, especially those in the second half of the book. We also make use of PyMC[11] for Markov chain Monte Carlo methods, and HealPy[12] for spherical coordinates and spherical harmonic transformations.

There are a number of other useful Python packages that are not used in this book. For example, Erin Sheldon's esutil package[13] includes a wide variety of handy utilities, focused primarily on Numerical Python, statistics, and file input/output. The CosmoloPy package[14] includes a set of functions for performing cosmological computations. Kapteyn[15] contains many useful routines for astronomical data manipulation, most notably a very complete set of tools for translating between various sky coordinate systems. The AstroPython site[16] acts as a community knowledge base for performing astronomy research with Python. Useful resources include a discussion forum, a collection of code snippets and various tutorials. AstroPy[17] is another effort in this direction, with a focus on community development of a single core Python package for astronomers.

1.4.2. Code Management with Git

Complex analyses of large data sets typically produce substantial amounts of special-purpose code. It is often easy to end up with an unmanageable collection of different software versions, or lose code due to computer failures. Additional management difficulties are present when multiple developers are working on the same code. Professional programmers address these and similar problems using code management tools. There are various freely available tools such as CVS, SVN, Bazaar, Mercurial, and Git. While they all differ a bit, their basic functionality is similar: they support collaborative development of software and the tracking of changes to software source code over time.

This book and the associated code are managed using Git. Installing[18] Git, using it for code management, and for distributing code, are all very user friendly and easy to learn.[19] Unlike CVS, Git can manage not only changes to files, but new files, deleted files, merged files, and entire file structures.[20] One of the most useful features of Git is its ability to set up a remote repository, so that code can be checked in and out from multiple computers. Even when a computer is not connected to a repository (e.g., in the event of a server outage, or when no internet connection is available), the local copy can still be modified and changes reported to the repository later. In the event of a disk failure, the remote repository can even be rebuilt from the local copy.

---

[11] http://pymc-devs.github.com/pymc/

[12] http://healpy.readthedocs.org

[13] http://code.google.com/p/esutil/

[14] http://roban.github.com/CosmoloPy/

[15] http://www.astro.rug.nl/software/kapteyn/

[16] http://www.astropython.org/

[17] see http://www.astropy.org/

[18] http://git-scm.com/

[19] For example, see http://www.github.com/

[20] For a Git manual, see http://progit.org/book/

Because of these features, Git has become the de facto standard code management tool in the Python community: most of the core Python packages listed above are managed with Git, using the website http://github.com to aid in collaboration. We strongly encourage you to consider using Git in your projects. You will not regret the time spent learning how to use it.

1.5. Description of Surveys and Data Sets Used in Examples

Many of the examples and applications in this book require realistic data sets in order to test their performance. There is an increasing amount of high-quality astronomical data freely available online. However, unless a person knows exactly where to look, and is familiar with database tools such as SQL (Structured Query Language,[21] for searching databases), finding suitable data sets can be very hard. For this reason, we have created a suite of data set loaders within the package AstroML. These loaders use an intuitive interface to download and manage large sets of astronomical data, which are used for the examples and plots throughout this text. In this section, we describe these data loading tools, list the data sets available through this interface, and show some examples of how to work with these data in Python.

1.5.1. AstroML Data Set Tools

Because of the size of these data sets, bundling them with the source code distribution would not be very practical. Instead, the data sets are maintained on a web page with http access via the data-set scripts in astroML.datasets. Each data set will be downloaded to your machine only when you first call the associated function. Once it is downloaded, the cached version will be used in all subsequent function calls.

For example, to work with the SDSS imaging photometry (see below), use the function fetch_imaging_sample. The function takes an optional string argument, data_home. When the function is called, it first checks the data_home directory to see if the data file has already been saved to disk (if data_home is not specified, then the default directory is $HOME/astroML_data/; alternatively, the $ASTROML_DATA environment variable can be set to specify the default location). If the data file is not present in the specified directory, it is automatically downloaded from the web and cached in this location.

The nice part about this interface is that the user does not need to remember whether the data has been downloaded and where it has been stored. Once the function is called, the data is returned whether it is already on disk or yet to be downloaded.

For a complete list of data set fetching functions, make sure AstroML is properly installed in your Python path, and open an IPython terminal and type

    In [1]: from astroML.datasets import<TAB>

The tab-completion feature of IPython will display the available data downloaders (see appendix A for more details on IPython).

---

[21]See, for example, http://en.wikipedia.org/wiki/SQL

1.5.2. Overview of Available Data Sets

Most of the astronomical data that we make available were obtained by the Sloan Digital Sky Survey[22] (SDSS), which operated in three phases starting in 1998. The SDSS used a dedicated 2.5 m telescope at the Apache Point Observatory, New Mexico, equipped with two special-purpose instruments, to obtain a large volume of imaging and spectroscopic data. For more details see [15]. The 120 MP camera (for details see [14]) imaged the sky in five photometric bands (u, g, r, i, and z; see appendix C for more details about astronomical flux measurements, and for a figure with the SDSS passbands). As a result of the first two phases of SDSS, Data Release 7 has publicly released photometry for 357 million unique sources detected in ∼12,000 deg^2 of sky[23] (the full sky is equivalent to ∼40,000 deg^2). For bright sources, the photometric precision is 0.01–0.02 mag (1–2% flux measurement errors), and the faint limit is $r \sim 22.5$. For more technical details about SDSS, see [1, 34, 42].

The SDSS imaging data were used to select a subset of sources for spectroscopic follow-up. A pair of spectrographs fed by optical fibers measured spectra for more than 600 galaxies, quasars and stars in each single observation. These spectra have wavelength coverage of 3800–9200 Å and a spectral resolving power of $R \sim 2000$. Data Release 7 includes about 1.6 million spectra, with about 900,000 galaxies, 120,000 quasars and 460,000 stars. The total volume of imaging and spectroscopic data products in the SDSS Data Release 7 is about 60 TB.

The second phase of the SDSS included many observations of the same patch of sky, dubbed "Stripe 82." This opens up a new dimension of astronomical data: the time domain. The Stripe 82 data have led to advances in the understanding of many time-varying phenomena, from asteroid orbits to variable stars to quasars and supernovas. The multiple observations have also been combined to provide a catalog of nonvarying stars with excellent photometric precision.

In addition to providing an unprecedented data set, the SDSS has revolutionized the public dissemination of astronomical data by providing exquisite portals for easy data access, search, analysis, and download. For professional purposes, the Catalog Archive Server (CAS[24]) and its SQL-based search engine is the most efficient way to get SDSS data. While detailed discussion of SQL is beyond the scope of this book,[25] we note that the SDSS site provides a very useful set of example queries[26] which can be quickly adapted to other problems.

Alongside the SDSS data, we also provide the Two Micron All Sky Survey (2MASS) photometry for stars from the SDSS Standard Star Catalog, described in [19]. 2MASS [32] used two 1.3 m telescopes to survey the entire sky in near-infrared light. The three 2MASS bands, spanning the wavelength range 1.2–2.2 μm (adjacent

---

[22]http://www.sdss.org

[23]http://www.sdss.org/dr7/

[24]http://cas.sdss.org/astrodr7/en/tools/search/sql.asp

[25]There are many available books about SQL since it is heavily used in industry and commerce. *Sams Teach Yourself SQL in 10 Minutes* by Forta (Sams Publishing) is a good start, although it took us more than 10 minutes to learn SQL; a more complete reference is *SQL in a Nutshell* by Kline, Kline, and Hunt (O'Reilly), and *The Art of SQL* by Faroult and Robson (O'Reilly) is a good choice for those already familiar with SQL.

[26]http://cas.sdss.org/astrodr7/en/help/docs/realquery.asp

to the SDSS wavelength range on the red side), are called J, H, and K_s ("s" in K_s stands for "short").

We provide several other data sets in addition to SDSS and 2MASS: the LINEAR database features time-domain observations of thousands of variable stars; the LIGO "Big Dog" data[27] is a *simulated* data set from a gravitational wave observatory; and the asteroid data file includes orbital data that come from a large variety of sources. For more details about these samples, see the detailed sections below.

We first describe tools and data sets for accessing SDSS imaging data for an arbitrary patch of sky, and for downloading an arbitrary SDSS spectrum. Several data sets specialized for the purposes of this book are described next and include galaxies with SDSS spectra, quasars with SDSS spectra, stars with SDSS spectra, a high-precision photometric catalog of SDSS standard stars, and a catalog of asteroids with known orbits and SDSS measurements.

Throughout the book, these data are supplemented by simulated data ranging from simple one-dimensional toy models to more accurate multidimensional representations of real data sets. The example code for each figure can be used to quickly reproduce these simulated data sets.

1.5.3. SDSS Imaging Data

The total volume of SDSS imaging data is measured in tens of terabytes and thus we will limit our example to a small (20 deg^2, or 0.05% of the sky) patch of sky. Data for a different patch size, or a different direction on the sky, can be easily obtained by minor modifications of the SQL query listed below.

We used the following SQL query (fully reprinted here to illustrate SDSS SQL queries) to assemble a catalog of ∼330,000 sources detected in SDSS images in the region bounded by $0° < \alpha < 10°$ and $-1° < \delta < 1°$ (α and δ are equatorial sky coordinates called the right ascension and declination).

```
SELECT
    round(p.ra,6) as ra, round(p.dec,6) as dec,
    p.run,                              --- comments are preceded by ---
    round(p.extinction_r,3) as rExtSFD, --- r band extinction from SFD
    round(p.modelMag_u,3) as uRaw,      --- ISM-uncorrected model mags
    round(p.modelMag_g,3) as gRaw,      --- rounding up model magnitudes
    round(p.modelMag_r,3) as rRaw,
    round(p.modelMag_i,3) as iRaw,
    round(p.modelMag_z,3) as zRaw,
    round(p.modelMagErr_u,3) as uErr,   --- errors are important!
    round(p.modelMagErr_g,3) as gErr,
    round(p.modelMagErr_r,3) as rErr,
    round(p.modelMagErr_i,3) as iErr,
    round(p.modelMagErr_z,3) as zErr,
    round(p.psfMag_u,3) as uRawPSF,     --- psf magnitudes
    round(p.psfMag_g,3) as gRawPSF,
    round(p.psfMag_r,3) as rRawPSF,
    round(p.psfMag_i,3) as iRawPSF,
    round(p.psfMag_z,3) as zRawPSF,
    round(p.psfMagErr_u,3) as upsfErr,
```

[27] See http://www.ligo.org/science/GW100916/

```
    round(p.psfMagErr_g,3) as gpsfErr,
    round(p.psfMagErr_r,3) as rpsfErr,
    round(p.psfMagErr_i,3) as ipsfErr,
    round(p.psfMagErr_z,3) as zpsfErr,
    p.type,                         --- tells if a source is resolved or not
    (case when (p.flags & '16') = 0 then 1 else 0 end) as ISOLATED --- useful
INTO mydb.SDSSimagingSample
FROM PhotoTag p
WHERE
    p.ra > 0.0 and p.ra < 10.0 and p.dec > -1 and p.dec < 1 --- 10x2 sq.deg.
    and (p.type = 3 OR p.type = 6) and    --- resolved and unresolved sources
    (p.flags & '4295229440') = 0 and --- '4295229440' is magic code for no
                            --- DEBLENDED_AS_MOVING or SATURATED objects
    p.mode = 1 and  --- PRIMARY objects only, which implies
                    --- !BRIGHT && (!BLENDED || NODEBLEND || nchild == 0)]
    p.modelMag_r < 22.5  --- adopted faint limit (same as about SDSS limit)
--- the end of query
```

This query can be copied verbatim into the SQL window at the CASJobs site[28] (the CASJobs tool is designed for jobs that can require long execution time and requires registration). After running it, you should have your own database called SDSSimagingSample available for download.

The above query selects objects from the PhotoTag table (which includes a subset of the most popular data columns from the main table PhotoObjAll). Detailed descriptions of all listed parameters in all the available tables can be found at the CAS site.[29] The subset of PhotoTag parameters returned by the above query includes positions, interstellar dust extinction in the *r* band (from [28]), and the five SDSS magnitudes with errors in two flavors. There are several types of magnitudes measured by SDSS (using different aperture weighting schemes) and the so-called model magnitudes work well for both unresolved (type=6, mostly stars and quasars) and resolved (type=3, mostly galaxies) sources. Nevertheless, the query also downloads the so-called psf (point spread function) magnitudes. For unresolved sources, the model and psf magnitudes are calibrated to be on average equal, while for resolved sources, model magnitudes are brighter (because the weighting profile is fit to the observed profile of a source and thus can be much wider than the psf, resulting in more contribution to the total flux than in the case of psf-based weights from the outer parts of the source). Therefore, the difference between psf and model magnitudes can be used to recognize resolved sources (indeed, this is the gist of the standard SDSS "star/galaxy" separator whose classification is reported as type in the above query). More details about various magnitude types, as well as other algorithmic and processing details, can be found at the SDSS site.[30]

The WHERE clause first limits the returned data to a 20 deg^2 patch of sky, and then uses several conditions to select unique stationary and well-measured sources above the chosen faint limit. The most mysterious part of this query is the use of processing flags. These 64-bit flags[31] are set by the SDSS photometric processing

---

[28]http://casjobs.sdss.org/CasJobs/

[29]See *Schema Browser* at http://skyserver.sdss3.org/dr8/en/help/browser/browser.asp

[30]http://www.sdss.org/dr7/algorithms/index.html

[31]http://www.sdss.org/dr7/products/catalogs/flags.html

pipeline *photo* [24] and indicate the status of each object, warn of possible problems with the image itself, and warn of possible problems in the measurement of various quantities associated with the object. The use of these flags is unavoidable when selecting a data set with reliable measurements.

To facilitate use of this data set, we have provided code in `astroML.datasets` to download and parse this data. To do this, you must import the function `fetch_imaging_sample`:[32]

```
In [1]: from astroML.datasets import\
        fetch_imaging_sample
In [2]: data = fetch_imaging_sample()
```

The first time this is called, the code will send an http request and download the data from the web. On subsequent calls, it will be loaded from local disk. The object returned is a *record array*, which is a data structure within NumPy designed for labeled data. Let us explore these data a bit:

```
In [3]: data.shape
Out[3]: (330753,)
```

We see that there are just over 330,000 objects in the data set. The names for each of the attributes of these objects are stored within the array data type, which can be accessed via the `dtype` attribute of `data`. The names of the columns can be accessed as follows:

```
In [4]: data.dtype.names[:5]
Out[4]: ('ra', 'dec', 'run', 'rExtSFD', 'uRaw')
```

We have printed only the first five names here using the array slice syntax `[:5]`. The data within each column can be accessed via the column name:

```
In [5]: data['ra'][:5]
Out[5]: array([ 0.358174,   0.358382,   0.357898,
0.35791 ,   0.358881])

In [6]: data['dec'][:5]
Out[6]: array([-0.508718, -0.551157, -0.570892,
        -0.426526, -0.505625])
```

Here we have printed the right ascension and declination (i.e., angular position on the sky) of the first five objects in the catalog. Utilizing Python's plotting package Matplotlib, we show a simple scatter plot of the colors and magnitudes of the first 5000 galaxies and the first 5000 stars from this sample. The result can be seen in

---

[32]Here and throughout we will assume the reader is using the IPython interface, which enables clean interactive plotting with Matplotlib. For more information, refer to appendix A.

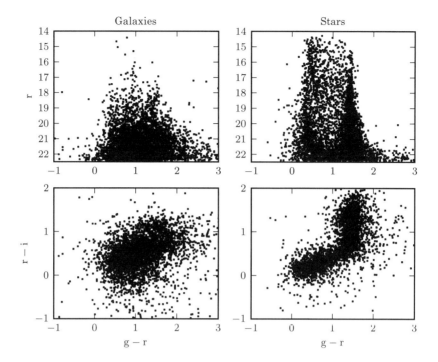

Figure 1.1. The r vs. $g - r$ color–magnitude diagrams and the $r - i$ vs. $g - r$ color–color diagrams for galaxies (left column) and stars (right column) from the SDSS imaging catalog. Only the first 5000 entries for each subset are shown in order to minimize the blending of points (various more sophisticated visualization methods are discussed in §1.6). This figure, and all the others in this book, can be easily reproduced using the astroML code freely downloadable from the supporting website.

figure 1.1. Note that as with all figures in this text, the Python code used to generate the figure can be viewed and downloaded on the book website.

Figure 1.1 suffers from a significant shortcoming: even with only 5000 points shown, the points blend together and obscure the details of the underlying structure. This blending becomes even worse when the full sample of 330,753 points is shown. Various visualization methods for alleviating this problem are discussed in §1.6. For the remainder of this section, we simply use relatively small samples to demonstrate how to access and plot data in the provided data sets.

1.5.4. Fetching and Displaying SDSS Spectra

While the above imaging data set has been downloaded in advance due to its size, it is also possible to access the SDSS database directly and in real time. In astroML.datasets, the function fetch_sdss_spectrum provides an interface to the FITS (Flexible Image Transport System; a standard file format in astronomy for manipulating images and tables[33]) files located on the SDSS spectral server. This

---

[33] See http://fits.gsfc.nasa.gov/iaufwg/iaufwg.html

operation is done in the background using the built-in Python module urllib2. For details on how this is accomplished, see the source code of `fetch_sdss_spectrum`.

The interface is very similar to those from other examples discussed in this chapter, except that in this case the function call must specify the parameters that uniquely identify an SDSS spectrum: the spectroscopic plate number, the fiber number on a given plate, and the date of observation (modified Julian date, abbreviated mjd). The returned object is a custom class which wraps the `pyfits` interface to the FITS data file.

```
In [1]: %pylab
Welcome to pylab, a matplotlib-based Python
  environment [backend: TkAgg].
For more information, type 'help(pylab)'.

In [2]: from astroML.datasets import\
        fetch_sdss_spectrum
In [3]: plate = 1615   # plate number of the spectrum
In [4]: mjd = 53166    # modified Julian date
In [5]: fiber = 513    # fiber ID number on a given
        plate
In [6]: data = fetch_sdss_spectrum(plate, mjd, fiber)
In [7]: ax = plt.axes()
In [8]: ax.plot(data.wavelength(), data.spectrum,
        '-k')
In [9]: ax.set_xlabel(r'$\lambda (\AA)$')
In [10]: ax.set_ylabel('Flux')
```

The resulting figure is shown in figure 1.2. Once the spectral data are loaded into Python, any desired postprocessing can be performed locally.

There is also a tool for determining the plate, mjd, and fiber numbers of spectra in a basic query. Here is an example, based on the spectroscopic galaxy data set described below.

```
In [1]: from astroML.datasets import tools
In [2]: target = tools.TARGET_GALAXY
# main galaxy sample
In [3]: plt, mjd, fib = tools.query_plate_mjd_fiber
        (5, primtarget=target)
In [4]: plt
Out[4]: array([266, 266, 266, 266, 266])

In [5]: mjd
Out[5]: array([51630, 51630, 51630, 51630, 51630])

In [6]: fib
Out[6]: array([27, 28, 30, 33, 35])
```

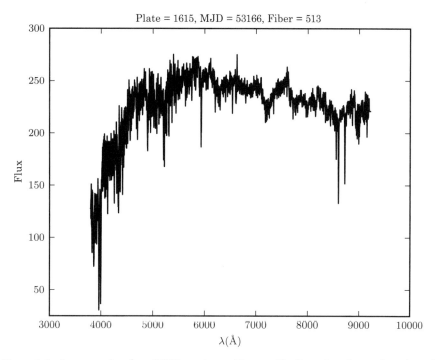

Figure 1.2. An example of an SDSS spectrum (the specific flux plotted as a function of wavelength) loaded from the SDSS SQL server in real time using Python tools provided here (this spectrum is uniquely described by SDSS parameters plate=1615, fiber=513, and mjd=53166).

Here we have asked for five objects, and received a list of five IDs. These could then be passed to the `fetch_sdss_spectrum` function to download and work with the spectral data directly. This function works by constructing a fairly simple SQL query and using urllib to send this query to the SDSS database, parsing the results into a NumPy array. It is provided as a simple example of the way SQL queries can be used with the SDSS database.

The plate and fiber numbers and mjd are listed in the next three data sets that are based on various SDSS spectroscopic samples. The corresponding spectra can be downloaded using `fetch_sdss_spectrum`, and processed as desired. An example of this can be found in the script `examples/datasets/compute_sdss_pca.py` within the `astroML` source code tree, which uses spectra to construct the spectral data set used in chapter 7.

1.5.5. Galaxies with SDSS Spectroscopic Data

During the main phase of the SDSS survey, the imaging data were used to select about a million galaxies for spectroscopic follow-up, including the main flux-limited sample (approximately $r < 18$; see the top-left panel in figure 1.1) and a smaller color-selected sample designed to include very luminous and distant galaxies (the so-called giant elliptical galaxies). Details about the selection of the galaxies for the spectroscopic follow-up can be found in [36].

In addition to parameters computed by the SDSS processing pipeline, such as redshift and emission-line strengths, a number of groups have developed post-processing algorithms and produced so-called "value-added" catalogs with additional scientifically interesting parameters, such as star-formation rate and stellar mass estimates. We have downloaded a catalog with some of the most interesting parameters for ∼660,000 galaxies using the query listed in appendix D submitted to the SDSS Data Release 8 database.

To facilitate use of this data set, in the AstroML package we have included a data set loading routine, which can be used as follows:

```
In [1]: from astroML.datasets import\
        fetch_sdss_specgals
In [2]: data = fetch_sdss_specgals()
In [3]: data.shape
Out[3]: (661598,)

In [4]: data.dtype.names[:5]
Out[4]: ('ra', 'dec', 'mjd', 'plate', 'fiberID')
```

As above, the resulting data is stored in a NumPy record array. We can use the data for the first 10,000 entries to create an example color–magnitude diagram, shown in figure 1.3.

```
In [5]: data = data[:10000]   # truncate data
In [6]: u = data['modelMag_u']
In [7]: r = data['modelMag_r']
In [8]: rPetro = data['petroMag_r']
In [9]: %pylab
Welcome to pylab, a matplotlib-based Python
environment [backend: TkAgg].
For more information, type 'help(pylab)'.

In [10]: ax = plt.axes()
In [11]: ax.scatter(u-r, rPetro, s=4, lw=0, c='k')
In [12]: ax.set_xlim(1, 4.5)
In [13]: ax.set_ylim(18.1, 13.5)
In [14]: ax.set_xlabel('$u - r$')
In [15]: ax.set_ylabel('$r_{petrosian}$')
```

Note that we used the Petrosian magnitudes for the magnitude axis and model magnitudes to construct the $u - r$ color; see [36] for details. Through squinted eyes, one can just make out a division at $u - r \approx 2.3$ between two classes of objects (see [2, 35] for an astrophysical discussion). Using the methods discussed in later chapters, we will be able to automate and quantify this sort of rough by-eye binary classification.

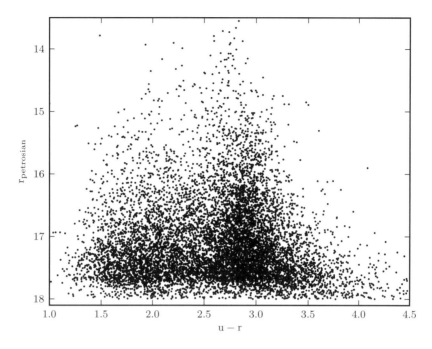

Figure 1.3. The r vs. $u - r$ color–magnitude diagram for the first 10,000 entries in the catalog of spectroscopically observed galaxies from the Sloan Digital Sky Survey (SDSS). Note two "clouds" of points with different morphologies separated by $u - r \approx 2.3$. The abrupt decrease of the point density for $r > 17.7$ (the bottom of the diagram) is due to the selection function for the spectroscopic galaxy sample from SDSS.

1.5.6. SDSS DR7 Quasar Catalog

The SDSS Data Release 7 (DR7) Quasar Catalog contains 105,783 spectroscopically confirmed quasars with highly reliable redshifts, and represents the largest available data set of its type. The construction and content of this catalog are described in detail in [29].

The function `astroML.datasets.fetch_dr7_quasar()` can be used to fetch these data as follows:

```
In [1]: from astroML.datasets import fetch_dr7_quasar
In [2]: data = fetch_dr7_quasar()
In [3]: data.shape
Out[3]: (105783,)

In [4]: data.dtype.names[:5]
Out[4]: ('sdssID', 'RA', 'dec', 'redshift', 'mag_u')
```

One interesting feature of quasars is the redshift dependence of their photometric colors. We can visualize this for the first 10,000 points in the data set as follows:

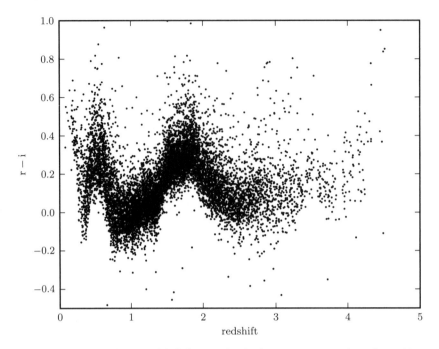

Figure 1.4. The $r - i$ color vs. redshift diagram for the first 10,000 entries from the SDSS Data Release 7 Quasar Catalog. The color variation is due to emission lines entering and exiting the r and i band wavelength windows.

```
In [5]:  data = data[:10000]
In [6]:  r = data['mag_r']
In [7]:  i = data['mag_i']
In [8]:  z = data['redshift']
In [9]:  %pylab
Welcome to pylab, a matplotlib-based Python
environment [backend: TkAgg].
For more information, type 'help(pylab)'.

In [10]: ax = plt.axes()
In [11]: ax.scatter(z, r - i, s=4, c='black',
linewidth=0)
In [12]: ax.set_xlim(0, 5)
In [13]: ax.set_ylim(-0.5, 1.0)
In [14]: ax.set_xlabel('redshift')
In [15]: ax.set_ylabel('r-i')
```

Figure 1.4 shows the resulting plot. The very clear structure in this diagram (and analogous diagrams for other colors) enables various algorithms for the photometric estimation of quasar redshifts, a type of problem discussed in detail in chapters 8–9.

1.5.7. SEGUE Stellar Parameters Pipeline Parameters

SDSS stellar spectra are of sufficient quality to provide robust and accurate values of the main stellar parameters, such as effective temperature, surface gravity, and metallicity (parametrized as [Fe/H]; this is the base 10 logarithm of the ratio of abundance of Fe atoms relative to H atoms, itself normalized by the corresponding ratio measured for the Sun, which is $\sim$ 0.02; i.e., [Fe/H]=0 for the Sun). These parameters are estimated using a variety of methods implemented in an automated pipeline called SSPP (SEGUE Stellar Parameters Pipeline); a detailed discussion of these methods and their performance can be found in [5] and references therein.

We have selected a subset of stars for which, in addition to [Fe/H], another measure of chemical composition, [α/Fe] (for details see [21]), is also available from SDSS Data Release 9. Note that Data Release 9 is the first release with publicly available [α/Fe] data. These measurements meaningfully increase the dimensionality of the available parameter space; together with the three spatial coordinates and the three velocity components (the radial component is measured from spectra, and the two tangential components from angular displacements on the sky called proper motion), the resulting space has eight dimensions. To ensure a clean sample, we have selected $\sim$330,000 stars from this catalog by applying various selection criteria that can be found in the documentation for function `fetch_sdss_sspp`.

The data set loader `fetch_sdss_sspp` for this catalog can be used as follows:

```
In [1]: from astroML.datasets import fetch_sdss_sspp
In [2]: data = fetch_sdss_sspp()
In [3]: data.shape
Out[3]: (327260,)
In [4]: data.dtype.names[:5]
Out[4]: ('ra', 'dec', 'Ar', 'upsf', 'uErr')
```

As above, we use a simple example plot to show how to work with the data. Astronomers often look at a plot of surface gravity vs. effective temperature because it is related to the famous luminosity vs. temperature Hertzsprung–Russell diagram which summarizes well the theories of stellar structure. The surface gravity is typically expressed in the cgs system (in units of cm/s^2), and its logarithm is used in analysis (for orientation, $\log g$ for the Sun is $\sim$4.44). As before, we plot only the first 10,000 entries, shown in figure 1.5.

```
In [5]: data = data[:10000]
In [6]: rpsf = data['rpsf']    # make some reasonable
         # cuts
In [7]: data = data[(rpsf > 15) & (rpsf < 19)]
In [8]: logg = data['logg']
In [9]: Teff = data['Teff']
In [10]: %pylab
Welcome to pylab, a matplotlib-based Python
environment [backend: TkAgg].
For more information, type 'help(pylab)'.
```

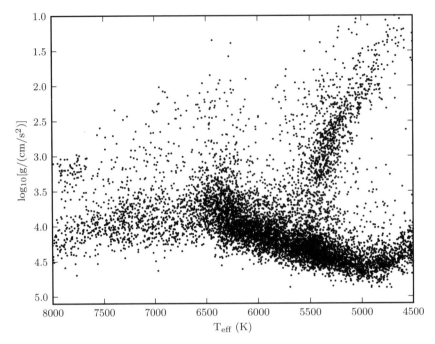

Figure 1.5. The surface gravity vs. effective temperature plot for the first 10,000 entries from the catalog of stars with SDSS spectra. The rich substructure reflects both stellar physics and the SDSS selection criteria for spectroscopic follow-up. The plume of points centered on $T_{\mathrm{eff}} \sim 5300$ K and $\log g \sim 3$ is dominated by red giant stars, and the locus of points with $T_{\mathrm{eff}} < 6500$ K and $\log g > 4.5$ is dominated by main sequence stars. Stars to the left from the main sequence locus are dominated by the so-called blue horizontal branch stars. The axes are plotted backward for ease of comparison with the classical Hertzsprung–Russell diagram: the luminosity of a star approximately increases upward in this diagram.

```
In [11]: ax = plt.axes()
In [12]: ax.scatter(Teff, logg, s=4, lw=0, c='k')
In [13]: ax.set_xlim(8000, 4500)
In [14]: ax.set_ylim(5.1, 1)
In [15]: ax.set_xlabel(r'$\mathrm{T_{eff}\ (K)}$')
In [16]: ax.set_ylabel(r'$\mathrm{log_{10}[g /
         (cm/s^2)]}$')
```

1.5.8. SDSS Standard Star Catalog from Stripe 82

In a much smaller area of ~300 deg², SDSS has obtained repeated imaging that enabled the construction of a more precise photometric catalog containing ~1 million stars (the precision comes from the averaging of typically over ten observations). These stars were selected as nonvariable point sources and have photometric precision better than 0.01 mag at the bright end (or about twice as good as single measurements). The size and photometric precision of this catalog make it a good choice for exploring various methods described in this book, such as stellar locus parametrization in the four-dimensional color space, and search for outliers. Further details about the construction of this catalog and its contents can be found in [19].

There are two versions of this catalog available from `astroML.datasets`. Both are accessed with the function `fetch_sdss_S82standards`. The first contains just the attributes measured by SDSS, while the second version includes a subset of stars cross-matched to 2MASS. This second version can be obtained by calling

```
fetch_sdss_S82standards(crossmatch_2mass = True).
```

The following shows how to fetch and plot the data:

```
In [1]: from astroML.datasets import\
        fetch_sdss_S82standards
In [2]: data = fetch_sdss_S82standards()
In [3]: data.shape
Out[3]: (1006849,)
In [4]: data.dtype.names[:5]
Out[4]: ('RA', 'DEC', 'RArms', 'DECrms', 'Ntot')
```

Again, we will create a simple color–color scatter plot of the first 10,000 entries, shown in figure 1.6.

```
In [5]: data = data[:10000]
In [6]: g = data['mmu_g']  # g-band mean magnitude
In [7]: r = data['mmu_r']  # r-band mean magnitude
In [8]: i = data['mmu_i']  # i-band mean magnitude
In [9]: %pylab
Welcome to pylab, a matplotlib-based Python
environment [backend: TkAgg].
For more information, type 'help(pylab)'.

In [10]: ax = plt.axes()
In [11]: ax.scatter(g - r, r - i, s=4, c='black',
         linewidth=0)
In [12]: ax.set_xlabel('g - r')
In [13]: ax.set_ylabel('r - i')
```

1.5.9. LINEAR Stellar Light Curves

The LINEAR project has been operated by the MIT Lincoln Laboratory since 1998 to discover and track near-Earth asteroids (the so-called "killer asteroids"). Its archive now contains approximately 6 million images of the sky, most of which are 5 MP images covering 2 deg^2. The LINEAR image archive contains a unique combination of sensitivity, sky coverage, and observational cadence (several hundred observations per object). A shortcoming of original reductions of LINEAR data is that its photometric calibration is fairly inaccurate because the effort was focused on

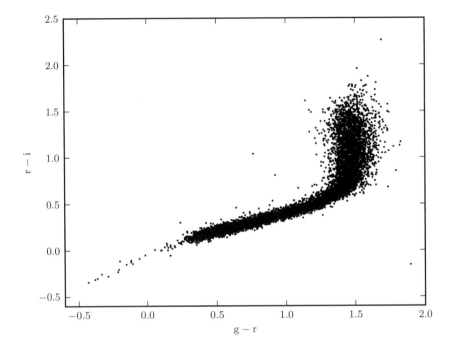

Figure 1.6. The $g - r$ vs. $r - i$ color–color diagram for the first 10,000 entries in the Stripe 82 Standard Star Catalog. The region with the highest point density is dominated by main sequence stars. The thin extension toward the lower-left corner is dominated by the so-called blue horizontal branch stars and white dwarf stars.

astrometric observations of asteroids. Here we use recalibrated LINEAR data from the sky region covered by SDSS which aided recalibration [30]. We focus on 7000 likely periodic variable stars. The full data set with 20 million light curves is publicly available.[34]

The loader for the LINEAR data set is `fetch_LINEAR_sample`. This data set contains light curves and associated catalog data for over 7000 objects:

```
In [1]: from astroML.datasets import\
        fetch_LINEAR_sample
In [2]: data = fetch_LINEAR_sample()
In [3]: gr = data.targets['gr']   # g-r color
In [4]: ri = data.targets['ri']   # r-i color
In [5]: logP = data.targets['LP1']
# log_10(period) in days
In [6]: gr.shape
Out[6]: (7010,)

In [7]: id = data.ids[2756]   # choose one id from the
        # sample
```

[34]The LINEAR Survey Photometric Database is available from https://astroweb.lanl.gov/lineardb/

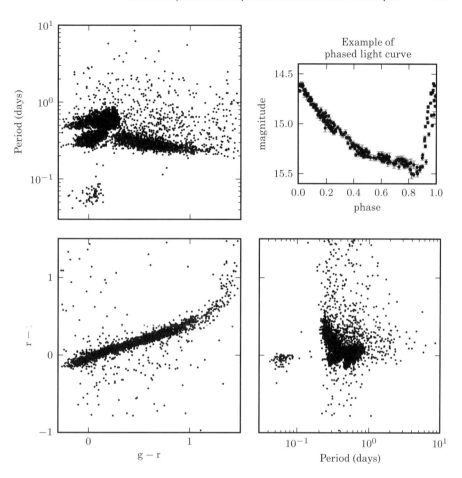

Figure 1.7. An example of the type of data available in the LINEAR data set. The scatter plots show the $g - r$ and $r - i$ colors, and the variability period determined using a Lomb–Scargle periodogram (for details see chapter 10). The upper-right panel shows a phased light curve for one of the over 7000 objects.

```
In [8]: id
Out[8]: 18527462

In [9]: t, mag, dmag = data[id].T
# access light curve data
In [10]: logP = data.get_target_parameter(id, 'LP1')
```

The somewhat cumbersome interface is due to the size of the data set: to avoid the overhead of loading all of the data when only a portion will be needed in any given script, the data are accessed through a class interface which loads the needed data on demand. Figure 1.7 shows a visualization of the data loaded in the example above.

1.5.10. SDSS Moving Object Catalog

SDSS, although primarily designed for observations of extragalactic objects, contributed significantly to studies of Solar system objects. It increased the number of asteroids with accurate five-color photometry by more than a factor of one hundred, and to a flux limit about one hundred times fainter than previous multicolor surveys. SDSS data for asteroids is collated and available as the Moving Object Catalog[35] (MOC). The 4th MOC lists astrometric and photometric data for ~472,000 Solar system objects. Of those, ~100,000 are unique objects with known orbital elements obtained by other surveys.

We can use the provided Python utilities to access the MOC data. The loader is called `fetch_moving_objects`.

```
In [1]: from astroML.datasets import\
        fetch_moving_objects
In [2]: data = fetch_moving_objects(Parker2008_cuts=
        True)
In [3]: data.shape
Out[3]: (33160,)

In [4]: data.dtype.names[:5]
Out[4]: ('moID', 'sdss_run', 'sdss_col', 'sdss_field',
        'sdss_obj')
```

As an example, we make a scatter plot of the orbital semimajor axis vs. the orbital inclination angle for the first 10,000 catalog entries (figure 1.8). Note that we have set a flag to make the data quality cuts used in [26] to increase the measurement quality for the resulting subsample. Additional details about this plot can be found in the same reference, and references therein.

```
In [5]: data = data[:10000]
In [6]: a = data['aprime']
In [7]: sini = data['sin_iprime']
In [8]: %pylab
Welcome to pylab, a matplotlib-based Python
environment [backend: TkAgg].
For more information, type 'help(pylab)'.

In [9]: ax = plt.axes()
In [10]: ax.scatter(a, sini, s=4, c='black',
        linewidth=0)
In [11]: ax.set_xlabel('Semi-major Axis (AU)')
In [12]: ax.set_ylabel('Sine of Inclination Angle')
```

[35]http://www.astro.washington.edu/users/ivezic/sdssmoc/sdssmoc.html

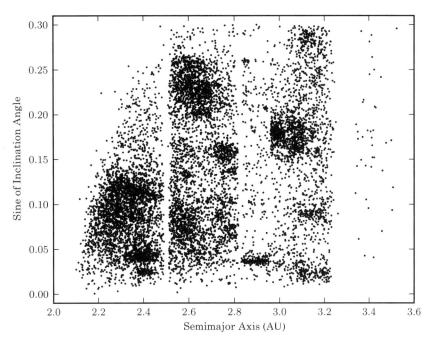

Figure 1.8. The orbital semimajor axis vs. the orbital inclination angle diagram for the first 10,000 catalog entries from the SDSS Moving Object Catalog (after applying several quality cuts). The gaps at approximately 2.5, 2.8, and 3.3 AU are called the Kirkwood gaps and are due to orbital resonances with Jupiter. The several distinct clumps are called asteroid families and represent remnants from collisions of larger asteroids.

1.6. Plotting and Visualizing the Data in This Book

Data visualization is an important part of scientific data analysis, both during exploratory analysis (e.g., to look for problems in data, searching for patterns, and informing quantitative hypothesis) and for the presentation of results. There are a number of books of varying quality written on this topic. An exceptional book is *The Visual Display of Quantitative Information* by Tufte [37], with excellent examples of both good and bad graphics, as well as clearly exposed design principles. Four of his principles that directly pertain to large data sets are (i) present many numbers in a small space, (ii) make large data sets coherent, (iii) reveal the data at several levels of detail, and (iv) encourage the eye to compare different pieces of data. For a recent review of high-dimensional data visualization in astronomy see [11].

1.6.1. Plotting Two-Dimensional Representations of Large Data Sets

The most fundamental quantity we typically want to visualize and understand is the distribution or density of the data. The simplest way to do this is via a scatter plot. When there are too many points to plot, individual points tend to blend together in dense regions of the plot. We must find an effective way to model the density. Note that, as we will see in the case of the histogram (§5.7.2), visualization of the density cannot be done ad hoc, that is, estimating the density is a statistical problem in

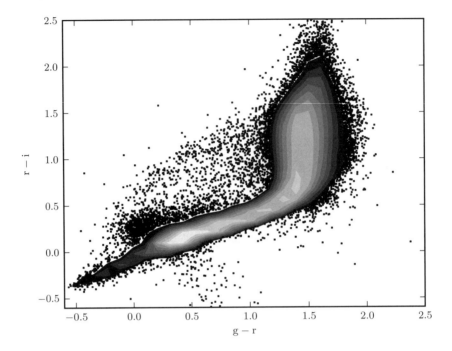

Figure 1.9. Scatter plot with contours over dense regions. This is a color–color diagram of the entire set of SDSS Stripe 82 standard stars; cf. figure 1.6.

itself—choices in simple visualizations of the density may undersmooth or oversmooth the data, misleading the analyst about its properties (density estimation methods are discussed in chapter 6).

A visualization method which addresses this blending limitation is the *contour plot*. Here the contours successfully show the distribution of dense regions, but at the cost of losing information in regions with only a few points. An elegant solution is to use contours for the high-density regions, and show individual points in low-density regions (due to Michael Strauss from Princeton University, who pioneered this approach with SDSS data). An example is shown in figure 1.9 (compare to the scatter plot of a subset of this data in figure 1.6).

Another method is to pixelize the plotted diagram and display the counts of points in each pixel (this "two-dimensional histogram" is known as a *Hess diagram* in astronomy, though this term is often used to refer specifically to color–magnitude plots visualized in this way). The counts can be displayed with different "stretch" (or mapping functions) in order to improve dynamic range (e.g., a logarithmic stretch). A Hess diagram for the color–color plot of the SDSS Stripe 82 standard stars is shown in figure 1.10.

Hess diagrams can be useful in other ways as well. Rather than simply displaying the count or density of points as a function of two parameters, one often desires to show the variation of a separate statistic or measurement. An example of this is shown in figure 1.11. The left panel shows the Hess diagram of the density of points as a function of temperature and surface gravity. The center panel shows a Hess diagram, except here the value in each pixel is the mean metallicity ([Fe/H]).

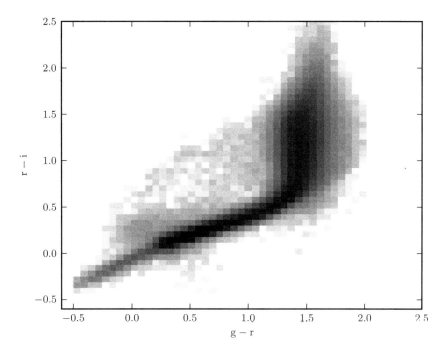

Figure 1.10. A Hess diagram of the $r - i$ vs. $g - r$ colors for the entire set of SDSS Stripe 82 standard stars. The pixels are colored with a logarithmic scaling; cf. figures 1.6 and 1.9.

The number density contours are overplotted for comparison. The grayscale color scheme in the middle panel can lead to the viewer missing fine changes in scale: for this reason, the right panel shows the same plot with a multicolor scale. This is one situation in which a multicolored scale allows better representation of information than a simple grayscale. Combining the counts and mean metallicity into a single plot provides much more information than the individual plots themselves.

Sometimes the quantity of interest is the density variation traced by a sample of points. If the number of points per required resolution element is very large, the simplest method is to use a Hess diagram. However, when points are sparsely sampled, or the density variation is large, it can happen that many pixels have low or vanishing counts. In such cases there are better methods than the Hess diagram where, in low-density regions, we might display a model for the density distribution as discussed, for example, in §6.1.1.

1.6.2. Plotting in Higher Dimensions

In the case of three-dimensional data sets (i.e., three vectors of length N, where N is the number of points), we have already seen examples of using color to encode a third component in a two-dimensional diagram. Sometimes we have four data vectors and would like to find out whether the position in one two-dimensional diagram is correlated with the position in another two-dimensional diagram. For example, we can ask whether two-dimensional color information for asteroids is correlated with their orbital semimajor axis and inclination [18], or whether the color and luminosity of galaxies are correlated with their position in a spectral emission-line diagram [33].

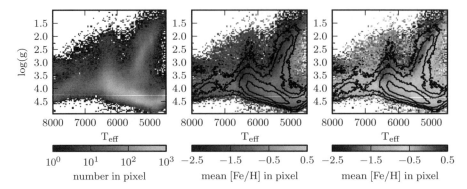

Figure 1.11. A Hess diagram of the number per pixel (left) and [Fe/H] metallicity (center, right) of SEGUE Stellar Parameters Pipeline stars. In the center and right panels, contours representing the number density are overplotted for comparison. These two panels show identical data, but compare a grayscale and multicolor plotting scheme. This is an example of a situation in which multiple colors are very helpful in distinguishing close metallicity levels. This is the same data as shown in figure 1.5. See color plate 1.

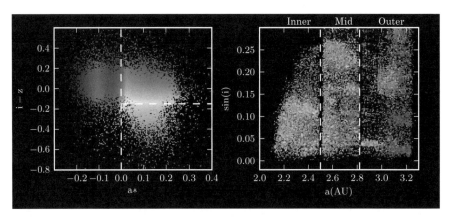

Figure 1.12. A multicolor scatter plot of the properties of asteroids from the SDSS Moving Object Catalog (cf. figure 1.8). The left panel shows observational markers of the chemical properties of the asteroids: two colors a^* and $i - z$. The right panel shows the orbital parameters: semimajor axis a vs. the sine of the inclination. The color of points in the right panel reflects their position in the left panel. See color plate 2.

Let us assume that the four data vectors are called (x, y, z, w). It is possible to define a continuous two-dimensional color palette that assigns a unique color to each data pair from, say, (z, w). Then we can plot the $x - y$ diagram with each symbol, or pixel, color coded according to this palette (of course, one would want to show the $z - w$ diagram, too). An example of this visualization method, based on [18], is shown in figure 1.12.

For higher-dimensional data, visualization can be very challenging. One possibility is to seek various low-dimensional projections which preserve certain "interesting" aspects of the data set. Several of these *dimensionality reduction* techniques are discussed in chapter 7.

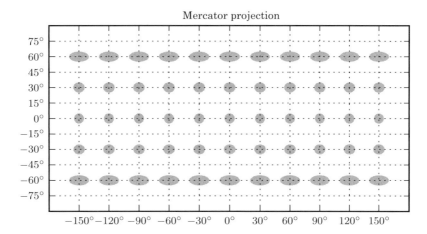

Figure 1.13. The Mercator projection. Shown are the projections of circles of constant radius $10°$ across the sky. Note that the area is not preserved by the Mercator projection: the projection increases the size of finite regions on the sphere, with a magnitude which increases at high latitudes.

1.6.3. Plotting Representations of Data on the Sky

Plotting the distributions or densities of sources as they would appear on the sky is an integral part of many large-scale analyses (including the analysis of the cosmic microwave background or the angular clustering of galaxies). The projection of various spherical coordinate systems (equatorial, ecliptic, galactic) to a plane is often used in astronomy, geography, and other sciences. There are a few dozen different projections that can be found in the literature, but only a few are widely used. There are always distortions associated with projecting a curved surface onto a plane, and various projections are constructed to preserve different properties (e.g., distance, angle, shape, area).

The Mercator projection is probably the most well known since it was used for several centuries for nautical purposes. The lines of constant true compass bearing (called loxodromes or rhumb lines) are straight line segments in this projection, hence its use in navigation. Unfortunately, it distorts the size of map features. For example, world maps in this projection can be easily recognized by the size of Greenland being about the same as the size of Africa (with the latter being much larger in reality). This can be seen from the sizes of the projected circles (called Tissot's indicatrix) in figure 1.13. Projections that preserve the feature size, known as equal-area projections, are more appropriate for use in astronomy, and here we review and illustrate a few of the most popular choices.

The Hammer and Aitoff projections are visually very similar. The former is an equal-area projection and the latter is an equal-distance projection. Sometimes, the Hammer projection is also referred to as the Hammer–Aitoff projection. They show an entire sphere centered on the equator and rescaled to cover twice as much equatorial distance as polar distance (see figure 1.14). For example, these projections were used for the all-sky maps produced by IRAS (the InfraRed Astronomy Satellite).

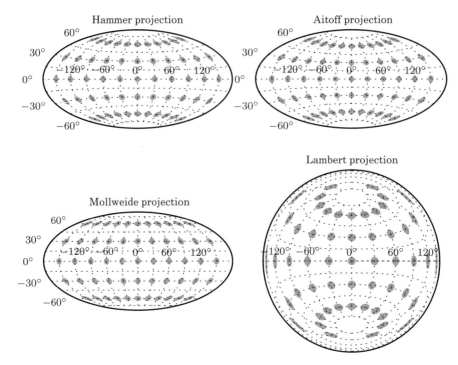

Figure 1.14. Four common full-sky projections. The shaded ellipses represent the distortion across the sky: each is projected from a circle of radius 10° on the sphere. The extent to which these are distorted and/or magnified shows the distortion inherent to the mapping.

The Mollweide projection is another equal-area projection, similar to the Hammer projection, except for straight parallels of latitude instead of the Hammer's curved parallels (developed by an astronomer). It is also known as the Babinet projection, elliptical projection, and homolographic (or homalographic) projection. This projection was used to visualize the WMAP (Wilkinson Microwave Anisotropy Probe) maps.

The Lambert azimuthal equal-area projection maps spherical coordinates to a disk. It is especially useful for projecting the two sky hemispheres into two disks.

In general, given spherical coordinates, (α, δ), the projected planar coordinates, (x, y), are computed using formulas for a particular projection. For example, for the Hammer projection planar coordinates can be computed from

$$x = \frac{2\sqrt{2}\cos(\delta)\sin(\alpha/2)}{\sqrt{1+\cos(\delta)\cos(\alpha/2)}} \tag{1.5}$$

and

$$y = \frac{\sqrt{2}\sin(\delta)}{\sqrt{1+\cos(\delta)\cos(\alpha/2)}}. \tag{1.6}$$

The inverse transformation can be computed as

$$\alpha = 2\arctan\left[\frac{zx}{2(2z^2-1)}\right] \tag{1.7}$$

and

$$\delta = \arcsin(zy), \tag{1.8}$$

where

$$z = \sqrt{1 - (x/4)^2 - (y/2)^2}. \tag{1.9}$$

These projections are available in Matplotlib by specifying the `projection` keyword when building the axis. See the source code associated with figure 1.14 for an example of how this can be accomplished in practice. For more obscure projections, the basemap toolkit,[36] an add-on to Matplotlib, has a more complete set of utilities. These are primarily geared toward visualization for earth sciences, but can be very useful for astronomical sky projections as well.

A concept related to spherical projection is pixelization of a spherical surface. One of the most useful tools is HEALPix (Hierarchical Equal Area isoLatitude Pixelization). HEALPix subdivides a sphere into equal-area pixels (which are not squares but rather curvilinear quadrilaterals). This tessellation is done hierarchically, with higher levels corresponding to smaller pixels. The lowest resolution partition includes 12 pixels, and each new level divides each pixel into four new ones (see figure 1.15). For example, to reach ~3 arcmin resolution, it takes about 12 million pixels. Pixels are distributed on lines of constant latitude, which simplifies and speeds up analysis based on spherical harmonics [12]. The HEALPix code (in IDL and Fortran 90) is publicly available from NASA.[37] A Python version, called HealPy is also available.[38] The lower panel of figure 1.15 shows an example of raw WMAP data, in a Mollweide projection using data in a HEALPix format.

1.7. How to Efficiently Use This Book

We hope that this book will be found useful both as formal course material, and as a self-study guide and reference book. Sufficient statistical background is provided in chapters 2–5 to enable a semester-long course on astronomical statistics (perhaps with one additional chapter from chapters 6–10 to make the course more data-analysis oriented). On the other hand, chapters 6–10 (together with supporting chapter 1) can enable a semester-long course on data mining and machine learning in astronomy. Unlike most textbooks, we do not provide specific exercises with answers. The main reason is that modern scientific data analysis is intimately intertwined with the writing and execution of efficient computer code, and we have designed this book as a practical text with that fact in mind. If a lecturer prefers problem assignments, we highly recommend exercises from Lup93 for a course on astronomical statistics.

A unique feature of this text is the free availability of example code to fetch relevant data sets and recreate every figure in each chapter of the book. This code

---

[36] http://matplotlib.github.com/basemap/

[37] Details about HEALPix are available from http://healpix.jpl.nasa.gov/

[38] http://healpy.readthedocs.org/en/latest/index.html

HEALPix Pixels (Mollweide)

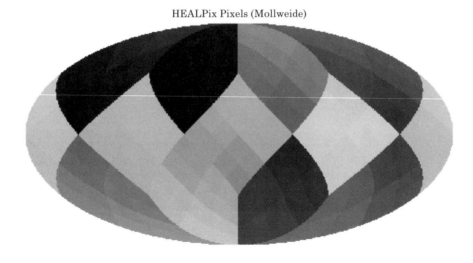

Raw WMAP data

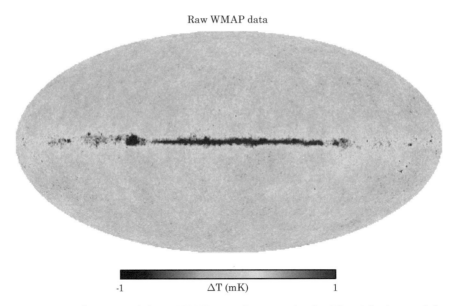

-1 ΔT (mK)   1

Figure 1.15. The top panel shows HEALPix pixels in nested order. The 12 fundamental sky divisions can be seen, as well as the hierarchical nature of the smaller pixels. This shows a pixelization with $n_{side} = 4$, that is, each of the 12 large regions has 4×4 pixels, for a total of 192 pixels. The lower panel shows a seven-year co-add of raw WMAP data, plotted using the HEALPix projection using the HealPy package. This particular realization has $n_{side} = 512$, for a total of 3,145,728 pixels. The pixels are roughly 6.8 arcminutes on a side. See color plate 3.

is available online at http://www.astroML.org, where the examples are organized by figure number. Additionally, throughout this text we include minimal code snippets which are meant to give a flavor of how various tools can be used. These snippets are not generally meant to be complete examples; this is the purpose of the online resources.

All code snippets in the book are set aside and appear like this. They will show some minimal code for purposes of illustration. For example, this is how to compute the cosine of a sequence of numbers:

```
import numpy as np
x = np.random.random(100)   # 100 numbers between 0
    # and 1
cos_x = np.cos(x)   # cosine of each element.
```

For more details on the essential modules for scientific computing in Python, see appendix A.

To take advantage of this book layout, we suggest downloading, examining, modifying, and experimenting with the source code used to create each figure in this text. In order to run these examples on your own machine, you need to install AstroML and its dependencies. A discussion of installation requirements can be found in appendix B, and on the AstroML website.

You can test the success of the installation by plotting one of the example figures from this chapter. For example, to plot figure 1.1, download the source code from `http://www.astroML.org/book_figures/chapter1/` and run the code. The data set will be downloaded and the code should generate a plot identical to figure 1.1. You can then modify the code: for example, rather than $g - r$ and $r - i$ colors, you may wish to see the diagram for $u - g$ and $i - z$ colors.

To get the most out of reading this book, we suggest the following interactive approach: When you come across a section which describes a technique or method which interests you, first find the associated figure on the website and copy the source code into a file which you can modify. Experiment with the code: run the code several times, modifying it to explore how variations of the input (e.g., number of points, number of features used or visualized, type of data) affect the results. See if you can find combinations of parameters that improve on the results shown in the book, or highlight the strengths and weaknesses of the method in question. Finally, you can use the code as a template for running a similar method on your own research data.

We hope that this interactive way of reading the text, working with the data firsthand, will give you the experience and insight needed to successfully apply data mining and statistical learning approaches to your own research, whether it is in astronomy or another data-intensive science.

References

[1] Abazajian, K. N., J. K. Adelman-McCarthy, M. A. Agüeros, and others (2009). The Seventh Data Release of the Sloan Digital Sky Survey. *ApJS 182*, 543–558.

[2] Baldry, I. K., K. Glazebrook, J. Brinkmann, and others (2004). Quantifying the bimodal color-magnitude distribution of galaxies. *ApJ 600*, 681–694.

[3] Ball, N. M. and R. J. Brunner (2010). Data mining and machine learning in astronomy. *International Journal of Modern Physics D 19*, 1049–1106.

[4] Barlow, R. (1989). *Statistics. A Guide to the Use of Statistical Methods in the Physical Sciences*. The Manchester Physics Series, New York: Wiley, 1989.

[5] Beers, T. C., Y. Lee, T. Sivarani, and others (2006). The SDSS-I Value-Added Catalog of stellar parameters and the SEGUE pipeline. *Mem.S.A.I. 77*, 1171.

[6] Bishop, C. M. (2006). *Pattern Recognition and Machine Learning*. Springer.

[7] Borne, K. (2009). Scientific data mining in astronomy. *ArXiv:astro-ph/0911.0505*.

[8] Borne, K., A. Accomazzi, J. Bloom, and others (2009). Astroinformatics: A 21st century approach to astronomy. In *Astro2010: The Astronomy and Astrophysics Decadal Survey. ArXiv:astro-ph/0909.3892*.

[9] Feigelson, E. D. and G. J. Babu (2012). *Modern Statistical Methods for Astronomy With R Applications*. Cambridge University Press.

[10] Feigelson, E. D. and G. Jogesh Babu (2012). Statistical methods for astronomy. *ArXiv:astro-ph/1205.2064*.

[11] Goodman, A. A. (2012). Principles of high-dimensional data visualization in astronomy. *Astronomische Nachrichten 333*, 505.

[12] Górski, K. M., E. Hivon, A. J. Banday, and others (2005). HEALPix: A framework for high-resolution discretization and fast analysis of data distributed on the sphere. *ApJ 622*, 759–771.

[13] Gregory, P. C. (2005). *Bayesian Logical Data Analysis for the Physical Sciences: A Comparative Approach with 'Mathematica' Support*. Cambridge University Press.

[14] Gunn, J. E., M. Carr, C. Rockosi, and others (1998). The Sloan Digital Sky Survey photometric camera. *AJ 116*, 3040–3081.

[15] Gunn, J. E., W. A. Siegmund, E. J. Mannery, and others (2006). The 2.5 m telescope of the Sloan Digital Sky Survey. *AJ 131*, 2332–2359.

[16] Hastie, T., R. Tibshirani, and J. Friedman (2009). *The Elements of Statistical Learning: Data Mining, Inference, and Prediction*. Springer.

[17] Hobson, M. P., A. H. Jaffe, A. R. Liddle, P. Mukherjee, and D. Parkinson (2010). *Bayesian Methods in Cosmology*. Cambridge: University Press.

[18] Ivezić, Ž., R. H. Lupton, M. Jurić, and others (2002). Color confirmation of asteroid families. *AJ 124*, 2943–2948.

[19] Ivezić, Ž., J. A. Smith, G. Miknaitis, and others (2007). Sloan Digital Sky Survey Standard Star Catalog for Stripe 82: The dawn of industrial 1% optical photometry. *AJ 134*, 973–998.

[20] Jaynes, E. T. (2003). *Probability Theory: The Logic of Science*. Cambridge University Press.

[21] Lee, Y. S., T. C. Beers, C. Allende Prieto, and others (2011). The SEGUE Stellar Parameter Pipeline. V. estimation of alpha-element abundance ratios from low-resolution SDSS/SEGUE stellar spectra. *AJ 141*, 90.

[22] Loredo, T. and the Astro/Info Working Group (2009). The astronomical information sciences: A keystone for 21st-century astronomy. Position paper for the Astro2010 Decadal Survey, # 34.

[23] Lupton, R. (1993). *Statistics in Theory and Practice*. Princeton University Press.

[24] Lupton, R. H., J. E. Gunn, Ž. Ivezić, and others (2001). The SDSS imaging pipelines. In F. R. Harnden Jr., F. A. Primini, and H. E. Payne (Ed.), *Astronomical Data Analysis Software and Systems X*, Volume 238 of *Astronomical Society of the Pacific Conference Series*, pp. 269.

[25] MacKay, D. J. C. (2010). *Information Theory, Inference, and Learning Algorithms*. Cambridge: University Press.

[26] Parker, A., Ž. Ivezić, M. Jurić, and others (2008). The size distributions of asteroid families in the SDSS Moving Object Catalog 4. *Icarus 198*, 138–155.

[27] Press, W. H., S. A. Teukolsky, W. T. Vetterling, and B. P. Flannery (1992). *Numerical Recipes in FORTRAN. The Art of Scientific Computing.* Cambridge: University Press.

[28] Schlegel, D. J., D. P. Finkbeiner, and M. Davis (1998). Maps of dust infrared emission for use in estimation of reddening and cosmic microwave background radiation foregrounds. *ApJ 500*, 525.

[29] Schneider, D. P., G. T. Richards, P. B. Hall, and others (2010). The Sloan Digital Sky Survey Quasar Catalog. V. Seventh Data Release. *AJ 139*, 2360–2373.

[30] Sesar, B., J. S. Stuart, Ž. Ivezić, and others (2011). Exploring the variable sky with LINEAR. I. photometric recalibration with the Sloan Digital Sky Survey. *AJ 142*, 190.

[31] Sivia, D. S. (2006). *Data Analysis: A Bayesian Tutorial.* Oxford University Press.

[32] Skrutskie, M. F., R. M. Cutri, R. Stiening, and others (2006). The Two Micron All Sky Survey (2MASS). *AJ 131*, 1163–1183.

[33] Smolčić, V., Ž. Ivezić, M. Gaćeša, and others (2006). The rest-frame optical colours of 99000 Sloan Digital Sky Survey galaxies. *MNRAS 371*, 121–137.

[34] Stoughton, C., R. H. Lupton, M. Bernardi, and others (2002). Sloan Digital Sky Survey: Early Data Release. *AJ 123*, 485–548.

[35] Strateva, I., Ž. Ivezić, G. R. Knapp, and others (2001). Color separation of galaxy types in the Sloan Digital Sky Survey imaging data. *AJ 122*, 1861–1874.

[36] Strauss, M. A., D. H. Weinberg, R. H. Lupton, and others (2002). Spectroscopic target selection in the Sloan Digital Sky Survey: The main galaxy sample. *AJ 124*, 1810–1824.

[37] Tufte, E. R. (2009). *The Visual Display of Quantitative Information.* Cheshire, Connecticut: Graphics Press.

[38] Wall, J. V. and C. R. Jenkins (2003). *Practical Statistics for Astronomers.* Cambridge University Press, 2003.

[39] Wasserman, L. (2010a). *All of Nonparametric Statistics.* Springer.

[40] Wasserman, L. (2010b). *All of Statistics: A Concise Course in Statistical Inference.* Springer.

[41] Way, M., J. Scargle, K. Ali, and A. Srivastava (2012). *Advances in Machine Learning and Data Mining for Astronomy.* Chapman and Hall/CRC Data Mining and Knowledge Discovery Series. Taylor and Francis.

[42] York, D. G., J. Adelman, J. E. Anderson, Jr., and others (2000). The Sloan Digital Sky Survey: Technical summary. *AJ 120*, 1579–1587.

2 Fast Computation on Massive Data Sets

"I do not fear computers. I fear the lack of them." (Isaac Asimov)

This chapter describes basic concepts and tools for tractably performing the computations described in the rest of this book. The need for fast algorithms for such analysis subroutines is becoming increasingly important as modern data sets are approaching billions of objects. With such data sets, even analysis operations whose computational cost is linearly proportional to the size of the data set present challenges, particularly since statistical analyses are inherently interactive processes, requiring that computations complete within some reasonable human attention span. For more sophisticated machine learning algorithms, the often worse-than-linear runtimes of straightforward implementations become quickly unbearable. In this chapter we will look at some techniques that can reduce such runtimes in a rigorous manner that does not sacrifice the accuracy of the analysis through unprincipled approximations. This is far more important than simply speeding up calculations: in practice, computational performance and statistical performance can be intimately linked. The ability of a researcher, within his or her effective time budget, to try more powerful models or to search parameter settings for each model in question, leads directly to better fits and predictions.

2.1. Data Types and Data Management Systems

2.1.1. Data Types

There are three basic data types considered in this book: continuous, ordinal, and nominal. Continuous data (or variables) are real numbers, such as results of quantitative measurements (e.g., temperature, length, time), and often have a measurement error associated with them (i.e., the values can be thought of as probability distributions quantitatively summarized by two numbers; a detailed discussion is presented in chapter 3). A subset of continuous variables are circular variables, where the lowest and highest data values are next to each other (angles on the sky, time in a day). These are of course very important in astronomy. Ordinal data (or ranked variables) have discrete values which can be ordered (e.g., the stellar spectral classification). For example, a continuous data set can be sorted and turned into a ranked list. Nominal data (or categorical data, or attributes) are descriptive and unordered, such as types of galaxies (spiral, elliptical, etc.). They are often nonnumeric.

This book is centered on data organized in tables. Each row corresponds to an object, and different columns correspond to various data values. These values are mostly real-valued measurements with uncertainties, though often ordinal and nominal data will be present, too.

2.1.2. Data Management Systems

Relational databases (or *Relational Database Management Systems*; RDBMSs) represent a technologically and commercially mature way to store and manage tabular data. RDBMSs are systems designed to serve SQL queries quickly—SQL supports queries typically having the form of concatenated *unidimensional* constraints. We will cover basic ideas for making such computations fast in §2.5.1. This is to be distinguished from truly *multidimensional* queries such as nearest-neighbor searches, which we will discuss in §2.5.2. In general, relational databases are not appropriate for multidimensional operations.

More recently, so-called "noSQL" data management systems have gained popular interest. The most popular is *Hadoop*,[1] an open-source system which was designed to perform text processing for the building of web search engines. Its basic representation of data is in terms of key-value pairs, which is a particularly good match for the sparsity of text data, but general enough to be useful for many types of data. Note especially that this approach is inefficient for tabular or array-based data. As of this writing, the data management aspect of Hadoop distributions is still fairly immature compared to RDBMSs. Hadoop distributions typically also come with a simple parallel computing engine.

Traditional database systems are not well set up for the needs of future large astronomical surveys, which involve the creation of large amounts of array-based data, with complex multidimensional relationships. There is ongoing research with the goal of developing efficient database architectures for this type of scientific analysis. One fairly new system with the potential to make a large impact is SciDB[2] [5, 43], a DBMS which is optimized for array-based data such as astronomical images. In fact, the creation of SciDB was inspired by the huge data storage and processing needs of LSST. The astronomical and astrophysical research communities are, at the time of this writing, just beginning to understand how this framework could enable more efficient research in their fields (see, e.g., [28]).

2.2. Analysis of Algorithmic Efficiency

A central mathematical tool for understanding and comparing the efficiency of algorithms is that of "big O" notation. This is simply a way to discuss the growth of an algorithm's runtime as a function of one or more variables of interest, often focused on N, the number of data points or records. See figure 2.1 for an example of

[1] http://hadoop.apache.org/
[2] http://www.scidb.org/

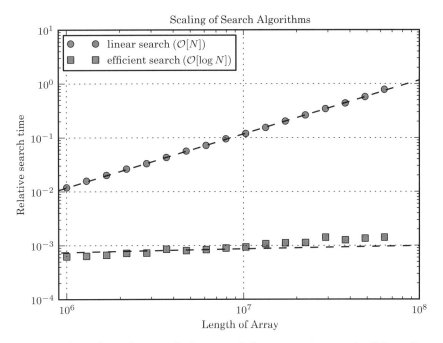

Figure 2.1. The scaling of two methods to search for an item in an ordered list: a linear method which performs a comparison on all N items, and a binary search which uses a more sophisticated algorithm. The theoretical scalings are shown by dashed lines.

the actual runtimes of two different algorithms which both compute the same thing, in this case a one-dimensional search. One exhibits a growth in runtime which is linear in the number of data points, or $\mathcal{O}(N)$, and one uses a smarter algorithm which exhibits growth which is logarithmic in the number of data points, or $\mathcal{O}(\log N)$.

Note that "big O" notation considers only the *order* of growth, that is, ignoring the constant factors in front of the function of N. It also measures "runtime" in terms of the number of certain designated key operations only, to keep things conveniently mathematically formalizable. Thus it is an abstraction which deliberately does not capture details such as whether the program was written in Python or any other language, whether the computer was a fast one or a slow one—those factors are captured in the constants that "big O" ignores, for the sake of mathematical analyzability. Generally speaking, when comparing two algorithms with different growth rates, all such constants eventually become unimportant when N becomes large enough. The order of growth of things other than runtime, such as memory usage, can of course be discussed. We can also consider variables other than N, such as the number of variables or dimensions D. We have given an informal definition of "big O" notation—a formal but accessible definition can be found in [8], an excellent introductory-level text on algorithms written for computer science PhD students.

Some machine learning methods (MLM) are more difficult to compute than others. In general, more accurate MLM are more difficult to compute. The naive or straightforward ways to implement such methods are often $\mathcal{O}(N^2)$ or even $\mathcal{O}(N^3)$ in runtime. However in recent years fast algorithms have been developed which can often compute various MLM as fast as $\mathcal{O}(N)$ or $\mathcal{O}(N \log N)$. We will highlight a few of the basic concepts behind such methods.

2.3. Seven Types of Computational Problem

There are a large number of statistical/machine learning methods described in this book. Making them run fast boils down to a number of different types of computational problems, including the following:

1. **Basic problems**: These include simple statistics, like means, variances, and covariance matrices. We also put basic one-dimensional sorts and range searches in this category. These are all typically simple to compute in the sense that they are $\mathcal{O}(N)$ or $\mathcal{O}(N \log N)$ at worst. We will discuss some key basic problems in §2.5.1.

2. **Generalized N-body problems**: These include virtually any problem involving distances or other similarities between (all or many) pairs (or higher-order n-tuples) of points, such as nearest-neighbor searches, correlation functions, or kernel density estimates. Such problems are typically $\mathcal{O}(N^2)$ or $\mathcal{O}(N^3)$ if computed straightforwardly, but more sophisticated algorithms are available (WSAS, [12]). We will discuss some such problems in §2.5.2.

3. **Linear algebraic problems**: These include all the standard problems of computational linear algebra, including linear systems, eigenvalue problems, and inverses. Assuming typical cases with $N \gg D$, these can be $\mathcal{O}(N)$ but in some cases the matrix of interest is $N \times N$, making the computation $\mathcal{O}(N^3)$. Some common examples where parameter fitting ends up being conveniently phrased in terms of linear algebra problems appear in dimensionality reduction (chapter 7) and linear regression (chapter 8).

4. **Optimization problems**: Optimization is the process of finding the minimum or maximum of a function. This class includes all the standard subclasses of optimization problems, from unconstrained to constrained, convex and nonconvex. Unconstrained optimizations can be fast (though somewhat indeterminate as they generally only lead to local optima), being $\mathcal{O}(N)$ for each of a number of iterations. Constrained optimizations, such as the quadratic programs required by nonlinear support vector machines (discussed in chapter 9) are $\mathcal{O}(N^3)$ in the worst case. Some optimization approaches beyond the widely used unconstrained optimization methods such as gradient descent or conjugate gradient are discussed in §4.4.3 on the expectation maximization algorithm for mixtures of Gaussians.

5. **Integration problems**: Integration arises heavily in the estimation of Bayesian models, and typically involves high-dimensional functions. Performing integration with high accuracy via quadrature has a computational complexity which is exponential in the dimensionality D. In §5.8 we describe the Markov chain Monte Carlo (MCMC) algorithm, which can be used for efficient high-dimensional integration and related computations.

6. **Graph-theoretic problems**: These problems involve traversals of graphs, as in probabilistic graphical models or nearest-neighbor graphs for manifold learning. The most difficult computations here are those involving discrete variables, in which the computational cost may be $\mathcal{O}(N)$ but is exponential in the number of interacting discrete variables among the D dimensions. Exponential computations are by far the most time consuming and generally must be avoided at all cost.

We will encounter graph-theoretic problems in manifold learning (e.g., IsoMap, §7.5.2) and clustering (e.g., Euclidean minimum spanning tree, §6.4.5).

7. **Alignment problems**: This class consists of various types of problem involving matchings between two or more data objects or data sets (known as "cross-matching" in astronomy). The worst-case cost is exponential in N, since in principle all possible assignments of points to each other must be evaluated to achieve the exact best-match solution. Cross-matching is closely related to N-point problems, and a tree-based cross-matching strategy is used in the source code of, for example, figures 1.6 and 6.12. A good introduction to the theory of cross-identification can be found in [6], or in chapter 7 of WSAS.

2.4. Seven Strategies for Speeding Things Up

What are the best general algorithmic strategies and concepts for accelerating these computations? Looking across the various problem types, they include the following:

1. **Trees**: Pointwise comparisons of data sets often require $\mathcal{O}(N^2)$ or worse. In these cases, tree implementations can help. These can be thought of as instances of the well-proven "divide-and-conquer" (or more precisely, "divide-conquer-merge") technique of discrete algorithm design. The basic idea is to chop up the space (at all scales), then prove that some parts can be ignored or approximated during the computation. This strategy is useful across a wide range of problems of interest in astronomy, and thus we will explain it in more detail beginning with §2.5.2, in the context of nearest-neighbor searches. For further reading on the strategy, in particular its general form involving one or more trees, see [12] and chapter 21 of WSAS. The conjecture that such algorithms bring naively $\mathcal{O}(N^2)$ problems to $\mathcal{O}(N)$ was shown in [36]. Examples of further successful application of tree-based techniques for astronomy problems beyond those described in WSAS include remarkable, fast algorithms for blind astrometric calibration [22] and asteroid trajectory tracking [20]. The difficult regime for such algorithms is where the intrinsic dimension of the data is high. In practice the intrinsic dimension (roughly, the dimension of the manifold upon which the points actually lie) of real data is generally much lower than its extrinsic dimension (the actual number of columns). Such correlation in columns allows for speedup even in the case of high-dimensional data (see §2.5.2 for further discussion).

2. **Subproblem reuse**: The techniques of "dynamic programming" and "memoization" refer to ways to avoid repeating work on subproblems that has already been performed, by storing solutions and recalling them if needed, typically in problems that are otherwise exponential in cost. The first represents the bottom-up approach to this, and the second the top-down (recursive) approach. Such principles are at the core of the IsoMap algorithm (§7.5.2) and the Bayesian blocks algorithm (§5.7.2, §10.3.5, [41]). As this approach relies on storage of previously calculated results, memory can be a bottleneck in more difficult problems such as combinatorial optimizations.

3. **Locality**: At times the speed of a computation is not limited by computation itself, but by data bandwidth. Locality is the idea of taking into account the typically large differences in latencies at the various levels of the memory hierarchy of a real computer system: network, disk, RAM, and multiple levels of cache. This is done

by rearranging the computation so that it avoids jumping around unnecessarily, keeping work localized in the fastest parts of the stack—for example, in RAM, as a window of the disk in a so-called "out-of-core" algorithm. Examples can be found throughout the design of the SDSS database, such as [21]. Locality is also an acute issue when programming GPUs. The power of a GPU is that it comes with a large number of processors, cheaply, but that is countered by the downside that its fast memory is generally very limited. Thus the need, again, to organize work so that computation stays within a small window on the data, much like the disk cache. Algorithms which exploit locality often require deep knowledge and exploitation of the characteristics of a specific computer system. The quick evolution of computer system particulars, where significant changes in the trade-off space can occur perhaps every few years, can make the value of the effort investment unclear.

4. **Streaming**: For extremely large data sets, algorithms which require access to the entire data set may not be tractable. Streaming, or *online*, algorithms rearrange computation so that it is decomposed into operations that can be done one data point at a time. This is most clearly appropriate when the data actually arrive in a stream, for example when new observations are obtained by instruments daily. In these cases it is desired that the model be updated to account for each data point as it comes in, rather than having to recalculate the statistic on all $N + 1$ points. Clever streaming approaches can yield very good approximations to statistics which in naive implementations require simultaneous access to the entire data set. A simple example of the streaming approach is the online computation of the mean, where each value may be discarded before accessing the next. More complicated examples exist as well: the well-known Sherman–Morrison formula (described in NumRec) represents an online way to compute a matrix inverse. Using this approach to approximate the optimization in a machine learning method that is normally done in batch is called *online learning* or *stochastic programming* depending on the theoretical lens employed. For a successful example of a streaming approach used in astronomy, see [7]. See [18, 31, 32] for examples of recent schemes with state-of-the-art theoretical and practical performance, for methods such as support vector machines (SVM) (§9.6) and LASSO regression (§8.3.2). The typical behavior of such algorithms is to quickly yield optimization solutions near the optimum achieved by exact batch optimization, possibly without even having touched all of the data yet. However, clear and accurate convergence can be a practical issue due to the stochastic nature of the approach; if the streaming algorithm can even achieve the same level of optimality, it can often take just as long, overall, as the batch solution.

5. **Function transforms**: When the function of interest is very costly to evaluate, one can often proceed by transforming the desired function into another space, or otherwise decomposing it into simpler functions to achieve a more tractable representation. Examples of this include the familiar Taylor series approximations and Fourier transforms. Examples of this approach can be found in Gaussian process regression (§8.10), kernel density estimation (§6.1.1), kernel regression (§8.5), and in dual-tree fast Gauss transforms [23], which employ the Hermite expansion in a fast-multipole-method-like [13] algorithm for efficiently computing sums of many Gaussians with very high accuracy. Techniques based on series expansions

are typically limited to low- to moderate-dimensional data due to the exponential dependence of the number of coefficients on the dimension.

6. **Sampling**: For large data sets, sampling can increase the efficiency by reducing the number N of data points on which a computation will be performed. A crude version of this is to simply randomly choose a subset. More intelligent schemes select a subset of the points that yield higher leverage or somehow more effectively represent the information in the whole data set, thus achieving higher accuracy in the resulting approximation. See [16, 24, 37] for examples of using trees to perform more accurate sampling than previous simpler sampling methods for principal component analysis (§7.3), kernel density estimation (§6.1.1), and nearest-neighbor search (§2.5.2), respectively. Sampling has the advantage that it is relatively insensitive to the dimension of the data, further motivating the augmentation of tree-based algorithms with Monte Carlo concepts for high-dimensional cases. Sampling schemes can yield significant speedups but come at the cost of relaxing strict approximation guarantees to merely probabilistic guarantees, and potentially adding nondeterminism to the results. *Active learning* can be regarded as a more guided special case of sampling in which points are chosen in order to maximally reduce error. Often the context is a prediction problem (say, classification), and the points are chosen from a set of candidate points for the analyst to label to create a training set, such that they are the best given the analyst's time budget for labeling. As the selection process itself is often computationally expensive, the goal is often to reduce human time rather than computation time, but formally the methodology is the same. An empirical study of various common alternative schemes is shown in [38]. An approach having the property of yielding unbiased estimates is shown in [40].

7. **Parallelism**: Parallelism is one of the first ideas that comes to mind for speeding up computations on massive data. It is more relevant than ever, with the advent of seas of computers for rent in the form of cloud computing, multicore processors now being the norm in even basic desktop workstations, and general-purpose GPUs. The idea is to speed up a computation by breaking it into parts which can be performed simultaneously by different processors. Many simple data-processing tasks can be parallelized via the MapReduce framework [9], implemented most notably in the Hadoop system,[3] which was originally designed for text preprocessing for search engines. Though easy to use for nonexperts in parallel computing, such frameworks are largely ineffective for machine learning methods, whose computations require a tighter coupling between machines. For a good introduction to parallel processing in an astronomical machine learning context, see chapter 27 of WSAS. The disadvantage of parallelism is the pragmatic relative difficulty of programming and debugging parallel codes, compared to the highly mature tools available for serial debugging. The speedups are also at best linear in the number of machines, while a better serial algorithm can yield several orders of magnitude in speedup on a single core (e.g., nine orders of magnitude in speedup were demonstrated in [12]). The ideal is of course to parallelize fast algorithms rather than brute-force algorithms, though that can be much more difficult. In any case, if the data cannot fit on one machine, parallelism may be

---

[3]http://hadoop.apache.org

the only option. For an example of state-of-the-art parallel multidimensional tree algorithms that can be used on very large data sets see [11] and [25]. For an example of the effective use of GPUs in astronomy, see [26].

8. **Problem transformation**: The final trick is, when all else fails, to change the problem itself. This can take a number of forms. One is to change the problem type into one which admits faster algorithms, for example reformulating an exponential-cost discrete graph problem to an approximation which can be cast in terms of continuous optimization, the central idea of variational methods for probability evaluations in graphical models [17]. Another is to reformulate a mathematical program (optimization problem) to create one which is easier to solve, for example via the Lagrangian dual in the optimization of support vector machines (§9.6) or lesser-known transformations for difference-of-convex functions or semidefinite relaxations in more challenging formulations of machine learning methods [14, 44]. A final option is to create a new machine learning method which maintains the statistical properties of interest while being inherently easier to compute. An example is the formulation of a decision-tree-like method for density estimation [33] which is nonparametric like KDE but inherits the $\mathcal{O}(N \log N)$ construction time and $\mathcal{O}(\log N)$ querying time of a tree.

2.5. Case Studies: Speedup Strategies in Practice

Examples of the above problem types and speedup strategies abound in the literature, and the state of the art is always evolving. For a good discussion of practical approaches to these, see the references listed above. In the remainder of this chapter we will offer a brief practical discussion of the first speedup strategy: tree-based approaches for searching, sorting, and multidimensional neighbors-based statistics, and their implementation in Python. We will begin with unidimensional searching and sorting, and move to a specific case of N-point problems, the nearest-neighbor search. This discussion will serve to contextualize the problems and principles discussed above, and also act as a starting point for thinking about efficient algorithmic approaches to difficult problems.

2.5.1. Unidimensional Searches and Other Problems

Searching and sorting algorithms are fundamental to many applications in machine learning and data mining, and are concepts that are covered in detail in many texts on computer algorithms (e.g., NumRec). Because of this, most data processing and statistics packages include some implementation of efficient sorts and searches, including various tools available in Python.

Searching

The SQL language of relational databases is good at expressing queries which are composed of unidimensional searches. An example is finding all objects whose brightness is between two numbers *and* whose size is less than a certain number. Each of the unidimensional searches within the query is often called a range search.

The typical approach for accelerating such range searches is a one-dimensional tree data structure called a *B-tree*. The idea of binary search is that of checking whether the value or range of interest is less than some pivot value, which determines which branch of the tree to follow. If the tree is balanced the search time becomes $\mathcal{O}(\log N)$.

A hash table data structure is an array, with input data mapped to array indices using a hash function. A simple example of hashing is the sorting of a data set in a pixelized grid. If one of the dimensions is x, and the pixel grid ranges from $x_{\min}$ to $x_{\max}$, with the pixel width Δ, then the (zero-based) pixel index can be computed as $\text{int}[(x - x_{\min})/\Delta]$, where the function *int* returns the integer part of a real number. The cost of computing a hash function must be small enough to make a hashing-based solution more efficient than alternative approaches, such as trees. The strength of hashing is that its lookup time can be independent of N, that is, $\mathcal{O}(1)$, or constant in time.

Hash tables can thus be more efficient than search trees in principle, but this is difficult to achieve in general, and thus search trees are most often used in real systems.

The Python package NumPy implements efficient array-based searching and hashing. Efficient searching can be accomplished via the function numpy.searchsorted, and scales as $\mathcal{O}(N \log N)$ (see figure 2.1, and the example code below). Basic hashing can be performed using numpy.histogram, and the multidimensional counterparts numpy.histogram2d and numpy.histogramdd. These functions are used throughout this text in the context of data visualization (see, e.g., figure 1.10).

Sorting

For sorting, the built-in Python function sorted and the more efficient array sorting function numpy.sort are useful tools. The most important thing to understand is the scaling of these functions with the size of the array. By default, numpy.sort uses a *quicksort* algorithm which scales as $\mathcal{O}(N \log N)$ (see figure 2.2). Quicksort is a good multipurpose sorting algorithm, and will be sufficient for most data analysis situations. Below there are some examples of searching and sorting using the tools available in Python. Note that we are using the IPython interpreter in order to have access to the commands %time and %timeit, which are examples of the "magic functions" made available by IPython (see appendix A).

```
In [1]: import numpy as np
In [2]: np.random.seed(0)
In [3]: x = np.random.rand(1E7)
In [4]: %time x.sort()    # time a single run
CPU times: user 1.80 s, sys: 0.00 s, total: 1.80 s
Wall time: 1.80 s
In [5]: print x
[   2.51678389e-08    1.63714365e-07
1.89048978e-07  ...,    9.99999814e-01
    9.99999837e-01    9.99999863e-01]
```

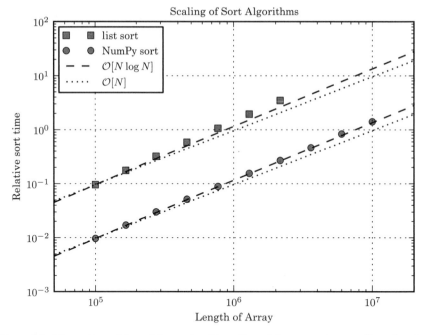

Figure 2.2. The scaling of the quicksort algorithm. Plotted for comparison are lines showing $\mathcal{O}(N)$ and $\mathcal{O}(N \log N)$ scaling. The quicksort algorithm falls along the $\mathcal{O}(N \log N)$ line, as expected.

This sorts the array in place, and accomplishes the task very quickly. The numpy package also has an efficient means of searching this sorted list for a desired value:

```
In [7]: np.searchsorted(x, 0.5)
Out[7]: 4998210
```

As expected, 0.5 falls very near the midpoint in the list of values. We can see the speed of this algorithm using IPython's % timeit functionality:

```
In [8]: %timeit np.searchsorted(x, 0.3)
100000 loops, best of 3: 2.37 us per loop
```

If you have an array of values and would like to sort by a particular column, the argsort function is the best bet:

```
In [9]: X = np.random.random((5, 3))
In [10]: np.set_printoptions(precision=2)
In [11]: print X
[[ 0.96  0.92  1.  ]
 [ 0.71  0.22  0.63]
 [ 0.34  0.82  0.97]
 [ 0.33  0.98  0.44]
```

```
    [ 0.95   0.33   0.73]]
In [12]: i_sort = np.argsort(X[:, 0])
In [13]: print X[i_sort]
[[ 0.33   0.98   0.44]
 [ 0.34   0.82   0.97]
 [ 0.71   0.22   0.63]
 [ 0.95   0.33   0.73]
 [ 0.96   0.92   1.  ]]
```

Here we have sorted the data by the first column, and the values in the second and third columns were rearranged in the same order. To sort each column independently, the sort function can be given an axis argument:

```
In [14]: X.sort(0)
In [15]: print X
[[ 0.33   0.22   0.44]
 [ 0.34   0.33   0.63]
 [ 0.71   0.82   0.73]
 [ 0.95   0.92   0.97]
 [ 0.96   0.98   1.  ]]
```

Now every column of X has been individually sorted, and the values in each row are no longer associated. Beware the difference of these last two operations! In the case where X represents an array of observations, sorting along every column like this will lead to a meaningless data set.

2.5.2. Multidimensional Searches and Other Operations

Just about any computational problem involving distances or other similarities between points falls into the class of generalized N-body problems mentioned in §2.3. The overall most efficient way to exactly/accurately perform such computations is via algorithms based on *multidimensional trees*. Such algorithms begin by building an indexing structure analogous to the B-trees which accelerate SQL queries. This indexing structure is built once for the lifetime of that particular data set (assuming the data set does not change), and thereafter can be used by fast algorithms that traverse it to perform a wide variety of computations. This fact should be emphasized: a single tree for a data set can be used to increase the efficiency of many different machine learning methods. We begin with one of the simplest generalized N-body problems, that of *nearest-neighbor search*.

Nearest-neighbor searches

The basic nearest-neighbor problem can be stated relatively easily. We are given an $N \times D$ matrix X representing N points (vectors) in D dimensions. The ith point in X is specified as the vector x_i, with $i = 1, \ldots, N$, and each x_i has D components, $x_{i,d}$, $d = 1, \ldots, D$. Given a query point x, we want to find the closest point in X under a given distance metric. For simplicity, we use the well-known Euclidean metric,

$$D(x, x_i) = \sqrt{\sum_{d=1}^{D}(x_d - x_{i,d})^2}. \qquad (2.1)$$

The goal of our computation is to find

$$x^* = \arg\min_i D(x, x_i). \qquad (2.2)$$

It is common to have to do this search for more than one query object at a time—in general there will be a set of query objects. This case is called the *all*-nearest-neighbor search. The special but common case where the query set is the same as the reference set, or the set over which we are searching, is called the *monochromatic case*, as opposed to the more general *bichromatic case* where the two sets are different. An example of a bichromatic case is the *cross-matching* of objects from two astronomical catalogs. We consider here the monochromatic case for simplicity: for each point x_i in X, find its nearest neighbor (or more generally, k nearest neighbors) in X (other than itself):

$$\forall_i, \quad x_i^* = \arg\min_j D(x_i, x_j). \qquad (2.3)$$

At first glance, this seems relatively straightforward: all we need is to compute the distances between every pair of points, and then choose the closest. This can be quickly written in Python:

```python
# file: easy_nearest_neighbor.py
import numpy as np

def easy_nn(X):
    N, D = X.shape
    neighbors = np.zeros(N, dtype=int)
    for i in range(N):
        # initialize closest distance to infinity
        j_closest = i
        d_closest = np.inf
        for j in range(N):
            # skip distance between a point and itself
            if i == j:
                continue
            d = np.sqrt(np.sum((X[i] - X[j]) ** 2))
            if d < d_closest:
                j_closest = j
                d_closest = d
        neighbors[i] = j_closest
    return neighbors
```

```
# IPython
In [1]: import numpy as np
In [2]: np.random.seed(0)
In [4]: from easy_nearest_neighbor import easy_nn
In [5]: X = np.random.random((10, 3))
# 10 points in 3 dimensions
In [6]: easy_nn(X)
Out[6]: array([3, 7, 6, 0, 3, 1, 2, 1, 3, 7])
In [7]: X = np.random.random((1000, 3))
# 1000 points in 3 dimensions
In [8]: %timeit easy_nn(X)
1 loops, best of 3: 18.3 s per loop
```

This naive algorithm is simple to code, but leads to *very* long computation times for large numbers of points. For N points, the computation time is $\mathcal{O}(N^2)$—if we increase the sample size by a factor of 10, the computation time will increase by approximately a factor of 100. In astronomical contexts, when the number of objects can number in the billions, this can quickly lead to problems.

Those familiar with Python and NumPy (and those who have read appendix A) will notice a glaring problem here: this uses loops instead of a vectorized implementation to compute the distances. We can vectorize this operation by observing the following identity:

$$\sum_k (X_{ik} - X_{jk})^2 = \sum_k \left[X_{ik}^2 - 2X_{ik}X_{jk} + X_{jk}^2 \right] \tag{2.4}$$

$$= \sum_k X_{ik}^2 - 2\sum_k X_{ik}X_{jk} + \sum_k X_{jk}^2 \tag{2.5}$$

$$= [XX^T]_{ii} + [XX^T]_{jj} - 2[XX^T]_{ij}, \tag{2.6}$$

where in the final line, we have written the sums in terms of matrix products. Now the entire operation can be reexpressed in terms of fast vectorized math:

```
# file: vectorized_nearest_neighbor.py
import numpy as np

def vectorized_nn(X):
    XXT = np.dot(X, X.T)
    Xii = XXT.diagonal()

    D = Xii - 2 * XXT + Xii[:, np.newaxis]

    # numpy.argsort returns sorted indices along a
    # given axis we'll take the second column
    # (index 1) because the first column corresponds
    # to the distance between each point and itself.
    return np.argsort(D, axis=1)[:, 1]
```

```
# IPython:
In [1]: import numpy as np
In [2]: np.random.seed(0)
In [3]: from vectorized_nearest_neighbor import
        vectorized_nn
In [4]: X = np.random.random((10, 3))
In [5]: vectorized_nn(X)
Out[5]: array([3, 7, 6, 0, 3, 1, 2, 1, 3, 7])
In [6]: X = np.random.random((1000, 3))
In [7]: %timeit vectorized_nn(X)
# timeit is a special feature of IPython
1 loops, best of 3: 139 ms per loop
```

Through vectorization, we have sped up our calculation by a factor of over 100. We have to be careful here, though. Our clever speed improvement does not come without cost. First, the vectorization requires a large amount of memory. For N points, we allocate not one but two $N \times N$ matrices. As N grows larger, then, the amount of memory used by this algorithm increases in proportion to N^2. Also note that, although we have increased the computational efficiency through vectorization, the algorithm still computes $\mathcal{O}(N^2)$ distance computations.

There is another disadvantage here as well. Because we are splitting the computation into separate parts, the machine floating-point precision can lead to unexpected results. Consider the following example:

```
In [1]: import numpy as np
In [2]: x = 1.0
In [3]: y = 0.0
In [4]: np.sqrt((x - y) ** 2)   # how we computed
        # non-vectorized distances
Out[4]: 1.0

In [5]: np.sqrt(x**2 + y**2 - 2*x*y)
# vectorized distances
Out[5]: 1.0

In [6]: x += 100000000
In [7]: y += 100000000
In [8]: np.sqrt((x - y) ** 2)   # non-vectorized
        # distances
Out[8]: 1.0

In [9]: np.sqrt(x**2 + y**2 - 2*x*y)
# vectorized distances
Out[9]: 0.0
```

The distance calculations in lines 4 and 8 correspond to the method used in the slow example above. The distance calculations in lines 5 and 9 correspond to our fast vectorized example, and line 9 leads to the wrong result. The reason for this is the floating-point precision of the computer. Because we are taking one very large number (x**2 + y**2) and subtracting another large number (2*x*y) which differs by only one part in 10^{16}, we suffer from roundoff error in line 9. Thus our efficient method, though faster than the initial implementation, suffers from three distinct disadvantages:

- Like the nonvectorized version, the computational efficiency still scales as $\mathcal{O}(N^2)$, which will be too slow for data sets of interest.
- Unlike the nonvectorized version, the memory use also scales as $\mathcal{O}(N^2)$, which may cause problems for large data sets.
- Roundoff error due to the vectorization tricks can cause incorrect results in some circumstances.

This sort of method is often called a *brute-force*, *exhaustive*, or *naive* search. It takes no shortcuts in its approach to the data: it simply evaluates and compares every possible option. The resulting algorithm is very easy to code and understand, but can lead to very slow computation as the data set grows larger. Fortunately, there are a variety of tree-based algorithms available which can improve on this.

Trees for increasing the efficiency of a search

There are a number of common types of multidimensional tree structures which can be used for a nearest-neighbor search.

Quad-trees and oct-trees The earliest multidimensional tree structures were *quad-trees* and *oct-trees*. These work in two and three dimensions, respectively. Oct-trees, in particular, have long been used in astrophysics within the context of N-body and smoothed particle hydrodynamics (SPH) simulations (see, e.g.,[42, 45]).

A quad-tree is a simple data structure used to arrange two-dimensional data, in which each tree node has exactly four children, representing its four quadrants. Each node is defined by four numbers: its left, right, top, and bottom extents.

Figure 2.3 shows a visualization of a quad-tree for some generated structured data. Notice how quickly the quad-tree narrows in on the structured parts of the data. Whole regions of the parameter space can be eliminated from a search in this way. This idea can be generalized to higher dimensions, using an oct-tree, so named because each node has up to eight children (representing the eight quadrants of a three-dimensional space).

By grouping the points in this manner, one can progressively constrain the distances between a test point and groups of points in a tree, using the bounding box of each group of points to provide a lower bound on the distance between the query point and any point in the group. If the lower bound is not better than the best-candidate nearest-neighbor distance the algorithm knows about so far, it can prune the group completely from the search, saving very large amounts of work.

The result is that under certain conditions the cost of the nearest-neighbor search reduces to $\mathcal{O}(\log N)$ for a single query point: a significant improvement over

Quad-tree Example

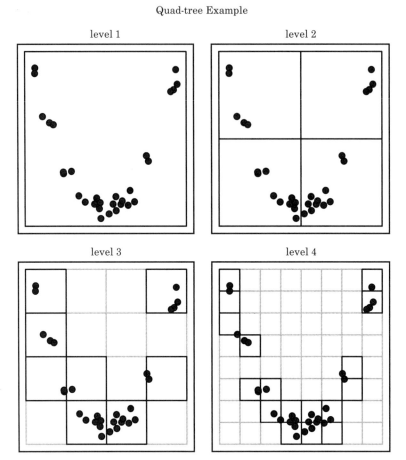

Figure 2.3. Example of a quad-tree.

brute force when N is large. The tree itself, the result of a one-time build operation, takes $\mathcal{O}(N \log N)$ time to construct. $\mathcal{O}(N \log N)$ is just a bit worse than $\mathcal{O}(N)$, and the constant in front of the build time is in practice very small, so tree construction is fast. All the multidimensional trees we will look at yield $\mathcal{O}(N \log N)$ build time and $\mathcal{O}(\log N)$ single-query search time under certain conditions, though they will still display different constants depending on the data. The dependence on the number of dimensions D also differs, as we discuss next.

kd-trees The quad-tree and oct-tree ideas above suggest a straightforward generalization to higher dimensions. In two dimensions, we build a tree with four children per node. In three dimensions, we build a tree with eight children per node. Perhaps in D dimensions, we should simply build a tree with 2^D children per node, and create a search algorithm similar to that of a quad- or oct-tree? A bit of calculation shows that this is infeasible: for even modest-sized values of D, the size of the tree quickly blows up. For example, if $D = 10$, each node would require $2^{10} = 1024$ children. This means that to go two levels down (i.e., to divide each dimension into four units) would already require over 10^6 nodes! The problem quickly gets out of hand as the

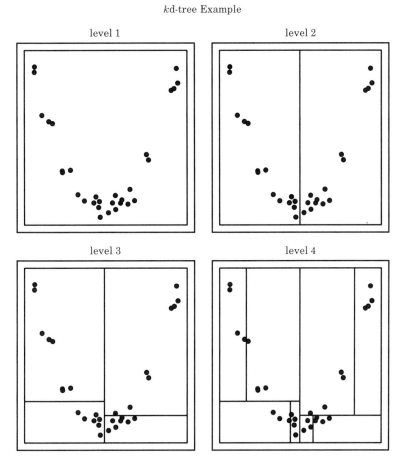

Figure 2.4. Example of a kd-tree.

dimension continues to increase. If we push the dimension up to $D = 100$, and make the dubious assumption that each node only requires 1 byte in memory, even a single level of the tree would require 10^{15} petabytes of storage. To put this in perspective, this is about ten billion times the estimated total volume of worldwide internet traffic in the year 2010. Evidently, this strategy is not going to work. This immense growth in the number of subdivisions of a space as the dimensionality of that space grows is one manifestation of the *curse of dimensionality* (see §7.1 for more details).

A solution to this problem is the kd-tree [2], so named as it is a k-dimensional generalization of the quad-tree and oct-tree. To get around the dimensionality issues discussed above, kd-trees are generally implemented as *binary* trees: that is, each node has two children. The top node of a kd-tree is a D-dimensional hyperrectangle which contains the entire data set. To create the subnodes, the volume is split into two regions along a single dimension, and this procedure is repeated recursively until the lowest nodes contain a specified number of points.

Figure 2.4 shows the kd-tree partition of a data set in two dimensions. Notice that, like the quad-tree in figure 2.3, the kd-tree partitions the space into rectilinear regions. Unlike the quad-tree, the kd-tree adapts the split points in order to better represent the data. Because of the binary nature of the kd-tree, it is suitable for higher-dimensional data.

A fast kd-tree implementation is available in SciPy, and can be used as follows:

```
In [1]:  import numpy as np
In [2]:  from scipy.spatial import cKDTree
In [3]:  np.random.seed(0)
In [4]:  X = np.random.random((1000, 3))
In [5]:  kdt = cKDTree(X) # build the KDTree
In [6]:  %timeit kdt.query(X, k=2) # query for two
            # neighbors
100 loops , best of 3: 7.95 ms per loop
```

The nearest neighbor of each point in a set of 1000 is found in just a few milliseconds: a factor of 20 improvement over the vectorized brute-force method for 1000 points. In general, as the number of points grows larger, the computation time will increase as $\mathcal{O}(\log N)$, for each of the N query points, for a total of $\mathcal{O}(N \log N)$.

```
In [7]:  X = np.random.random ((100000, 3))
In [8]:  kdt = cKDTree(X)
In [9]:  %timeit kdt.query(X, k=2) # query for two
            # neighbors
1 loops , best of 3: 949 ms per loop
```

A factor of 100 more points leads to a factor of 120 increase in computational cost, which is consistent with our prediction of $\mathcal{O}(N \log N)$. How does this compare to the brute-force vectorized method? Well, if brute-force searches truly scale as $\mathcal{O}(N^2)$, then we would expect the computation to take $100^2 \times 139$ ms, which comes to around 23 minutes, compared to a few seconds for a tree-based method!

Even as kd-trees solve the scaling and dimensionality issues discussed above, they are still subject to a fundamental weakness, at least in principle. Because the kd-tree relies on rectilinear splitting of the data space, it also falls subject to the curse of dimensionality. To see why, imagine building a kd-tree on points in a D-dimensional space. Because the kd-tree splits along a single dimension in each level, one must go D levels deep before each dimension has been split. For D relatively small, this does not pose a problem. But for, say, $D = 100$, this means that we must create $2^{100} \approx 10^{30}$ nodes in order to split each dimension once! This is a clear limitation of kd-trees in high dimensions. One would expect that, for N points in D dimensions, a kd-tree will lose efficiency when $D \gg \log_2 N$. As a result, other types of trees sometimes do better than kd-trees, such as the ones we describe below.

Ball-trees Ball-trees [29, 30] make use of an intuitive fact: if x_1 is far from x_2 and x_2 is near x_3, then x_1 is also far from x_3. This intuition is a reflection of the triangle inequality,

$$D(x_1, x_2) + D(x_2, x_3) \leq D(x_1, x_3), \qquad (2.7)$$

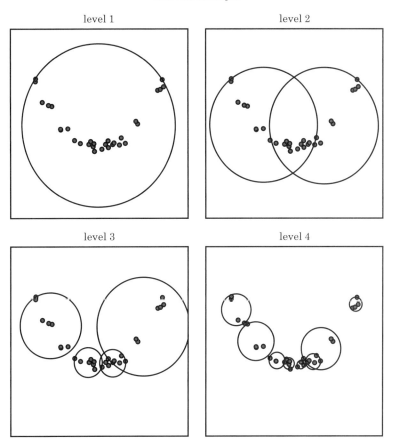

Figure 2.5. Example of a ball-tree.

which can be proven relatively easily for Euclidean distances, and also applies to any of a large number of other *distance metrics* which may be useful in certain applications.

Ball-trees represent a way to address the flaw of kd-trees applied to high-dimensional structured data. Rather than building rectilinear nodes in D dimensions, ball-tree construction builds hyperspherical nodes. Each node is defined by a centroid c_i and a radius r_i, such that the distance $D(y, c_i) \leq r_i$ for every point y contained in the node. With this construction, given a point **x** outside the node, it is straightforward to show from the triangle inequality (eq. 2.7) that

$$[D(\mathbf{x}, \mathbf{c}_i) - r_i] \leq D(\mathbf{x}, \mathbf{y}) \leq [D(\mathbf{x}, \mathbf{c}_i) + r_i] \qquad (2.8)$$

for any point **y** in the node. Using this fact, a neighbor search can proceed quickly by eliminating large parts of the data set from a query through a single distance computation.

Figure 2.5 shows an example of a ball-tree in two dimensions. Comparing to the kd-tree example in figure 2.4 (which uses the same data set), one can see that the ball-tree nodes more quickly converge on the nonlinear structure of the data set. This efficiency allows the ball-tree to be much more efficient than the kd-tree in high dimensions in some cases.

There is a fast ball-tree algorithm included in the package Scikit-learn, which can be used as follows (compare to the kd-tree example used above):

```
In [1]: import numpy as np
In [2]: from sklearn.neighbors import BallTree
In [3]: np.random.seed(0)
In [4]: X = np.random.random((1000, 3))
In [5]: bt = BallTree(X)   # build the Ball Tree
In [6]: %timeit bt.query(X, k=2)
# query for two neighbors
100 loops, best of 3: 5.7 ms per loop
In [7]: X = np.random.random((100000, 3))
In [8]: bt = BallTree(X)
In [9]: %timeit bt.query(X, k=2)
# query for two neighbors
1 loops, best of 3: 1.2 s per loop
```

We see that in low dimensions, the ball-tree and kd-tree have comparable computational complexity. As the number of dimensions increases, the ball-tree can outperform the kd-tree, but the actual performance depends highly on the internal structure or intrinsic dimensionality of the data (see below).

Other trees Many other fast methods for tree-based nearest-neighbor searches have been developed, too numerous to cover here. Cover trees [3] represent an interesting nonbinary kind of ball-tree that, by construction, allows a theoretical proof of the usual desired $\mathcal{O}(\log N)$ single-query search time under mild assumptions (unlike kd-trees and ball trees in which it is difficult to provide such hard runtime guarantees beyond simple cases such as uniformly distributed random data). The idea of deriving orthogonal directions from the data (see §7.3) for building trees is an old one—some modern twists are shown in [24, 27]. Maximum margin trees have been shown to perform better than the well-known tree structures in the setting of time-sensitive searches, that is, where the goal is to return the best nearest-neighbor answer within a bounded time budget [35]. In astronomy, for two-dimensional data on a sphere, a structure based on hexagons has been shown to be effective [21]. Trees have also been developed for other kinds of similarities between points, beyond distances—for example cosine trees [16] and cone trees [34], for dot products. There are in fact hundreds or possibly thousands of proposed data structures for nearest-neighbor searches. A number of useful references can be found in [4, 10, 39].

Intrinsic dimensionality Thankfully, despite the curse of dimensionality, there is a nice property that real data sets almost always have. In reality the dimensions are generally *correlated*, meaning that the "true" or "intrinsic" dimensionality of the data (in some sense the true number of degrees of freedom) is often much smaller than the simple number of columns, or "extrinsic" dimension. The real performance of multidimensional trees is subject to this intrinsic dimensionality, which unfortunately is not known in advance, and itself requires an expensive (typically $\mathcal{O}(N^2)$) algorithm to estimate. Krauthgamer and Lee [19] describe one notion of intrinsic dimensionality, the *expansion constant*, and examples of how the runtime of algorithms of interest in this book depend on it are shown in [3, 36].

The result is that kd-trees, while seemingly not a compelling idea in high dimensions, can often work well in up to hundreds or even thousands of dimensions, depending on the structure of the data and the number of data points— sometimes much better than ball-trees. So the story is not as clear as one might hope or think. So which tree should you use? Unfortunately, at this stage of research in computational geometry there is little theory to help answer the question without just trying a number of different structures and seeing which works the best.

Approximate neighbor methods The degradation in the speedup provided by tree-based algorithms in high dimensions has motivated the use of sampling methods in conjunction with trees. The locality-based hashing scheme [1], based on the simple idea of random projections, is one recent approach. It is based on the idea of returning points that lie within a fixed factor of the true distance to the query point. However, Hammsersley in 1950 [15] showed that distances numerically occupy a narrowing range as the dimension increases: this makes it clear that such an approximation notion in high dimensions will eventually accept any point as a candidate for the true nearest neighbor. A newer approximation notion [37] guarantees that the returned neighbor found is within a certain distance in *rank order*, for example, in the closest 1% of the points, without actually computing all the distances. This retains meaning even in the face of arbitrary dimension.

In general we caution that the design of practical fast algorithms can be subtle, in particular requiring that substantial thought be given to any mathematical notion of approximation that is employed. Another example of an important subtlety in this regard occurs in the computation of approximate kernel summations, in which it is recommended to bound the *relative* error (the error as a percentage of the true sum) rather than the more straightforward and typical practice of bounding the absolute error (see [23]).

References

[1]   Andoni, A. and P. Indyk (2008). Near-optimal hashing algorithms for approximate nearest neighbor in high dimensions. *Communications of the ACM 51*(1), 117–122.

[2]   Bentley, J. L. (1975). Multidimensional binary search trees used for associative searching. *Commun. ACM 18*, 509–517.

[3]   Beygelzimer, A., S. Kakade, and J. Langford (2006). Cover trees for nearest neighbor. In *ICML*, pp. 97–104.

[4]   Bhatia, N. and Vandana (2010). Survey of nearest neighbor techniques. *Journal of Computer Science 8*(2), 4.

[5]   Brown, P. G. (2010). Overview of SciDB: Large scale array storage, processing and analysis. In *Proceedings of the 2010 ACM SIGMOD International Conference on Management of data*, SIGMOD '10, New York, NY, USA, pp. 963–968. ACM.

[6]   Budavári, T. and A. S. Szalay (2008). Probabilistic cross-identification of astronomical sources. *ApJ 679*, 301–309.

[7]   Budavári, T., V. Wild, A. S. Szalay, L. Dobos, and C.-W. Yip (2009). Reliable eigenspectra for new generation surveys. *MNRAS 394*, 1496–1502.

[8]   Cormen, T. H., C. E. Leiserson, and R. L. Rivest (2001). *Introduction to Algorithms*. MIT Press.

[9]   Dean, J. and S. Ghemawat (2008). MapReduce: Simplified data processing on large clusters. *Commun. ACM 51*(1), 107–113.

[10]   Dhanabal, S. and D. S. Chandramathi (2011). Article: A review of various k-nearest neighbor query processing techniques. *International Journal of Computer Applications 31*(7), 14–22. Foundation of Computer Science, New York, USA.

[11]   Gardner, J. P., A. Connolly, and C. McBride (2007). Enabling rapid development of parallel tree search applications. In *Proceedings of the 5th IEEE Workshop on Challenges of Large Applications in Distributed Environments*, CLADE '07, New York, NY, USA, pp. 1–10. ACM.

[12]   Gray, A. and A. Moore (2000). 'N-Body' problems in statistical learning. In *Advances in Neural Information Processing Systems 13*, pp. 521–527. MIT Press.

[13]   Greengard, L. and V. Rokhlin (1987). A fast algorithm for particle simulations. *Journal of Computational Physics 73*.

[14]   Guan, W. and A. G. Gray (2013). Sparse fractional-norm support vector machine via DC programming. *Computational Statistics and Data Analysis (CSDA) (in press)*.

[15]   Hammersley, J. M. (1950). The distribution of distance in a hypersphere. *Annals of Mathematical Statistics 21*(3), 447–452.

[16]   Holmes, M., A. G. Gray, and C. Isbell (2009). QUIC-SVD: Fast SVD using cosine trees. In *Advances in Neural Information Processing Systems (NIPS) (Dec 2008)*. MIT Press.

[17]   Jordan, M. I., Z. Ghahramani, T. S. Jaakkola, and L. K. Saul (1999). An introduction to variational methods for graphical models. *Mach. Learn. 37*(2), 183–233.

[18]   Juditsky, A., G. Lan, A. Nemirovski, and A. Shapiro (2009). Robust stochastic approximation approach to stochastic programming. *SIAM Journal on Optimization 19*(4), 1574–1609.

[19]   Krauthgamer, R. and J. R. Lee (2004). Navigating nets: simple algorithms for proximity search. In *Proceedings of the Fifteenth Annual ACM-SIAM Symposium on Discrete Algorithms*, SODA '04, pp. 798–807. Society for Industrial and Applied Mathematics.

[20]   Kubica, J., L. Denneau, Jr., A. Moore, R. Jedicke, and A. Connolly (2007). Efficient algorithms for large-scale asteroid discovery. In R. A. Shaw, F. Hill, and D. J. Bell (Eds.), *Astronomical Data Analysis Software and Systems XVI*, Volume 376 of *Astronomical Society of the Pacific Conference Series*, pp. 395–404.

[21]   Kunszt, P., A. S. Szalay, and A. Thakar (2000). The hierarchical triangular mesh. In *Mining the Sky: Proc. of the MPA/ESO/MPE Workshop*, pp. 631–637.

[22]   Lang, D., D. W. Hogg, K. Mierle, M. Blanton, and S. Roweis (2010). Astrometry.net: Blind astrometric calibration of arbitrary astronomical images. *Astronomical Journal 137*, 1782–2800.

[23]   Lee, D. and A. G. Gray (2006). Faster Gaussian summation: Theory and experiment. In *In Proceedings of the Twenty-second Conference on Uncertainty in Artificial Intelligence*.

[24]   Lee, D. and A. G. Gray (2009). Fast high-dimensional kernel summations using the Monte Carlo multipole method. In *Advances in Neural Information Processing Systems (NIPS) (Dec 2008)*. MIT Press.

[25]   Lee, D., R. Vuduc, and A. G. Gray (2012). A distributed kernel summation framework for general-dimension machine learning. In *SIAM International Conference on Data Mining (SDM)*.

[26]   Lee, M. and T. Budavari (2012). Fast cross-matching of astronomical catalogs on GPUs. In *GPU Technology Conference (GTC)*.

[27]   Liaw, Y.-C., M.-L. Leou, and C.-M. Wu (2010). Fast exact k nearest neighbors search using an orthogonal search tree. *Pattern Recognition 43*(6), 2351–2358.

[28]   Malon, D., P. van Gemmeren, and J. Weinstein (2012). An exploration of SciDB in the context of emerging technologies for data stores in particle physics and cosmology. *Journal of Physics Conference Series 368*(1), 012021.

[29]   Moore, A. (2000). The Anchors Hierarchy: Using the triangle inequality to survive high dimensional data. In *Proceedings of the Sixteenth Conference on Uncertainty in Artificial Intelligence*, pp. 397–405. Morgan Kaufmann.

[30]   Omohundro, S. M. (1989). Five balltree construction algorithms. Technical report. International Computer Science Institute, TR-89-063.

[31]   Ouyang, H. and A. G. Gray (2012a). NASA: Achieving lower regrets and faster rates via adaptive stepsizes. In *ACM SIGKDD International Conference on Knowledge Discovery and Data Mining (KDD)*.

[32]   Ouyang, H. and A. G. Gray (2012b). Stochastic smoothing for nonsmooth minimizations: Accelerating SGD by exploiting structure. In *International Conference on Machine Learning (ICML)*.

[33]   Ram, P. and A. G. Gray (2011). Density estimation trees. In *ACM SIGKDD International Conference on Knowledge Discovery and Data Mining (KDD)*.

[34]   Ram, P. and A. G. Gray (2012a). Maximum inner-product search using cone trees. In *ACM SIGKDD International Conference on Knowledge Discovery and Data Mining (KDD)*.

[35]   Ram, P. and A. G. Gray (2012b). Nearest-neighbor search on a time budget via max-margin trees. In *SIAM International Conference on Data Mining (SDM)*.

[36]   Ram, P., D. Lee, W. March, and A. G. Gray (2010). Linear-time algorithms for pairwise statistical problems. In *Advances in Neural Information Processing Systems (NIPS) (Dec 2009)*. MIT Press.

[37]   Ram, P., D. Lee, H. Ouyang, and A. G. Gray (2010). Rank-approximate nearest neighbor search: Retaining meaning and speed in high dimensions. In *Advances in Neural Information Processing Systems (NIPS) (Dec 2009)*. MIT Press.

[38]   Richards, J. W., D. L. Starr, H. Brink, A. A. Miller, J. S. Bloom, and others (2012). Active learning to overcome sample selection bias: Application to photometric variable star classification. *ApJ 744*, 192.

[39]   Samet, H. (1990). *The Design and Analysis of Spatial Data Structures*. Addison-Wesley Longman.

[40]   Sastry, R. and A. G. Gray (2012). UPAL: Unbiased pool based active learning. In *Conference on Artificial Intelligence and Statistics (AISTATS)*.

[41]   Scargle, J. D., J. P. Norris, B. Jackson, and J. Chiang (2012). Studies in astronomical time series analysis. VI. Bayesian block representations. *ArXiv:astro-ph/1207.5578*.

[42]   Springel, V., N. Yoshida, and S. D. M. White (2001). GADGET: A code for collisionless and gasdynamical cosmological simulations. *New Astr. 6*, 79–117.

[43] Stonebraker, M., P. Brown, A. Poliakov, and S. Raman (2011). The architecture of SciDB. In *SSDBM'11: Proceedings of the 23rd International Conference on Scientific and Statistical Database Management*, SSDBM'11, Berlin, Heidelberg, pp. 1–16. Springer.

[44] Vasiloglou, N., A. G. Gray, and D. Anderson (2008). Scalable semidefinite manifold learning. In *IEEE International Workshop on Machine Learning For Signal Processing (MLSP)*.

[45] Wadsley, J. W., J. Stadel, and T. Quinn (2004). Gasoline: A flexible, parallel implementation of TreeSPH. *New Astr. 9*, 137–158.

PART II
Statistical Frameworks and Exploratory Data Analysis

3 Probability and Statistical Distributions

"There are three kinds of lies: lies, damned lies, and statistics." (popularized by Mark Twain)
"In ancient times they had no statistics so they had to fall back on lies." (Stephen Leacock)

The main purpose of this chapter is to review notation and basic concepts in probability and statistics. The coverage of various topics cannot be complete, and it is aimed at concepts needed to understand material covered in the book. For an in-depth discussion of probability and statistics, please refer to numerous readily available textbooks, such as Bar89, Lup93, WJ03, Wass10, mentioned in §1.3.

The chapter starts with a brief overview of probability and random variables, then it reviews the most common univariate and multivariate distribution functions, and correlation coefficients. We also summarize the central limit theorem and discuss how to generate mock samples (random number generation) for a given distribution function.

Notation

Notation in probability and statistics is highly variable and ambiguous, and can make things confusing all on its own (and even more so in data mining publications!). We try to minimize the notational clutter, though this is not always possible. We have already introduced some notation in §1.2. For example, lowercase letters are used for probability density (differential distribution) functions (pdf), and the corresponding uppercase letter for their cumulative counterpart (cdf), for example, $h(x)$ and $H(x)$.

We have been able to simplify our nomenclature by ignoring some annoying difficulties, particularly in the case of continuous values. In reality, we cannot talk about the probability of x taking on a specific real-number value as being anything other than zero. The product $h(x)\,dx$ gives a probability that the value x would fall in a dx wide interval around x, but we will not explicitly write dx (see Wass10 for a clear treatment in this regard).

We shall use p for probability whenever possible, both for the probability of a single event and for the probability density functions (pdf).

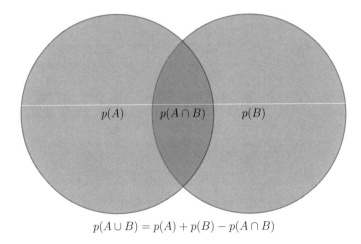

$$p(A \cup B) = p(A) + p(B) - p(A \cap B)$$

Figure 3.1. A representation of the sum of probabilities in eq. 3.1.

3.1. Brief Overview of Probability and Random Variables

3.1.1. Probability Axioms

Given an event A, such as the outcome of a coin toss, we assign it a real number $p(A)$, called the *probability* of A. As discussed above, $p(A)$ could also correspond to a probability that a value of x falls in a dx wide interval around x. To qualify as a probability, $p(A)$ must satisfy three Kolmogorov axioms:

1. $p(A) \geq 0$ for each A.
2. $p(\Omega) = 1$, where Ω is a set of all possible outcomes.
3. If $A_1, A_2, \ldots$ are disjoint events, then $p\left(\bigcup_{i=1}^{\infty} A_i\right) = \sum_{i=1}^{\infty} p(A_i)$, where $\bigcup$ stands for "union."

As a consequence of these axioms, several useful rules can be derived. The probability that the union of two events, A and B, will happen is given by the sum rule,

$$p(A \cup B) = p(A) + p(B) - p(A \cap B), \tag{3.1}$$

where $\cap$ stands for "intersection." That is, the probability that either A or B will happen is the sum of their respective probabilities minus the probability that *both* A and B will happen (this rule avoids the double counting of $p(A \cap B)$ and is easy to understand graphically: see figure 3.1).

If the complement of event A is $\overline{A}$, then

$$p(A) + p(\overline{A}) = 1. \tag{3.2}$$

The probability that both A and B will happen is equal to

$$p(A \cap B) = p(A|B)\, p(B) = p(B|A)\, p(A). \tag{3.3}$$

Here "|" is pronounced "given" and $p(A|B)$ is the probability of event A given that (conditional on) B is true. We discuss conditional probabilities in more detail in §3.1.3.

If events B_i, $i = 1, \ldots, N$ are disjoint and their union is the set of all possible outcomes, then

$$p(A) = \sum_i p(A \cap B_i) = \sum_i p(A|B_i)\, p(B_i). \tag{3.4}$$

This expression is known as the law of total probability. Conditional probabilities also satisfy the law of total probability. Assuming that an event C is not mutually exclusive with A or any of B_i, then

$$p(A|C) = \sum_i p(A|C \cap B_i)\, p(B_i|C). \tag{3.5}$$

Cox derived the same probability rules starting from a different set of axioms than Kolmogorov [2]. Cox's derivation is used to justify the so-called "logical" interpretation of probability and the use of Bayesian probability theory (for an illuminating discussion, see chapters 1 and 2 in Jay03). To eliminate possible confusion in later chapters, note that both the Kolmogorov and Cox axioms result in essentially the same probabilistic framework.

The difference between classical inference and Bayesian inference is fundamentally in the interpretation of the resulting probabilities (discussed in detail in chapters 4 and 5). Briefly, classical statistical inference is concerned with $p(A)$, interpreted as the long-term outcome, or frequency with which A occurs (or would occur) in identical repeats of an experiment, and events are restricted to propositions about random variables (see below). Bayesian inference is concerned with $p(A|B)$, interpreted as the plausibility of a proposition A, conditional on the truth of B, and A and B can be any logical proposition (i.e., they are not restricted to propositions about random variables).

3.1.2. Random Variables

A random, or stochastic, variable is, roughly speaking, a variable whose value results from the measurement of a quantity that is subject to random variations. Unlike normal mathematical variables, a random variable can take on a set of possible different values, each with an associated probability. It is customary in the statistics literature to use capital letters for random variables, and a lowercase letter for a particular realization of random variables (called random variates). We shall use lowercase letters for both.

There are two main types of random variables: discrete and continuous. The outcomes of discrete random variables form a countable set, while the outcomes of continuous random variables usually map on to the real number set (though one can define mapping to the complex plane, or use matrices instead of real numbers, etc.). The function which ascribes a probability value to each outcome of the random variable is the probability density function (pdf).

Independent identically distributed (iid) random variables are drawn from the same distribution and are independent. Two random variables, x and y, are *independent* if and only if

$$p(x, y) = p(x)\, p(y) \tag{3.6}$$

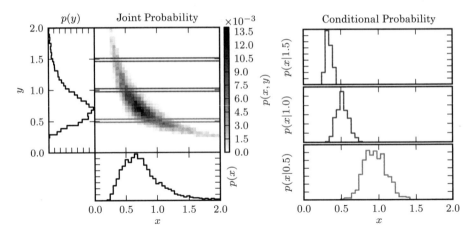

Figure 3.2. An example of a two-dimensional probability distribution. The color-coded panel shows $p(x, y)$. The two panels to the left and below show marginal distributions in x and y (see eq. 3.8). The three panels to the right show the conditional probability distributions $p(x|y)$ (see eq. 3.7) for three different values of y (as marked in the left panel).

for all values x and y. In other words, the knowledge of the value of x tells us nothing about the value of y.

The *data* are specific ("measured") values of random variables. We will refer to measured values as x_i, and to the set of all N measurements as $\{x_i\}$.

3.1.3. Conditional Probability and Bayes' Rule

When two continuous random variables are not independent, it follows from eq. 3.3 that

$$p(x, y) = p(x|y)\, p(y) = p(y|x)\, p(x). \tag{3.7}$$

The *marginal probability function* is defined as

$$p(x) = \int p(x, y)\, dy, \tag{3.8}$$

and analogously for $p(y)$. Note that complete knowledge of the conditional pdf $p(y|x)$, and the marginal probability $p(x)$, is sufficient to fully reconstruct $p(x, y)$ (the same is true with x and y reversed).

By combining eqs. 3.7 and 3.8, we get a continuous version of the law of total probability,

$$p(x) = \int p(x|y)p(y)\, dy. \tag{3.9}$$

An example of a two-dimensional probability distribution is shown in figure 3.2, together with corresponding marginal and conditional probability distributions. Note that the conditional probability distributions $p(x|y = y_0)$ are simply one-dimensional "slices" through the two-dimensional image $p(x, y)$ at given values of y_0, and then divided (renormalized) by the value of the marginal distribution $p(y)$ at $y = y_0$. As a result of this renormalization, the integral of $p(x|y)$ (over x) is unity.

Eqs. 3.7 and 3.9 can be combined to yield *Bayes' rule*:

$$p(y|x) = \frac{p(x|y)p(y)}{p(x)} = \frac{p(x|y)p(y)}{\int p(x|y)p(y)\,dy}. \tag{3.10}$$

Bayes' rule relates conditional and marginal probabilities to each other. In the case of a discrete random variable, y_j, with M possible values, the integral in eq. 3.10 becomes a sum:

$$p(y_j|x) = \frac{p(x|y_j)p(y_j)}{p(x)} = \frac{p(x|y_j)p(y_j)}{\sum_{j=1}^{M} p(x|y_j)p(y_j)}. \tag{3.11}$$

Bayes' rule follows from a straightforward application of the rules of probability and is by no means controversial. It represents the foundation of Bayesian statistics, which has been a very controversial subject until recently. We briefly note here that it is not the rule itself that has caused controversy, but rather its application. Bayesian methods are discussed in detail in chapter 5.

We shall illustrate the use of marginal and conditional probabilities, and of Bayes' rule, with a simple example.

Example: the Monty Hall problem

The following problem illustrates how different probabilistic inferences can be derived about the same physical system depending on the available prior information. There are $N=1000$ boxes, of which 999 are empty and one contains some "prize." You choose a box at random; the probability that it contains the prize is 1/1000. This box remains closed. The probability that any one of other 999 boxes contains the prize is also 1/1000, and the probability that the box with the prize is among those 999 boxes is 999/1000. Then another person who *knows which box contains the prize* opens 998 empty boxes chosen from the 999 remaining boxes (i.e., the box you chose is "set aside"). It is important to emphasize that these 998 boxes are *not* selected randomly from the set of 999 boxes you did not choose—instead, they are selected as empty boxes. So, the remaining 999th box is almost certain to contain the prize; the probability is 999/1000 because there is a chance of only 1 in 1000 that the prize is in the box you chose initially, and the probabilities for the two unopened boxes must add up to 1. Alternatively, before 998 empty boxes were opened, the probability that the 999 boxes contained the prize was 999/1000. Given that all but one were demonstrated to be empty, the last 999th box now contains the prize with the same probability. If you were offered to switch the box you initially chose with other unopened box, you would increase the chances of getting the prize by a factor of 999 (from 1/1000 to 999/1000). On the other hand, if a third person walked in and had to choose one of the two remaining unopened boxes, but *without knowing* that initially there were 1000 boxes, nor which one you initially chose, he or she would pick the box with the prize with a probability of 1/2. The difference in expected outcomes is due to different prior information, and it nicely illustrates that the probabilities we assign to events reflect the state of our knowledge.

This problem, first discussed in a slightly different form in 1959 by Martin Gardner in his "Mathematical Games" column [3] in *Scientific American* (the "Three Prisoner Problem"), sounds nearly trivial and uncontroversial. Nevertheless, when the same mathematical problem was publicized for the case of $N = 3$ by Marilyn vos Savant in her newspaper column in 1990 [6], it generated an amazing amount of controversy. Here is a transcript of her column:

> Suppose you're on a game show, and you're given the choice of three doors. Behind one door is a car, behind the others, goats. You pick a door, say #1, and the host, who knows what's behind the doors, opens another door, say #3, which has a goat. He says to you, "Do you want to pick door #2?" Is it to your advantage to switch your choice of doors?

vos Savant also provided the correct answer to her question (also known as "the Monty Hall problem"): you should switch the doors because it increases your chance of getting the car from 1/3 to 2/3. After her column was published, vos Savant received thousands of letters from readers, including many academics and mathematicians,[1] all claiming that vos Savant's answer is wrong and that the probability is 1/2 for both unopened doors. But as we know from the less confusing case with large N discussed above (this is why we started with the $N = 1000$ version), vos Savant was right and the unhappy readers were wrong. Nevertheless, if you side with her readers, you may wish to write a little computer simulation of this game and you will change your mind (as did about half of her readers). Indeed, vos Savant called on math teachers to perform experiments with playing cards in their classrooms—they *experimentally* verified that it pays to switch! Subsequently, the problem was featured in a 2011 episode of the pop-science television series *MythBusters*, where the hosts reached the same conclusion.

Here is a formal derivation of the solution using Bayes' rule. H_i is the hypothesis that the prize is in the ith box, and $p(H_i|I) = 1/N$ is its prior probability given background information I. Without a loss of generality, the box chosen initially can be enumerated as the first box. The "data" that $N - 2$ boxes, all but the first box and the kth box ($k > 1$), are empty is d_k (i.e., d_k says that the kth box remains closed). The probability that the prize is in the first box, given I and k, can be evaluated using Bayes' rule (see eq. 3.11),

$$p(H_1|d_k, I) = \frac{p(d_k|H_1, I)p(H_1|I)}{p(d_k|I)}. \tag{3.12}$$

The probability that the kth box remains unopened given that H_1 is true, $p(d_k|H_1, I)$, is $1/(N-1)$ because this box is randomly chosen from $N-1$ boxes. The denominator can be expanded using the law of total probability,

$$p(d_k|I) = \sum_{i=1}^{N} p(d_k|H_i, I)p(H_i|I). \tag{3.13}$$

[1] For amusing reading, check out http://www.marilynvossavant.com/articles/gameshow.html

The probability that the kth box stays unopened, given that H_i is true, is

$$p(d_k|H_i, I) = \begin{cases} 1 \text{ for } k = i, \\ 0 \text{ otherwise,} \end{cases} \tag{3.14}$$

except when $i = 1$ (see above). This reduces the sum to only two terms:

$$p(d_k|I) = p(d_k|H_1, I)p(H_1|I) + p(d_k|H_k, I)p(H_k|I) = \frac{1}{(N-1)N} + \frac{1}{N} = \frac{1}{N-1}. \tag{3.15}$$

This result might appear to agree with our intuition because there are $N - 1$ ways to choose one box out of $N - 1$ boxes, but this interpretation is not correct: the kth box is not chosen randomly in the case when the prize is not in the first box, but instead it *must* be chosen (the second term in the sum above). Hence, from eq. 3.12, the probability that the prize is in the first (initially chosen) box is

$$p(H_1|d_k, I) = \frac{\frac{1}{(N-1)}\frac{1}{N}}{\frac{1}{(N-1)}} = \frac{1}{N}. \tag{3.16}$$

It is easy to show that $p(H_k|d_k, I) = (N - 1)/N$. Note that $p(H_1|d_k, I)$ is equal to the prior probability $p(H_1|I)$; that is, the opening of $N - 2$ empty boxes (data d_k) did not improve our knowledge of the content of the first box (but it did improve our knowledge of the content of the kth box by a factor of $N - 1$).

Example: 2×2 contingency table

A number of illustrative examples about the use of conditional probabilities exist for the simple case of two discrete variables that can have two different values, yielding four possible outcomes. We shall use a medical test here: one variable is the result of a test for some disease, T, and the test can be negative (0) or positive (1); the second variable is the health state of the patient, or the presence of disease D: the patient can have a disease (1) or not (0). There are four possible combinations in this sample space: $T = 0, D = 0$; $T = 0, D = 1$; $T = 1, D = 0$; and $T = 1, D = 1$. Let us assume that we know their probabilities. If the patient is healthy ($T = 0$), the probability for the test being positive (a false positive) is $p(T = 1|D = 0) = \epsilon_{fP}$, where ϵ_{fP} is (typically) a small number, and obviously $p(T = 0|D = 0) = 1 - \epsilon_{fP}$. If the patient has the disease ($T = 1$), the probability for the test being negative (a false negative) is $p(T = 0|D = 1) = \epsilon_{fN}$, and $p(T = 1|D = 1) = 1 - \epsilon_{fN}$. For a visual summary, see figure 3.3. Let us assume that we also know that the prior probability (in the absence of any testing, for example, based on some large population studies unrelated to our test) for the disease in question is $p(D = 1) = \epsilon_D$, where ϵ_D is a small number (of course, $p(D = 0) = 1 - \epsilon_D$).

Assume now that our patient took the test and it came out positive ($T = 1$). What is the probability that our patient has contracted the disease, $p(D = 1|T = 1)$?

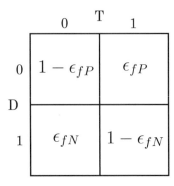

Figure 3.3. A contingency table showing $p(T|D)$.

Using Bayes' rule (eq. 3.11), we have

$$p(D = 1|T = 1) = \frac{p(T = 1|D = 1)\, p(D = 1)}{p(T = 1|D = 0)\, p(D = 0) + p(T = 1|D = 1)\, p(D = 1)},$$
(3.17)

and given our assumptions,

$$p(D = 1|T = 1) = \frac{\epsilon_D - \epsilon_{fN}\,\epsilon_D}{\epsilon_D + \epsilon_{fP} - [\epsilon_D\,(\epsilon_{fP} + \epsilon_{fN})]}.$$
(3.18)

For simplicity, let us ignore second-order terms since all ϵ parameters are presumably small, and thus

$$p(D = 1|T = 1) = \frac{\epsilon_D}{\epsilon_D + \epsilon_{fP}}.$$
(3.19)

This is an interesting result: we can only reliably diagnose a disease (i.e., $p(D = 1|T = 1) \sim 1$) if $\epsilon_{fP} \ll \epsilon_D$. For rare diseases, the test must have an exceedingly low false-positive rate! On the other hand, if $\epsilon_{fP} \gg \epsilon_D$, then $p(D = 1|T = 1) \sim \epsilon_D/\epsilon_{fP} \ll 1$ and the testing does not produce conclusive evidence. The false-negative rate is not quantitatively important as long as it is not much larger than the other two parameters. Therefore, when being tested, it is good to ask about the test's false-positive rate.

If this example is a bit confusing, consider a sample of 1000 tested people, with $\epsilon_D = 0.01$ and $\epsilon_{fP} = 0.02$. Of those 1000 people, we expect that 10 of them have the disease and all, assuming a small ϵ_{fN}, will have a positive test. However, an additional $\sim$20 people will be selected due to a false-positive result, and we will end up with a group of 30 people who tested positively. The chance to pick a person with the disease will thus be 1/3.

An identical computation applies to a jury deciding whether a DNA match is sufficient to declare a murder suspect guilty (with all the consequences of such a verdict). In order for a positive test outcome to represent conclusive evidence, the applied DNA test must have a false-positive rate much lower than the probability of randomly picking the true murderer on the street. The larger the effective community

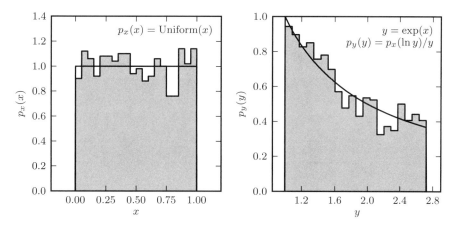

Figure 3.4. An example of transforming a uniform distribution. In the left panel, x is sampled from a uniform distribution of unit width centered on $x = 0.5$ ($\mu = 0$ and $W = 1$; see §3.3.1). In the right panel, the distribution is transformed via $y = \exp(x)$. The form of the resulting pdf is computed from eq. 3.20.

from which a murder suspect is taken (or DNA database), the better the DNA test must be to convincingly reach a guilty verdict.

These contingency tables are simple examples of the concepts which underlie *model selection* and *hypothesis testing*, which will be discussed in more detail in §4.6.

3.1.4. Transformations of Random Variables

Any function of a random variable is itself a random variable. It is a common case in practice that we measure the value of some variable x, but the interesting final result is a function $y(x)$. If we know the probability density distribution $p(x)$, where x is a random variable, what is the distribution $p(y)$, where $y = \Phi(x)$ (with $x = \Phi^{-1}(y)$)? It is easy to show that

$$p(y) = p\left[\Phi^{-1}(y)\right] \left|\frac{d\Phi^{-1}(y)}{dy}\right|. \tag{3.20}$$

For example, if $y = \Phi(x) = \exp(x)$, then $x = \Phi^{-1}(y) = \ln(y)$. If $p(x) = 1$ for $0 \le x \le 1$ and 0 otherwise (a uniform distribution), eq. 3.20 leads to $p(y) = 1/y$, with $1 \le y \le e$. That is, a uniform distribution of x is transformed into a nonuniform distribution of y (see figure 3.4).

Note that cumulative statistics, such as the median, do not change their order under monotonic transformations (e.g., given $\{x_i\}$, the median of x and the median of $\exp(x)$ correspond to the same data point).

If some value of x, say x_0, is determined with an uncertainty σ_x, then we can use a Taylor series expansion to estimate the uncertainty in y, say σ_y, at $y_0 = \Phi(x_0)$ as

$$\sigma_y = \left|\frac{d\Phi(x)}{dx}\right|_0 \sigma_x, \tag{3.21}$$

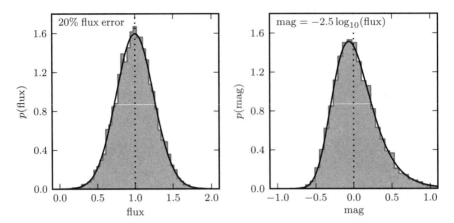

Figure 3.5. An example of Gaussian flux errors becoming non-Gaussian magnitude errors. The dotted line shows the location of the mean flux; note that this is not coincident with the peak of the magnitude distribution.

where the derivative is evaluated at x_0. While often used, this approach can produce misleading results when it is insufficient to keep only the first term in the Taylor series. For example, if the flux measurements follow a Gaussian distribution with a relative accuracy of a few percent, then the corresponding distribution of astronomical magnitudes (the logarithm of flux; see appendix C) is close to a Gaussian distribution. However, if the relative flux accuracy is 20% (corresponding to the so-called "5σ" detection limit), then the distribution of magnitudes is skewed and non-Gaussian (see figure 3.5). Furthermore, the mean magnitude is not equal to the logarithm of the mean flux (but the medians still correspond to each other!).

3.2. Descriptive Statistics

An arbitrary distribution function $h(x)$ can be characterized by its "location" parameters, "scale" or "width" parameters, and (typically dimensionless) "shape" parameters. As discussed below, these parameters, called descriptive statistics, can describe both various analytic distribution functions, as well as being determined directly from data (i.e., from our estimate of $h(x)$, which we named $f(x)$). When these parameters are based on $h(x)$, we talk about *population* statistics; when based on a finite-size data set, they are called *sample* statistics.

3.2.1. Definitions of Descriptive Statistics

Here are definitions for some of the more useful descriptive statistics:

- Arithmetic mean (also known as the expectation value),

$$\mu = E(x) = \int_{-\infty}^{\infty} x h(x)\, dx \qquad (3.22)$$

- Variance,

$$V = \int_{-\infty}^{\infty} (x - \mu)^2 h(x)\,dx \tag{3.23}$$

- Standard deviation,

$$\sigma = \sqrt{V} \tag{3.24}$$

- Skewness,

$$\Sigma = \int_{-\infty}^{\infty} \left(\frac{x - \mu}{\sigma}\right)^3 h(x)\,dx \tag{3.25}$$

- Kurtosis,

$$K = \int_{-\infty}^{\infty} \left(\frac{x - \mu}{\sigma}\right)^4 h(x)\,dx \; - \; 3 \tag{3.26}$$

- Absolute deviation about d,

$$\delta = \int_{-\infty}^{\infty} |x - d| h(x)\,dx \tag{3.27}$$

- Mode (or the most probable value in case of unimodal functions), x_m,

$$\left(\frac{dh(x)}{dx}\right)_{x_m} = 0 \tag{3.28}$$

- $p\%$ quantiles (p is called a percentile), q_p,

$$\frac{p}{100} = \int_{-\infty}^{q_p} h(x)\,dx \tag{3.29}$$

Although this list may seem to contain (too) many quantities, remember that they are trying to capture the behavior of a completely general function $h(x)$. The variance, skewness, and kurtosis are related to the kth central moments (with $k = 2, 3, 4$) defined analogously to the variance (the variance is identical to the second central moment). The skewness and kurtosis are measures of the distribution shape, and will be discussed in more detail when introducing specific distributions below. Distributions that have a long tail toward x larger than the "central location" have positive skewness, and symmetric distributions have no skewness. The kurtosis is defined relative to the Gaussian distribution (thus it is adjusted by the "3" in eq. 3.26), with highly peaked ("leptokurtic") distributions having positive kurtosis, and flat-topped ("platykurtic") distributions

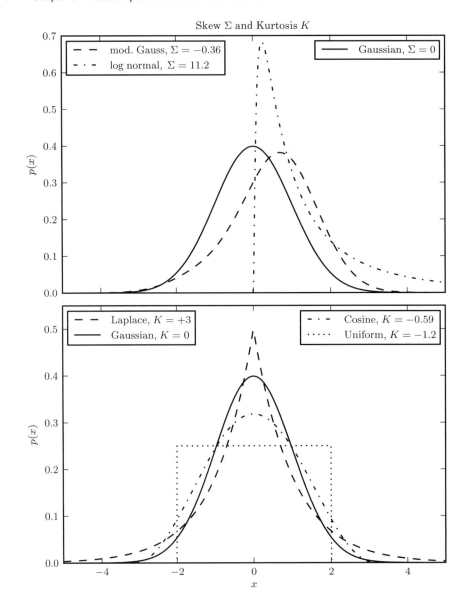

Figure 3.6. An example of distributions with different skewness Σ (top panel) and kurtosis K (bottom panel). The modified Gaussian in the upper panel is a normal distribution multiplied by a Gram–Charlier series (see eq. 4.70), with $a_0 = 2$, $a_1 = 1$, and $a_2 = 0.5$. The log-normal has $\sigma = 1.2$.

having negative kurtosis (see figure 3.6). The higher the distribution's moment, the harder it is to estimate it with small samples, and furthermore, there is more sensitivity to outliers (less robustness). For this reason, higher-order moments, such as skewness and kurtosis should be used with caution when samples are small.

The above statistical functions are among the many built into NumPy and SciPy. Useful functions to know about are numpy.mean, numpy.median, numpy.var, numpy.percentile, numpy.std, scipy.stats.skew, scipy.stats.kurtosis, and scipy.stats.mode. For example, to compute the quantiles of a one-dimensional array x, use the following:

```
import numpy as np
x = np.random.random(100) # 100 random numbers
q25, q50, q75 = np.percentile(x, [25, 50, 75])
```

For more information, see the NumPy and SciPy documentation of the above functions.

The absolute deviation about the mean (i.e., $d = \bar{x}$) is also called the mean deviation. When taken about the median, the absolute deviation is minimized. The most often used quantiles are the median, q_{50}, and the first and third quartile, q_{25} and q_{75}. The difference between the third and the first quartiles is called the interquartile range. A very useful relationship between the mode, the median and the mean, valid for mildly non-Gaussian distributions (see problem 2 in Lup93 for an elegant proof based on Gram–Charlier series[2]) is

$$x_m = 3\,q_{50} - 2\,\mu. \tag{3.30}$$

For example, this relationship is valid exactly for the Poisson distribution.

Note that some distributions do not have finite variance, such as the Cauchy distribution discussed below (§3.3.5). Obviously, when the distribution's variance is infinite (i.e., the tails of $h(x)$ do not decrease faster than x^{-3} for large $|x|$), the skewness and kurtosis will diverge as well.

3.2.2. Data-Based Estimates of Descriptive Statistics

Any of these quantities can be estimated directly from data, in which case they are called sample statistics (instead of population statistics). However, in this case we also need to be careful about the uncertainties of these estimates. Hereafter, assume that we are given N measurements, x_i, $i = 1, \ldots, N$, abbreviated as $\{x_i\}$. We will ignore for a moment the fact that measurements must have some uncertainty of their own (errors); alternatively, we can assume that x_i are measured much more accurately than the range of observed values (i.e., $f(x)$ reflects some "physics" rather than measurement errors). Of course, later in the book we shall relax this assumption.

In general, when estimating the above quantities for a sample of N measurements, the integral $\int_{-\infty}^{\infty} g(x)h(x)\,dx$ becomes proportional to the sum $\sum_i^N g(x_i)$, with the constant of proportionality $\sim(1/N)$. For example, the *sample arithmetic*

[2]The Gram–Charlier series is a convenient way to describe distribution functions that do not deviate strongly from a Gaussian distribution. The series is based on the product of a Gaussian distribution and the sum of the Hermite polynomials (see §4.7.4).

mean, $\bar{x}$, and the *sample standard deviation, s*, can be computed via standard formulas,

$$\bar{x} = \frac{1}{N} \sum_{i=1}^{N} x_i \tag{3.31}$$

and

$$s = \sqrt{\frac{1}{N-1} \sum_{i=1}^{N} (x_i - \bar{x})^2}. \tag{3.32}$$

The reason for the $(N-1)$ term instead of the naively expected N in the second expression is related to the fact that $\bar{x}$ is also determined from data (we discuss this subtle fact and the underlying statistical justification for the $(N-1)$ term in more detail in §5.6.1). With N replaced by $N-1$ (the so-called Bessel's correction), the sample variance (i.e., s^2) becomes unbiased (and the sample standard deviation given by expression 3.32 becomes a less biased, but on average still underestimated, estimator of the true standard deviation; for a Gaussian distribution, the underestimation varies from 20% for $N = 2$, to 3% for $N = 10$ and is less than 1% for $N > 30$). Similar factors that are just a bit different from N, and become N for large N, also appear when computing the skewness and kurtosis. What a "large N" means depends on a particular case and preset level of accuracy, but generally this transition occurs somewhere between $N = 10$ and $N = 100$ (in a different context, such as the definition of a "massive" data set, the transition may occur at N of the order of a million, or even a billion, again depending on the problem at hand).

We use different symbols in the above two equations ($\bar{x}$ and s) than in eqs. 3.22 and 3.24 (μ and σ) because the latter represent the "truth" (they are definitions based on the true $h(x)$, whatever it may be), and the former are simply *estimators* of that truth based on a *finite-size* sample ($\hat{x}$ is often used instead of $\bar{x}$). These estimators have a variance and a bias, and often they are judged by comparing their mean squared errors,

$$\text{MSE} = V + \text{bias}^2, \tag{3.33}$$

where V is the variance, and *the bias is defined as the expectation value of the difference between the estimator and its true (population) value*. Estimators whose variance and bias vanish as the sample size goes to infinity are called *consistent estimators*. An estimator can be unbiased but not consistent: as a simple example, consider taking the first measured value as an estimator of the mean value. This is unbiased, but its variance does not decrease with the sample size.

Obviously, we should also know the uncertainty in our estimators for μ ($\bar{x}$) and σ (s; note that s is *not* an uncertainty estimate for $\bar{x}$—this is a common misconception!). A detailed discussion of what exactly "uncertainty" means in this context, and how to derive the following expressions, can be found in chapter 5. Briefly, when N is large (at least 10 or so), and if the variance of $h(x)$ is finite, we expect from the central limit theorem (see below) that $\bar{x}$ and s will be distributed around their values given by eqs. 3.31 and 3.32 according to Gaussian distributions with the widths (standard errors) equal to

$$\sigma_{\bar{x}} = \frac{s}{\sqrt{N}}, \tag{3.34}$$

which is called *the standard error of the mean*, and

$$\sigma_s = \frac{s}{\sqrt{2(N-1)}} = \frac{1}{\sqrt{2}} \sqrt{\frac{N}{N-1}} \, \sigma_{\bar{x}}. \tag{3.35}$$

The first expression is also valid when the standard deviation for parent population is known a priori (i.e., it is not determined from data using eq. 3.32). Note that for large N, the uncertainty of the location parameter is about 40% larger than the uncertainty of the scale parameter ($\sigma_{\bar{x}} \sim \sqrt{2}\,\sigma_s$). Note also that for small N, σ_s is not much smaller than s itself. The implication is that $s < 0$ is allowed according to the standard interpretation of "error bars" that implicitly assumes a Gaussian distribution! We shall return to this seemingly puzzling result in chapter 5 (§5.6.1), where an expression to be used instead of eq. 3.35 for small N (< 10) is derived.

Estimators can be compared in terms of their *efficiency*, which measures how large a sample is required to obtain a given accuracy. For example, the median determined from data drawn from a Gaussian distribution shows a scatter around the true location parameter (μ in eq. 1.4) larger by a factor of $\sqrt{\pi/2} \sim 1.253$ than the scatter of the mean value (see eq. 3.37 below). Since the scatter decreases with $1/\sqrt{N}$, the efficiency of the mean is $\pi/2$ times larger than the efficiency of the median. The smallest attainable variance for an unbiased estimator is called the *minimum variance bound (MVB)* and such an estimator is called the *minimum variance unbiased estimator (MVUE)*. We shall discuss in more detail how to determine the MVB in §4.2. Methods for estimating the bias and variance of various estimators are further discussed in §4.5 on bootstrap and jackknife methods. An estimator is *asymptotically normal* if its distribution around the true value approaches a Gaussian distribution for large sample size, with variance decreasing proportionally to $1/N$.

For the case of real data, which can have spurious measurement values (often, and hereafter, called "outliers"), quantiles offer a more robust method for determining location and scale parameters than the mean and standard deviation. For example, the median is a much more *robust* estimator of the location than the mean, and the interquartile range ($q_{75} - q_{25}$) is a more *robust* estimator of the scale parameter than the standard deviation. This means that the median and interquartile range are much less affected by the presence of outliers than the mean and standard deviation. It is easy to see why: if you take 25% of your measurements that are larger than q_{75} and arbitrarily modify them by adding a large number to all of them (or multiply them all by a large number, or different large numbers), both the mean and the standard deviation will be severely affected, while the median and the interquartile range will remain unchanged. Furthermore, even in the absence of outliers, for some distributions that do not have finite variance, such as the Cauchy distribution, the median and the interquartile range are the best choices for estimating location and scale parameters. Often, the interquartile range is renormalized so that the width estimator, σ_G, becomes an unbiased estimator of σ

for a *perfect*[3] Gaussian distribution (see §3.3.2 for the origin of the factor 0.7413),

$$\sigma_G = 0.7413 \, (q_{75} - q_{25}). \tag{3.36}$$

There is, however, a price to pay for this robustness. For example, we already discussed that the efficiency of the median as a location estimator is poorer than that for the mean in the case of a Gaussian distribution. An additional downside is that it is much easier to compute the mean than the median for large samples; although the efficient algorithms described in §2.5.1 make this downside somewhat moot. In practice, one is often willing to pay the price of ~25% larger errors for the median than for the mean (assuming nearly Gaussian distributions) to avoid the possibility of catastrophic failures due to outliers.

AstroML provides a convenience routine for calculating σ_G:

```
import numpy as np
from astroML import stats
x = np.random.normal(size=1000)   # 1000 normally
    # distributed points
stats.sigmaG(x)
1.0302378533978402
```

A very useful result is the following expression for computing standard error, σ_{qp}, for an arbitrary quantile q_p (valid for large N; see Lup93 for a derivation):

$$\sigma_{qp} = \frac{1}{h_p} \sqrt{\frac{p(1-p)}{N}}, \tag{3.37}$$

where h_p is the value of the probability distribution function at the pth percentile (e.g., for the median, $p = 0.5$). Unfortunately, σ_{qp} depends on the underlying $h(x)$. In the case of a Gaussian distribution, it is easy to derive that the standard error for the median is

$$\sigma_{q50} = s \sqrt{\frac{\pi}{2N}}, \tag{3.38}$$

with $h_{50} = 1/(s\sqrt{2\pi})$ and $s \sim \sigma$ in the limit of large N, as mentioned above. Similarly, the standard error for σ_G (eq. 3.36) is $1.06s/\sqrt{N}$, or about 50% larger than σ_s (eq. 3.35). The coefficient (1.06) is derived assuming that q_{25} and q_{75} are

---

[3]A real mathematician would probably laugh at placing the adjective *perfect* in front of "Gaussian" here. What we have in mind is a habit, especially common among astronomers, to (mis)use the word Gaussian for any distribution that even remotely resembles a bell curve, even when outliers are present. Our statement about the scatter of the median being larger than the scatter of the mean is not correct in such cases.

not correlated, which is not true for small samples (which can have scatter in σ_G up to 10–15% larger, even if drawn from a Gaussian distribution).

We proceed with an overview of the most common distributions used in data mining and machine learning applications. These distributions are called $p(x|I)$ to distinguish their mathematical definitions (equivalent to $h(x)$ introduced earlier) from $f(x)$ estimated from data. All the parameters that quantitatively describe the distribution $p(x)$ are "hidden" in I (for "information"), and $p(x|I)\,dx$ is interpreted as the probability of having a value of the random variable between x and $x + dx$. All $p(x|I)$ are normalized according to eq. 1.2.

3.3. Common Univariate Distribution Functions

Much of statistics is based on the analysis of distribution functions for random variables, and most textbooks cover this topic in great detail. From a large number of distributions introduced by statisticians, a relatively small number cover the lion's share of practical applications. We introduce them and provide their basic properties here, without pretending to cover all the relevant aspects. For an in-depth understanding, please consult statistics textbooks (e.g., Lup93, Wass10).

3.3.1. The Uniform Distribution

The uniform distribution is implemented in `scipy.stats.uniform`:

```
from scipy import stats
dist = stats.uniform(0, 2)  # left edge = 0,
        # width = 2
r = dist.rvs(10)    # ten random draws
p = dist.pdf(1)  # pdf evaluated at x=1
```

The uniform distribution (also known as "top-hat" and "box") is described by

$$p(x|\mu, W) = \frac{1}{W} \text{ for } |x - \mu| \leq \frac{W}{2}, \tag{3.39}$$

and 0 otherwise, where W is the width of the "box" (see figure 3.7). Obviously, the mean and the median are equal to μ. Although simple, the uniform distribution often appears in practical problems (e.g., the distribution of the position of a photon detected in a finite-size pixel) and a well-known result is that

$$\sigma = \frac{W}{\sqrt{12}} \sim 0.3W, \tag{3.40}$$

where σ is computed using eq. 3.24. Since the uniform distribution is symmetric, its skewness is 0, and it is easy to show that its kurtosis is -1.2 (i.e., platykurtic as we would expect). Note that we can arbitrarily vary the location of this distribution

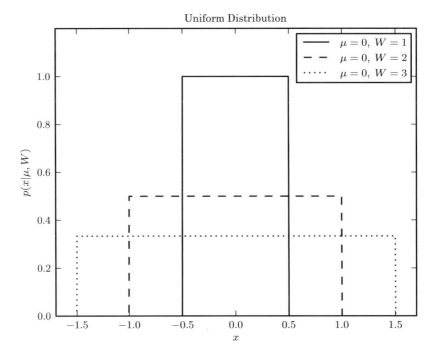

Figure 3.7. Examples of a uniform distribution (see eq. 3.39).

along the x-axis, and multiply x by an arbitrary scale factor, without any impact on the distribution's *shape*. That is, if we define

$$z = \frac{x - \mu}{W}, \tag{3.41}$$

then

$$p(z) = 1 \text{ for } |z| \leq \frac{1}{2}, \text{ and } p(z) = 0 \text{ for } |z| > \frac{1}{2}. \tag{3.42}$$

This independence of the distribution's shape on shift and normalization of the random variable x is a general result, and illustrates the meaning of the "location" and "scale" (or "width") parameters. An analogous "shift and rescale" transformation can be seen in the definitions of skewness and kurtosis (eqs. 3.25 and 3.26).

3.3.2. The Gaussian (normal) Distribution

The normal distribution is implemented in `scipy.stats.norm`:

```
from scipy import stats
dist = stats.norm(0, 1)    # mean = 0, stdev = 1
r = dist.rvs(10)   # ten random draws
p = dist.pdf(0) # pdf evaluated at x=0
```

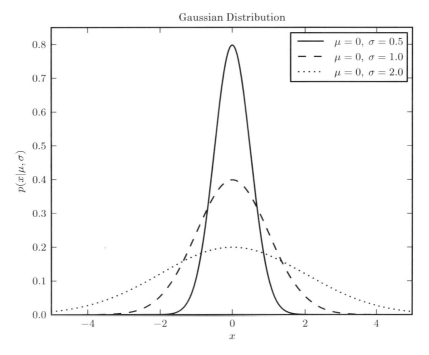

Figure 3.8. Examples of a Gaussian distribution (see eq. 3.43).

We have already introduced the Gaussian distribution in eq. 1.4, repeated here for completeness,

$$p(x|\mu, \sigma) = \frac{1}{\sigma\sqrt{2\pi}} \exp\left(\frac{-(x-\mu)^2}{2\sigma^2}\right). \tag{3.43}$$

The Gaussian distribution is also called the normal distribution, $\mathcal{N}(\mu, \sigma)$. The standard deviation σ is often replaced by the variance σ^2 and the distribution is then referred to as $\mathcal{N}(\mu, \sigma^2)$, but here we uniformly adopt the first form. Since the Gaussian distribution is symmetric (see figure 3.8), its skewness is 0, and by the definition of kurtosis, its kurtosis is 0 as well (kurtosis is defined so that this is the case).

The Gaussian distribution has two main properties that make it special. First, it lends itself to analytic treatment in many cases; most notably, a convolution of two Gaussian distributions is also Gaussian (it's hard to believe, but computers haven't existed forever). The convolution of a function $f(x)$ with a function $g(x)$ (both assumed real functions) is defined as

$$(f \star g)(x) = \int_{-\infty}^{\infty} f(x')\,g(x-x')\,dx' = \int_{-\infty}^{\infty} f(x-x')\,g(x')\,dx'. \tag{3.44}$$

In particular, the convolution of a Gaussian distribution $\mathcal{N}(\mu_o, \sigma_o)$ (e.g., an intrinsic distribution we are trying to measure) with a Gaussian distribution $\mathcal{N}(b, \sigma_e)$ (i.e., Gaussian error distribution with bias b and random error σ_e) produces parameters for the resulting Gaussian[4] $\mathcal{N}(\mu_C, \sigma_C)$ given by

$$\mu_C = (\mu_o + b) \quad \text{and} \quad \sigma_C = \left(\sigma_o^2 + \sigma_e^2\right)^{1/2}. \tag{3.45}$$

Similarly, the Fourier transform of a Gaussian is also a Gaussian (see §10.2.2). Another unique feature of the Gaussian distribution is that the sample mean and the sample variance are independent.

Second, the central limit theorem (see below) tells us that the mean of samples drawn from an almost arbitrary distribution will follow a Gaussian distribution. Hence, much of statistics and most classical results are based on an underlying assumption that a Gaussian distribution is applicable, although often it is not. For example, the ubiquitous least-squares method for fitting linear models to data is invalid when the measurement errors do not follow a Gaussian distribution (see chapters 5 and 8). Another interesting aspect of the Gaussian distribution is its "information content," which we discuss in chapter 5.

The cumulative distribution function for a Gaussian distribution,

$$P(x|\mu, \sigma) = \frac{1}{\sigma\sqrt{2\pi}} \int_{-\infty}^{x} \exp\left(\frac{-(x' - \mu)^2}{2\sigma^2}\right) dx', \tag{3.46}$$

cannot be evaluated in closed form in terms of elementary functions, and usually it is expressed in terms of the *Gauss error function*,

$$\text{erf}(z) = \frac{2}{\sqrt{\pi}} \int_{0}^{z} \exp\left(-t^2\right) dt. \tag{3.47}$$

Tables and computer algorithms for evaluating $\text{erf}(z)$ are readily available (note that $\text{erf}(\infty) = 1$). With the aid of the error function,

$$P(x|\mu, \sigma) = \frac{1}{2}\left(1 \pm \text{erf}\left(\frac{|x - \mu|}{\sigma\sqrt{2}}\right)\right), \tag{3.48}$$

with the plus sign for $x > \mu$ and the minus sign otherwise.

The Gauss error function is available in `scipy.special`:

```
>>> from scipy.special import erf
>>> erf(1)
0.84270079294971478
```

[4]Note that the product of two Gaussians retains a Gaussian shape but its integral is *not* unity. Furthermore, the location and scale parameters are *very different* from those given by eq. 3.45. For example, the location parameter is a weighted sum of the two input location parameters (see §5.2.1 and eq. 5.20 for a detailed discussion).

Note that the integral of $p(x|\mu, \sigma)$ given by eq. 3.43 between two arbitrary integration limits, a and b, can be obtained as the difference of the two integrals $P(b|\mu, \sigma)$ and $P(a|\mu, \sigma)$. As a special case, the integral for $a = \mu - M\sigma$ and $b = \mu + M\sigma$ ("$\pm M\sigma$" ranges around μ) is equal to $\text{erf}(M/\sqrt{2})$. The values for $M = 1, 2,$ and 3 are 0.682, 0.954, and 0.997. The incidence level of "one in a million" corresponds to $M = 4.9$ and "one in a billion" to $M = 6.1$. Similarly, the interquartile range for the Gaussian distribution can be expressed as

$$q_{75} - q_{25} = \sigma \, 2 \sqrt{2} \, \text{erf}^{-1}(0.5) \approx 1.349 \, \sigma, \tag{3.49}$$

which explains the coefficient in eq. 3.36.

If x follows a Gaussian distribution $\mathcal{N}(\mu, \sigma)$, then $y = \exp(x)$ has a log-normal distribution. The log-normal distribution arises when the variable is a product of many independent positive variables (and the central limit theorem is applicable to the sum of their logarithms). The mean value of y is $\exp(\mu + \sigma^2/2)$, the mode is $\exp(\mu - \sigma^2)$, and the median is $\exp(\mu)$. Note that the mean value of y is *not* equal to $\exp(\mu)$, with the discrepancy increasing with the σ^2/μ ratio; this fact is related to our discussion of the failing Taylor series expansion at the end of §3.1.4.

3.3.3. The Binomial Distribution

The binomial distribution is implemented in `scipy.stats.binomial`:

```
from scipy import stats
dist = stats.binom(20, 0.7)  # N = 20, b = 0.7
r = dist.rvs(10)   # ten random draws
p = dist.pmf(8)  # prob. evaluated at k=8
```

Unlike the Gaussian distribution, which describes the distribution of a continuous variable, the binomial distribution describes the distribution of a variable that can take only two discrete values (say, 0 or 1, or success vs. failure, or an event happening or not). If the probability of success is b, then the distribution of a discrete variable k (an integer number, unlike x which is a real number) that measures how many times success occurred in N trials (i.e., measurements), is given by

$$p(k|b, N) = \frac{N!}{k!(N-k)!} \, b^k \, (1-b)^{N-k} \tag{3.50}$$

(see figure 3.9). The special case of $N = 1$ is also known as a Bernoulli distribution. This result can be understood as the probability of k consecutive successes followed by $(N - k)$ consecutive failures (the last two terms), multiplied by the number of different permutations of such a draw (the first term). The mean (i.e., the expected number of successes) for the binomial distribution is $\bar{k} = bN$, and its standard

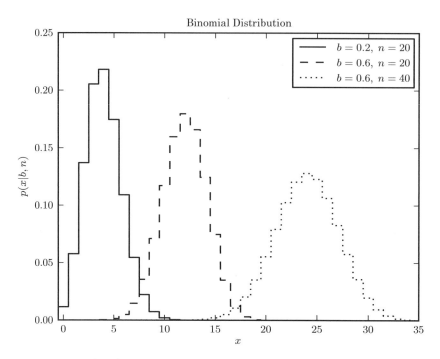

Figure 3.9. Examples of a binomial distribution (see eq. 3.50).

deviation is

$$\sigma_k = [N\,b\,(1-b)]^{1/2}\,. \tag{3.51}$$

A common example of a process following a binomial distribution is the flipping of a coin. If the coin is fair, then $b = 0.5$, and success can be defined as obtaining a head (or a tail). For a real coin tossed N times, with k successes, we can ask what is our best estimate of b, called b^0, and its uncertainty given these data, σ_b. If N is large, we can simply determine $b^0 = k/N$, with its uncertainty (standard error) following from eq. 3.51 ($\sigma_b = \sigma_k/N$), and *assume* that the probability distribution for the true value of b is given by a Gaussian distribution $\mathcal{N}(b^0, \sigma_b)$. For example, it is easy to compute that it takes as many as 10^4 tosses of a fair coin to convince yourself that it is indeed fair within 1% accuracy ($b = 0.500 \pm 0.005$). How to solve this problem in a general case, without having to assume a Gaussian error distribution, will be discussed in chapter 5.

Coin flips are assumed to be independent events (i.e., the outcome of one toss does not influence the outcome of another toss). In the case of drawing a ball from a bucket full of red and white balls, we will get a binomial distribution if the bucket is infinitely large (so-called drawing with replacement). If instead the bucket is of finite size, then $p(k|b, N)$ is described by a hypergeometric distribution, which we will not use here.

A related distribution is the multinomial distribution, which is a generalization of the binomial distribution. It describes the distribution of a variable that can have more than two discrete values, say M values, and the probability of each value is different and given by b_m, $b = 1, \ldots, M$. The multinomial distribution describes the distribution of M discrete variables k_m which count how many times the value indexed by m occurred in n trials.

> The function `multinomial` in the `numpy.random` submodule implements random draws from a multinomial distribution:
>
> ```
> from numpy.random import multinomial
> vals = multinomial(10, pvals=[0.2, 0.3, 0.5])
> # pvals sum to 1
> ```

3.3.4. The Poisson Distribution

> The Poisson distribution is implemented in `scipy.stats.poisson`:
>
> ```
> from scipy import stats
> dist = stats.poisson(5)   # mu = 5
> r = dist.rvs(10)   # ten random draws
> p = dist.pmf(3) # prob. evaluated at k=3
> ```

The Poisson distribution is a special case of the binomial distribution and thus it also describes the distribution of a discrete variable. The classic example of this distribution, and an early application, is analysis of the chance of a Prussian cavalryman being killed by the kick of a horse. If the number of trials, N, for a binomial distribution goes to infinity such that the probability of success, $p = k/N$, stays fixed, then the distribution of the number of successes, k, is controlled by $\mu = pN$ and given by

$$p(k|\mu) = \frac{\mu^k \exp(-\mu)}{k!} \tag{3.52}$$

(see figure 3.10). The mean (or expectation) value is μ, and it fully describes a Poisson distribution: the mode (most probable value) is $(\mu - 1)$, the standard deviation is $\sqrt{\mu}$, the skewness is $1/\sqrt{\mu}$, and the kurtosis is $1/\mu$. As μ increases, both the skewness and kurtosis decrease, and thus the Poisson distribution becomes more and more similar to a Gaussian distribution, $\mathcal{N}(\mu, \sqrt{\mu})$ (and remember that the Poisson distribution is a limiting case of the binomial distribution). Interestingly, although the Poisson distribution morphs into a Gaussian distribution for large μ, the expectation value of the difference between the mean and median does *not* become 0, but rather 1/6. Because of this transition to a Gaussian for large μ, the Poisson distribution is sometimes called the "law of small numbers." Sometimes it is also called the "law of rare events" but we emphasize that μ need *not* be a small number—the adjective "rare" comes from the fact that only a small fraction of a large number of trials N results in success ("p is small, not μ").

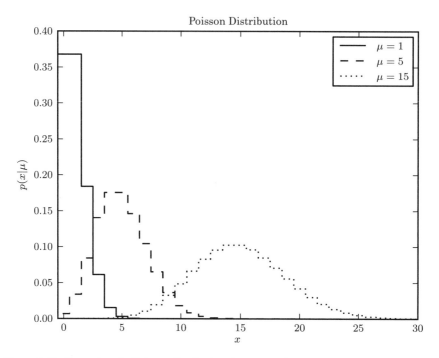

Figure 3.10. Examples of a Poisson distribution (see eq. 3.52).

The Poisson distribution is especially important in astronomy because it describes the distribution of the number of photons counted in a given interval. Even when it is replaced in practice by a Gaussian distribution for large μ, its Poissonian origin can be recognized from the relationship $\sigma^2 = \mu$ (the converse is not true).

3.3.5. The Cauchy (Lorentzian) Distribution

The Cauchy distribution is implemented in `scipy.stats.cauchy`:

```
from scipy import stats
dist = stats.cauchy(0, 1)  # mu = 0, gamma = 1
r = dist.rvs(10)   # ten random draws
p = dist.pdf(3) # pdf evaluated at x=3
```

Let us now return to distributions of a continuous variable and consider a distribution whose behavior is markedly different from any of those discussed above: the Cauchy, or Lorentzian, distribution,

$$p(x|\mu, \gamma) = \frac{1}{\pi \gamma} \left(\frac{\gamma^2}{\gamma^2 + (x - \mu)^2} \right). \tag{3.53}$$

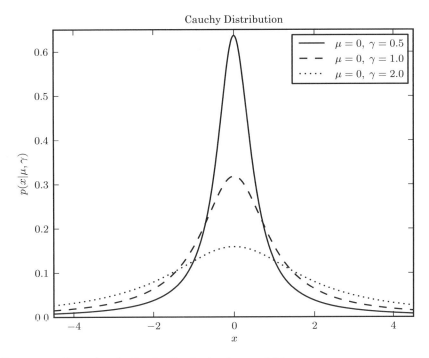

Figure 3.11. Examples of a Cauchy distribution (see eq. 3.53).

It is a symmetric distribution described by the location parameter μ and the scale parameter γ, and its median and mode are equal to μ (see figure 3.11). Because its tails decrease as slowly as x^{-2} for large $|x|$, the mean, variance, standard deviation, and higher moments do not exist. Therefore, given a set of measured x_i drawn from the Cauchy distribution, the location and scale parameters *cannot* be estimated by computing the mean value and standard deviation using standard expressions given by eqs. 3.31 and 3.32. To clarify, one can always compute a mean value for a set of numbers x_i, but this mean value will have a large scatter around μ, and furthermore, this scatter will *not* decrease with the sample size (see figure 3.12). Indeed, the mean values for many independent samples will themselves follow a Cauchy distribution. Nevertheless, μ and γ *can* be estimated as the median value and interquartile range for $\{x_i\}$. The interquartile range for the Cauchy distribution is equal to 2γ and thus

$$\sigma_G = 1.483\gamma. \tag{3.54}$$

We will see in chapter 5 how to justify this approach in a general case.

We note that the *ratio* of two independent standard normal variables ($z = (x - \mu)/\sigma$, with z drawn from $\mathcal{N}(0, 1)$) follows a Cauchy distribution with $\mu = 0$ and $\gamma = 1$. Therefore, in cases when the quantity of interest is obtained as a ratio of two other measured quantities, assuming that it is distributed as a Gaussian is a really bad idea if the quantity in the denominator has a finite chance of taking on a zero value. Furthermore, using the mean value to determine its location parameter (i.e., to get a "best" value implied by the measurements) will not achieve the "$1/\sqrt{N}$" error reduction. For a general case of the ratio of two random variables drawn from two

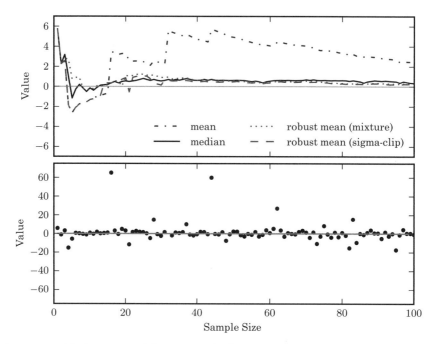

Figure 3.12. The bottom panel shows a sample of N points drawn from a Cauchy distribution with $\mu = 0$ and $\gamma = 2$. The top panel shows the sample median, sample mean, and two robust estimates of the location parameter (see text) as a function of the sample size (only points to the left from a given sample size are used). Note that the sample mean is not a good estimator of the distribution's location parameter. Though the mean appears to converge as N increases, this is deceiving: because of the large tails in the Cauchy distribution, there is always a high likelihood of a far-flung point affecting the sample mean. This behavior is markedly different from a Gaussian distribution where the probability of such "outliers" is much smaller.

different Gaussian distributions, $x = \mathcal{N}(\mu_2, \sigma_2)/\mathcal{N}(\mu_1, \sigma_1)$, the distribution of x is much more complex than the Cauchy distribution (it follows the so-called Hinkley distribution, which we will not discuss here).

Figure 3.12 also shows the results of two other robust procedures for estimating the location parameter that are often used in practice. The "clipped mean" approach computes the mean and the standard deviation for a sample (using σ_G is more robust than using eq. 3.32) and then excludes all points further than $K\sigma$ from the mean, with typically $K = 3$. This procedure is applied iteratively and typically converges very fast (in the case of a Gaussian distribution, only about 1 in 300 points is excluded). Another approach is to model the distribution as the sum of two Gaussian distributions, using methods discussed in §4.4. One of the two components is introduced to model the non-Gaussian behavior of the tails and its width is set to $K\sigma$, again with typically $K = 3$. As illustrated in figure 3.12, the location parameters estimated by these methods are similar to that of the median. In general, the performance of these methods depends on the actual distribution whose location parameter is being estimated. To summarize, when estimating the location parameter for a distribution that is not guaranteed to be Gaussian, one should use a more robust estimator than the plain sample mean given by eq. 3.31. The 3σ clipping and the median are the simplest alternatives, but ultimately full modeling of the underlying distribution in cases when additional information is available will yield the best results.

3.3.6. The Exponential (Laplace) Distribution

> The (two-sided) Laplace distribution is implemented in `scipy.stats.cauchy`, while the one-sided exponential distribution is implemented in `scipy.stats.expon`:

```
from scipy import stats
dist = stats.laplace(0, 0.5)   # mu = 0, delta = 0.5
r = dist.rvs(10)   # ten random draws
p = dist.pdf(3)   # pdf evaluated at x=3
```

Similarly to a Gaussian distribution, the exponential distribution is also fully specified by only two parameters,

$$p(x|\mu, \Delta) = \frac{1}{2\Delta} \exp\left(\frac{-|x - \mu|}{\Delta}\right). \tag{3.55}$$

We note that often the exponential distribution is defined only for $x > 0$, in which case it is called the one-sided exponential distribution, and that the above expression is often called the double exponential, or Laplace distribution (see figure 3.13). The formulation adopted here is easier to use in the context of modeling error distributions than a one-sided exponential distribution. The simplest case of a one-sided exponential distribution is $p(x|\tau) = \tau^{-1} \exp(-x/\tau)$, with both the expectation (mean) value and standard deviation equal to τ. This distribution describes the time between two successive events which occur continuously and independently at a constant rate (such as photons arriving at a detector); the number of such events during some fixed time interval, T, is given by the Poisson distribution with $\mu = T/\tau$ (see eq. 3.52).

Since the Laplace distribution is symmetric around μ, its mean, mode, and median are μ, and its skewness is 0. The standard deviation is

$$\sigma = \sqrt{2}\,\Delta \approx 1.414\,\Delta, \tag{3.56}$$

and the equivalent Gaussian width estimator determined from the interquartile range ($q_{75} - q_{25} = 2(\ln 2)\Delta$) is (see eq. 3.36)

$$\sigma_G = 1.028\,\Delta. \tag{3.57}$$

Note that σ is larger than σ_G ($\sigma \approx 1.38\sigma_G$), and thus their comparison can be used to detect deviations from a Gaussian distribution toward an exponential distribution. In addition, the fractions of objects in the distribution tails are vastly

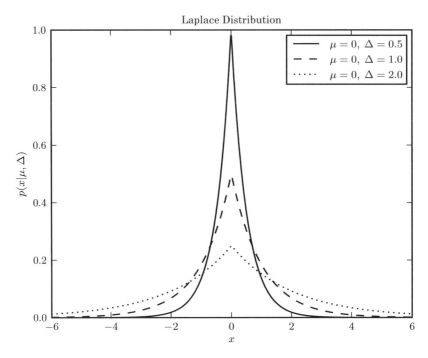

Figure 3.13. Examples of a Laplace distribution (see eq. 3.55).

different. While for the Gaussian distribution, $|x_i - \mu| > 5\sigma$ happens in fewer than one per million cases, for the exponential distribution it happens about once in a thousand cases. The kurtosis is yet another quantity that can distinguish between the Gaussian and exponential distributions; the kurtosis for the latter is 3 (see figure 3.6). Methods for comparing a measured distribution to hypothetical distributions, such as Gaussian and exponential, are discussed later in chapter 4 (§4.7) and in chapter 5.

3.3.7. The χ^2 Distribution

The χ^2 distribution is implemented in `scipy.stats.chi2`:

```
from scipy import stats
dist = stats.chi2(5)   # k = 5
r = dist.rvs(10)   # ten random draws
p = dist.pdf(1)  # pdf evaluated at x=1
```

The χ^2 distribution is one of the most important distributions in statistics. If $\{x_i\}$ are drawn from a Gaussian distribution and we define $z_i = (x_i - \mu)/\sigma$, then the sum of its squares, $Q = \sum_{i=1}^{N} z_i^2$, follows a χ^2 distribution with $k = N$ degrees of freedom

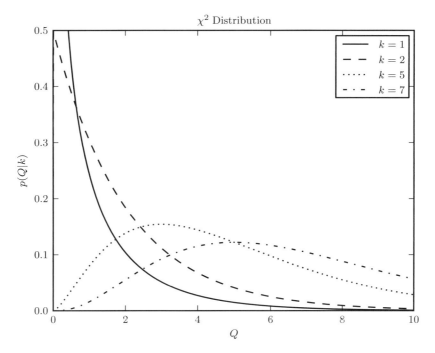

Figure 3.14. Examples of a χ^2 distribution (see eq. 3.58).

(see figure 3.14),

$$p(Q|k) \equiv \chi^2(Q|k) = \frac{1}{2^{k/2}\Gamma(k/2)} \, Q^{k/2-1} \, \exp(-Q/2) \text{ for } Q > 0, \quad (3.58)$$

where Γ is the gamma function (for positive integers k, $\Gamma(k) \equiv (k-1)!$, and it has a closed-form expression at the half integers relevant here). In other words, the distribution of Q values depends only on the sample size, and not on the actual values of μ and σ. The importance of the χ^2 distribution will become apparent when discussing the maximum likelihood method in chapter 4.

The gamma function and log-gamma function are two of the many functions available in scipy.special:

```
>>> from scipy import special
>>> special.gamma(5)
24.0
>>> special.gammaln(100)   # returns log(gamma(100))
359.1342053695754
```

TABLE 3.1.

Summary of the most useful descriptive statistics for some common distributions of a continuous variable. For the definitions of listed quantities, see eqs. 3.22–3.29 and 3.36. The symbol N/A is used in cases where a simple expression does not exist, or when the quantity does not exist.

Distribution	Parameters	$\bar{x}$	q_{50}	x_m	σ	σ_G	Σ	K
Gaussian	μ, σ	μ	μ	μ	σ	σ	0	0
Uniform	μ, W	μ	μ	N/A	$W/\sqrt{12}$	$0.371W$	0	-1.2
Exponential	μ, Δ	μ	μ	μ	$\sqrt{2}\Delta$	1.028Δ	0	3
Poisson	μ	μ	$\mu - 1/3$	$\mu - 1$	$\sqrt{\mu}$	N/A	$1/\sqrt{\mu}$	$1/\mu$
Cauchy	μ, γ	N/A	μ	μ	N/A	1.483γ	N/A	N/A
χ^2_{dof}	k	1	$(1 - 2/9k)^3$	max $(0, 1 - 2/k)$	$\sqrt{2/k}$	N/A	$\sqrt{8/k}$	$12/k$

It is convenient to define the χ^2 distribution *per degree of freedom* as

$$\chi^2_{\mathrm{dof}}(Q|k) \equiv \chi^2(Q/k|k). \tag{3.59}$$

The mean value for χ^2_{dof} is 1, the median is approximately equal to $(1 - 2/9k)^3$, and the standard deviation is $\sqrt{2/k}$. The skewness for χ^2_{dof} is $\sqrt{8/k}$, and its kurtosis is $12/k$. Therefore, as k increases, χ^2_{dof} tends to $\mathcal{N}(1, \sqrt{2/k})$. For example, if $k = 200$, the probability that the value of Q/k is in the 0.9–1.1 interval is 0.68. On the other hand, when $k = 20,000$, the interval containing the same probability is 0.99–1.01. The quantiles for the χ^2_{dof} distribution do not have a closed-form expression.

Note that the sum Q computed for a set of x_i is very sensitive to outliers (i.e., points many σ "away" from μ). A single spurious measurement has the potential of significantly influencing analysis based on χ^2. How to deal with this problem in a general case is discussed in §4.3.

Descriptive statistics for the above distributions are summarized in table 3.1.

We review five more distributions below that frequently arise when analyzing data and comparing empirical distributions.

3.3.8. Student's t Distribution

Student's t distribution is implemented in `scipy.stats.t`:

```
from scipy import stats
dist = stats.t(5)    # k = 5
r = dist.rvs(10)     # ten random draws
p = dist.pdf(4)      # pdf evaluated at x=4
```

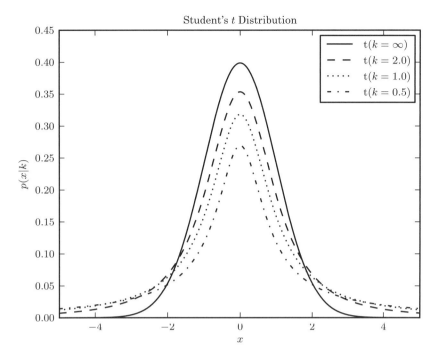

Figure 3.15. Examples of Student's t distribution (see eq. 3.60).

Student's t distribution has the probability density function

$$p(x|k) = \frac{\Gamma(\frac{k+1}{2})}{\sqrt{\pi k}\,\Gamma(\frac{k}{2})} \left(1 + \frac{x^2}{k}\right)^{-\frac{k+1}{2}}, \qquad (3.60)$$

where k is the number of degrees of freedom (see figure 3.15). Note that for $k = 2$, this distribution is a Cauchy distribution with $\mu = 0$ and $\gamma = 1$ (i.e., $p(x) = \pi^{-1}(1 + x^2)^{-1}$). Student's t distribution[5] is symmetric and bell shaped, but with heavier tails than for a Gaussian distribution. The mean, median, and mode for Student's t distribution are all zero when $k > 1$ and undefined for $k = 1$. When $k > 2$, the standard deviation is $\sqrt{k/(k-2)}$; when $k > 3$, skewness is 0; and when $k > 4$, kurtosis is $6/(k-4)$. For large k, Student's t distribution tends to $\mathcal{N}(0, 1)$.

Given a sample of N measurements, $\{x_i\}$, drawn from a Gaussian distribution $\mathcal{N}(\mu, \sigma)$, the variable

$$t = \frac{\overline{x} - \mu}{s/\sqrt{N}}, \qquad (3.61)$$

where $\overline{x}$ and s are given by eqs. 3.31 and 3.32, respectively, follows Student's t distribution with $k = N - 1$ degrees of freedom. Note that, although there is some

---

[5]It seems that just about every book on statistics must mention that Student was the pen name of W. S. Gosset, and that Gosset worked for the Guinness brewery, possibly because making beer is more fun than practicing statistics.

similarity between t defined here and Q from the χ^2 distribution, the main difference is that the definition of t is based on data-based *estimates* $\overline{x}$ and s, while the χ^2 statistic is based on *true* values μ and σ. Note that, analogously to the χ^2 distribution, we can repeatedly draw $\{x_i\}$ from Gaussian distributions with *different* μ and σ, and the corresponding t values for each sample will follow *identical* distributions as long as N stays unchanged.

The ratio of a standard normal variable and a variable drawn from the χ^2 distribution follows Student's t distribution. Although often approximated as a Gaussian distribution, the mean of a sample follows Student's t distribution (because s in the denominator is only an estimate of the true σ) and this difference may matter when samples are small. Student's t distribution also arises when comparing means of two samples. We discuss these results in more detail in chapter 5.

3.3.9. Fisher's F Distribution

Fisher's F distribution is implemented in `scipy.stats.f`:

```
from scipy import stats
dist = stats.f(2, 3)   # d1 = 2, d2 = 3
r = dist.rvs(10)   # ten random draws
p = dist.pdf(1) # pdf evaluated at x=1
```

Fisher's F distribution has the probability density function for $x \geq 0$, and parameters $d_1 > 0$ and $d_2 > 0$,

$$p(x|d_1, d_2) = C \left(1 + \frac{d_1}{d_2}x\right)^{-\frac{d_1+d_2}{2}} x^{\frac{d_1}{2}-1}, \tag{3.62}$$

where the normalization constant is

$$C = \frac{1}{B(d_1/2, d_2/2)} \left(\frac{d_1}{d_2}\right)^{d_1/2}, \tag{3.63}$$

and B is the beta function. Depending on d_1 and d_2, the appearance of the F distribution can greatly vary (see figure 3.16). When both d_1 and d_2 are large, the mean value is $d_2/(d_2-2) \approx 1$, and the standard deviation is $\sigma = \sqrt{2(d_1 + d_2)/(d_1 d_2)}$. Fisher's F distribution describes the distribution of the ratio of two independent χ^2_{dof} variables with d_1 and d_2 degrees of freedom, and is useful when comparing the standard deviations of two samples. Also, if x_1 and x_2 are two independent random variables drawn from the Cauchy distribution with location parameter μ, then the ratio $|x_1 - \mu|/|x_2 - \mu|$ follows Fisher's F distribution with $d_1 = d_2 = 2$.

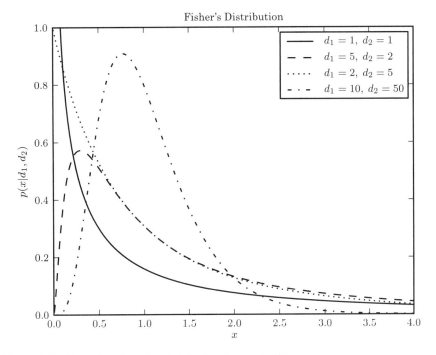

Figure 3.16. Examples of a Fisher distribution (see eq. 3.62).

3.3.10. The Beta Distribution

The beta distribution is implemented in `scipy.stats.beta`:

```
from scipy import stats
dist = stats.beta(0.5, 1.5)  # alpha = 0.5,
      # beta = 1.5
r = dist.rvs(10)  # ten random draws
p = dist.pdf(0.6) # pdf evaluated at x=0.6
```

The beta distribution is defined for $0 < x < 1$ and described by two parameters, $\alpha > 0$ and $\beta > 0$, as

$$p(x|\alpha, \beta) = \frac{\Gamma(\alpha + \beta)}{\Gamma(\alpha)\Gamma(\beta)} \, x^{\alpha-1} \, (1 - x)^{\beta-1}, \tag{3.64}$$

where Γ is the gamma function. The mean value for the beta distribution is $\alpha/(\alpha+\beta)$. Various combinations of the parameters α and β can result in very different shapes (see figure 3.17). This property makes it very useful in Bayesian analysis, as discussed in §5.6.2. In particular, the beta distribution is the conjugate prior for the binomial distribution (see §5.2.3).

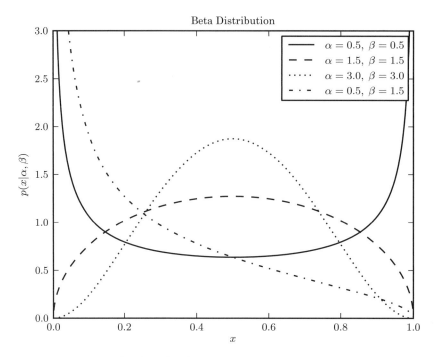

Figure 3.17. Examples of the beta distribution (see eq. 3.64).

3.3.11. The Gamma Distribution

The gamma distribution is implemented in `scipy.stats.gamma`:

```
from scipy import stats
dist = stats.gamma(1, 0, 2)  # k = 1, loc = 0,
        # theta = 2
r = dist.rvs(10)  # ten random draws
p = dist.pdf(1) # pdf evaluated at x=1
```

The gamma distribution is defined for $0 < x < \infty$, and is described by two parameters, a shape parameter k and a scale parameter θ (see figure 3.18). The probability distribution function is given by

$$p(x|k, \theta) = \frac{1}{\theta^k} \frac{x^{k-1} e^{-x/\theta}}{\Gamma(k)}, \qquad (3.65)$$

where $\Gamma(k)$ is the gamma function.

The gamma distribution is useful in Bayesian statistics, as it is a conjugate prior to several distributions including the exponential (Laplace) distribution, and the Poisson distribution (see §5.2.3).

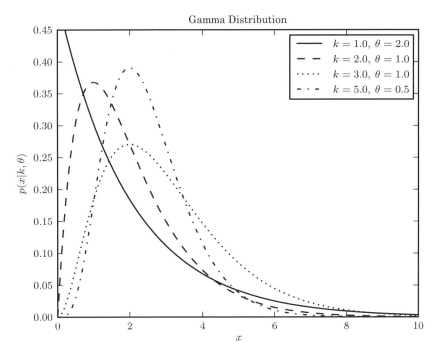

Figure 3.18. Examples of a gamma distribution (see eq. 3.65).

3.3.12. The Weibull Distribution

The Weibull distribution is implemented in `scipy.stats.dweibull`:

```
from scipy import stats
dist = stats.dweibull(1, 0, 2)   # k = 1, loc = 0,
        # lambda = 2
r = dist.rvs(10)   # ten random draws
p = dist.pdf(1) # pdf evaluated at x=1
```

The Weibull distribution is defined for $x \geq 0$ and described by the shape parameter k and the scale parameter λ as

$$p(x|k, \lambda) = \frac{k}{\lambda} \left(\frac{x}{\lambda}\right)^{k-1} e^{-(x/\lambda)^k} \tag{3.66}$$

(see figure 3.19). The mean value is $\lambda \, \Gamma(1 + 1/k)$ and the median is $\lambda \, (\ln 2)^{1/k}$. The shape parameter k can be used to smoothly interpolate between the exponential distribution (corresponding to $k = 1$; see §3.3.6) and the Rayleigh distribution ($k = 2$; see §3.3.7). As k tends to infinity, the Weibull distribution morphs into a Dirac δ function. If x is uniformly distributed on the interval $(0, 1)$, then the random

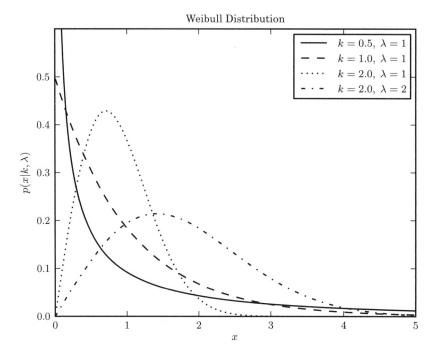

Figure 3.19. Examples of a Weibull distribution (see eq. 3.66).

variable $y = \lambda(-\ln x)^{1/k}$ is distributed as the Weibull distribution with parameters k and λ.

The Weibull distribution is often encountered in physics and engineering because it provides a good description of a random failure process with a variable rate, wind behavior, distribution of extreme values, and the size distribution of particles. For example, if x is the time to failure for a device with the failure rate proportional to a power of time, t^m, then x follows the Weibull distribution with $k = m + 1$. The distribution of the wind speed at a given location typically follows the Weibull distribution with $k \approx 2$ (the $k = 2$ case arises when two components of a two-dimensional vector are uncorrelated and follow Gaussian distributions with equal variances).

In an engineering context, an elegant method was developed to assess whether data $\{x_i\}$ was drawn from the Weibull distribution. The method utilizes the fact that the cumulative distribution function for the Weibull distribution is very simple:

$$H_W(x) = 1 - e^{-(x/\lambda)^k}. \tag{3.67}$$

The cumulative distribution function based on data, $F(x)$, is used to define $z = \ln(-\ln(1 - F(x)))$ and z is plotted as a function of $\ln x$. For the Weibull distribution, the z vs. $\ln x$ plot is a straight line with a slope equal to k and intercept equal to $-k \ln \lambda$. For small samples, it is difficult to distinguish the Weibull and log-normal distributions.

3.4. The Central Limit Theorem

The central limit theorem provides the theoretical foundation for the practice of repeated measurements in order to improve the accuracy of the final result. Given an *arbitrary* distribution $h(x)$, characterized by its mean μ and standard deviation σ, the central limit theorem says that the mean of N values x_i drawn from that distribution will approximately follow a Gaussian distribution $\mathcal{N}(\mu, \sigma/\sqrt{N})$, with the approximation accuracy improving with N. This is a remarkable result since the details of the distribution $h(x)$ are not specified—we can "average" our measurements (i.e., compute their mean value using eq. 3.31) and expect the $1/\sqrt{N}$ improvement in accuracy *regardless of details in our measuring apparatus!* The underlying reason why the central limit theorem can make such a far-reaching statement is the strong assumption about $h(x)$: it must have a standard deviation and thus its tails must fall off faster than $1/x^2$ for large x. As more measurements are combined, the tails will be "clipped" and eventually (for large N) the mean will follow a Gaussian distribution (it is easy to prove this theorem using standard tools from statistics such as characteristic functions; e.g., see Lup93). Alternatively, it can be shown that the resulting Gaussian distribution rises as the result of many consecutive convolutions (e.g., see Greg05). An illustration of the central limit theorem in action, using a uniform distribution for $h(x)$, is shown in figure 3.20.

However, there are cases when the central limit theorem *cannot* be invoked! We already discussed the Cauchy distribution, which does not have a well-defined mean or standard deviation, and thus the central limit theorem is *not applicable* (recall figure 3.12). In other words, if we repeatedly draw N values x_i from a Cauchy distribution and compute their mean value, the resulting distribution of these mean values will *not* follow a Gaussian distribution (it will follow the Cauchy distribution, and will have an infinite variance). If we decide to use the mean of measured values to estimate the location parameter μ, we will *not* gain the $\sqrt{N}$ improvement in accuracy promised by the central limit theorem. Instead, we need to compute the median and interquartile range for x_i, which are unbiased estimators of the location and scale parameters for the Cauchy distribution. Of course, the reason why the central limit theorem is not applicable to the Cauchy distribution is its extended tails that decrease only as x^{-2}.

We mention in passing the weak law of large numbers (also known as Bernoulli's theorem): the sample mean converges to the distribution mean as the sample size increases. Again, for distributions with ill-defined variance, such as the Cauchy distribution, *the weak law of large numbers breaks down.*

In another extreme case of tail behavior, we have the uniform distribution which does not even have tails (cf. §3.3.1). If we repeatedly draw N values x_i from a uniform distribution described by its mean μ and width W, the distribution of their mean value $\overline{x}$ will be centered on μ, as expected from the central limit theorem. In addition, the uncertainty of our estimate for the location parameter μ will decrease proportionally to $1/\sqrt{N}$, again in agreement with the central limit theorem. However, using the mean to estimate μ is *not* the best option here, and indeed μ can *be estimated with an accuracy that improves as $1/N$, that is, faster than $1/\sqrt{N}$.*

How is this arguably surprising result possible? Given the uniform distribution described by eq. 3.39, a value x_i that happens to be larger than μ rules out all

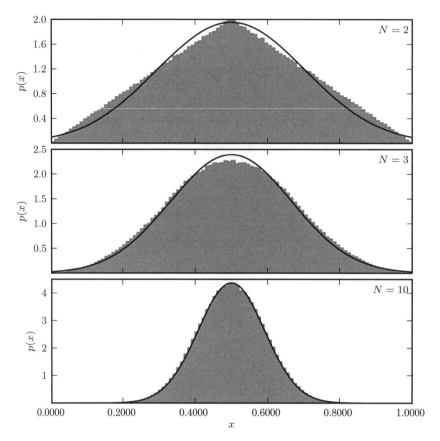

Figure 3.20. An illustration of the central limit theorem. The histogram in each panel shows the distribution of the mean value of N random variables drawn from the (0, 1) range (a uniform distribution with $\mu = 0.5$ and $W = 1$; see eq. 3.39). The distribution for $N = 2$ has a triangular shape and as N increases it becomes increasingly similar to a Gaussian, in agreement with the central limit theorem. The predicted normal distribution with $\mu = 0.5$ and $\sigma = 1/\sqrt{12N}$ is shown by the line. Already for $N = 10$, the "observed" distribution is essentially the same as the predicted distribution.

values $\mu < x_i - W/2$. This strong conclusion is of course the result of the sharp edges of the uniform distribution. The strongest constraint on μ comes from the extremal value of x_i and thus we know that $\mu > \max(x_i) - W/2$. Analogously, we know that $\mu < \min(x_i) + W/2$ (of course, it must be true that $\max(x_i) \leq W/2$ and $\min(x_i) \geq -W/2$). Therefore, given N values x_i, the allowed range for μ is $\max(x_i) - W/2 < \mu < \min(x_i) + W/2$, with a uniform probability distribution for μ within that range. The best estimate for μ is then in the middle of the range,

$$\tilde{\mu} = \frac{\min(x_i) + \max(x_i)}{2}, \tag{3.68}$$

and the standard deviation of this estimate (note that the scatter of this estimate around the true value μ is not Gaussian) is the width of the allowed interval, R,

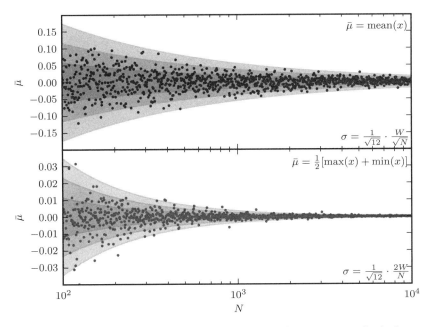

Figure 3.21. A comparison of the sample-size dependence of two estimators for the location parameter of a uniform distribution, with the sample size ranging from $N = 100$ to $N = 10{,}000$. The estimator in the top panel is the sample mean, and the estimator in the bottom panel is the mean value of two extreme values. The theoretical 1σ, 2σ, and 3σ contours are shown for comparison. When using the sample mean to estimate the location parameter, the uncertainty decreases proportionally to $1/\sqrt{N}$, and when using the mean of two extreme values as $1/N$. Note different vertical scales for the two panels.

divided by $\sqrt{12}$ (cf. eq. 3.40). In addition, the best estimate for W is given by

$$\tilde{W} = [\max(x_i) - \min(x_i)]\,\frac{N}{N-2}. \tag{3.69}$$

What is the width of the allowed interval, $R = (\max(x_i) - \min(x_i) - W)$? By considering the distribution of extreme values of x_i, it can be shown that the expectation values are $E[\min(x_i)] = (\mu - W/2 + W/N)$ and $E[\max(x_i)] = (\mu + W/2 - W/N)$. These results can be easily understood: if N values x_i are uniformly scattered within a box of width W, then the two extreme points will be on average $\sim W/N$ away from the box edges. Therefore, the width of the allowed range for μ is $R = 2W/N$, and $\tilde{\mu}$ is an unbiased estimator of μ with a standard deviation of

$$\sigma_{\tilde{\mu}} = \frac{2W}{\sqrt{12}\,N}. \tag{3.70}$$

While the mean value of x_i is also an unbiased estimator of μ, $\tilde{\mu}$ is a much more *efficient* estimator: the ratio of the two uncertainties is $2/\sqrt{N}$ and $\tilde{\mu}$ wins for $N > 2$. The different behavior of these two estimators is illustrated in figure 3.21.

In summary, while the central limit theorem is of course valid for the uniform distribution, the mean of x_i is not the most efficient estimator of the location parameter μ. Due to the absence of tails, the distribution of extreme values of x_i provides the most efficient estimator, $\tilde{\mu}$, which improves with the sample size as fast as $1/N$.

The Cauchy distribution and the uniform distribution are vivid examples of cases where taking the mean of measured values is not an appropriate procedure for estimating the location parameter. What do we do in a general case when the optimal procedure is not known? We will see in chapter 5 that maximum likelihood and Bayesian methods offer an elegant general answer to this question (see §5.6.4).

3.5. Bivariate and Multivariate Distribution Functions

3.5.1. Two-Dimensional (Bivariate) Distributions

All the distribution functions discussed so far are one-dimensional: they describe the distribution of N measured values x_i. Let us now consider the case when two values are measured in each instance: x_i and y_i. Let us assume that they are drawn from a two-dimensional distribution described by $h(x, y)$, with $\int_{-\infty}^{\infty} dx \int_{-\infty}^{\infty} h(x, y) \, dy = 1$. The distribution $h(x, y)$ should be interpreted as giving the probability that x is between x and $x + dx$ and that y is between y and $y + dy$.

In analogy with eq. 3.23, the two variances are defined as

$$V_x = \int_{-\infty}^{\infty} \int_{-\infty}^{\infty} (x - \mu_x)^2 \, h(x, y) \, dx \, dy \tag{3.71}$$

and

$$V_y = \int_{-\infty}^{\infty} \int_{-\infty}^{\infty} (y - \mu_y)^2 \, h(x, y) \, dx \, dy, \tag{3.72}$$

where the mean values are defined as

$$\mu_x = \int_{-\infty}^{\infty} \int_{-\infty}^{\infty} x \, h(x, y) \, dx \, dy \tag{3.73}$$

and analogously for μ_y. In addition, the covariance of x and y, which is a measure of the dependence of the two variables on each other, is defined as

$$V_{xy} = \int_{-\infty}^{\infty} \int_{-\infty}^{\infty} (x - \mu_x)(y - \mu_y) \, h(x, y) \, dx \, dy. \tag{3.74}$$

Sometimes, Cov(x,y) is used instead of V_{xy}. For later convenience, we define $\sigma_x = \sqrt{V_x}$, $\sigma_y = \sqrt{V_y}$, and $\sigma_{xy} = V_{xy}$ (note that there is no square root; i.e., the unit for σ_{xy} is the square of the unit for σ_x and σ_y). A very useful related result is that the variance of the sum $z = x + y$ is

$$V_z = V_x + V_y + 2\,V_{xy}. \tag{3.75}$$

When x and y are uncorrelated ($V_{xy} = 0$), the variance of their sum is equal to the sum of their variances. For $w = x - y$,

$$V_w = V_x + V_y - 2\,V_{xy}. \tag{3.76}$$

In the two-dimensional case, it is important to distinguish the marginal distribution of one variable, for example, here for x:

$$m(x) = \int_{-\infty}^{\infty} h(x, y)\,dy, \tag{3.77}$$

from the two-dimensional distribution evaluated at a given $y = y_o$, $h(x, y_o)$ (and analogously for y). The former is generally wider than the latter, as will be illustrated below using a Gaussian example. Furthermore, while $m(x)$ is a properly normalized probability distribution ($\int_{-\infty}^{\infty} m(x)\,dx = 1$), $h(x, y = y_o)$ is not (recall the discussion in §3.1.3).

If $\sigma_{xy} = 0$, then x and y are uncorrelated and we can treat them separately as two independent one-dimensional distributions. Here "independence" means that whatever range we impose on one of the two variables, the distribution of the other one remains unchanged. More formally, we can describe the underlying two-dimensional probability distribution function as the product of two functions that each depends on only one variable:

$$h(x, y) = h_x(x)\,h_y(y). \tag{3.78}$$

Note that in this special case, marginal distributions are identical to h_x and h_y, and $p(x|y = y_o)$ is the same as $h_x(x)$ except for different normalization.

3.5.2. Bivariate Gaussian Distributions

A generalization of the Gaussian distribution to the two-dimensional case is given by

$$p(x, y | \mu_x, \mu_y, \sigma_x, \sigma_y, \sigma_{xy}) = \frac{1}{2\pi \sigma_x \sigma_y \sqrt{1 - \rho^2}} \exp\left(\frac{-z^2}{2(1 - \rho^2)}\right), \tag{3.79}$$

where

$$z^2 = \frac{(x - \mu_x)^2}{\sigma_x^2} + \frac{(y - \mu_y)^2}{\sigma_y^2} - 2\rho\,\frac{(x - \mu_x)(y - \mu_y)}{\sigma_x \sigma_y}, \tag{3.80}$$

and the (dimensionless) correlation coefficient between x and y is defined as

$$\rho = \frac{\sigma_{xy}}{\sigma_x \sigma_y} \tag{3.81}$$

(see figure 3.22). For perfectly correlated variables such that $y = ax + b$, $\rho = a/|a| \equiv \text{sign}(a)$, and for uncorrelated variables, $\rho = 0$. The *population* correlation coefficient ρ is directly related to Pearson's *sample* correlation coefficient r discussed in §3.6.

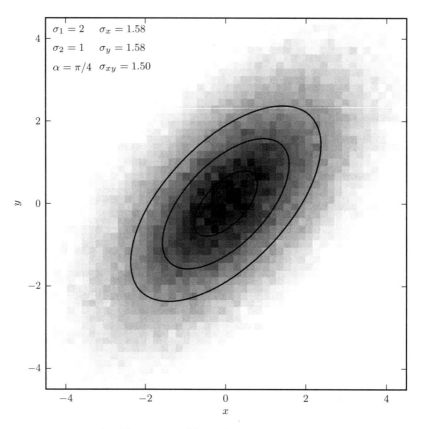

Figure 3.22. An example of data generated from a bivariate Gaussian distribution. The shaded pixels are a Hess diagram showing the density of points at each position.

The contours in the (x, y) plane defined by $p(x, y|\mu_x, \mu_y, \sigma_x, \sigma_y, \sigma_{xy}) =$ constant are ellipses centered on $(x = \mu_x, y = \mu_y)$, and the angle α (defined for $-\pi/2 \leq \alpha \leq \pi/2$) between the x-axis and the ellipses' major axis is given by

$$\tan(2\alpha) = 2\, \rho \, \frac{\sigma_x \sigma_y}{\sigma_x^2 - \sigma_y^2} = 2 \frac{\sigma_{xy}}{\sigma_x^2 - \sigma_y^2}. \tag{3.82}$$

When the (x, y) coordinate system is rotated by an angle α around the point $(x = \mu_x, y = \mu_y)$,

$$P_1 = (x - \mu_x) \cos\alpha + (y - \mu_y) \sin\alpha,$$
$$P_2 = -(x - \mu_x) \sin\alpha + (y - \mu_y) \cos\alpha, \tag{3.83}$$

the correlation between the two new variables P_1 and P_2 disappears, and the two widths are

$$\sigma_{1,2}^2 = \frac{\sigma_x^2 + \sigma_y^2}{2} \pm \sqrt{\left(\frac{\sigma_x^2 - \sigma_y^2}{2}\right)^2 + \sigma_{xy}^2}. \tag{3.84}$$

The coordinate axes P_1 and P_2 are called the principal axes, and σ_1 and σ_2 represent the minimum and maximum widths obtainable for any rotation of the coordinate axes. In this coordinate system where the correlation vanishes, the bivariate Gaussian is the product of two univariate Gaussians (see eq. 3.78). We shall discuss a multidimensional extension of this idea (principal component analysis) in chapter 7.

Alternatively, starting from the principal axes frame, we can compute

$$\sigma_x = \sqrt{\sigma_1^2 \cos^2 \alpha + \sigma_2^2 \sin^2 \alpha}, \tag{3.85}$$

$$\sigma_y = \sqrt{\sigma_1^2 \sin^2 \alpha + \sigma_2^2 \cos^2 \alpha}, \tag{3.86}$$

and ($\sigma_1 \geq \sigma_2$ by definition)

$$\sigma_{xy} = (\sigma_1^2 - \sigma_2^2) \sin \alpha \cos \alpha. \tag{3.87}$$

Note that σ_{xy}, and thus the correlation coefficient ρ, vanish for both $\alpha = 0$ and $\alpha = \pi/2$, and have maximum values for $\pi/4$. By inverting eq. 3.83, we get

$$x = \mu_x + P_1 \cos \alpha - P_2 \sin \alpha,$$
$$y = \mu_y + P_1 \sin \alpha + P_2 \cos \alpha. \tag{3.88}$$

These expressions are very useful when generating mock samples based on bivariate Gaussians (see §3.7).

The marginal distribution of the y variable is given by

$$m(y|I) = \int_{-\infty}^{\infty} p(x, y|I) \, dx = \frac{1}{\sigma_y \sqrt{2\pi}} \exp \left(\frac{-(y - \mu_y)^2}{2\sigma_y^2} \right), \tag{3.89}$$

where we used shorthand $I = (\mu_x, \mu_y, \sigma_x, \sigma_y, \sigma_{xy})$, and analogously for $m(x)$. Note that $m(y|I)$ does not depend on μ_x, σ_x, and σ_{xy}, and it is equal to $\mathcal{N}(\mu_y, \sigma_y)$. Let us compare $m(y|I)$ to $p(x, y|I)$ evaluated for the most probable x,

$$p(x = \mu_x, y|I) = \frac{1}{\sigma_x \sqrt{2\pi}} \frac{1}{\sigma_* \sqrt{2\pi}} \exp \left(\frac{-(y - \mu_y)^2}{2\sigma_*^2} \right) = \frac{1}{\sigma_x \sqrt{2\pi}} \mathcal{N}(\mu_y, \sigma_*), \tag{3.90}$$

where

$$\sigma_* = \sigma_y \sqrt{1 - \rho^2} \leq \sigma_y. \tag{3.91}$$

Since $\sigma_* \leq \sigma_y$, $p(x = \mu_x, y|I)$ is narrower than $m(y|I)$, reflecting the fact that the latter carries additional uncertainty due to unknown (marginalized) x. It is generally true that $p(x, y|I)$ evaluated for any fixed value of x will be proportional to a Gaussian with the width equal to σ_* (and centered on the P_1-axis). In other words, eq. 3.79 can be used to "predict" the value of y for an arbitrary x when $\mu_x, \mu_y, \sigma_x, \sigma_y$, and σ_{xy} are estimated from a given data set.

In the next section we discuss how to estimate the parameters of a bivariate Gaussian $(\mu_x, \mu_y, \sigma_1, \sigma_2, \alpha)$ using a set of points (x_i, y_i) whose uncertainties are negligible compared to σ_1 and σ_2. We shall return to this topic when discussing regression methods in chapter 8, including the fitting of linear models to a set of points (x_i, y_i) whose measurement uncertainties (i.e., *not* their distribution) are described by an analog of eq. 3.79.

3.5.3. A Robust Estimate of a Bivariate Gaussian Distribution from Data

AstroML provides a routine for both the robust and nonrobust estimates of the parameters for a bivariate normal distribution:

```
# assume x and y are pre-defined data arrays
from astroML.stats import fit_bivariate_normal
mean, sigma1, sigma2, alpha =\
        fit_bivariate_
    normal(x, y)
```

For further examples, see the source code associated with figure 3.23.

A bivariate Gaussian distribution is often encountered in practice when dealing with two-dimensional problems, and typically we need to estimate its parameters using data vectors x and y. Analogously to the one-dimensional case, where we can estimate parameters μ and σ as $\overline{x}$ and s using eqs. 3.31 and 3.32, here we can estimate the five parameters $(\overline{x}, \overline{y}, s_x, s_y, s_{xy})$ using similar equations that correspond to eqs. 3.71–3.74. In particular, the correlation coefficient is estimated using Pearson's sample correlation coefficient, r (eq. 3.102, discussed in §3.6). The principal axes can be easily found with α estimated using

$$\tan(2\alpha) = 2\frac{s_x s_y}{s_x^2 - s_y^2}\, r, \tag{3.92}$$

where for simplicity we use the same symbol for both population and sample values of α.

When working with real data sets that often have outliers (i.e., a small fraction of points are drawn from a significantly different distribution than for the majority of the sample), eq. 3.92 can result in grossly incorrect values of α because of the impact of outliers on s_x, s_y, and r. A good example is the measurement of the velocity ellipsoid for a given population of stars, when another population with vastly different kinematics contaminates the sample (e.g., halo vs. disk stars). A simple and efficient remedy is to use the median instead of the mean, and to use the interquartile range to estimate variances.

While it is straightforward to estimate s_x and s_y from the interquartile range (see eq. 3.36), it is not so for s_{xy}, or equivalently, r. To robustly estimate r, we can use the

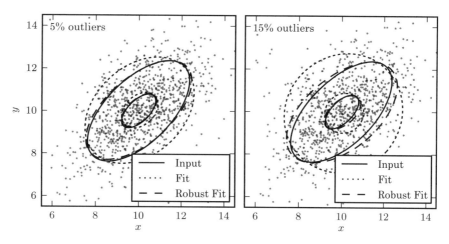

Figure 3.23. An example of computing the components of a bivariate Gaussian using a sample with 1000 data values (points), with two levels of contamination. The core of the distribution is a bivariate Gaussian with $(\mu_x, \mu_y, \sigma_1, \sigma_2, \alpha) = (10, 10, 2, 1, 45°)$. The "contaminating" subsample contributes 5% (left) and 15% (right) of points centered on the same (μ_x, μ_y), and with $\sigma_1 = \sigma_2 = 5$. Ellipses show the 1σ and 3σ contours. The solid lines correspond to the input distribution. The thin dotted lines show the nonrobust estimate, and the dashed lines show the robust estimate of the best-fit distribution parameters (see §3.5.3 for details).

following identity for the correlation coefficient (for details and references, see [5]):

$$\rho = \frac{V_u - V_w}{V_u + V_w},\tag{3.93}$$

where V stands for variance, and transformed coordinates are defined as $(\mathrm{Cov}(u,w) = 0)$

$$u = \frac{\sqrt{2}}{2}\left(\frac{x}{\sigma_x} + \frac{y}{\sigma_y}\right)\tag{3.94}$$

and

$$w = \frac{\sqrt{2}}{2}\left(\frac{x}{\sigma_x} - \frac{y}{\sigma_y}\right).\tag{3.95}$$

By substituting the robust estimator σ_G^2 in place of the variance V in eq. 3.93, we can compute a robust estimate of r, and in turn a robust estimate of the principal axis angle α. Error estimates for r and α can be easily obtained using the bootstrap and jackknife methods discussed in §4.5. Figure 3.23 illustrates how this approach helps when the sample is contaminated by outliers. For example, when the fraction of contaminating outliers is 15%, the best-fit α determined using the nonrobust method is grossly incorrect, while the robust best fit still recognizes the orientation of the distribution's core. Even when outliers contribute only 5% of the sample, the robust estimate of σ_2/σ_1 is much closer to the input value.

3.5.4. Multivariate Gaussian Distributions

The function `multivariate_normal` in the module `numpy.random` implements random samples from a multivariate Gaussian distribution:

```
>>> import numpy as np
>>> mu = [1, 2]
>>> cov = [[1, 0.2],
...        [0.2, 3]]
>>> np.random.multivariate_normal(mu, cov)
array([ 0.03438156, -2.60831303])
```

This was a two-dimensional example, but the function can handle any number of dimensions.

Analogously to the two-dimensional (bivariate) distribution given by eq. 3.79, the Gaussian distribution can be extended to multivariate Gaussian distributions in an arbitrary number of dimensions. Instead of introducing new variables by name, as we did by adding y to x in the bivariate case, we introduce a vector variable $\mathbf{x}$ (i.e., instead of a scalar variable x). We use M for the problem dimensionality ($M = 2$ for the bivariate case) and thus the vector $\mathbf{x}$ has M components. In the one-dimensional case, the variable x has N values x_i. In the multivariate case, each of M components of $\mathbf{x}$, let us call them x^j, $j = 1, \ldots, M$, has N values denoted by x_i^j. With the aid of linear algebra, results from the preceding section can be expressed in terms of matrices, and then trivially extended to an arbitrary number of dimensions. The notation introduced here will be extensively used in later chapters.

The argument of the exponential function in eq. 3.79 can be rewritten as

$$\text{arg} = -\frac{1}{2}\left(\alpha x^2 + \beta y^2 + 2\gamma xy\right), \tag{3.96}$$

with σ_x, σ_y, and σ_{xy} expressed as functions of α, β, and γ (e.g., $\sigma_x^2 = \beta/(\alpha\beta - \gamma^2)$), and the distribution is centered on the origin for simplicity (we could replace $\mathbf{x}$ by $\mathbf{x} - \overline{\mathbf{x}}$, where $\overline{\mathbf{x}}$ is the vector of mean values, if need be). This form lends itself better to matrix notation:

$$p(\mathbf{x}|I) = \frac{1}{(2\pi)^{M/2}\sqrt{\det(\mathbf{C})}} \exp\left(-\frac{1}{2}\mathbf{x}^T\mathbf{H}\mathbf{x}\right), \tag{3.97}$$

where $\mathbf{x}$ is a column vector, $\mathbf{x}^T$ is its transposed row vector, $\mathbf{C}$ is the covariance matrix and $\mathbf{H}$ is equal to the inverse of the covariance matrix, $\mathbf{C}^{-1}$ (note that $\mathbf{H}$ is a symmetric matrix with positive eigenvalues).

Analogously to eq. 3.74, the elements of the covariance matrix $\mathbf{C}$ are given by

$$C_{kj} = \int_{-\infty}^{\infty} x^k x^j \, p(\mathbf{x}|I) \, d^M x. \tag{3.98}$$

The argument of the exponential function can be expanded in component form as

$$\mathbf{x}^T \mathbf{H} \mathbf{x} = \sum_{k=1}^{M} \sum_{j=1}^{M} H_{kj} x^k x^j. \tag{3.99}$$

For example, in the bivariate case discussed above we had $\mathbf{x}^T = (x, y)$ and

$$H_{11} = \alpha, \ H_{22} = \beta, \ H_{12} = H_{21} = \gamma, \tag{3.100}$$

with

$$\det(\mathbf{C}) = \sigma_x^2 \sigma_y^2 - \sigma_{xy}^2. \tag{3.101}$$

Equivalently, starting with $C_{11} = \sigma_x^2$, $C_{12} = C_{21} = \sigma_{xy}$, and $C_{22} = \sigma_y^2$, and using $\mathbf{H} = \mathbf{C}^{-1}$ and eq. 3.97, it is straightforward to recover eq. 3.79.

Similarly to eq. 3.89, when multivariate $p(\mathbf{x}|I)$ is marginalized over all but one dimension, the result is a univariate (one-dimensional) Gaussian distribution.

3.6. Correlation Coefficients

Several correlation tests are implemented in `scipy.stats`, including spearmanr, kendalltau, and pearsonr:

```
from scipy import stats
x, y = np.random.random((2, 100))   # two random
        # arrays
corr_coeff, p_value = stats.pearsonr(x, y)
rho, p_value = stats.spearmanr(x, y)
tau, p_value = stats.kendalltau(x, y)
```

We have already introduced the covariance of x and y (see eq. 3.74) and the correlation coefficient (see eq. 3.81) as measures of the dependence of the two variables on each other. Here we extend our discussion to the interpretation of the sample correlation coefficient.

Given two data sets of equal size N, $\{x_i\}$ and $\{y_i\}$, Pearson's sample correlation coefficient is

$$r = \frac{\sum_{i=1}^{N}(x_i - \bar{x})(y_i - \bar{y})}{\sqrt{\sum_{i=1}^{N}(x_i - \bar{x})^2}\sqrt{\sum_{i=1}^{N}(y_i - \bar{y})^2}}, \tag{3.102}$$

with $-1 \leq r \leq 1$. For uncorrelated variables, $r = 0$. If the pairs (x_i, y_i) are drawn from two uncorrelated univariate Gaussian distributions (i.e., the population correlation coefficient $\rho = 0$), then the distribution of r follows Student's t distribution with $k = N - 2$ degrees of freedom and

$$t = r \sqrt{\frac{N-2}{1-r^2}}. \tag{3.103}$$

Given this known distribution, a measured value of r can be transformed into the significance of the statement that $\{x_i\}$ and $\{y_i\}$ are correlated. For example, if $N=10$, the probability that a value of r as large as 0.72 would arise by chance is 1% (the one-sided 99% confidence level for Student's t distribution with $k = 8$ degrees of freedom is $t = 2.896$). We shall return to such an analysis in §4.6 on hypothesis testing.

When the sample is drawn from a bivariate Gaussian distribution with a nonvanishing population correlation coefficient ρ, the Fisher transformation can be used to estimate the confidence interval for ρ from the measured value r. The distribution of F,

$$F(r) = \frac{1}{2} \ln \left(\frac{1+r}{1-r} \right), \tag{3.104}$$

approximately follows a Gaussian distribution with the mean $\mu_F = F(\rho)$ and a standard deviation $\sigma_F = (N - 3)^{-1/2}$. For the above sample with $N = 10$ and $r = 0.72$, this approximate approach gives a significance level of 0.8% when $\rho = 0$ (instead of the exact value of 1%).

Pearson's correlation coefficient has two main deficiencies. First, the measurement errors for $\{x_i\}$ and $\{y_i\}$ are not used. As they become very large, the significance of any given value of r will decrease (if errors are much larger than the actual ranges of data values, the evidence for correlation vanishes). The case of nonnegligible errors is discussed in chapter 8. Second, Pearson's correlation coefficient is sensitive to Gaussian outliers (recall the discussion in the previous section) and, more generally, the distribution of r does not follow Student's t distribution if $\{x_i\}$ and $\{y_i\}$ are not drawn from a bivariate Gaussian distribution. In such cases, nonparametric correlation tests, discussed next, are a better option.

3.6.1. Nonparametric Correlation Tests

The two best known nonparametric (distribution-free) correlation tests are based on Spearman's correlation coefficient and Kendall's correlation coefficient. These two correlation coefficients are themselves strongly correlated ($r = 0.98$ in the case of Gaussian distributions), though their magnitude is not equal (see below). Traditionally, Spearman's correlation coefficient is used more often because it is easier to compute, but Kendall's correlation coefficient is gaining popularity because it approaches normality faster than Spearman's correlation coefficient.

Both Spearman's and Kendall's correlation coefficients are based on the concept of ranks. To get ranks, sort a data set $\{x_i\}$ in ascending order. The index i of a value x_i in the sorted data set is its rank, R_i^x. Since most data sets of interest here involve continuous variables, we will ignore the possibility that some values could be equal, but there are also methods to handle ties in rank (see Lup93). The main advantage of ranks is that their distribution is known: each value $(1, \ldots, N)$ occurs exactly once and this fact can be used to derive useful results such as

$$\sum_{i=1}^{N} R_i = \frac{N(N+1)}{2} \tag{3.105}$$

and

$$\sum_{i=1}^{N} (R_i)^2 = \frac{N(N+1)(2N+1)}{6}. \tag{3.106}$$

Spearman's correlation coefficient is defined analogously to Pearson's coefficient, with ranks used instead of the actual data values,

$$r_S = \frac{\sum_{i=1}^{N}(R_i^x - \overline{R^x})(R_i^y - \overline{R^y})}{\sqrt{\sum_{i=1}^{N}(R_i^x - \overline{R^x})^2}\sqrt{\sum_{i=1}^{N}(R_i^y - \overline{R^y})^2}}, \tag{3.107}$$

or alternatively (see Lup93 for a brief derivation),

$$r_S = 1 - \frac{6}{N(N^2 - 1)} \sum_{i=1}^{N} \left(R_i^x - R_i^y\right)^2. \tag{3.108}$$

The distribution of r_S for uncorrelated variables is the same as given by eq. 3.103, with Pearson's r replaced by r_S.

Kendall's correlation coefficient is based on a comparison of two ranks, and does not use their actual difference (i.e., the difference $R_i^x - R_i^y$ in expression above). If $\{x_i\}$ and $\{y_i\}$ are not correlated, a comparison of two pairs of values j and k, defined by $R_j^x = R_j^y$ and $R_k^x = R_k^y$ will produce similar numbers of *concordant pairs*, defined by $(x_j - x_k)(y_j - y_k) > 0$, and *discordant pairs*, defined by $(x_j - x_k)(y_j - y_k) < 0$. The condition for concordant pairs corresponds to requiring that *both* differences have the same sign, and opposite signs for discordant pairs. For a perfect correlation (anticorrelation), all $N(N-1)/2$ possible pairs will be concordant (discordant).

Hence, to get Kendall's correlation coefficient, τ, count the number of concordant pairs, N_c, and the number of discordant pairs, N_d,

$$\tau = 2 \frac{N_c - N_d}{N(N-1)} \tag{3.109}$$

(note that $-1 \leq \tau \leq 1$). Kendall's τ can be interpreted as the probability that the two data sets are in the same order minus the probability that they are not in the same order.

For small N, the significance of τ can be found in tabulated form. When $N > 10$, the distribution of Kendall's τ, for the case of *no correlation*, can be approximated as a Gaussian with $\mu = 0$ and width

$$\sigma_\tau = \left[\frac{2(2N+5)}{9N(N-1)} \right]^{1/2}. \tag{3.110}$$

This expression can be used to find a significance level corresponding to a given τ, that is, the probability that such a large value would arise by chance in the case of no correlation. Note, however, that Kendall's τ is not an estimator of ρ in the general case.

When $\{x_i\}$ and $\{y_i\}$ are correlated with a true correlation coefficient ρ, then the distributions of measured Spearman's and Kendall's correlation coefficients become harder to describe. It can be shown that for a bivariate Gaussian distribution of x and y with a correlation coefficient ρ, the expectation value for Kendall's τ is

$$\overline{\tau} = \frac{2}{\pi} \sin^{-1}(\rho) \tag{3.111}$$

(see [7] for a derivation, and for a more general expression for τ in the presence of noise). Note that τ offers an unbiased estimator of the population value, while r_S does not (see Lup93). In practice, a good method for placing a confidence estimate on the measured correlation coefficient is the bootstrap method (see §4.5). An example, shown in figure 3.24 compares the distribution of Pearson's, Spearman's, and Kendall's correlation coefficients for the sample shown in figure 3.23. As is evident, Pearson's correlation coefficient is *very sensitive* to outliers!

The efficiency of Kendall's τ relative to Pearson's correlation coefficient for a bivariate Gaussian distribution is greater than 90%, and can exceed it by large factors for non-Gaussian distributions (the method of so-called normal scores can be used to raise the efficiency to 100% in the case of a Gaussian distribution). Therefore, Kendall's τ is a good general choice for measuring the correlation of any two data sets.

The computation of N_c and N_d needed for Kendall's τ by direct evaluation of $(x_j - x_k)(y_j - y_k)$ is an $\mathcal{O}(N^2)$ algorithm. In the case of large samples, more sophisticated $\mathcal{O}(N \log N)$ algorithms are available in the literature (e.g., [1]).

3.7. Random Number Generation for Arbitrary Distributions

The distributions in `scipy.stats.distributions` each have a method called `rvs`, which implements a pseudorandom sample from the distribution (see examples in the above sections). In addition, the module `numpy.random` implements samplers for a number of distributions. For example, to select five random integers between 0 and 10:

```
>>> import numpy as np
>>> np.random.random_integers(0, 10, 5)
array([7, 5, 1, 1, 6])
```

For a full list of available distributions, see the documentation of `numpy.random` and of `scipy.stats`.

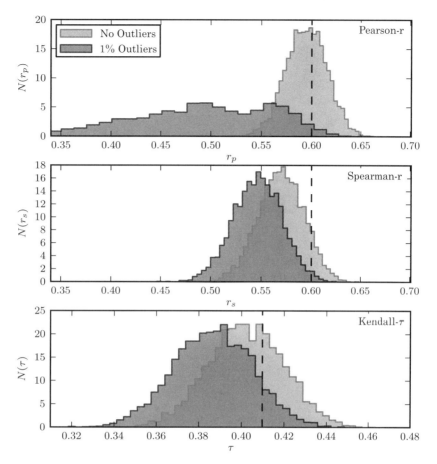

Figure 3.24. Bootstrap estimates of the distribution of Pearson's, Spearman's, and Kendall's correlation coefficients based on 2000 resamplings of the 1000 points shown in figure 3.23. The true values are shown by the dashed lines. It is clear that Pearson's correlation coefficient is not robust to contamination.

Numerical simulations of the measurement process are often the only way to understand complicated selection effects and resulting biases. These approaches are often called Monte Carlo simulations (or modeling) and the resulting artificial (as opposed to real measurements) samples are called Monte Carlo or mock samples. Monte Carlo simulations require a sample drawn from a specified distribution function, such as the analytic examples introduced earlier in this chapter, or given as a lookup table. The simplest case is the uniform distribution function (see eq. 3.39), and it is implemented in practically all programming languages. For example, module `random` in Python returns a random (really pseudorandom since computers are deterministic creatures) floating-point number greater than or equal to 0 and less than 1, called a uniform deviate. The `random` submodule of NumPy provides some more sophisticated random number generation, and can be much faster than the random number generation built into Python, especially when generating large random arrays.

When "random" is used without a qualification, it usually means a uniform deviate. The mathematical background of such random number generators (and pitfalls associated with specific implementation schemes, including strong correlation between successive values) is concisely discussed in NumRec. Both the Python and NumPy random number generators are based on the Mersenne twister algorithm [4], which is one of the most extensively tested random number generators available. Although many distribution functions are already implemented in Python (in the `random` module) and in NumPy and SciPy (in the `numpy.random` and `scipy.stats` modules), it is often useful to know how to use a uniform deviate generator to generate a simulated (mock) sample drawn from an arbitrary distribution.

In the one-dimensional case, the solution is exceedingly simple and is called the transformation method. Given a differential distribution function $f(x)$, its cumulative distribution function $F(x)$ given by eq. 1.1 can be used to choose a specific value of x as follows. First use a uniform deviate generator to choose a value $0 \leq y \leq 1$, and then choose x such that $F(x) = y$. If $f(x)$ is hard to integrate, or given in a tabular form, or $F(x)$ is hard to invert, an appropriate numerical integration scheme can be used to produce a lookup table for $F(x)$. An example of "cloning" 100,000 data values following the same distribution as 10,000 "measured" values using table interpolation is given in figure 3.25. This particular implementation uses a cubic spline interpolation to approximate the inverse of the observed cumulative distribution $F(x)$. Though slightly more involved, this approach is much faster than the simple selection/rejection method (see NumRec for details). Unfortunately, this rank-based approach cannot be extended to higher dimensions. We will return to the subject of cloning a general multidimensional distribution in §6.3.2.

In multidimensional cases, and when the distribution is separable (i.e., it is equal to the product of independent one-dimensional distributions, e.g., as given for the two-dimensional case by eq. 3.6), one can generate the distribution of each random deviate using a one-dimensional prescription. When the multidimensional distribution is not separable, one needs to consider marginal distributions. For example, in a two-dimensional case $h(x, y)$, one would first draw the value of x using the marginal distribution given by eq. 3.77. Given this x, say x_o, the value of y, say y_o, would be generated using the properly normalized one-dimensional cumulative conditional probability distribution in the y direction,

$$H(y|x_o) = \frac{\int_{-\infty}^{y} h(x_o, y') \, dy'}{\int_{-\infty}^{\infty} h(x_o, y') \, dy'}. \tag{3.112}$$

In higher dimensions, x_o and y_o would be kept fixed, and the properly normalized cumulative distributions of other variables would be used to generate their values.

In the special case of multivariate Gaussian distributions (see §3.5), mock samples can be simply generated in the space of principal axes, and then the values can be "rotated" to the appropriate coordinate system (recall the discussion in §3.5.2). For example, two independent sets of values η_1 and η_2 can be drawn from an $\mathcal{N}(0, 1)$ distribution, and then x and y coordinates can be obtained using the transformations (cf. eq. 3.88)

$$x = \mu_x + \eta_1 \sigma_1 \cos \alpha - \eta_2 \sigma_2 \sin \alpha \tag{3.113}$$

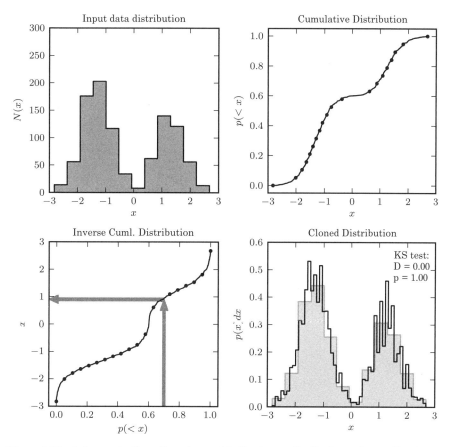

Figure 3.25. A demonstration of how to empirically clone a distribution, using a spline interpolation to approximate the inverse of the observed cumulative distribution. This allows us to
nonparametrically select new random samples approximating an observed distribution. First
the list of points is sorted, and the rank of each point is used to approximate the cumulative
distribution (upper right). Flipping the axes gives the inverse cumulative distribution on a
regular grid (lower left). After performing a cubic spline fit to the inverse distribution, a
uniformly sampled x value maps to a y value which approximates the observed pdf. The lower-
right panel shows the result. The K-S test (see §4.7.2) indicates that the samples are consistent
with being drawn from the same distribution. This method, while fast and effective, cannot be
easily extended to multiple dimensions.

and

$$y = \mu_y + \eta_1 \, \sigma_1 \sin \alpha + \eta_2 \, \sigma_2 \cos \alpha. \tag{3.114}$$

The generalization to higher dimensions is discussed in §3.5.4.

The cloning of an arbitrary high-dimensional distribution is possible if one can
sufficiently model the density of the generating distribution. We will return to this
problem within the context of density estimation routines: see §6.3.2.

References

[1]   Christensen, D. (2005). Fast algorithms for the calculation of Kendall's τ. *Computational Statistics 20*, 51–62.

[2]   Cox, R. T. (1946). Probability, frequency, and reasonable expectation. *Am. Jour. Phys. 14*, 1–13.

[3]   Gardner, M. (1959). Mathematical games. *Scientific American 201*, 180–182.

[4]   Matsumoto, M. and T. Nishimura (1998). Mersenne twister: A 623-dimensionally equidistributed uniform pseudo-random number generator. *ACM Trans. Model. Comput. Simul. 8*(1), 3–30.

[5]   Shevlyakov, G. and P. Smirnov (2011). Robust estimation of the correlation coefficient: An attempt of survey. *Austrian Journal of Statistics 40*, 147–156.

[6]   Vos Savant, M. (1990). Game show problem. *Parade Magazine*, 16.

[7]   Xu, W., Y. Hou, Y. S. Hung, and Y. Zou (2010). Comparison of Spearman's rho and Kendall's tau in normal and contaminated normal models. *ArXiv:cs/1011.2009*.

4 Classical Statistical Inference

"Frequentist or Bayesian? Flip a coin. (And observe the long-term outcome.)" (Frequentists)

This chapter introduces the main concepts of *statistical inference*, or drawing conclusions from data. There are three main types of inference:

- **Point estimation**: What is the best estimate for a model parameter θ, based on the available data?
- **Confidence estimation**: How confident should we be in our point estimate?
- **Hypothesis testing**: Are data at hand consistent with a given hypothesis or model?

There are two major statistical paradigms which address the statistical inference questions: the classical, or *frequentist* paradigm, and the *Bayesian* paradigm (despite the often-used adjective "classical," historically the frequentist paradigm was developed after the Bayesian paradigm). While most of statistics and machine learning is based on the classical paradigm, Bayesian techniques are being embraced by the statistical and scientific communities at an ever-increasing pace. These two paradigms are sufficiently different that we discuss them in separate chapters.

In this chapter we begin with a short comparison of classical and Bayesian paradigms, and then discuss the three main types of statistical inference from the classical point of view. The following chapter attempts to follow the same structure as this chapter, but from the Bayesian point of view. The topics covered in these two chapters complete the foundations for the remaining chapters on exploratory data analysis and data-based prediction methods.

4.1. Classical vs. Bayesian Statistical Inference

We start with a brief summary of the basic philosophical differences between the two approaches. A more detailed discussion of their practical pros and cons is presented in §5.9. We follow the balanced perspective set forth by Wasserman (Wass10; see §1.3), an expert in both Bayesian and frequentist theory.

Classical or frequentist statistics is based on these tenets:

- Probabilities refer to *relative frequencies of events*. They are objective properties of the real world.
- Parameters (such as the fraction of coin flips, for a certain coin, that are heads) are *fixed, unknown constants*. Because they are not fluctuating, probability statements about parameters are meaningless.
- Statistical procedures should have well-defined long-run frequency properties. For example, a 95% confidence interval should bracket the true value of the parameter with a limiting frequency of at least 95%.

In contrast, Bayesian inference takes this stance:

- Probability describes the degree of subjective belief, not the limiting frequency. Probability statements can be made about things *other than data*, including model parameters and models themselves.
- Inferences about a parameter are made by producing its probability distribution—*this distribution quantifies the uncertainty of our knowledge about that parameter*. Various point estimates, such as expectation value, may then be readily extracted from this distribution.

Note that both are equally concerned with uncertainties about estimates. The main difference is whether one is allowed, or not, to discuss the "probability" of some aspect of the fixed universe having a certain value. The choice between the two paradigms is in some sense a philosophical subtlety, but also has very real practical consequences. In terms of philosophy, there is a certain symmetry and thus elegance in the Bayesian construction giving a "grand unified theory" feel to it that appeals naturally to the sensibilities of many scientists.

In terms of pragmatism, the Bayesian approach is accompanied by significant difficulties, particularly when it comes to computation. Despite much advancement in the past few decades, these challenges should not be underestimated. In addition, the results of classical statistics are often the same as Bayesian results, and still represent a standard for reporting scientific results. Therefore, despite many strong advantages of the Bayesian approach, classical statistical procedures cannot be simply and fully replaced by Bayesian results. Indeed, maximum likelihood analysis, described next, is a major concept in both paradigms.

4.2. Maximum Likelihood Estimation (MLE)

We will start with *maximum likelihood estimation* (MLE), a common special case of the larger frequentist world. It is a good starting point because Bayesian estimation in the next chapter builds directly on top of the apparatus of maximum likelihood. The frequentist world, however, is not tied to the likelihood (as Bayesian estimation is), and we will also visit additional ideas outside of likelihood-based approaches later in this section (§4.2.8).

4.2.1. The Likelihood Function

The starting point for both MLE and Bayesian estimation is the *likelihood* of the data. The data likelihood represents a quantitative description of our measuring process. The concept was introduced by Gauss and Laplace, and popularized by Fisher in the first half of the last century.[1]

Given a known, or assumed, behavior of our measuring apparatus (or, statistically speaking, the distribution from which our sample was drawn), we can compute the probability, or likelihood, of observing any given value. For example, if we know or assume that our data $\{x_i\}$ are drawn from an $\mathcal{N}(\mu, \sigma)$ parent distribution, then the likelihood of a given value x_i is given by eq. 1.4. Assuming that individual values are independent (e.g., they would not be independent in the case of drawing from a small parent sample without replacement, such as the second drawing from an urn that initially contains a red and a white ball), the likelihood of the entire data set, L, is then the product of likelihoods for each particular value,

$$L \equiv p(\{x_i\}|M(\boldsymbol{\theta})) = \prod_{i=1}^{n} p(x_i|M(\boldsymbol{\theta})), \tag{4.1}$$

where M stands for a model (i.e., our understanding of the measurement process, or assumptions about it). In general, the model M includes k model parameters θ_p, $p = 1, \ldots, k$, abbreviated as the vector $\boldsymbol{\theta}$ with components θ_p. We will sometimes write just M, or just $\boldsymbol{\theta}$ and θ in one-dimensional cases, in place of $M(\boldsymbol{\theta})$, depending on what we are trying to emphasize. Instead of the specific one-dimensional data set $\{x_i\}$, we will often use D for data in general cases.

Although $L \equiv p(\{x_i\}|M(\boldsymbol{\theta}))$ can be read as "the probability of the data given the model," note that L is not a true (properly normalized) pdf. The likelihood of a given single value x_i is given by a true pdf (e.g., by eq. 1.4), but the product of such functions is no longer normalized to 1. The data likelihood L can take on extremely small values when a data set is large, and its logarithm is often used instead in practice, as we will see in the practical examples below.

The likelihood can be considered both as a function of the data and as a function of the model. When computing the likelihood of some data value x, L is a function of x for some fixed model parameters. Given some fixed data set, it can be considered as a function of the model parameters instead. These parameters can then be varied to maximize the likelihood of observing this specific data set, as described next. Note that this concept of likelihood maximization does *not* imply that likelihoods can be interpreted as probabilities for parameters $\boldsymbol{\theta}$. To do so, a full Bayesian analysis is required, as discussed in the next chapter.

4.2.2. The Maximum Likelihood Approach

Maximum likelihood estimation consists of the following conceptual steps:

1. The formulation of the data likelihood for some model M, $p(D|M)$, which amounts to an assumption about how the data are generated. This step is

---

[1] Fisher's first paper on this subject, and his first mathematical paper ever, was published in 1912 while he was still an undergraduate student.

crucial and the accuracy of the resulting inferences is strongly affected by the quality of this assumption (i.e., how well our model describes the actual data generation process). Models are typically described using a set of model parameters, $\boldsymbol{\theta}$, i.e., the model is $M(\boldsymbol{\theta})$.

2. Search for the best model parameters ($\boldsymbol{\theta}$) which maximize $p(D|M)$. This search yields the MLE *point estimates*, $\boldsymbol{\theta}^0$ (i.e., we obtain k estimates, θ_p^0, $p = 1, \ldots, k$).

3. Determination of the confidence region for model parameters, $\boldsymbol{\theta}^0$. Such a *confidence estimate* can be obtained analytically in MLE by doing mathematical derivations specific to the model chosen, but can also be done numerically for arbitrary models using general frequentist techniques, such as bootstrap, jackknife, and cross-validation, described later. Since the bootstrap can simulate draws of samples from the true underlying distribution of the data, various descriptive statistics can be computed on such samples to examine the uncertainties surrounding the data and our estimators based on that data.

4. Perform *hypothesis tests* as needed to make other conclusions about models and point estimates.

While these steps represent a blueprint for the frequentist approach in general, the likelihood is just one of many possible so-called objective functions (also called fitness functions, or cost functions); other possibilities are explored briefly in §4.2.8. An example of how to perform MLE in practice is described next, using a Gaussian likelihood.

4.2.3. The MLE Applied to a Homoscedastic Gaussian Likelihood

We will now solve a simple problem where we have a set of N measurements, $\{x_i\}$, of, say, the length of a rod. The measurement errors are known to be Gaussian, and we will consider two cases: here we analyze a case where all measurements have the same known error, σ (homoscedastic errors). In §4.2.6 we will analyze a case when each measurement has a different known error σ_i (heteroscedastic errors). The goal of our analysis in both cases is to find the maximum likelihood estimate for the length of the rod, and its confidence interval.

The likelihood for obtaining data $D = \{x_i\}$ given the rod length μ and measurement error σ is

$$L \equiv p(\{x_i\}|\mu, \sigma) = \prod_{i=1}^{N} \frac{1}{\sqrt{2\pi}\sigma} \exp\left(\frac{-(x_i - \mu)^2}{2\sigma}\right). \tag{4.2}$$

Here there is only one model parameter, that is, $k = 1$ and $\theta_1 = \mu$. We can find the maximum likelihood estimate, μ^0, as the value of μ that maximizes L, as follows.

The *log-likelihood function* is defined by $\ln L \equiv \ln[L(\boldsymbol{\theta})]$. Its maximum occurs at the same place as that of the likelihood function, and the same is true of the likelihood function times any constant. Thus we shall often ignore constants in the likelihood function and work with its logarithm. The value of the model parameter μ that maximizes lnL can be determined using the condition

$$\left.\frac{d\,\text{lnL}(\mu)}{d\mu}\right|_{\mu^0} \equiv 0. \tag{4.3}$$

For our Gaussian example, we get from eq. 4.2,

$$\text{lnL}(\mu) = \text{constant} - \sum_{i=1}^{N} \frac{(x_i - \mu)^2}{2\sigma^2}. \tag{4.4}$$

This particularly simple result is a direct consequence of the Gaussian error distribution and admits an analytic solution for MLE of μ derived from eq. 4.3,

$$\mu^0 = \frac{1}{N} \sum_{i=1}^{N} x_i. \tag{4.5}$$

That is, μ^0 is simply the arithmetic mean of all measurements (cf. eq. 3.31).

We started this example by assuming that σ is a measurement error and that there is a unique value of the length of a rod that we are trying to estimate. An alternative interpretation is that there is an intrinsic Gaussian distribution of rod lengths, given by σ, and the measurement errors are negligible compared to σ. We can combine these two possibilities by invoking convolution results for Gaussians as long as measurement errors are homoscedastic. Using eq. 3.45, we can interpret σ as the intrinsic distribution width and the measurement error added in quadrature. This simple result is valid only in the case of homoscedastic errors and the generalization to heteroscedastic errors is discussed below.

4.2.4. Properties of Maximum Likelihood Estimators

Maximum likelihood estimators have several optimality properties, under certain assumptions. The critical assumption is that the data truly come from the specified model class (e.g., they really are drawn from a Gaussian, if that is the model used). Additional assumptions include some relatively mild *regularity conditions*, which amount to various smoothness conditions, certain derivatives existing, etc. (see Lup93 for detailed discussion).

Maximum likelihood estimators have the following properties:

- They are *consistent* estimators; that is, they can be proven to converge to the true parameter value as the number of data points increases.
- They are *asymptotically normal* estimators. The distribution of the parameter estimate, as the number of data points increases to infinity, approaches a normal distribution, centered at the MLE, with a certain spread. This spread can often be easily calculated and used as a confidence band around the estimate, as discussed below (see eq. 4.7).
- They asymptotically achieve the theoretical minimum possible variance, called the Cramér–Rao bound. In other words, they achieve the best possible error given the data at hand; that is, no other estimator can do better in terms of efficiently using each data point to reduce the total error of the estimate (see eq. 3.33).

4.2.5. The MLE Confidence Intervals

Given an MLE, such as μ^0 above, how do we determine its uncertainty? The asymptotic normality of MLE is invoked to demonstrate that the error matrix can be computed from the covariance matrix as

$$\sigma_{jk} = \left(-\frac{d^2 \ln L}{d\theta_j \, d\theta_k} \bigg|_{\theta=\theta_0} \right)^{-1/2}. \tag{4.6}$$

This result is derived by expanding lnL in a Taylor series and retaining terms up to second order (essentially, lnL is approximated by a parabola, or an ellipsoidal surface in multidimensional cases, around its maximum). If this expansion is exact (as is the case for a Gaussian error distribution, see below), then eq. 4.6 is exact. In general, this is not the case and the likelihood surface can significantly deviate from a smooth elliptical surface. Furthermore, it often happens in practice that the likelihood surface is multimodal. It is always a good idea to visualize the likelihood surface when in doubt (see examples in §5.6).

The above expression is related to *expected Fisher information*, which is defined as the expectation value of the second derivative of $-\ln L$ with respect to θ (or the Fisher information matrix when θ is a vector). The inverse of the Fisher information gives a lower bound on the variance of any unbiased estimator of θ; this limit is known as the Cramér–Rao lower bound (for detailed discussion, see [9]). When evaluated at θ_0, such as in eq. 4.6, it is called *observed Fisher information* (for large samples, expected and observed Fisher information become asymptotically equal). Detailed discussion of the connection between Fisher information and confidence intervals, including asymptotic normality properties, is available in Wass10.

The diagonal elements, σ_{ii}, correspond to marginal error bars for parameters θ_i (in analogy with eq. 3.8). If $\sigma_{jk} = 0$ for $j \neq k$, then the inferred values of parameters are uncorrelated (i.e., the error in one parameter does not have an effect on other parameters; if somehow we were given the true value of one of the parameters, our estimates of the remaining parameters would not improve). In this case, the σ_{ii} are the direct analogs of error bars in one-dimensional problems.

It often happens that $\sigma_{jk} \neq 0$ when $j \neq k$. In this case, errors for parameters θ_j and θ_k are correlated (e.g., in the two-dimensional case, the likelihood surface is a bivariate Gaussian with principal axes that are not aligned with coordinate axes; see §3.5.2). This correlation tells us that some combinations of parameters are better determined than others. We can use eq. 3.84 for a two-dimensional case, and the results from §3.5.4 for a general case, to compute these parameter combinations (i.e., the principal axes of the error ellipse) and their uncertainties. When all diagonal elements are much larger than the off-diagonal elements, some combinations of parameters may be measured with a better accuracy than individual parameters.

Returning to our Gaussian example, the uncertainty of the mean is

$$\sigma_\mu = \left(-\frac{d^2 \ln L(\mu)}{d\mu^2} \bigg|_{\mu^0} \right)^{-1/2} = \frac{\sigma}{\sqrt{N}}, \tag{4.7}$$

which is the same as eq. 3.34. Note again that σ in this example was assumed to be constant and known.

4.2.6. The MLE Applied to a Heteroscedastic Gaussian Likelihood

The computation in the heteroscedastic case proceeds analogously to that in the homoscedastic case. The log-likelihood is

$$\ln L = \text{constant} - \sum_{i=1}^{N} \frac{(x_i - \mu)^2}{2\sigma_i^2} \tag{4.8}$$

and again we can derive an analytic solution for the maximum likelihood estimator of μ:

$$\mu^0 = \frac{\sum_i^N w_i x_i}{\sum_i^N w_i}, \tag{4.9}$$

with weights $w_i = \sigma_i^{-2}$. That is, μ^0 is simply a weighted arithmetic mean of all measurements. When all σ_i are equal, eq. 4.9 reduces to the standard result given by eq. 3.31 and eq. 4.5.

Using eq. 4.6, the uncertainty of μ^0 is

$$\sigma_\mu = \left(\sum_{i=1}^{N} \frac{1}{\sigma_i^2} \right)^{-1/2} = \left(\sum_{i=1}^{N} w_i \right)^{-1/2}. \tag{4.10}$$

Again, when all σ_i are equal to σ, this reduces to the standard result given by eqs. 3.34 and 4.7.

We can generalize these results to the case when the measured quantity follows an intrinsic Gaussian distribution with *known* width σ_o, and measurement errors are also known and given as e_i. In this case, eq. 3.45 tells us that we can use the above results with $\sigma_i = (\sigma_o^2 + e_i^2)^{1/2}$. We will discuss important differences in cases when σ_o is not known when analyzing examples of Bayesian analysis in §5.6.1. Those examples will also shed more light on how to treat multidimensional likelihoods, as well as complex problems which need to be solved using numerical techniques.

4.2.7. The MLE in the Case of Truncated and Censored Data

Often the measuring apparatus does not sample the measured range of variable x with equal probability. The probability of drawing a measurement x is quantified using the selection probability, or selection function, $S(x)$. When $S(x) = 0$ for $x > x_{max}$ (analogously for $x < x_{min}$), the data set is *truncated* and we know nothing about sources with $x > x_{max}$ (not even whether they exist or not). A related but different concept is *censored* data sets, where a measurement of an *existing* source was attempted, but the value is outside of some known interval (a familiar astronomical case is an "upper limit" for flux measurement when we look for, e.g., an X-ray source in an optical image of the same region on the sky but do not find it).

We will now revisit the Gaussian example and discuss how to account for data truncation using the MLE approach. For simplicity, we will assume that the selection function is unity for $x_{min} \leq x \leq x_{max}$ and $S(x)=0$ otherwise. The treatment of a more complicated selection function is discussed in §4.9, and censored data are discussed in the context of regression in §8.1.

The key point when accounting for truncated data is that the data likelihood of a single datum must be a properly normalized pdf. The fact that data are truncated enters analysis through a renormalization constant. In the case of a Gaussian error distribution (we assume that σ is known), the likelihood for a single data point is

$$p(x_i|\mu, \sigma, x_{min}, x_{max}) = C(\mu, x_{min}, x_{max})\frac{1}{\sqrt{2\pi}\sigma} \exp\left(\frac{-(x_i - \mu)^2}{2\sigma}\right), \qquad (4.11)$$

where the renormalization constant is evaluated as

$$C(\mu, \sigma, x_{min}, x_{max}) = (P(x_{max}|\mu, \sigma) - P(x_{min}|\mu, \sigma))^{-1} \qquad (4.12)$$

with the cumulative distribution function for Gaussian, P, given by eq. 3.48.

The log-likelihood is

$$\ln L(\mu) = \text{constant} - \sum_{i=1}^{N} \frac{(x_i - \mu)^2}{2\sigma^2} + N \ln\left[C(\mu, \sigma, x_{min}, x_{max})\right]. \qquad (4.13)$$

The first two terms on the right-hand side are identical to those in eq. 4.4, and the third term accounts for truncation. Note that the third term does not depend on data because x_{min} and x_{max} are the same for all points; it would be straightforward to incorporate varying x_{min} and x_{max} (i.e., data points with different "selection limits"). Because the Gauss error function (see eq. 3.47) is needed to evaluate the renormalization constant, there is no simple closed-form expression for μ^0 in this case.

For an illustrative numerical example, let us consider a heavily truncated Gaussian and the following estimation problem. In the country Utopia, graduate schools are so good that students from the country of Karpathia prefer to study in Utopia. They are admitted to a graduate school if they pass a number of tests, which we will assume leads to a hard lower limit of 120 for the IQ score of Karpathian students. We also know that IQ follows a normal distribution centered at 100 with a standard deviation of 15. If you meet a Karpathian student with an IQ of S_{IQ}, what is your best estimate for the mean IQ of Karpathia?

The answer can be easily obtained by finding the maximum of $\ln L(\mu)$ given by eq. 4.13, evaluated for $x_{min} = 120$, $x_{max} = \infty$, $N = 1$, $x_1 = S_{IQ}$, and $\sigma = 15$. Whether the MLE for μ, μ^0, is larger or smaller than 100 depends on the exact value of S_{IQ}. It is easy to show (e.g., using a mock sample) that the mean value for an $\mathcal{N}(100, 15)$ Gaussian truncated at $x_{min} = 120$ is ~127. If S_{IQ} is smaller than 127, the implied mean IQ for Karpathia is smaller than 100, and conversely if $S_{IQ} > 127$, then $\mu^0 > 100$. For example, if $S_{IQ} = 140$, the MLE for μ is 130. For a sample of two students with an IQ of 120 and 140, $\mu^0 = 107$, with uncertainty $\sigma_\mu = 20$. For an arbitrary number of students, their mean IQ must be greater than 127 to obtain $\mu^0 > 100$, and the sample size must be considerable to, for example, reject the hypothesis (see §4.6) that the mean IQ of Karpathia is smaller (or larger) than 100. If all your Karpathian friends have an IQ around 120, bunched next to the selection threshold, it is likely that the mean IQ in Karpathia is below 100! Therefore, if you run into a smart Karpathian, do not automatically assume that all Karpathians have high IQs on average because it could be due to selection effects. Note that if you had a large sample of Karpathian students, you could bin their IQ scores and fit a Gaussian (the data would only constrain the tail of the Gaussian). Such regression methods are discussed in chapter 8. However, as this example shows, there is no need to bin your data, except perhaps for visualization purposes.

4.2.8. Beyond the Likelihood: Other Cost Functions and Robustness

Maximum likelihood represents perhaps the most common choice of the so-called "cost function" (or objective function) within the frequentist paradigm, but not the only one. Here the cost function quantifies some "cost" associated with parameter estimation. The expectation value of the cost function is called "risk" and can be minimized to obtain best-fit parameters.

The mean integrated square error (MISE), defined as

$$\text{MISE} = \int_{\infty}^{+\infty} [f(x) - h(x)]^2 \, dx, \qquad (4.14)$$

is an often-used form of risk; it shows how "close" is our empirical estimate $f(x)$ to the true pdf $h(x)$. The MISE is based on a cost function given by the mean square error, also known as the L_2 norm. A cost function that minimizes absolute deviation is called the L_1 norm. As shown in examples earlier in this section, the MLE applied to a Gaussian likelihood leads to an L_2 cost function (see eq. 4.4). If data instead followed the Laplace (exponential) distribution (see §3.3.6), the MLE would yield an L_1 cost function.

There are many other possible cost functions and often they represent a distinctive feature of a given algorithm. Some cost functions are specifically designed to be robust to outliers, and can thus be useful when analyzing contaminated data (see §8.9 for some examples). The concept of a cost function is especially important in cases where it is hard to formalize the likelihood function, because an optimal solution can still be found by minimizing the corresponding risk. We will address cost functions in more detail when discussing various methods in chapters 6–10.

4.3. The Goodness of Fit and Model Selection

When using maximum likelihood methods, the MLE approach estimates the "best-fit" model parameters and gives us their uncertainties, but it does not tell us how good the fit is. For example, the results given in §4.2.3 and §4.2.6 will tell us the best-fit parameters of a Gaussian, but what if our data was not drawn from a Gaussian distribution? If we select another model, say a Laplace distribution, how do we compare the two possibilities? This comparison becomes even more involved when models have a varying number of model parameters. For example, we know that a fifth-order polynomial fit will always be a better fit to data than a straight-line fit, but do the data really support such a sophisticated model?

4.3.1. The Goodness of Fit for a Model

Using the best-fit parameters, we can compute the maximum value of the likelihood from eq. 4.1, which we will call L^0. Assuming that our model is correct, we can ask how likely it is that this particular value would have arisen by chance. If it is very unlikely to obtain L^0, or $\ln L^0$, by randomly drawing data from the implied best-fit distribution, then the best-fit model is not a good description of the data. Evidently, we need to be able to predict the distribution of L, or equivalently $\ln L$.

For the case of the Gaussian likelihood, we can rewrite eq. 4.4 as

$$\ln L = \text{constant} - \frac{1}{2} \sum_{i=1}^{N} z_i^2 = \text{constant} - \frac{1}{2}\chi^2, \tag{4.15}$$

where $z_i = (x_i - \mu)/\sigma$. Therefore, the distribution of $\ln L$ can be determined from the χ^2 distribution with $N - k$ degrees of freedom (see §3.3.7), where k is the number of model parameters determined from data (in this example $k = 1$ because μ is determined from data and σ was assumed fixed). The distribution of χ^2 does not depend on the actual values of μ and σ; the expectation value for the χ^2 distribution is $N - k$ and its standard deviation is $\sqrt{2(N-k)}$. For a "good fit," we expect that χ^2 *per degree of freedom,*

$$\chi_{\text{dof}}^2 = \frac{1}{N-k} \sum_{i=1}^{N} z_i^2 \approx 1. \tag{4.16}$$

If instead $(\chi_{\text{dof}}^2 - 1)$ is many times larger than $\sqrt{2/(N-k)}$, it is unlikely that the data were generated by the assumed model. Note, however, that outliers may significantly increase χ_{dof}^2. The likelihood of a particular value of χ_{dof}^2 for a given number of degrees of freedom can be found in tables or evaluated using the function `scipy.stats.chi2`.

As an example, consider the simple case of the luminosity of a single star being measured multiple times (figure 4.1). Our model is that of a star with no intrinsic luminosity variation. If the model and measurement errors are consistent, this will lead to χ_{dof}^2 close to 1. Overestimating the measurement errors can lead to an improbably low χ_{dof}^2, while underestimating the measurement errors can lead to an improbably high χ_{dof}^2. A high χ_{dof}^2 may also indicate that the model is insufficient to fit the data: for example, if the star has intrinsic variation which is either periodic (e.g., in the so-called RR-Lyrae-type variable stars) or stochastic (e.g., active M dwarf stars). In this case, accounting for this variability in the model can lead to a better fit to the data. We will explore these options in later chapters. Because the number of samples is large ($N = 50$), the χ^2 distribution is approximately Gaussian: to aid in evaluating the fits, figure 4.1 reports the deviation in σ for each fit.

The probability that a certain maximum likelihood value L^0 might have arisen by chance can be evaluated using the χ^2 distribution only when the likelihood is Gaussian. When the likelihood is not Gaussian (e.g., when analyzing small count data which follows the Poisson distribution), L^0 is still a measure of how well a model fits the data. Different models, assuming that they have the same number of free parameters, can be ranked in terms of L^0. For example, we could derive the best-fit estimates of a Laplace distribution using MLE, and compare the resulting L^0 to the value obtained for a Gaussian distribution.

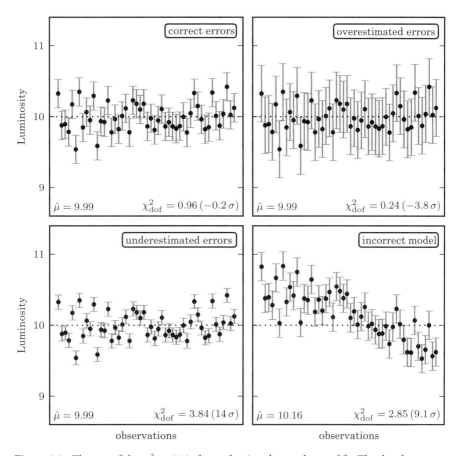

Figure 4.1. The use of the χ^2 statistic for evaluating the goodness of fit. The data here are a series of observations of the luminosity of a star, with known error bars. Our model assumes that the brightness of the star does not vary; that is, all the scatter in the data is due to measurement error. $\chi^2_{\rm dof} \approx 1$ indicates that the model fits the data well (upper-left panel). $\chi^2_{\rm dof}$ much smaller than 1 (upper-right panel) is an indication that the errors are overestimated. $\chi^2_{\rm dof}$ much larger than 1 is an indication either that the errors are underestimated (lower-left panel) or that the model is not a good description of the data (lower-right panel). In this last case, it is clear from the data that the star's luminosity is varying with time: this situation will be treated more fully in chapter 10.

Note, however, that L^0 by itself does not tell us how well a model fits the data. That is, we do not know in general if a particular value of L^0 is consistent with simply arising by chance, as opposed to a model being inadequate. To quantify this probability, we need to know the expected distribution of L^0, as given by the χ^2 distribution in the special case of Gaussian likelihood.

4.3.2. Model Comparison

Given the maximum likelihood for a set of models, $L^0(M)$, the model with the largest value provides the best description of the data. However, this is not necessarily the best model overall when models have different numbers of free parameters.

A "scoring" system also needs to take into account the model complexity and "penalize" models for additional parameters not supported by data. In the Bayesian framework, there is a unique scoring system based on the posterior model probability, as discussed in detail in §5.4.3. Here we limit discussion to classical methods.

There are several methods for comparing models in classical statistics. For example, we discuss the cross-validation technique and bias–variance trade-off (based on mean square error, as discussed above in the context of cost functions) in the context of regression in chapter 8. A popular general classical method for model comparison is the Aikake information criterion (AIC). The AIC is a simple approach based on an asymptotic approximation; the preferred approach to model comparison for the highest accuracy is cross-validation (discussed in §8.11.1), which is based on only the finite data at hand rather than approximations based on infinite data. Nonetheless the AIC is easy to use, and often effective for simple models.

The AIC (the version corrected for small samples, see [15]) is computed as

$$\text{AIC} \equiv -2\ln\left(L^0(M)\right) + 2k + \frac{2k(k+1)}{N-k-1}, \tag{4.17}$$

where k is the number of model parameters and N is the number of data points. The AIC is related to methods based on bias–variance trade-off (see Wass10), and can be derived using information theory (see HTF09).

Under the assumption of normality, the first term is equal to the model's χ^2 (up to a constant). When multiple models are compared, the one with the smallest AIC is the best model to select. If the models are equally successful in describing the data (they have the same value of $L^0(M)$), then the model with fewer free parameters wins. A closely related concept to AIC is the Bayesian information criterion (BIC), introduced in §5.4.3. We will discuss an example of model selection using AIC and BIC criteria in the following section.

4.4. ML Applied to Gaussian Mixtures: The Expectation Maximization Algorithm

Data likelihood can be a complex function of many parameters that often does not admit an easy analytic solution for MLE. In such cases, numerical methods such as those described in §5.8, are used to obtain model parameters and their uncertainties.

A special case of a fairly complex likelihood which can still be maximized using a relatively simple and straightforward numerical method is a mixture of Gaussians. We first describe the model and the resulting data likelihood function, and then discuss the expectation maximization algorithm for maximizing the likelihood.

4.4.1. Gaussian Mixture Model

The likelihood of a datum x_i for a Gaussian mixture model is given by

$$p(x_i|\boldsymbol{\theta}) = \sum_{j=1}^{M} \alpha_j \, \mathcal{N}(\mu_j, \sigma_j), \tag{4.18}$$

where dependence on x_i comes via a Gaussian $\mathcal{N}(\mu_j, \sigma_j)$. The vector of parameters $\boldsymbol{\theta}$ that need to be estimated for a given data set $\{x_i\}$ includes normalization factors for each Gaussian, α_j, and its parameters μ_j and σ_j. It is assumed that the data have negligible uncertainties (e.g., compared to the smallest σ_j), and that M is given. We shall see below how to relax both of these assumptions. Given that the likelihood for a single datum must be a true pdf, α_j must satisfy the normalization constraint

$$\sum_{j=1}^{M} \alpha_j = 1. \tag{4.19}$$

The log-likelihood for the whole data set is then

$$\ln L = \sum_{i=1}^{N} \ln \left[\sum_{j=1}^{M} \alpha_j \, \mathcal{N}(\mu_j, \sigma_j) \right] \tag{4.20}$$

and needs to be maximized as a function of $k = (3M - 1)$ parameters (not $3M$ because of the constraint given by eq. 4.19).

An attempt to derive constraints on these parameters by setting partial derivatives of $\ln L$ with respect to each parameter to zero would result in a complex system of $(3M - 1)$ nonlinear equations and would not bring us much closer to the solution (the problem is that in eq. 4.20 the logarithm is taken of the whole sum over j classes, unlike in the case of a single Gaussian where taking the logarithm of the exponential function results in a simple quadratic function). We could also attempt to simply find the maximum of $\ln L$ through an exhaustive search of the parameter space. However, even when M is small, such an exhaustive search would be too time consuming, and for large M it becomes impossible. For example, if the search grid for each parameter included only 10 values (typically insufficient to achieve the required parameter accuracy), even with a relatively small $M = 5$, we would have to evaluate the function given by eq. 4.20 about 10^{14} times!

A practical solution for maximizing $\ln L$ is to use the Levenberg–Marquardt algorithm, which combines gradient descent and Gauss–Newton optimization (see NumRec). Another possibility is to use Markov chain Monte Carlo methods, discussed in detail in §5.8. However, a much faster procedure is available, especially for the case of large M, based on the concept of hidden variables, as described in the next section.

4.4.2. Class Labels and Hidden Variables

The likelihood given by eq. 4.18 can be interpreted using the concept of "hidden" (or missing) variables. If M Gaussian components are interpreted as different "classes," which means that a particular datum x_i was generated by one and only one of the individual Gaussian components, then the index j is called a "class label." The hidden variable here is the class label j responsible for generating each x_i. If we knew the class label for each datum, then this maximization problem would be trivial and equivalent to examples based on a single Gaussian distribution discussed in the previous section. That is, all the data could be sorted into M subsamples according to their class label. The fraction of points in each subsample would be an estimator of α_j, while μ_j and σ_j could be trivially obtained using eqs. 3.31 and 3.32. In a more general case when the probability function for each class is described by a non-Gaussian function, eqs. 3.31 and 3.32 cannot be used, but given that we know the class labels the problem can still be solved and corresponds to the so-called naive Bayesian classifier discussed in §9.3.2.

Since the class labels are not known, for each data value we can only determine the probability that it was generated by class j (sometimes called responsibility, e.g., HTF09). Given x_i, this probability can be obtained for each class using Bayes' rule (see eq. 3.10),

$$p(j|x_i) = \frac{\alpha_j \, \mathcal{N}(\mu_j, \sigma_j)}{\sum_{j=1}^{M} \alpha_j \, \mathcal{N}(\mu_j, \sigma_j)}. \tag{4.21}$$

The class probability $p(j|x_i)$ is small when x_i is not within "a few" σ_j from μ_j (assuming that x_i is close to some other mixture component). Of course, $\sum_{j=1}^{M} p(j|x_i) = 1$. This probabilistic class assignment is directly related to the hypothesis testing concept introduced in §4.6, and will be discussed in more detail in chapter 9.

4.4.3. The Basics of the Expectation Maximization Algorithm

Of course, we do not have to interpret eq. 4.18 in terms of classes and hidden variables. After all, lnL is just a scalar function that needs to be maximized. However, this interpretation leads to an algorithm, called the expectation maximization (EM) algorithm, which can be used to make this maximization fast and straightforward in practice. The EM algorithm was introduced by Dempster, Laird, and Rubin in 1977 ([7]), and since then many books have been written about its various aspects (for a good short tutorial, see [20]).

The key ingredient of the iterative EM algorithm is the assumption that the class probability $p(j|x_i)$ is known and fixed in each iteration (for a justification based on conditional probabilities, see [20] and HTF09). The EM algorithm is not limited to Gaussian mixtures, so instead of $\mathcal{N}(\mu_j, \sigma_j)$ in eq. 4.18, let us use a more general pdf for each component, $p_j(x_i|\boldsymbol{\theta})$ (for notational simplicity, we do not explicitly account for the fact that p_j includes only a subset of all $\boldsymbol{\theta}$ parameters, e.g., only μ_j and σ_j are relevant for the jth Gaussian component). By analogy with eq. 4.20, the log-likelihood is

$$\text{lnL} = \sum_{i=1}^{N} \ln \left[\sum_{j=1}^{M} \alpha_j \, p_j(x_i|\boldsymbol{\theta}) \right]. \tag{4.22}$$

We can take a partial derivative of lnL with respect to the parameter θ_j,

$$\frac{\partial \text{lnL}}{\partial \theta_j} = \sum_{i=1}^{N} \frac{\alpha_j}{\sum_{j=1}^{M} \alpha_j \, p_j(x_i|\boldsymbol{\theta})} \left[\frac{\partial p_j(x_i|\boldsymbol{\theta})}{\partial \theta_j} \right] \tag{4.23}$$

and motivated by eq. 4.21, rewrite it as

$$\frac{\partial \ln L}{\partial \theta_j} = \sum_{i=1}^{N} \left[\frac{\alpha_j \, p_j(x_i|\boldsymbol{\theta})}{\sum_{j=1}^{M} \alpha_j \, p_j(x_i|\boldsymbol{\theta})} \right] \left[\frac{1}{p_j(x_i|\boldsymbol{\theta})} \frac{\partial p_j(x_i|\boldsymbol{\theta})}{\partial \theta_j} \right]. \qquad (4.24)$$

Although this equation looks horrendous, it can be greatly simplified. The first term corresponds to the class probability given by eq. 4.21. Because it will be fixed in a given iteration, we introduce a shorthand $w_{ij} = p(j|x_i)$. The second term is the partial derivative of $\ln[p_j(x_i|\boldsymbol{\theta})]$. When $p_j(x_i|\boldsymbol{\theta})$ is Gaussian, it leads to particularly simple constraints for model parameters because now we take the logarithm of the exponential function *before* taking the derivative. Therefore,

$$\frac{\partial \ln L}{\partial \theta_j} = -\sum_{i=1}^{N} w_{ij} \frac{\partial}{\partial \theta_j} \left[\ln \sigma_j + \frac{(x_i - \mu_j)^2}{2\sigma_j^2} \right], \qquad (4.25)$$

where θ_j now corresponds to μ_j or σ_j. By setting the derivatives of $\ln L$ with respect to μ_j and σ_j to zero, we get the estimators (this derivation is discussed in more detail in §5.6.1)

$$\mu_j = \frac{\sum_{i=1}^{N} w_{ij} x_i}{\sum_{i=1}^{N} w_{ij}}, \qquad (4.26)$$

$$\sigma_j^2 = \frac{\sum_{i=1}^{N} w_{ij} (x_i - \mu_j)^2}{\sum_{i=1}^{N} w_{ij}}, \qquad (4.27)$$

and from the normalization constraint,

$$\alpha_j = \frac{1}{N} \sum_{i=1}^{N} w_{ij}. \qquad (4.28)$$

These expressions and eq. 4.21 form the basis of the iterative EM algorithm in the case of Gaussian mixtures. Starting with a guess for w_{ij}, the values of α_j, μ_j, and σ_j are estimated using eqs. 4.26–4.28. This is the "maximization" *M-step* which brings the parameters closer toward the local maximum. In the subsequent "expectation" *E-step*, w_{ij} are updated using eq. 4.21. The algorithm is not sensitive to the initial guess of parameter values. For example, setting all σ_j to the sample standard deviation, all α_j to $1/M$, and randomly drawing μ_j from the observed $\{x_i\}$ values, typically works well in practice (see HTF09).

Treating w_{ij} as constants during the M-step may sound ad hoc, and the whole EM algorithm might look like a heuristic method. After all, the above derivation does not guarantee that the algorithm will converge. Nevertheless, the EM algorithm has a rigorous foundation and it is provable that it will indeed find a local maximum of $\ln L$ for a wide class of likelihood functions (for discussion and references, see [20, 25]). In practice, however, the EM algorithm may fail due to numerical difficulties, especially when the available data are sparsely distributed, in the case of outliers, and if some data points are repeated (see [1]).

Scikit-learn contains an EM algorithm for fitting N-dimensional mixtures of Gaussians:

```
>>> import numpy as np
>>> from sklearn.mixture import GMM
>>> X = np.random.normal(size=(100, 1)) # 100 points
        # in 1 dim
>>> model = GMM(2) # two components
>>> model.fit(X)
>>> model.means_ # the locations of the best-fit
    # components
array([[-0.05786756],
       [ 0.69668864]])
```

See the source code of figure 4.2 for a further example. Multidimensional Gaussian mixture models are also discussed in the context of clustering and density estimation: see §6.3.

How to choose the number of classes?

We have assumed in the above discussion of the EM algorithm that the number of classes in a mixture, M, is known. As M is increased, the description of the data set $\{x_i\}$ using a mixture model will steadily improve. On the other hand, a very large M is undesired—after all, $M = N$ will assign a mixture component to each point in a data set. How do we choose M in practice?

Selecting an optimal M for a mixture model is a case of model selection discussed in §4.3. Essentially, we evaluate multiple models and score them according to some metric to get the best M. Additional detailed discussion of this important topic from the Bayesian viewpoint is presented in §5.4. A basic example of this is shown in figure 4.2, where the AIC and BIC are used to choose the optimal number of components to represent a simulated data set generated using a mixture of three Gaussian distributions. Using these metrics, the correct optimal $M = 3$ is readily recognized.

The EM algorithm as a classification tool

The right panel in figure 4.2 shows the class probability for the optimal model ($M = 3$) as a function of x (cf. eq. 4.21). These results can be used to probabilistically assign all measured values $\{x_i\}$ to one of the three classes (mixture components). There is no unique way to deterministically assign a class to each of the data points because there are unknown hidden parameters. In practice, the so-called completeness vs. contamination trade-off plays a major role in selecting classification thresholds (for a detailed discussion see §4.6).

Results analogous to the example shown in figure 4.2 can be obtained in multidimensional cases, where the mixture involves multivariate Gaussian distributions

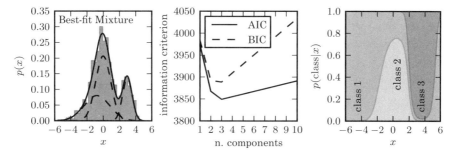

Figure 4.2. Example of a one-dimensional Gaussian mixture model with three components. The left panel shows a histogram of the data, along with the best-fit model for a mixture with three components. The center panel shows the model selection criteria AIC (see §4.3) and BIC (see §5.4) as a function of the number of components. Both are minimized for a three-component model. The right panel shows the probability that a given point is drawn from each class as a function of its position. For a given x value, the vertical extent of each region is proportional to that probability. Note that extreme values are most likely to belong to class 1.

discussed in §3.5.4. Here too an optimal model can be used to assign a probabilistic classification to each data point, and this and other classification methods are discussed in detail in chapter 9.

How to account for measurement errors?

In the above discussion of the EM algorithm, it was assumed that measurement errors for $\{x_i\}$ are negligible when compared to the smallest component width, σ_j. However, in practice this assumption is often not acceptable and the best-fit σ_j that are "broadened" by measurement errors are biased estimates of "intrinsic" widths (e.g., when measuring the widths of spectral lines). How can we account for errors in x_i, given as e_i?

We will limit our discussion to Gaussian mixtures, and assume that measurement uncertainties, as quantified by e_i, follow a Gaussian distribution. In the case of homoscedastic errors, where all $e_i = e$, we can make use of the fact that the convolution of two Gaussians is a Gaussian (see eq. 3.45) and obtain intrinsic widths as

$$\sigma_j^* = (\sigma_j^2 - e^2)^{1/2}. \tag{4.29}$$

This "poor-man's" correction procedure fails in the heteroscedastic case. Furthermore, due to uncertainties in the best-fit values, it is entirely possible that the best-fit value of σ_j may turn out to be smaller than e.

A remedy is to account for measurement errors already in the model description: we can replace σ_j in eq. 4.18 by $(\sigma_j^2 + e_i^2)^{1/2}$, where the σ_j now correspond to the intrinsic widths of each class. However, these new class pdfs do not admit simple explicit prescriptions for the maximization step given by eqs. 4.26–4.27 because they are no longer Gaussian (see §5.6.1 for a related discussion).

Following the same derivation steps, the new prescriptions for the M-step are now

$$\mu_j = \frac{\sum_{i=1}^{N} \frac{w_{ij}}{\sigma_j^2 + e_i^2} x_i}{\sum_{i=1}^{N} \frac{w_{ij}}{\sigma_j^2 + e_i^2}} \tag{4.30}$$

and

$$\sum_{i=1}^{N} \frac{w_{ij}}{\sigma_j^2 + e_i^2} = \sum_{i=1}^{N} \frac{w_{ij}}{(\sigma_j^2 + e_i^2)^2} (x_i - \mu_j)^2. \tag{4.31}$$

Compared to eqs. 4.26–4.27, σ_j is now "coupled" to e_i and cannot be moved outside of the sum, which prevents a few cancelations that led to simple forms when $e_i = 0$. Update rules for μ_j and α_j are still explicit and would require only a minor modification of specific implementation. The main difficulty, which prevents the use of standard EM routines for performing the M-step, is eq. 4.31, because the update rule for σ_j is not explicit anymore. Nevertheless, it still provides a rule for updating σ_j, which can be found by numerically solving eq. 4.31. We discuss a very similar problem and provide a numerical example in §5.6.1. An expectation maximization approach to Gaussian mixture models in the presence of errors is also discussed in chapter 6 (under the heading *extreme deconvolution* [5]) in the context of clustering and density estimation.

Non-Gaussian mixture models

The EM algorithm is not confined to Gaussian mixtures. As eq. 4.24 shows, the basic premise of the method can be derived for any mixture model. In addition to various useful properties discussed in §3.3.2, a major benefit of Gaussian pdfs is the very simple set of explicit equations (eqs. 4.26–4.28) for updating model parameters. When other pdfs are used, a variety of techniques are proposed in the literature for implementation of the maximization M-step. For cases where Gaussian mixtures are insufficient descriptors of data, we recommend consulting abundant and easily accessible literature on the various forms of the EM algorithm.

4.5. Confidence Estimates: the Bootstrap and the Jackknife

Most standard expressions for computing confidence limits for estimated parameters are based on fairly strong assumptions, such as Gaussianity and large samples. Fortunately, there are two alternative methods for computing confidence limits that are general, powerful, and easy to implement. Compared to the rest of statistics, they are relatively new and are made possible by the advent of cheap computing power. Both rely on resampling of the data set $\{x_i\}$.

Our data set $\{x_i\}$ is drawn from some distribution function $h(x)$. If we knew $h(x)$ perfectly well, we could compute any statistic without uncertainty (e.g., we could draw a large sample from $h(x)$ and, e.g., compute the mean). However, we do not know $h(x)$, and the best we can do are computations which rely on various estimates of $h(x)$ derived from the data, which we call here $f(x)$. Bootstrapping is based on the approximation (see Lup93)

$$f(x) = \frac{1}{N} \sum_{i=1}^{N} \delta(x - x_i), \tag{4.32}$$

where $\delta(x)$ is the Dirac δ function. The function $f(x)$ maximizes the probability of obtaining observed data values (f is a maximum likelihood estimator of h; for a discussion of bootstrap from the Bayesian viewpoint, see HTF09). We can now pretend that $f(x)$ is actually $h(x)$, and use it to perform various computations. For example, we could use eq. 4.32 to estimate the mean and its uncertainty.

When determining parameter uncertainties in practice, we use eq. 4.32 to draw an almost arbitrary number of new data sets. There are $N!$ possible distinct samples of size N and the probability that a new data set is identical to the original data set is $N!/N^n$ (even for a small N, this probability is small, e.g., for $N = 10$ it is only 0.00036). In other words, we draw from the observed data set with replacement: select N new index values j from the range $i = 1, \ldots, N$, and this is your new sample (some values from $\{x_i\}$ can appear twice or more times in the resampled data set). This resampling is done B times, and the resulting B data sets are used to compute the statistic of interest B times. The distribution of these values maps the uncertainty of the statistics of interest and can be used to estimate its bias and standard error, as well as other statistics.

The bootstrap method was proposed by Efron in 1979 [8]; more information can be found in [13] and references therein. The bootstrap method described above is called the nonparametric bootstrap; there is also the parametric bootstrap method which draws samples from the best-fit model. According to Wall and Jenkins, Efron named this method after "the image of lifting oneself up by one's own bootstraps." The nonparametric bootstrap method is especially useful when errors for individual data values are not independent (e.g., cumulative histogram, or two-point correlation function; see §6.5). Nevertheless, astronomers sometimes misuse the bootstrap idea (see [17]) and ignore nontrivial implementation considerations in complex problems; for example, the treatment in NumRec is misleadingly simple (for a detailed discussion of recent developments in bootstrap methodology and an excellent reference list, see [6]).

Figure 4.3 illustrates an application of the bootstrap method for estimating uncertainty in the standard deviation and σ_G (see eq. 3.36). The data sample has $N = 1000$ values and it was drawn from a Gaussian distribution with $\mu = 0$ and $\sigma = 1$. It was resampled 10,000 times and the histograms in figure 4.3 show the distribution of the resulting σ and σ_G. We can see that the bootstrap estimates of uncertainties are in good agreement with the values computed using eqs. 3.35 and 3.37.

The jackknife method, invented by Tukey in 1958, is similar in spirit to the bootstrap method (according to Lup93, the name jackknife implies robustness and general applicability). Rather than drawing a data set of the same size as the original data set during the resampling step, one or more observations are left unused when computing the statistic of interest. Let us call this statistic α with its value computed from the full data set α_N. Assuming that one observation (data value) is removed when resampling, we can form N such data sets, and compute a statistic of interest,

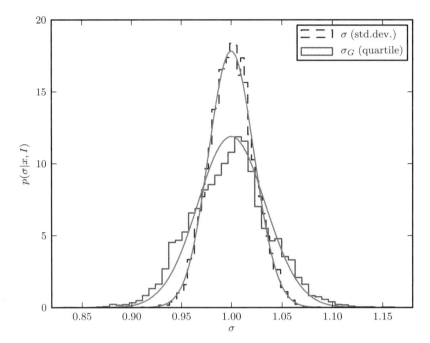

Figure 4.3. The bootstrap uncertainty estimates for the sample standard deviation σ (dashed line; see eq. 3.32) and σ_G (solid line; see eq. 3.36). The sample consists of $N = 1000$ values drawn from a Gaussian distribution with $\mu = 0$ and $\sigma = 1$. The bootstrap estimates are based on 10,000 samples. The thin lines show Gaussians with the widths determined as $s/\sqrt{2(N-1)}$ (eq. 3.35) for σ and $1.06s/\sqrt{N}$ (eq. 3.37) for σ_G.

α_i^*, for each of them. It can be shown that in the case of a single observation removed from the data set, a bias-corrected jackknife estimate of α can be computed as (see Lup93 for a concise derivation)

$$\alpha^J = \alpha_N + \Delta\alpha, \tag{4.33}$$

where the jackknife correction is

$$\Delta\alpha = (N-1)\left(\alpha_N - \frac{1}{N}\sum_{i=1}^{N}\alpha_i^*\right). \tag{4.34}$$

For estimators which are asymptotically normal, the standard error for a jackknife estimate α^J is

$$\sigma_\alpha = \sqrt{\frac{1}{N(N-1)}\sum_{i=1}^{N}\left[N\alpha_N - \alpha^J - (N-1)\alpha_i^*\right]^2}. \tag{4.35}$$

The confidence limits for α can be computed using Student's t distribution (see §3.3.8) with $t = (\alpha - \alpha^J)/\sigma_\alpha$ and $N - 1$ degrees of freedom. The jackknife standard error is more reliable than the jackknife bias correction because it is based on a

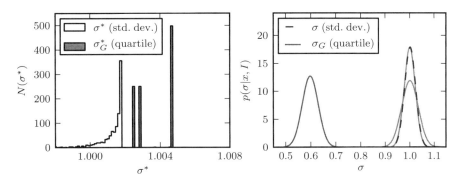

Figure 4.4. The jackknife uncertainty estimates for the width of a Gaussian distribution. This example uses the same data as figure 4.3. The upper panel shows a histogram of the widths determined using the sample standard deviation, and using the interquartile range. The lower panel shows the corrected jackknife estimates (eqs. 4.33 and 4.35) for the two methods. The gray lines show the theoretical results, given by eq. 3.35 for σ and eq. 3.37 for σ_G. The result for σ matches the theoretical result almost exactly, but note the failure of the jackknife to correctly estimate σ_G (see the text for a discussion of this result).

simpler approximation (see Lup93 for a detailed discussion). For the sample whose bootstrap uncertainty estimates for σ and σ_G are shown in figure 4.3, the jackknife method (eq. 4.35) gives similar widths as with the bootstrap method. Note, however, that the bias correction estimate for σ_G given by eq. 4.34 is completely unreliable (see figure 4.4). This failure is a general problem with the standard jackknife method, which performs well for smooth differential statistics such as the mean and standard deviation, but does not perform well for medians, quantiles, and other rank-based statistics. For these sorts of statistics, a jackknife implementation that removes more than one observation can overcome this problem. The reason for this failure becomes apparent upon examination of the upper panel of figure 4.4: for σ_G, the vast majority of jackknife samples yield one of three discrete values! Because quartiles are insensitive to the removal of outliers, all samples created by the removal of a point larger than q_{75} lead to precisely the same estimate. The same is true for removal of any point smaller than q_{25}, and for any point in the range $q_{25} < x < q_{75}$. Because of this, the jackknife cannot accurately sample the error distribution, which leads to a gross misestimate of the result.

Should one use bootstrap or jackknife in practice? Although based on different approximations, they typically produce similar results for smooth statistics, especially for large samples. Jackknife estimates are usually easier to calculate, easier to apply to complex sampling schemes, and they also automatically remove bias. However, bootstrap is better for computing confidence intervals because it does not involve the assumption of asymptotic normality (i.e., it maps out the shape of the distribution). Note that bootstrap gives slightly different results even if the data set is fixed (because random resampling is performed), while jackknife gives repeatable results for a given data set (because all possible permutations are used). Of course, when feasible, it is prudent to use both bootstrap and jackknife and critically compare their results. Both methods should be used with caution when N is small. Again, before applying bootstrap to complex problems, consult the specialist literature (a good entry point is [6]).

Cross-validation and bootstrap aggregating (bagging) are methods closely related to jackknife and bootstrap. They are used in regression and classification contexts, and are discussed in §8.11 and §9.7, respectively.

AstroML contains some routines for performing basic nonparametric jackknife and bootstrap: `astroML.resample.bootstrap` and `astroML.resample.jackknife`.

```
>>> import numpy as np
>>> from astroML.resample import jackknife
>>> x = np.random.normal(loc=0, scale=1, size=1000)
>>> jackknife(x, np.std, kwargs=dict(ddof=1, axis=1))
(1.01, 0.02)
```

The standard deviation is found to be 1.01 ± 0.02. For more examples of the use of bootstrap and jackknife methods, see the source code of figure 4.3.

4.6. Hypothesis Testing

A common problem in statistics is to ask whether a given sample is consistent with some hypothesis. For example, we might be interested in whether a measured value x_i, or the whole set $\{x_i\}$, is consistent with being drawn from a Gaussian distribution $\mathcal{N}(\mu, \sigma)$. Here $\mathcal{N}(\mu, \sigma)$ is our *null hypothesis*, typically corresponding to a "no effect" case, and we are trying to *reject it* in order to demonstrate that we measured some effect. A good example from astronomy is the source detection in images with substantial background (e.g., atmospheric sky brightness in optical images). Because the background fluctuates, the contribution of the source flux to a particular image resolution element must be substantially larger than the background fluctuation to represent a robust detection. Here, the null hypothesis is that the measured brightness in a given resolution element is due to background, and when we can reject it, we have a source detection. It is always assumed that we know how to compute the probability of a given outcome from the null hypothesis: for example, given the cumulative distribution function, $0 \leq H_0(x) \leq 1$ (see eq. 1.1), the probability that we would get a value at least as large as x_i is $p(x > x_i) = 1 - H(x_i)$, and is called the *p value*. Typically, a threshold *p* value is adopted, called *the significance level* α, and the null hypothesis is rejected when $p \leq \alpha$ (e.g., if $\alpha = 0.05$ and $p < 0.05$, the null hypothesis is rejected at a 0.05 significance level). If we fail to reject a hypothesis, it does not mean that we proved its correctness because it may be that our sample is simply not large enough to detect an effect.

For example, if we flip a coin 10 times and get 8 tails, should we reject the hypothesis that the coin is fair? If it is indeed fair, the binomial distribution (eq. 3.50) predicts that the probability of 8 or more tails is 0.054 and thus we cannot reject the null hypothesis at the 0.05 significance level. We shall return to this coin-flip example when discussing Bayesian methods in chapter 5.

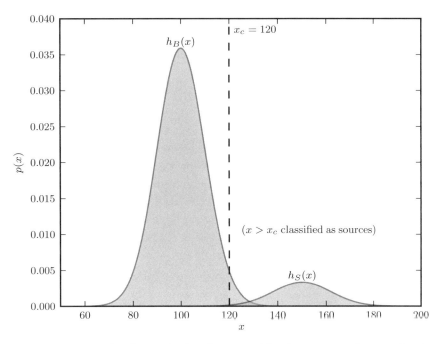

Figure 4.5. An example of a simple classification problem between two Gaussian distributions. Given a value of x, we need to assign that measurement to one of the two distributions (background vs. source). The cut at $x_c = 120$ leads to very few Type II errors (i.e., *false negatives*: points from the distribution h_S with $x < x_c$ being classified as background), but this comes at the cost of a significant number of Type I errors (i.e., *false positives*: points from the distribution h_B with $x > x_c$ being classified as sources).

When performing these tests, we are bound to make two types of errors, which statisticians memorably call *Type I and Type II errors* (Jerzy Neyman and Egon Pearson introduced this notation in 1933). Type I errors are cases when the null hypothesis is true but incorrectly rejected. In the context of source detection, these errors represent spurious sources, or more generally, false positives (see figure 4.5). The false-positive probability when testing a single datum is limited by the adopted significance level α (the case of multiple tests is discussed in the next section). Cases when the null hypothesis is false, but it is not rejected are called Type II errors (missed sources, or false negatives). The false-negative probability when testing a single datum is usually called β, and is related to *the power of a test* as $(1 - \beta)$. Hypothesis testing is intimately related to comparisons of distributions, as discussed below. The classical approach to hypothesis testing is not identical to the Bayesian approach, and we shall return to this topic in chapter 5 (see §5.4).

4.6.1. Simple Classification and Completeness vs. Contamination Trade-Off

As the significance level α is decreased (the criterion for rejecting the null hypothesis becomes more conservative), the number of false positives decreases and the number of false negatives increases. Therefore, there is a trade-off to be made to find an optimal value of α, which depends on the relative importance of false negatives and positives in a particular problem. For example, if the null hypothesis is "my parachute is good," we are more concerned about false negatives (it's bad but we accept it as good) than about false positives (it's good but we reject it as bad) because the former can kill us while the latter presumably has less dire consequences (what α would make you feel safe in this case?). On the other hand, if the null hypothesis is "this undergraduate student would do great in graduate school," then accepting a bad student (false positive) is arguably less harmful than rejecting a truly good student (false negative).

When many instances of hypothesis testing are performed, a process called *multiple hypothesis testing*, the fraction of false positives can significantly exceed the value of α. The fraction of false positives depends not only on α and the number of data points, but also on the number of true positives (the latter is proportional to the number of instances when an alternative hypothesis is true). We shall illustrate these trade-offs with an example.

Often the underlying distribution from which data $\{x_i\}$ were drawn, $h(x)$, is a sum of two populations

$$h(x) = (1 - a)\, h_B(x) + a\, h_S(x), \tag{4.36}$$

where a is the relative normalization factor (we assume that integrals of h_B and h_S are normalized to unity). In this example there is not only a null hypothesis (B, for background), but also a specific alternative hypothesis (S, for source). Given $\{x_i\}$, for example counts obtained with a measuring apparatus, we want to assign to each individual measurement x_i the probability that it belongs to population S, $p_S(x_i)$ (of course, $p_B(x_i) = 1 - p_S(x_i)$ as there are only these two possibilities). Recall that the size of sample $\{x_i\}$ is N, and thus the number of *true* sources in the sample is Na. A simplified version of this problem is *classification*, where we assign the class S or B without retaining the knowledge of the actual probability p_S. In order for classification based on x to be possible at all, obviously $h_B(x)$ and $h_S(x)$ must be different (how to "measure" the difference between two distributions is discussed in §4.7).

If we choose a classification boundary value x_c, then the *expected number* of spurious sources (false positives or Type I errors) in the classified sample is

$$n_{\text{spurious}} = N(1 - a)\, \alpha = N(1 - a) \int_{x_c}^{\infty} h_B(x)\, dx, \tag{4.37}$$

and the number of missed sources (false negatives or Type II errors) is

$$n_{\text{missed}} = N a\, \beta = N a \int_0^{x_c} h_S(x)\, dx. \tag{4.38}$$

The number of instances classified as a source, that is, instances when the null hypothesis is rejected, is

$$n_{\text{source}} = N a - n_{\text{missed}} + n_{\text{spurious}}. \tag{4.39}$$

The sample *completeness* (also called sensitivity and recall rate in the statistics literature) is defined as

$$\eta = \frac{Na - n_{\text{missed}}}{Na} = 1 - \int_0^{x_c} h_S(x)\, dx,$$

(4.40)

with $0 \leq \eta \leq 1$, and the sample *contamination* is defined as

$$\epsilon = \frac{n_{\text{spurious}}}{n_{\text{source}}},$$

(4.41)

with $0 \leq \epsilon \leq 1$ (the $1 - \epsilon$ rate is sometimes called classification efficiency). The sample contamination is also called the *false discovery rate* (FDR). As x_c increases, the sample contamination decreases (good), but at the same time completeness decreases too (bad). This trade-off can be analyzed using the so-called *receiver operating characteristic* (ROC) curve which typically plots the fraction of true positives vs. the fraction of true negatives (see HTF09). In astronomy, ROC curves are often plotted as expected completeness vs. contamination (or sometimes efficiency) rate. The position along the ROC curve is parametrized by x_c (i.e., by the classification rule). The area under the ROC curve, sometimes called the c statistic, can be used to quantify overall performance of the classification method (see §9.8).

Note that the sample completeness involves neither N nor a, but the sample contamination depends on a: obviously, for $a = 1$ we get $\epsilon = 0$ (there can be no contamination if only sources exist), but more concerningly, $\lim_{a \to 0}(\epsilon)=1$. In other words, for small a, that is, when the true fraction of instances with a false null hypothesis is small, we can have a large sample contamination even when x_c corresponds to a very small value of α. How do we choose an optimal value of x_c?

To have a concrete example, we will use the source detection example and assume that $h_B(x) = \mathcal{N}(\mu = 100, \sigma = 10)$ and $h_S(x) = \mathcal{N}(\mu = 150, \sigma = 12)$, with $a = 0.1$ and $N = 10^6$ (say, an image with 1000 by 1000 resolution elements; the x values correspond to the sum of background and source counts). For illustration, see figure 4.5. If we naively choose $x_c = 120$ (a "2σ cut" away from the mean for h_B, corresponding to a Type I error probability of $\alpha = 0.024$), 21,600 values will be incorrectly classified as a source. With $x_c = 120$, the sample completeness is 0.994 and 99,400 values are correctly classified as a source. Although the Type I error rate is only 0.024, the sample contamination is $21{,}600/(21{,}600+99{,}400) = 0.179$, or over 7 times higher! Of course, this result that $\epsilon \gg \alpha$ is a consequence of the fact that the true population contains 9 times as many background values as it contains sources ($a = 0.1$).

In order to decrease the expected contamination level ϵ, we need to increase x_c, but the optimal value depends on a. Since a is often unknown in practice, choosing x_c is not straightforward. A simple practical method for choosing the optimal value of x_c for a given desired ϵ (or FDR) was proposed by Benjamini and Hochberg [2].

The Benjamini and Hochberg method assumes that measurements can be described by eq. 4.36 and makes an additional assumption that $h_B(x)$ is known (e.g., when a is small, it is possible to isolate a portion of an image to measure background count distribution). Given $h_B(x)$, and its cumulative counterpart $H_B(x)$, it is possible to assign a p value to each value in $\{x_i\}$ as $p_i = 1 - H_B(x_i)$, and sort the sample so that p_i are increasing. If all $\{x_i\}$ values were drawn from $h_B(x)$, the differential distribution of these p_i values would be a uniform distribution by construction, and its cumulative distribution, $1 \leq C_i \leq N$, would increase linearly as

$$C_i^B = N \, p_i. \tag{4.42}$$

Instead, for Na cases in the adopted model the null hypothesis is false and they will result in an excess of small p_i values; hence, the observed cumulative distribution, $C_i = C(p_i) = i$, will have values much larger than C_i^B for small p. Benjamini and Hochberg realized that this fact can be used to find a classification threshold p_c (and the corresponding x_c and its index $i_c = C(p_c)$; recall that the sample is sorted by p_i) that guarantees that the sample contamination ϵ is below some desired value ϵ_0. Their proposal for finding p_c is very elegant and does not require involved computations: assume that the null hypothesis is rejected for all values $p_i \leq p_c$, resulting in a subsample of $i_c = C(p_c)$ values (i.e., these i_c values are selected as sources). The number of cases when the null hypothesis was actually true and falsely rejected is $(1 - a)Np_c < Np_c$, and thus the contamination rate is

$$\epsilon = \frac{(1 - a)Np_c}{i_c} < \frac{Np_c}{i_c}. \tag{4.43}$$

Therefore, the threshold value must satisfy

$$i_c < N \frac{p_c}{\epsilon_0}. \tag{4.44}$$

This condition corresponds to the intersection of the measured $C_i(p_i)$ curve and the straight line $C = Np/\epsilon_0$, and the algorithm is simply to find the largest value of i ($= C_i$) which satisfies eq. 4.44 (see figure 4.6). Note that $p_c < \epsilon_0$ because $i_c < N$; if one naively adopted $p_c = \epsilon_0$, the resulting expected sample contamination would be a factor of N/i_c larger than ϵ_0.

The Benjamini and Hochberg algorithm is conservative because it assumes that $(1-a) \approx 1$ when deriving the upper bound on ϵ. Hence, the resulting contamination rate ϵ is a factor of $(1 - a)$ smaller than the maximum allowed value ϵ_0. If we knew a, we could increase i_c given by eq. 4.44 by a factor of $1/(1 - a)$, and thus increase the sample completeness, while still guaranteeing that the sample contamination does not exceed ϵ_0.

In cases where one can assume that large values of p (say, for $p > 0.5$) are dominated by the null hypothesis, which is often the case, the cumulative distribution is

$$C(p_i) = Na + N(1 - a) \, p_i \quad \text{for } p_i > 0.5, \tag{4.45}$$

with, say, $C(p_i = 0.5) = C_{0.5}$. Given that the slope of this line is $2(N - C_{0.5})$, the number of cases when the null hypothesis is true but falsely rejected can be estimated as $2(N - C_{0.5})p_c$. This estimate amounts to scaling $h_B(x)$ to fit the observed

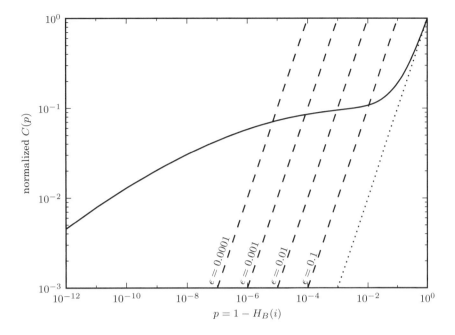

Figure 4.6. Illustration of the Benjamini and Hochberg method for 10^6 points drawn from the distribution shown in figure 4.5. The solid line shows the cumulative distribution of observed p values, normalized by the sample size. The dashed lines show the cutoff for various limits on contamination rate ϵ computed using eq. 4.44 (the accepted measurements are those with p smaller than that corresponding to the intersection of solid and dashed curves). The dotted line shows how the distribution would look in the absence of sources. The value of the cumulative distribution at $p = 0.5$ is 0.55, and yields a correction factor $\lambda = 1.11$ (see eq. 4.46).

distribution, or equivalently, estimating $(1 - a)$ as

$$\lambda^{-1} \equiv 1 - a = 2 \left(1 - \frac{C_{0.5}}{N} \right). \tag{4.46}$$

Thus, the Benjamini and Hochberg method can be improved by multiplying i_c by λ, yielding the sample completeness increased by a factor λ.

4.7. Comparison of Distributions

We often ask whether two samples are drawn from the same distribution, or equivalently whether two sets of measurements imply a difference in the measured quantity. A similar question is whether a sample is consistent with being drawn from some known distribution (while real samples are always finite, the second question is the same as the first one when one of the samples is considered as infinitely large). In general, obtaining answers to these questions can be very complicated. First, what do we mean by "the same distribution"? Distributions can be described by their location, scale, and shape. When the distribution shape is assumed known, for example when we know for one or another reason that the sample is drawn from a Gaussian distribution, the problem is greatly simplified to the consideration of only two parameters (location and scale, μ and σ from $\mathcal{N}(\mu, \sigma)$). Second, we might be interested in only one of these two parameters; for example, do two sets of measurements with different measurement errors imply the same mean value (e.g., two experimental groups measure the mass of the same elementary particle, or the same planet, using different methods).

Depending on data type (discrete vs. continuous random variables) and what we can assume (or not) about the underlying distributions, and the specific question we ask, we can use different statistical tests. The underlying idea of statistical tests is to use data to compute an appropriate statistic, and then compare the resulting data-based value to its expected distribution. The expected distribution is evaluated by *assuming that the null hypothesis is true*, as discussed in the preceding section. When this expected distribution implies that the data-based value is unlikely to have arisen from it by chance (i.e., the corresponding p value is small), the null hypothesis is rejected with some threshold probability α, typically 0.05 or 0.01 ($p < \alpha$). For example, if the null hypothesis is that our datum came from the $\mathcal{N}(0, 1)$ distribution, then $x = 3$ corresponds to $p = 0.003$ (see §3.3.2). Note again that $p > \alpha$ does *not* mean that the hypothesis is *proven* to be correct!

The number of various statistical tests in the literature is overwhelming and their applicability is often hard to discern. We describe here only a few of the most important tests, and further discuss hypothesis testing and distribution comparison in the Bayesian context in chapter 5.

4.7.1. Regression toward the Mean

Before proceeding with statistical tests for comparing distributions, we point out a simple statistical selection effect that is sometimes ignored and leads to spurious conclusions.

If two instances of a data set $\{x_i\}$ are drawn from some distribution, the mean difference between the matched values (i.e., the ith value from the first set and the ith value from the second set) will be zero. However, if we use one data set to select a subsample for comparison, the mean difference may become biased. For example, if we subselect the lowest quartile from the first data set, then the mean difference between the second and the first data set will be larger than zero.

Although this subselection step may sound like a contrived procedure, there are documented cases where the impact of a procedure designed to improve students' test scores was judged by applying it only to the worst performing students. Given that there is always some randomness (measurement error) in testing scores, these preselected students would have improved their scores without any intervention. This effect is called "regression toward the mean": if a random variable is extreme on its first measurement, it will tend to be closer to the population mean on a second measurement. In an astronomical context, a common related tale states that weather conditions observed at a telescope site today are, typically, not as good as those that would have been inferred from the prior measurements made during the site selection process.

Therefore, when selecting a subsample for further study, or a control sample for comparison analysis, one has to worry about various statistical selection effects. Going back to the above example with student test scores, a proper assessment of a new educational procedure should be based on a randomly selected subsample of students who will undertake it.

4.7.2. Nonparametric Methods for Comparing Distributions

When the distributions are not known, tests are called nonparametric, or distribution-free tests. The most popular nonparametric test is the Kolmogorov–Smirnov (K-S) test, which compares the cumulative distribution function, $F(x)$, for two samples, $\{x1_i\}$, $i = 1, \ldots, N_1$ and $\{x2_i\}$, $i = 1, \ldots, N_2$ (see eq. 1.1 for definitions; we sort the sample and divide the rank (recall §3.6.1) of x_i by the sample size to get $F(x_i)$; $F(x)$ is a step function that increases by $1/N$ at each data point; note that $0 \leq F(x) \leq 1$).

The K-S test and its variations can be performed in Python using the routines kstest, ks_2samp, and ksone from the module scipy.stats:

```
>>> import numpy as np
>>> from scipy import stats
>>> vals = np.random.normal(loc=0, scale=1,
                            size=1000)
>>> stats.kstest(vals, "norm")
(0.0255, 0.529)
```

The D value is 0.0255, and the p value is 0.529. For more examples of these statistics, see the SciPy documentation, and the source code for figure 4.7.

The K-S test is based on the following statistic which measures the maximum distance of the two cumulative distributions $F_1(x1)$ and $F_2(x2)$:

$$D = \max |F_1(x1) - F_2(x2)| \tag{4.47}$$

($0 \leq D \leq 1$; we note that other statistics could be used to measure the difference between F_1 and F_2, e.g., the integrated square error). The key question is how often would the value of D computed from the data arise by chance if the two samples were drawn from the *same* distribution (the null hypothesis in this case). Surprisingly, this question has a well-defined answer even when we know nothing about the underlying distribution. Kolmogorov showed in 1933 (and Smirnov published tables with the numerical results in 1948) that the probability of obtaining by chance a value of D larger than the measured value is given by the function

$$Q_{KS}(\lambda) = 2 \sum_{k=1}^{\infty} (-1)^{k-1} e^{-2k^2\lambda^2}, \tag{4.48}$$

where the argument λ can be accurately described by the following approximation (as shown by Stephens in 1970; see discussion in NumRec):

$$\lambda = \left(0.12 + \sqrt{n_e} + \frac{0.11}{\sqrt{n_e}} \right) D, \tag{4.49}$$

where the "effective" number of data points is computed from

$$n_e = \frac{N_1 \, N_2}{N_1 + N_2}. \tag{4.50}$$

Note that for large n_e, $\lambda \approx \sqrt{n_e} D$. If the probability that a given value of D is due to chance is very small (e.g., 0.01 or 0.05), we can reject the null hypothesis that the two samples were drawn from the same underlying distribution.

For n_e greater than about 10 or so, we can bypass eq. 4.48 and use the following simple approximation to evaluate D corresponding to a given probability α of obtaining a value at least that large:

$$D_{KS} = \frac{C(\alpha)}{\sqrt{n_e}}, \tag{4.51}$$

where $C(\alpha)$ is the critical value of the Kolmogorov distribution with $C(\alpha = 0.05) = 1.36$ and $C(\alpha = 0.01) = 1.63$. Note that the ability to reject the null hypothesis (if it is really false) increases with $\sqrt{n_e}$. For example, if $n_e = 100$, then $D > D_{KS} = 0.163$ would arise by chance in only 1% of all trials. If the actual data-based value is indeed 0.163, we can reject the null hypothesis that the data were drawn from the same (unknown) distribution, with our decision being correct in 99 out of 100 cases.

We can also use the K-S test to ask, "Is the measured $f(x)$ consistent with a known reference distribution function $h(x)$?" (When $h(x)$ is a Gaussian distribution with known parameters, it is more efficient to use the parametric tests described in the next section.) This case is called the "one-sample" K-S test, as opposed to the "two-sample" K-S test discussed above. In this case, $N_1 = N$ and $N_2 = \infty$, and thus $n_e = N$. Again, a small value of Q_{KS} (or $D > D_{KS}$) indicates that it is unlikely, at the given confidence level set by α, that the data summarized by $f(x)$ were drawn from $h(x)$.

The K-S test is sensitive to the location, the scale, and the shape of the underlying distribution(s) and, because it is based on cumulative distributions, it is invariant to reparametrization of x (we would obtain the same conclusion if, for example, we used $\ln x$ instead of x). The main strength but also the main weakness of the K-S test is its ignorance about the underlying distribution. For example, the test is insensitive to details in the differential distribution function (e.g., narrow regions where it drops to zero), and more sensitive near the center of the distribution than at the tails (the K-S test is not the best choice for distinguishing samples drawn from Gaussian and exponential distributions; see §4.7.4).

For an example of the two-sample K-S test, refer to figure 3.25, where it is used to confirm that two random samples are drawn from the same underlying data set. For an example of the one-sample K-S test, refer to figure 4.7, where it is compared to other tests of Gaussianity.

A simple test related to the K-S test was developed by Kuiper to treat distributions defined on a circle. It is based on the statistic

$$D^* = \max\{F_1(x1) - F_2(x2)\} + \max\{F_2(x1) - F_1(x2)\}. \tag{4.52}$$

As is evident, this statistic considers both positive and negative differences between two distributions (D from the K-S test is equal to the greater of the two terms).

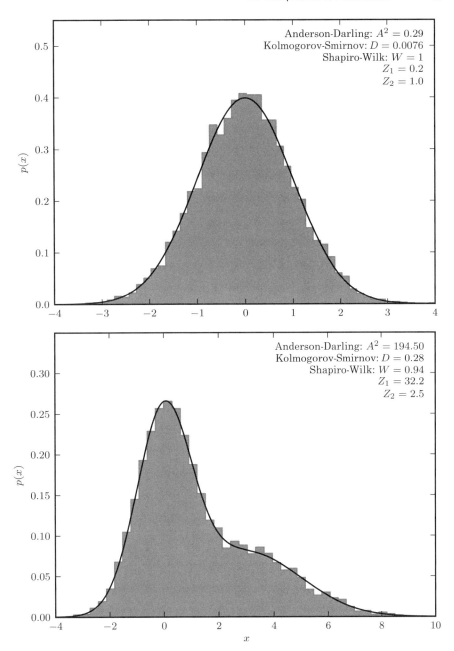

Figure 4.7. The results of the Anderson–Darling test, the Kolmogorov–Smirnov test, and the Shapiro–Wilk test when applied to a sample of 10,000 values drawn from a normal distribution (upper panel) and from a combination of two Gaussian distributions (lower panel).

For distributions defined on a circle (i.e., $0° < x < 360°$), the value of D^* is invariant to where exactly the origin ($x = 0°$) is placed. Hence, the Kuiper test is a good test for comparing the longitude distributions of two astronomical samples. By analogy with the K-S test,

$$Q_{\text{Kuiper}}(\lambda) = 2 \sum_{k=1}^{\infty} (4k^2\lambda^2 - 1) \, e^{-2k^2\lambda^2}, \tag{4.53}$$

with

$$\lambda = \left(0.155 + \sqrt{n_e} + \frac{0.24}{\sqrt{n_e}}\right) D^*. \tag{4.54}$$

The K-S test is not the only option for nonparametric comparison of distributions. The Cramér–von Mises criterion, the Watson test, and the Anderson–Darling test, to name but a few, are similar in spirit to the K-S test, but consider somewhat different statistics. For example, the Anderson–Darling test is more sensitive to differences in the tails of the two distributions than the K-S test. A practical difficulty with these other statistics is that a simple summary of their behavior, such as given by eq. 4.48 for the K-S test, is not readily available. We discuss a very simple test for detecting non-Gaussian behavior in the tails of a distribution in §4.7.4.

A somewhat similar quantity that is also based on the cumulative distribution function is the Gini coefficient (developed by Corrado Gini in 1912). It measures the deviation of a given cumulative distribution ($F(x)$, defined for $x_{\min} \leq x \leq x_{\max}$) from that expected for a uniform distribution:

$$G = 1 - 2 \int_{x_{\min}}^{x_{\max}} F(x) \, dx. \tag{4.55}$$

When $F(x)$ corresponds to a uniform differential distribution, $G = 0$, and $G \leq 1$ always. The Gini coefficient is *not* a statistical test, but we mention it here for reference because it is commonly used in classification (see §9.7.1), in economics and related fields (usually to quantify income inequality), and sometimes confused with a statistical test.

The U test and the Wilcoxon test

The U test and Wilcoxon test are implemented in `mannwhitneyu` and `ranksums` (i.e., Wilcoxon rank-sum test) within the `scipy.stats` module:

```
>>> import numpy as np
>>> from scipy import stats
>>> x, y = np.random.normal(0, 1, size=(2, 1000))
>>> stats.mannwhitneyu(x, y)
(487678.0, 0.1699)
```

The U test result is close to the expected $N_1 N_2/2$, indicating that the two samples are drawn from the same distribution. For more information, see the SciPy documentation.

Nonparametric methods for comparing distributions, for example, the K-S test, are often sensitive to more than a single distribution property, such as the location or scale parameters. Often, we are interested in differences in only a particular statistic, such as the mean value, and do not care about others. There are several widely used nonparametric tests for such cases. They are analogous to the better-known classical parametric tests, the t test and the paired t test (which assume Gaussian distributions and are described below), and are based on the ranks of data points, rather than on their values.

The U test, or the Mann–Whitney–Wilcoxon test (or the Wilcoxon rank-sum test, not to be confused with the Wilcoxon signed-rank test described below) is a nonparametric test for testing whether two data sets are drawn from distributions with different location parameters (if these distributions are known to be Gaussian, the standard classical test is called the t test, described in §4.7.6). The sensitivity of the U test is dominated by a difference in medians of the two tested distributions.

The U statistic is determined using the ranks for the full sample obtained by concatenating the two data sets and sorting them, while retaining the information about which data set a value came from. To compute the U statistic, take each value from sample 1 and count the number of observations in sample 2 that have a smaller rank (in the case of identical values, take half a count). The sum of these counts is U, and the minimum of the values with the samples reversed is used to assess the significance. For cases with more than about 20 points per sample, the U statistic for sample 1 can be more easily computed as

$$U_1 = R_1 - \frac{N_1(N_1 - 1)}{2}, \tag{4.56}$$

where R_1 is the sum of ranks for sample 1, and analogously for sample 2. The adopted U statistic is the smaller of the two (note that $U_1 + U_2 = N_1 N_2$, which can be used to check computations). The behavior of U for large samples can be well approximated with a Gaussian distribution, $\mathcal{N}(\mu_U, \sigma_U)$, of variable

$$z = \frac{U - \mu_U}{\sigma_U}, \tag{4.57}$$

with

$$\mu_U = \frac{N_1 N_2}{2} \tag{4.58}$$

and

$$\sigma_U = \sqrt{\frac{N_1 N_2 (N_1 + N_2 + 1)}{12}}. \tag{4.59}$$

For small data sets, consult the literature or use one of the numerous and widely available statistical programs.

A special case of comparing the means of two data sets is when the data sets have the same size ($N_1 = N_2 = N$) and data points are paired. For example, the two data sets could correspond to the same sample measured twice, "before" and "after" something that could have affected the values, and we are testing for evidence of a change in mean values. The nonparametric test that can be used to compare means of two arbitrary distributions is the Wilcoxon signed-rank test. The test is based on differences $y_i = x1_i - x2_i$, and the values with $y_i = 0$ are excluded, yielding the new sample size $m \leq N$. The sample is ordered by $|y_i|$, resulting in the rank R_i for each pair, and each pair is assigned $\Phi_i = 1$ if $x1_i > x2_i$ and 0 otherwise. The Wilcoxon signed-ranked statistic is then

$$W_+ = \sum_i^m \Phi_i R_i, \tag{4.60}$$

that is, all the ranks with $y_i > 0$ are summed. Analogously, W_- is the sum of all the ranks with $y_i < 0$, and the statistic T is the smaller of the two. For small values of m, the significance of T can be found in tables. For m larger than about 20, the behavior of T can be well approximated with a Gaussian distribution, $\mathcal{N}(\mu_T, \sigma_T)$, of the variable

$$z = \frac{T - \mu_T}{\sigma_T}, \tag{4.61}$$

with

$$\mu_T = \frac{N(2N+1)}{2} \tag{4.62}$$

and

$$\sigma_T = N\sqrt{\frac{(2N+1)}{12}}. \tag{4.63}$$

The Wilcoxon signed-rank test can be performed with the function `scipy.stats.wilcoxon`:

```
import numpy as np
from scipy import stats
x, y = np.random.normal(0, 1, size=(2, 1000))
T, p = stats.wilcoxon(x, y)
```

See the documentation of the `wilcoxon` function for more details.

4.7.3. Comparison of Two-Dimensional Distributions

There is no direct analog of the K-S test for multidimensional distributions because cumulative probability distribution is not well defined in more than one dimension. Nevertheless, it is possible to use a method similar to the K-S test, though not as straightforward (developed by Peacock in 1983, and Fasano and Franceschini in 1987; see §14.7 in NumRec), as follows.

Given two sets of points, $\{x_i^A, y_i^A\}, i = 1, \ldots, N_A$ and $\{x_i^B, y_i^B\}, i = 1, \ldots, N_B$, define four quadrants centered on the point (x_j^A, y_j^A) and compute the fraction of data points from each data set in each quadrant. Record the maximum difference (among the four quadrants) between the fractions for data sets A and B. Repeat for all points from sample A to get the overall maximum difference, D_A, and repeat the whole procedure for sample B. The final statistic is then $D = (D_A + D_B)/2$.

Although it is not strictly true that the distribution of D is independent of the details of the underlying distributions, Fasano and Franceschini showed that its variation is captured well by the coefficient of correlation, ρ (see eq. 3.81). Using simulated samples, they derived the following behavior (analogous to eq. 4.49 from the one-dimensional K-S test):

$$\lambda = \frac{\sqrt{n_e}\, D}{1 + (0.25 - 0.75/\sqrt{n_e})\sqrt{1 - \rho^2}}. \tag{4.64}$$

This value of λ can be used with eq. 4.48 to compute the significance level of D when $n_e > 20$.

4.7.4. Is My Distribution Really Gaussian?

When asking, "Is the measured $f(x)$ consistent with a known reference distribution function $h(x)$?", a few standard statistical tests can be used when we know, or can assume, that both $h(x)$ and $f(x)$ are Gaussian distributions. These tests are at least as efficient as any nonparametric test, and thus are the preferred option. Of course, in order to use them reliably we need to first convince ourselves (and others!) that our $f(x)$ is consistent with being a Gaussian.

Given a data set $\{x_i\}$, we would like to know whether we can reject the null hypothesis (see §4.6) that $\{x_i\}$ was drawn from a Gaussian distribution. Here we are not asking for specific values of the location and scale parameters, but only whether the *shape* of the distribution is Gaussian. In general, deviations from a Gaussian distribution could be due to nonzero skewness, nonzero kurtosis (i.e., thicker symmetric or asymmetric tails), or more complex combinations of such deviations. Numerous tests are available in statistical literature which have varying sensitivity to different deviations. For example, the difference between the mean and the median for a given data set is sensitive to nonzero skewness, but has no sensitivity whatsoever to changes in kurtosis. Therefore, if one is trying to detect a difference between the Gaussian $\mathcal{N}(\mu = 4, \sigma = 2)$ and the Poisson distribution with $\mu = 4$, the difference between the mean and the median might be a good test (0 vs. 1/6 for large samples), but it will not catch the difference between a Gaussian and an exponential distribution no matter what the size of the sample.

As already discussed in §4.6, a common feature of most tests is to predict the distribution of their chosen statistic under the assumption that the null hypothesis is true. An added complexity is whether the test uses any parameter estimates derived from data. Given the large number of tests, we limit our discussion here to only a few of them, and refer the reader to the voluminous literature on statistical tests in case a particular problem does not lend itself to these tests.

The first test is the Anderson–Darling test, specialized to the case of a Gaussian distribution. The test is based on the statistic

$$A^2 = -N - \frac{1}{N}\sum_{i=1}^{N}[(2i - 1)\ln(F_i) + (2N - 2i + 1)\ln(1 - F_i)], \tag{4.65}$$

TABLE 4.1.
The values of the Anderson–Darling statistic A^2 corresponding to significance level p.

μ and σ from data?	$p = 0.05$	$p = 0.01$
μ no, σ no	2.49	3.86
μ yes, σ no	1.11	1.57
μ no, σ yes	2.32	3.69
μ yes, σ yes	0.79	1.09

where F_i is the ith value of the cumulative distribution function of z_i, which is defined as

$$z_i = \frac{x_i - \mu}{\sigma}, \qquad (4.66)$$

and assumed to be in ascending order. In this expression, either one or both of μ and σ can be known, or determined from data $\{x_i\}$. Depending on which parameters are determined from data, the statistical behavior of A^2 varies. Furthermore, if *both* μ and σ are determined from data (using eqs. 3.31 and 3.32), then A^2 needs to be multiplied by $(1 + 4/N - 25/N^2)$. The specialization to a Gaussian distribution enters when predicting the detailed statistical behavior of A^2, and its values for a few common significance levels (p) are listed in table 4.1. The values corresponding to other significance levels, as well as the statistical behavior of A^2 in the case of distributions other than Gaussian can be computed with simple numerical simulations (see the example below).

---

scipy.stats.anderson implements the Anderson–Darling test:

```
>>> import numpy as np
>>> from scipy import stats
>>> x = np.random.normal(0, 1, size=1000)
>>> A, crit, sig = stats.anderson(x, 'norm')
>>> A
0.54728
```

See the source code of figure 4.7 for a more detailed example.

Of course, the K-S test can also be used to detect a difference between $f(x)$ and $\mathcal{N}(\mu, \sigma)$. A difficulty arises if μ and σ are determined from the same data set: in this case the behavior of Q_{KS} is different from that given by eq. 4.48 and has only been determined using Monte Carlo simulations (and is known as the Lilliefors distribution [16]).

The third common test for detecting non-Gaussianity in $\{x_i\}$ is the Shapiro–Wilk test. It is implemented in a number of statistical programs, and details about this test can be found in [23]. Its statistic is based on both data values, x_i, and data ranks, R_i (see §3.6.1):

$$W = \frac{\left(\sum_{i=1}^{N} a_i R_i\right)^2}{\sum_{i=1}^{N} (x_i - \bar{x})^2},$$

(4.67)

where constants a_i encode the expected values of the order statistics for random variables sampled from the standard normal distribution (the test's null hypothesis). The Shapiro–Wilk test is very sensitive to non-Gaussian tails of the distribution ("outliers"), but not as much to detailed departures from Gaussianity in the distribution's core. Tables summarizing the statistical behavior of the W statistic can be found in [11].

The Shapiro–Wilk test is implemented in `scipy.stats.shapiro`:

```
>>> import numpy as np
>>> from scipy import stats
>>> x = np.random.normal(0, 1, 1000)
>>> stats.shapiro(x)
(0.9975, 0.1495)
```

A value of W close to 1 indicates that the data is indeed Gaussian. For more information, see the documentation of the function `shapiro`.

Often the main departure from Gaussianity is due to so-called "catastrophic outliers," or largely discrepant values many σ away from μ. For example, the overwhelming majority of measurements of fluxes of objects in an astronomical image may follow a Gaussian distribution, but, for just a few of them, unrecognized cosmic rays could have had a major impact on flux extraction. A simple method to detect the presence of such outliers is to compare the sample standard deviation s (eq. 3.32) and σ_G (eq. 3.36). Even when the outlier fraction is tiny, the ratio s/σ_G can become significantly large. When $N > 100$, for a Gaussian distribution (i.e., for the null hypothesis), this ratio follows a nearly Gaussian distribution with $\mu \sim 1$ and with $\sigma \sim 0.92/\sqrt{N}$. For example, if you measure $s/\sigma_G = 1.3$ using a sample with $N = 100$, then you can state that the probability of such a large value appearing by chance is less than 1%, and reject the null hypothesis that your sample was drawn from a Gaussian distribution. Another useful result is that the difference of the mean and the median drawn from a Gaussian distribution also follows a nearly Gaussian distribution with $\mu \sim 0$ and $\sigma \sim 0.76 s/\sqrt{N}$. Therefore, when $N > 100$ we can define two simple statistics based on the measured values of (μ, q_{50}, s, and σ_G) that both measure departures in terms of Gaussian-like "sigma":

$$Z_1 = 1.3 \frac{|\mu - q_{50}|}{s} \sqrt{N}$$

(4.68)

and

$$Z_2 = 1.1 \left| \frac{s}{\sigma_G} - 1 \right| \sqrt{N}. \tag{4.69}$$

Of course, these and similar results for the statistical behavior of various statistics can be easily derived using Monte Carlo samples (see §3.7).

Figure 4.7 shows the results of these tests when applied to samples of $N = 10{,}000$ values selected from a Gaussian distribution and from a mixture of two Gaussian distributions. To summarize, for data that depart from a Gaussian distribution, we expect the Anderson–Darling A^2 statistic to be much larger than 1 (see table 4.1), the K-S D statistic (see eq. 4.47 and 4.51) to be much larger than $1/\sqrt{N}$, the Shapiro–Wilk W statistic to be smaller than 1, and Z_1 and Z_2 to be larger than several σ. All these tests correctly identify the first data set as being normally distributed, and the second data set as departing from normality.

In cases when our empirical distribution fails the tests for Gaussianity, but there is no strong motivation for choosing an alternative specific distribution, a good approach for modeling non-Gaussianity is to adopt the Gram–Charlier series,

$$h(x) = \mathcal{N}(\mu, \sigma) \sum_{k=0}^{\infty} a_k H_k(z), \tag{4.70}$$

where $z = (x - \mu)/\sigma$, and $H_k(z)$ are the Hermite polynomials ($H_0 = 1$, $H_1 = z$, $H_2 = z^2 - 1$, $H_3 = z^3 - 3z$, etc.). For "nearly Gaussian" distributions, even the first few terms of the series provide a good description of $h(x)$ (see figure 3.6 for an example of using the Gram–Charlier series to generate a skewed distribution). A related expansion, the Edgeworth series, uses derivatives of $h(x)$ to derive "correction" factors for a Gaussian distribution.

4.7.5. Is My Distribution Bimodal?

It happens frequently in practice that we want to test a hypothesis that the data were drawn from a unimodal distribution (e.g., in the context of studying bimodal color distribution of galaxies, bimodal distribution of radio emission from quasars, or the kinematic structure of the Galaxy's halo). Answering this question can become quite involved and we discuss it in chapter 5 (see §5.7.3).

4.7.6. Parametric Methods for Comparing Distributions

Given a sample $\{x_i\}$ that does not fail any test for Gaussianity, one can use a few standard statistical tests for comparing means and variances. They are more efficient (they require smaller samples to reject the null hypothesis) than nonparametric tests, but often by much less than a factor of 2, and for good nonparametric tests close to 1 (e.g., the efficiency of the U test compared to the t test described below is as high as 0.95). Hence, nonparametric tests are generally the preferable option to classical tests which assume Gaussian distributions. Nevertheless, because of their ubiquitous presence in practice and literature, we briefly summarize the two most important classical tests. As before, we assume that we are given two samples, $\{x1_i\}$ with $i = 1, \ldots, N_1$, and $\{x2_i\}$ with $i = 1, \ldots, N_2$.

Comparison of Gaussian means using the *t* test

Variants of the *t* test can be computed using the routines `ttest_rel`, `ttest_ind`, and `ttest_1samp`, available in the module `scipy.stats`:

```
>>> import numpy as np
>>> from scipy import stats
>>> x, y = np.random.normal(size=(2, 1000))
>>> t, p = stats.ttest_ind(x, y)
```

See the documentation of the above SciPy functions for more details.

If the only question we are asking is whether our data $\{x1_i\}$ and $\{x2_i\}$ were drawn from two Gaussian distributions with a different μ but the same σ, and we were given σ, the answer would be simple. We would first compute the mean values for both samples, $\overline{x1}$ and $\overline{x2}$, using eq. 3.31, and their standard errors, $\sigma_{\overline{x1}} = \sigma/\sqrt{N_1}$ and analogously for $\sigma_{\overline{x2}}$, and then ask how large is the difference $\Delta = \overline{x1} - \overline{x2}$ in terms of its expected scatter, $\sigma_\Delta = \sigma\sqrt{1/N_1^2 + 1/N_2^2}$: $M_\sigma = \Delta/\sigma_\Delta$. The probability that the observed value of M would arise by chance is given by the Gauss error function (see §3.3.2) as $p = 1 - \mathrm{erf}(M/\sqrt{2})$. For example, for $M = 3$, $p = 0.003$.

If we do *not* know σ, but need to estimate it from data (with possibly different values for the two samples, s_1 and s_2; see eq. 3.32), then the ratio $M_s = \Delta/s_\Delta$, where $s_\Delta = \sqrt{s_1^2/N_1 + s_2^2/N_2}$, can no longer be described by a Gaussian distribution! Instead, it follows Student's t distribution (see the discussion in §5.6.1). The number of degrees of freedom depends on whether we assume that the two underlying distributions from which the samples were drawn have the same variances or not. If we can make this assumption then the relevant statistic (corresponding to M_s) is

$$t = \frac{\overline{x1} - \overline{x2}}{s_D}, \tag{4.71}$$

where

$$s_D = \sqrt{s_{12}^2 \left(\frac{1}{N_1} + \frac{1}{N_2} \right)} \tag{4.72}$$

is an estimate of the standard error of the difference of the means, and

$$s_{12} = \sqrt{\frac{(N_1 - 1)s_1^2 + (N_2 - 1)s_2^2}{N_1 + N_2 - 2}} \tag{4.73}$$

is an estimator of the common standard deviation of the two samples. The number of degrees of freedom is $k = (N_1 + N_2 - 2)$. Hence, instead of looking up the significance of $M_\sigma = \Delta/\sigma_\Delta$ using the Gaussian distribution $\mathcal{N}(0, 1)$, we use the significance corresponding to t and Student's t distribution with k degrees of freedom. For very large samples, this procedure tends to the simple case with known σ described in the first paragraph because Student's t distribution tends to a Gaussian distribution (in other words, s converges to σ).

If we cannot assume that the two underlying distributions from which the samples were drawn have the same variances, then the appropriate test is called Welch's t test and the number of degrees of freedom is determined using the Welch–Satterthwaite equation (however, see §5.6.1 for the Bayesian approach). For formulas and implementation, see NumRec.

A special case of comparing the means of two data sets is when the data sets have the same size ($N_1 = N_2 = N$) and each pair of data points has the same σ, but the value of σ is not the same for all pairs (recall the difference between the nonparametric U and the Wilcoxon tests). In this case, the t test for paired samples should be used. The expression 4.71 is still valid, but eq. 4.72 needs to be modified as

$$s_D = \sqrt{\frac{(N_1 - 1)s_1^2 + (N_2 - 1)s_2^2 - 2\text{Cov}_{12}}{N}}, \tag{4.74}$$

where the covariance between the two samples is

$$\text{Cov}_{12} = \frac{1}{N-1}\sum_{i=1}^{N}(x1_i - \overline{x1})(x2_i - \overline{x2}). \tag{4.75}$$

Here the pairs of data points from the two samples need to be properly arranged when summing, and the number of degrees of freedom is $N - 1$.

Comparison of Gaussian variances using the F test

The F test can be computed using the routine `scipy.stats.f_oneway`:

```
>>> import numpy as np
>>> from scipy import stats
>>> x, y = np.random.normal(size=(2, 1000))
>>> F, p = stats.f_oneway(x, y)
```

See the SciPy documentation for more details.

The F test is used to compare the variances of two samples, $\{x1_i\}$ and $\{x2_i\}$, drawn from two unspecified Gaussian distributions. The null hypothesis is that the variances of two samples are equal, and the statistic is based on the ratio of the sample variances (see eq. 3.32),

$$F = \frac{s_1^2}{s_2^2}, \tag{4.76}$$

where F follows Fisher's F distribution with $d_1 = N_1 - 1$ and $d_2 = N_2 - 1$ (see §3.3.9). Situations when we are interested in only knowing whether $\sigma_1 < \sigma_2$ or $\sigma_2 < \sigma_1$ are treated by appropriately using the left and right tails of Fisher's F distribution.

We will conclude this section by quoting Wall and Jenkins: "The application of efficient statistical procedure has power; but the application of common sense has more." We will see in the next chapter that the Bayesian approach provides a transparent mathematical framework for quantifying our common sense.

4.8. Nonparametric Modeling and Histograms

When there is no strong motivation for adopting a parametrized description (typically an analytic function with free parameters) of a data set, nonparametric methods offer an alternative approach. Somewhat confusingly, "nonparametric" does not mean that there are no parameters. For example, one of the simplest nonparametric methods to analyze a one-dimensional data set is a histogram. To construct a histogram, we need to specify bin boundaries, and we implicitly assume that the estimated distribution function is piecewise constant within each bin. Therefore, here too there are parameters to be determined—the value of the distribution function in each bin. However, there is no specific distribution class, such as the set of all possible Gaussians, or Laplacians, but rather a general set of *distribution-free* models, called the Sobolev space. The Sobolev space includes all functions, $h(x)$, that satisfy some smoothness criteria, such as

$$\int [h''(x)]^2 dx < \infty. \tag{4.77}$$

This constraint, for example, excludes all functions with infinite spikes. Formally, a method is nonparametric if it provides a distribution function estimate $f(x)$ that approaches the true distribution $h(x)$ with enough data, for any $h(x)$ in a class of functions with relatively weak assumptions, such as the Sobolev space above.

Nonparametric methods play a central role in modern machine learning. They provide the highest possible predictive accuracies, as they can model any shape of distribution, down to the finest detail which still has predictive power, though they typically come at a higher computational cost than more traditional multivariate statistical methods. In addition, it is harder to interpret the results of nonparametric methods than those of parametric models.

Nonparametric methods are discussed extensively in the rest of this book, including methods such as nonparametric correction for the selection function in the context of luminosity function estimation (§4.9), kernel density estimation (§6.1.1), and decision trees (§9.7). In this chapter, we only briefly discuss one-dimensional histograms.

4.8.1. Histograms

A histogram can fit virtually any shape of distribution, given enough bins. This is the key—while each bin can be thought of as a simple constant estimator of the density in that bin, the overall histogram is a piecewise constant estimator which can be thought of as having a tuning parameter—the number of bins. When the number of data points is small, the number of bins should somehow be small, as there is not enough information to warrant many bins. As the number of data points grows, the number of bins should also grow to capture the increasing amount of detail in the distribution's shape that having more data points allows. This is a general feature of nonparametric methods—they are composed of simple pieces, and the number of pieces grows with the number of data points.

Getting the number of bins right is clearly critical. Pragmatically, it can easily make the difference between concluding that a distribution has a single mode or that it has two modes. Intuitively, we expect that a large bin width will destroy fine-scale features in the data distribution, while a small width will result in increased counting noise per bin. We emphasize that it is *not* necessary to bin the data before estimating model parameters. A simple example is the case of data drawn from a Gaussian distribution. We can estimate its parameters μ and σ using eqs. 3.31 and 3.32 without ever binning the data. This is a general result that will be discussed in the context of arbitrary distributions in chapter 5. Nevertheless, binning can allow us to visualize our data and explore various features in order to motivate the model selection.

We will now look at a few rules of thumb for the surprisingly subtle question of choosing the critical bin width, based on frequentist analyses. The gold standard for frequentist bin width selection is cross-validation, which is more computationally intensive. This topic is discussed in §6.1.1, in the context of a generalization of histograms (kernel density estimation). However, because histograms are so useful as quick data visualization tools, simple rules of thumb are useful to have in order to avoid large or complex computations.

Various proposed methods for choosing optimal bin width typically suggest a value proportional to some estimate of the distribution's scale, and decreasing with the sample size. The most popular choice is "Scott's rule" which prescribes a bin width

$$\Delta_b = \frac{3.5\sigma}{N^{1/3}}, \tag{4.78}$$

where σ is the sample standard deviation, and N is the sample size. This rule asymptotically minimizes the mean integrated square error (see eq. 4.14) and assumes that the underlying distribution is Gaussian; see [22]. An attempt to generalize this rule to non-Gaussian distributions is the Freedman–Diaconis rule,

$$\Delta_b = \frac{2(q_{75} - q_{25})}{N^{1/3}} = \frac{2.7\sigma_G}{N^{1/3}}, \tag{4.79}$$

which estimates the scale ("spread") of the distribution from its interquartile range (see [12]). In the case of a Gaussian distribution, Scott's bin width is 30% larger than the Freedman–Diaconis bin width. Some rules use the extremes of observed values to estimate the scale of the distribution, which is clearly inferior to using the interquartile range when outliers are present.

Although the Freedman–Diaconis rule attempts to account for non-Gaussian distributions, it is too simple to distinguish, for example, multimodal and unimodal distributions that have the same σ_G. The main reason why finding the optimal bin size is not straightforward is that the result depends on both the actual data distribution and the choice of metric (such as the mean square error) to be optimized.

The interpretation of binned data essentially represents a model fit, where the model is a piecewise constant function. Different bin widths correspond to different models, and choosing the best bin width amounts to the selection of the best model. The model selection is a topic discussed in detail in chapter 5 on Bayesian statistical inference, and in that context we will describe a powerful method that is cognizant of the detailed properties of a given data distribution. We will also compare these three different rules using multimodal and unimodal distributions (see §5.7.2, in particular figure 5.20).

NumPy and Matplotlib contain powerful tools for creating histograms in one dimension or multiple dimensions. The Matplotlib command pylab.hist is the easiest way to plot a histogram:

```
In [1]: %pylab
In [2]: import numpy as np
In [3]: x = np.random.normal(size=1000)
In [4]: plt.hist(x, bins=50)
```

For more details, see the source code for the many figures in this chapter which show histograms. For computing but not plotting a histogram, the functions numpy.histogram, numpy.histogram2d, and numpy.histogramdd provide optimized implementations:

```
In [5]: counts, bins = np.histogram(x, bins=50)
```

The above rules of thumb for choosing bin widths are implemented in the submodule astroML.density_estimation, using the functions knuth_bin_width, scotts_bin_width, and freedman_bin_width. There is also a pylab-like interface for simple histogramming:

```
In [6]: from astroML.plotting import hist
In [7]: hist(x, bins='freedman') # can also choose
        # 'knuth' or 'scott'
```

The hist function in AstroML operates just like the hist function in Matplotlib, but can optionally use one of the above routines to choose the binning. For more details see the source code associated with figure 5.20, and the associated discussion in §5.7.2.

4.8.2. How to Determine the Histogram Errors?

Assuming that we have selected a bin size, Δ_b, the N values of x_i are sorted into M bins, with the count in each bin $n_k, k = 1, \ldots, M$. If we want to express the results as a properly normalized $f(x)$, with the values f_k in each bin, then it is customary to adopt

$$f_k = \frac{n_k}{\Delta_b N}.$$ (4.80)

The unit for f_k is the inverse of the unit for x_i.

Each estimate of f_k comes with some uncertainty. It is customary to assign "error bars" for each n_k equal to $\sqrt{n_k}$ and thus the uncertainty of f_k is

$$\sigma_k = \frac{\sqrt{n_k}}{\Delta_b N}.$$ (4.81)

This practice assumes that n_k are scattered around the true values in each bin (μ) according to a Gaussian distribution, and that error bars enclose the 68% confidence range for the true value. However, when counts are low this assumption of Gaussianity breaks down and the Poisson distribution should be used instead. For example, according to the Gaussian distribution, negative values of μ have nonvanishing probability for small n_k (if $n_k = 1$, this probability is 16%). This is clearly wrong since in counting experiments, $\mu \geq 0$. Indeed, if $n_k \geq 1$, then even $\mu = 0$ is clearly ruled out. Note also that $n_k = 0$ does not necessarily imply that $\mu = 0$: even if $\mu = 1$, counts will be zero in $1/e \approx 37\%$ of cases. Another problem is that the range $n_k \pm \sigma_k$ does not correspond to the 68% confidence interval for true μ when n_k is small. These issues are important when fitting models to small count data (assuming that the available data are already binned). This idea is explored in a Bayesian context in §5.6.6.

4.9. Selection Effects and Luminosity Function Estimation

We have already discussed truncated and censored data sets in §4.2.7. We now consider these effects in more detail and introduce a nonparametric method for correcting the effects of the selection function on the inferred properties of the underlying pdf.

When the selection probability, or selection function $S(x)$, is known (often based on analysis of simulated data sets) and finite, we can use it to correct our estimate $f(x)$. The correction is trivial in the strictly one-dimensional case: the implied true distribution $h(x)$ is obtained from the observed $f(x)$ as

$$h(x) = \frac{f(x)}{S(x)}.$$ (4.82)

When additional observables are available, they might carry additional information about the behavior of the selection function, $S(x)$. One of the most important examples in astronomy is the case of flux-limited samples, as follows.

Assume that in addition to x, we also measure a quantity y, and that our selection function is such that $S(x) = 1$ for $0 \leq y \leq y_{max}(x)$, and $S(x) = 0$ for $y > y_{max}(x)$, with $x_{min} \leq x \leq x_{max}$. Here, the observable y may, or may not, be related to (correlated with) observable x, and the $y \geq 0$ assumption is

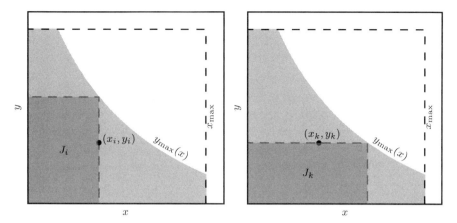

Figure 4.8. Illustration for the definition of a truncated data set, and for the comparable or associated subset used by the Lynden-Bell C^- method. The sample is limited by $x < x_{max}$ and $y < y_{max}(x)$ (light-shaded area). Associated sets J_i and J_k are shown by the dark-shaded area.

added for simplicity and without a loss of generality. In an astronomical context, x can be thought of as luminosity, L, (or absolute magnitude), and y as distance (or redshift in the cosmological context). The differential distribution of luminosity (probability density function) is called the luminosity function. In this example, and for noncosmological distances, we can compute $y_{max}(x) = (x/(4\pi F_{min}))^{1/2}$, where F_{min} is the smallest flux that our measuring apparatus can detect (or that we imposed on the sample during analysis); for illustration see figure 4.8. The observed distribution of x values is in general different from the distribution we would observe when $S(x) = 1$ for $y \leq (x_{max}/(4\pi F_{min}))^{1/2}$, that is, when the "missing" region, defined by $y_{max}(x) < y \leq (x_{max}/(4\pi F_{min}))^{1/2} = y_{max}(x_{max})$, is not excluded. If the two-dimensional probability density is $n(x, y)$, then the latter is given by

$$h(x) = \int_0^{y_{max}(x_{max})} n(x, y)\, dy, \tag{4.83}$$

and the observed distribution corresponds to

$$f(x) = \int_0^{y_{max}(x)} n(x, y)\, dy. \tag{4.84}$$

As is evident, the dependence of $n(x, y)$ on y directly affects the difference between $f(x)$ and $h(x)$. Therefore, in order to obtain an estimate of $h(x)$ based on measurements of $f(x)$ (the luminosity function in the example above), we need to estimate $n(x, y)$ first. Using the same example, $n(x, y)$ is the probability density function per unit luminosity *and* unit distance (or equivalently volume). Of course, there is no guarantee that the luminosity function is the same for near and far distances, that is, $n(x, y)$ need not be a separable function of x and y.

Let us formulate the problem as follows. Given a set of measured pairs (x_i, y_i), with $i = 1, \ldots, N$, and *known* relation $y_{max}(x)$, estimate the two-dimensional distribution, $n(x, y)$, from which the sample was drawn. Assume that measurement errors for both x and y are negligible compared to their observed ranges, that x is measured within a range defined by $x_{\min}$ and $x_{\max}$, and that the selection function is 1 for $0 \leq y \leq y_{\max}(x)$ and $x_{\min} \leq x \leq x_{\max}$, and 0 otherwise (for illustration, see figure 4.8).

In general, this problem can be solved by fitting some predefined (assumed) function to the data (i.e., determining a set of best-fit parameters), or in a nonparametric way. The former approach is typically implemented using maximum likelihood methods [4] as discussed in §4.2.2. An elegant nonparametric solution to this mathematical problem was developed by Lynden-Bell [18], and shown to be equivalent or better than other nonparametric methods by Petrosian [19]. In particular, Lynden-Bell's solution, dubbed the C^- method, is superior to the most famous nonparametric method, the $1/V_{\max}$ estimator of Schmidt [21]. Lynden-Bell's method belongs to methods known in statistical literature as the product-limit estimators (the most famous example is the Kaplan–Meier estimator for estimating the survival function; for example, the time until failure of a certain device).

4.9.1. Lynden-Bell's C^- Method

Lynden-Bell's C-minus method is implemented in the package astroML.lumfunc, using the functions Cminus, binned_Cminus, and bootstrap_Cminus. For data arrays x and y, with associated limits xmax and ymax, the call looks like this:

```
from astroML.lumfunc import Cminus
Nx, Ny, cuml_x, cuml_y = Cminus(x, y, xmax, ymax)
```

For details on the use of these functions, refer to the documentation and to the source code for figures 4.9 and 4.10.

Lynden-Bell's nonparametric C^- method can be applied to the above problem when the distributions along the two coordinates x and y are uncorrelated, that is, when we can assume that the bivariate distribution $n(x, y)$ is separable:

$$n(x, y) = \Psi(x)\, \rho(y). \tag{4.85}$$

Therefore, before using the C^- method we need to demonstrate that this assumption is valid.

Following Lynden-Bell, the basic steps for testing that the bivariate distribution $n(x, y)$ is separable are the following:

1. Define a *comparable* or *associated* set for each object i such that $J_i = \{j : x_j < x_i, y_j < y_{\max}(x_i)\}$; this is the largest x-limited and y-limited data subset for object i, with N_i elements (see the left panel of figure 4.8).
2. Sort the set J_i by y_j; this gives us the rank R_j for each object (ranging from 1 to N_i).
3. Define the rank R_i for object i in *its* associated set: this is essentially the number of objects with $y < y_i$ in set J_i.

4. Now, if x and y are truly independent, R_i must be distributed *uniformly* between 0 and N_i; in this case, it is trivial to determine the expectation value and variance for R_i: $E(R_i) = E_i = N_i/2$ and $V(R_i) = V_i = N_i^2/12$. We can define the statistic

$$\tau = \frac{\sum_i (R_i - E_i)}{\sqrt{\sum_i V_i}}.$$ (4.86)

If $\tau < 1$, then x and y are uncorrelated at $\sim 1\sigma$ level (this step appears similar to Schmidt's $V/V_{\max}$ test discussed below; nevertheless, they are fundamentally different because $V/V_{\max}$ tests the hypothesis of a uniform distribution in the y direction, while the statistic τ tests the hypothesis of uncorrelated x and y).

Assuming that $\tau < 1$, it is straightforward to show, using relatively simple probability integral analysis (e.g., see the appendix in [10], as well as the original Lynden-Bell paper [18]), how to determine cumulative distribution functions. The cumulative distributions are defined as

$$\Phi(x) = \int_{-\infty}^{x} \Psi(x')\,dx'$$ (4.87)

and

$$\Sigma(y) = \int_{-\infty}^{y} \rho(y')\,dy'.$$ (4.88)

Then,

$$\Phi(x_i) = \Phi(x_1) \prod_{k=2}^{i} (1 + 1/N_k),$$ (4.89)

where it is assumed that x_i are sorted ($x_1 \le x_k \le x_N$). Analogously, if M_k is the number of objects in a set defined by $J_k = \{j : y_j < y_k, y_{\max}(x_j) > y_k\}$ (see the right panel of figure 4.8), then

$$\Sigma(y_j) = \Sigma(y_1) \prod_{k=2}^{j} (1 + 1/M_k).$$ (4.90)

Note that both $\Phi(x_j)$ and $\Sigma(y_j)$ are defined on nonuniform grids with N values, corresponding to the N measured values. Essentially, the C^- method assumes a piecewise constant model for $\Phi(x)$ and $\Sigma(y)$ between data points (equivalently, differential distributions are modeled as Dirac δ functions at the position of each data point). As shown by Petrosian, $\Phi(x)$ and $\Sigma(y)$ represent an optimal data summary [19].

The differential distributions $\Psi(x)$ and $\rho(y)$ can be obtained by binning cumulative distributions in the relevant axis; the statistical noise (errors) for both quantities can be estimated as described in §4.8.2, or using bootstrap (§4.5).

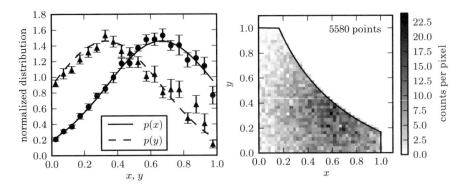

Figure 4.9. An example of using Lynden-Bell's C^- method to estimate a bivariate distribution from a truncated sample. The lines in the left panel show the true one-dimensional distributions of x and y (truncated Gaussian distributions). The two-dimensional distribution is assumed to be separable; see eq. 4.85. A realization of the distribution is shown in the right panel, with a truncation given by the solid line. The points in the left panel are computed from the truncated data set using the C^- method, with error bars from 20 bootstrap resamples.

An approximate normalization can be obtained by requiring that the total predicted number of objects is equal to their observed number.

We first illustrate the C^- method using a toy model where the answer is known; see figure 4.9. The input distributions are recovered to within uncertainties estimated using bootstrap resampling. A realistic example is based on two samples of galaxies with SDSS spectra (see §1.5.5). A flux-limited sample of galaxies with an r-band magnitude cut of $r < 17.7$ is selected from the redshift range $0.08 < z < 0.12$, and separated into blue and red subsamples using the color boundary $u-r = 2.22$. These color-selected subsamples closely correspond to spiral and elliptical galaxies and are expected to have different luminosity distributions [24]. Absolute magnitudes were computed from the distance modulus based on the spectroscopic redshift, assuming WMAP cosmology (see the source code of figure 4.10 for details). For simplicity, we ignore K corrections, whose effects should be very small for this redshift range (for a more rigorous treatment, see [3]). As expected, the difference in luminosity functions is easily discernible in figure 4.10. Due to the large sample size, statistical uncertainties are very small. True uncertainties are dominated by systematic errors because we did not take evolutionary and K corrections into account; we assumed that the bivariate distribution is separable, and we assumed that the selection function is unity. For a more detailed analysis and discussion of the luminosity function of SDSS galaxies, see [4].

It is instructive to compare the results of the C^- method with the results obtained using the $1/V_{max}$ method [21]. The latter assumes that the observed sources are uniformly distributed in probed volume, and multiplies the counts in each x bin j by a correction factor that takes into account the fraction of volume accessible to each measured source. With x corresponding to distance, and assuming that volume scales as the cube of distance (this assumption is not correct at cosmological distances),

$$S_j = \sum_i \left(\frac{x_i}{x_{max}(j)} \right)^3 , \tag{4.91}$$

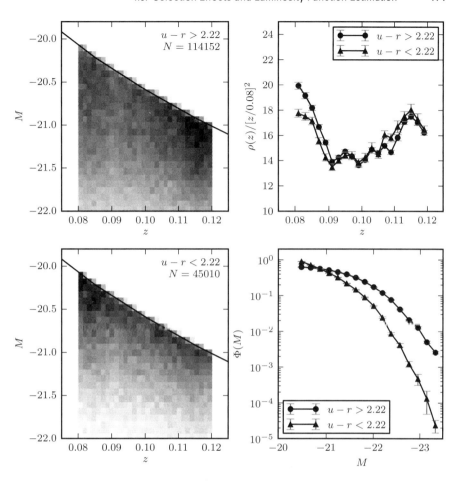

Figure 4.10. An example of computing the luminosity function for two $u - r$ color-selected subsamples of SDSS galaxies using Lynden-Bell's C^- method. The galaxies are selected from the SDSS spectroscopic sample, with redshift in the range $0.08 < z < 0.12$ and flux limited to $r < 17.7$. The left panels show the distribution of sources as a function of redshift and absolute magnitude. The distribution $p(z, M) = \rho(z)\Phi(m)$ is obtained using Lynden-Bell's method, with errors determined by 20 bootstrap resamples. The results are shown in the right panels. For the redshift distribution, we multiply the result by z^2 for clarity. Note that the most luminous galaxies belong to the photometrically red subsample, as discernible in the bottom-right panel.

where the sum is over all x_i measurements from y (luminosity) bin j, and the maximum distance $x_{max}(j)$ is defined by $y_j = y_{max}[x_{max}(j)]$. Given S_j, h_j is determined from f_j using eq. 4.82. Effectively, each measurement contributes more than a single count, proportionally to $1/x_i^3$. This correction procedure is correct only if there is no variation of the underlying distribution with distance. Lynden-Bell's C^- method is more versatile because it can treat cases when the underlying distribution varies with distance (as long as this variation does not depend on the other coordinate).

Complicated selection function

In practical problems, the selection function is often more complicated than given by the sharp boundary at $y_{max}(x)$. A generalization of the C^- method to the case of an arbitrary selection function, $S(x, y)$, is described in [10]. First define a generalized comparable set $J_i = \{j : x_j > x_i\}$, and then generalize N_i to the quantity

$$T_i = \sum_{j=1}^{N_i} \frac{S(x_i, y_j)}{S(x_j, y_j)},\tag{4.92}$$

with a redefined rank

$$R_i = \sum_{j=1}^{N_i} \frac{S(x_i, y_j)}{S(x_j, y_j)},\tag{4.93}$$

for $y_j < y_i$. It follows that $E(R_i) = T_i/2$ and $V(R_i) = T_i^2/12$, as in the case of a simple selection function.

4.9.2. A Flexible Method of Estimating Luminosity Functions

What can we do when measurement errors for x and y are not negligible, or when the bivariate distribution we want to infer is not a separable function of x and y? A powerful and completely general Bayesian approach is described by Kelly, Fan, and Vestergaard in [14]. They model the function $n(x, y)$ as a mixture of Gaussian functions (see §4.4). Although this approach is formally parametric, it is essentially as flexible as nonparametric methods.

4.10. Summary

In this chapter we have reviewed classical or frequentist techniques used for data modeling, estimating confidence intervals, and hypothesis testing. In the following chapter, we will build upon this toolkit by considering Bayesian methods. A combination of ideas from these two chapters will form the basis of the machine learning and data mining techniques presented in part III of the book.

References

[1] Archambeau, C., J. Lee, and M. Verleysen (2003). On convergence problems of the EM algorithm for finite Gaussian mixtures. In *European Symposium on Artificial Neural Networks*, pp. 99–106.

[2] Benjamini, Y. and Y. Hochberg (1995). Controlling the false discovery rate: a practical and powerful approach to multiple testing. *J. R. Stat. Soc. 57*, 289.

[3] Blanton, M. R., J. Brinkmann, I. Csabai, and others (2003). Estimating fixed-frame galaxy magnitudes in the Sloan Digital Sky Survey. *AJ 125*, 2348–2360.

[4] Blanton, M. R., D. W. Hogg, N. A. Bahcall, and others (2003). The galaxy luminosity function and luminosity density at redshift z = 0.1. *ApJ 592*, 819–838.

[5] Bovy, J., D. W. Hogg, and S. T. Roweis (2011). Extreme deconvolution: Inferring complete distribution functions from noisy, heterogeneous and incomplete observations. *Annals of Applied Statistics 5*, 1657–1677.

[6] Davison, A. C., D. V. Hinkley, and G. Young (2003). Recent developments in bootstrap methodology. *Statistical Science 18*, 141–57.

[7] Dempster, A. P., N. M. Laird, and D. Rubin (1977). Maximum-likelihood from incomplete data via the EM algorithm. *J. R. Stat. Soc. Ser. B 39*, 1–38.

[8] Efron, B. (1979). Bootstrap methods: Another look at the jackknife. *Ann. Statistics 7*, 1–26.

[9] Efron, B. and D. V. Hinkley (1978). Assessing the accuracy of the maximum likelihood estimator: Observed versus expected Fisher information. *Biometrika 65*, 457–87.

[10] Fan, X., M. A. Strauss, D. P. Schneider, and others (2001). High-redshift quasars found in Sloan Digital Sky Survey commissioning data. IV. Luminosity function from the Fall Equatorial Stripe Sample. *AJ 121*, 54–65.

[11] Franklin, J. (1972). *Biometrica Tables for Statisticians.* Cambridge University Press.

[12] Freedman, D. and P. Diaconis (1981). On the histogram as a density estimator: L2 theory. *Zeitschrift für Wahrscheinlichkeitstheorie und verwandte Gebiete 57*, 453–476.

[13] Hastie, T., R. Tibshirani, and J. Friedman (1993). *An Introduction to the Boostrap.* Chapman and Hall/CRC.

[14] Kelly, B. C., X. Fan, and M. Vestergaard (2008). A flexible method of estimating luminosity functions. *ApJ 682*, 874–895.

[15] Liddle, A. R. (2007). Information criteria for astrophysical model selection. *MNRAS 377*, L74–L78.

[16] Lilliefors, H. (1967). On the Kolmogorov-Smirnov test for normality with mean and variance unknown. *Journal of the American Statistical Association 62*, 399–402.

[17] Loredo, T. J. (2012). Bayesian astrostatistics: A backward look to the future. *ArXiv:astro-ph/1208.3036.*

[18] Lynden-Bell, D. (1971). A method of allowing for known observational selection in small samples applied to 3CR quasars. *MNRAS 155*, 95.

[19] Petrosian, V. (1992). Luminosity function of flux-limited samples. In E. D. Feigelson and G. J. Babu (Eds.), *Statistical Challenges in Modern Astronomy*, pp. 173–200.

[20] Roche, A. (2011). EM algorithm and variants: An informal tutorial. *ArXiv:statistics/1105.1476.*

[21] Schmidt, M. (1968). Space distribution and luminosity functions of quasi-stellar radio sources. *ApJ 151*, 393.

[22] Scott, D. W. (1979). On optimal and data-based histograms. *Biometrika 66*, 605–610.

[23] Shapiro, S. S. and M. B. Wilk (1965). An analysis of variance test for normality (complete samples). *Biometrika 52*, 591–61.

[24] Strateva, I., Ž. Ivezić, G. R. Knapp, and others (2001). Color separation of galaxy types in the Sloan Digital Sky Survey imaging data. *AJ 122*, 1861–1874.

[25] Wu, C. (1983). On the convergence properties of the EM algorithm. *The Annals of Statistics 11*, 94–103.

5 Bayesian Statistical Inference

"The Bayesian approach is the numerical realization of common sense." (Bayesians)

We have already addressed the main philosophical differences between classical and Bayesian statistical inferences in §4.1. In this chapter, we introduce the most important aspects of Bayesian statistical inference and techniques for performing such calculations in practice. We first review the basic steps in Bayesian inference in §5.1–5.4, and then illustrate them with several examples in §5.6–5.7. Numerical techniques for solving complex problems are discussed in §5.8, and the last section provides a summary of pros and cons for classical and Bayesian methods.

Let us briefly note a few historical facts. The Reverend Thomas Bayes (1702–1761) was a British amateur mathematician who wrote a manuscript on how to combine an initial belief with new data to arrive at an improved belief. The manuscript was published posthumously in 1763 and gave rise to the name Bayesian statistics. However, the first renowned mathematician to popularize Bayesian methodology was Pierre Simon Laplace, who rediscovered (1774) and greatly clarified Bayes' principle. He applied the principle to a variety of contemporary problems in astronomy, physics, population statistics, and even jurisprudence. One of the most famous results is his estimate of the mass of Saturn and its uncertainty, which remain consistent with the best measurements of today.

Despite Laplace's fame, Bayesian analysis did not secure a permanent place in science. Instead, classical frequentist statistics was adopted as the norm (this could be at least in part due to the practical difficulties of performing full Bayesian calculations without the aid of computers). Much of Laplace's Bayesian analysis was ignored until the early twentieth century when Harold Jeffreys reinterpreted Laplace's work with much clarity. Yet, even Jeffreys' work was not fully comprehended until around 1960, when it took off thanks to vocal proponents such as de Finetti, Savage, Wald, and Jaynes, and of course, the advent of computing technology. Today, a vast amount of literature exists on various Bayesian topics, including the two books by Jaynes and Gregory listed in §1.3. For a very informative popular book about the resurgence of Bayesian methods, see [26].

5.1. Introduction to the Bayesian Method

The basic premise of the Bayesian method is that probability statements are not limited to data, but can be made for model parameters and models themselves. Inferences are made by producing probability density functions (pdfs); most notably, model parameters are treated as random variables.

The Bayesian method has gained wide acceptance over the last few decades, in part due to maturing development of its philosophical and technical foundations, and in part due to the ability to actually perform the required computations. The Bayesian method yields optimal results, given all the available and explicitly declared information, assuming, of course, that all of the supplied information is correct. Even so, it is not without its own pitfalls, as discussed at the end of this chapter.

Classical and Bayesian techniques share an important ingredient: the data likelihood function introduced in §4.2. In classical statistics, the data likelihood function is used to find model parameters that yield the highest data likelihood. Yet, the likelihood function cannot be interpreted as a probability density function for model parameters. Indeed, the pdf for a model parameter is not even a valid concept in classical statistics. The Bayesian method extends the concept of the data likelihood function by adding extra, so-called *prior*, information to the analysis, and assigning pdfs to all model parameters and models themselves.

We use the following simple example for motivating the inclusion of prior information (problem 48 in Lup93). In this example, the maximum likelihood approach might be thought to give us an unsatisfying answer, whereas the addition of "side" or prior information can improve the inference.

Imagine you arrive at a bus stop, and observe that the bus arrives t minutes later (it is assumed that you had no knowledge about the bus schedule). What is the mean time between two successive buses, τ, if the buses keep a regular schedule? It is easy to derive an intuitive answer. The wait time is distributed uniformly in the interval $0 \leq t \leq \tau$, and on average you would wait for $t = \tau/2$ minutes. Rearranging this gives $\tau = 2t$, which agrees with intuition.

What does the maximum likelihood approach give? The probability that you will wait t minutes (the likelihood of data) is given by the uniform distribution

$$p(t|\tau) = \begin{cases} 1/\tau & \text{if } 0 \leq t \leq \tau, \\ 0 & \text{otherwise.} \end{cases} \tag{5.1}$$

Because we only observe a single point, the data likelihood (eq. 4.1) is simply equal to this probability. The maximum likelihood, then, corresponds to the smallest possible τ such that $t \leq \tau$: this is satisfied by $\tau = t$ and not $\tau = 2t$ as we expected! Computing the expectation value or the median for τ does not help either because the resulting integrals diverge. These puzzling results are resolved by the use of appropriate prior information, as discussed in the next section. We shall see several other examples where the addition of extra information changes the results we would get from the maximum likelihood approach.

The Bayesian method is not, however, motivated by the differences in results between maximum likelihood and Bayesian techniques. These differences are often negligible, especially when the data sets are large. Rather, the Bayesian method is motivated by its ability to provide a full probabilistic framework for data analysis. One of the most important aspects of Bayesian analysis is the ability to straight-forwardly incorporate unknown or uninteresting model parameters—the so-called *nuisance parameters*—in data analysis. But let us start with basics and introduce the Bayesian framework step by step. After reviewing the basic analysis steps, we shall illustrate them with practical examples collected in §5.6–5.7.

5.1.1. The Essence of the Bayesian Idea

We already introduced Bayes' rule (eqs. 3.10 and 3.11) in §3.1.3. Bayes' rule is simply a mathematical identity following from a straightforward application of the rules of probability and thus is not controversial in and of itself. The frequentist vs. Bayesian controversy sets in when we apply Bayes' rule to the likelihood function $p(D|M)$ to obtain Bayes' theorem:

$$p(M|D) = \frac{p(D|M)\, p(M)}{p(D)},$$ (5.2)

where D stands for data, and M stands for model. Bayes' theorem quantifies the rule for "combining an initial belief with new data to arrive at an improved belief" and says that "improved belief" is proportional to the product of "initial belief" and the probability that "initial belief" generated the observed data.

To be more precise, let us explicitly acknowledge the presence of prior informa-tion I and the fact that models are typically described by parameters whose values we want to estimate from data:

$$p(M, \boldsymbol{\theta}|D, I) = \frac{p(D|M, \boldsymbol{\theta}, I)\, p(M, \boldsymbol{\theta}|I)}{p(D|I)}.$$ (5.3)

In general, as we saw in the likelihood function in §4.2.1, the model M includes k model parameters θ_p, $p = 1, \ldots, k$, abbreviated as vector $\boldsymbol{\theta}$ with components θ_p. Strictly speaking, the vector $\boldsymbol{\theta}$ should be labeled by M since different models may be described by different parameters. We will sometimes write just M or $\boldsymbol{\theta}$ (θ in one-dimensional cases), depending on what we are trying to emphasize.

The result $p(M, \boldsymbol{\theta}|D, I)$ is called the *posterior* pdf for model M *and* parameters $\boldsymbol{\theta}$, given data D and other prior information I. This term is a $(k + 1)$-dimensional pdf in the space spanned by k model parameters and the model index M. The term $p(D|M, \boldsymbol{\theta}, I)$ is, as before, the *likelihood* of data *given* some model M and given some fixed values of parameters $\boldsymbol{\theta}$ describing it, and all other prior information I. The term $p(M, \boldsymbol{\theta}|I)$ is the a priori joint probability for model M and its parameters $\boldsymbol{\theta}$ in the absence of any of the data used to compute likelihood, and is often simply called the *prior*. The term prior is understood logically and not temporally: despite its name, measurements D that enter into the calculation of data likelihood may be collected before information that is used to construct prior $p(M, \boldsymbol{\theta}|I)$. The prior can be expanded as

$$p(M, \boldsymbol{\theta}|I) = p(\boldsymbol{\theta}|M, I)\, p(M|I),$$ (5.4)

and in parameter estimation problems we need only specify $p(\boldsymbol{\theta}|M, I)$. In the context of the model selection, however, we need the full prior.

The term $p(D|I)$ is the *probability of data*, or the prior predictive probability for D. It provides proper normalization for the posterior pdf and usually it is not explicitly computed when estimating model parameters: rather, $p(M, \boldsymbol{\theta}|D, I)$ for a given M is simply renormalized so that its integral over all model parameters $\boldsymbol{\theta}$ is unity. The integral of the prior $p(\boldsymbol{\theta}|M, I)$ over all parameters should also be unity, but for the same reason, calculations of the posterior pdf are often done with an arbitrary normalization. An important exception is model selection discussed below, where the correct normalization of the product $p(D|M, \boldsymbol{\theta}, I)\,p(\boldsymbol{\theta}|M, I)$ is crucial.

5.1.2. Discussion

Why is this interpretation controversial? If we want to write "$p(M, \boldsymbol{\theta}|D, I)$," we must somehow acknowledge that it is not a probability in the same sense as $p(D|M, \boldsymbol{\theta}, I)$. The latter can be easily understood in terms of the long-term frequency of events, for example as the long-term probability of observing heads a certain number of times given a certain physical coin (hence the term "frequentist"). If we measure the mass of a planet, or an apple, or an elementary particle, this mass is not a random number; it is what it is and cannot have a distribution! To acknowledge this, we accept that when using the Bayesian formalism, $p(M, \boldsymbol{\theta}|D, I)$ corresponds to the state of our *knowledge* (i.e., belief) about a model and its parameters, given data D and prior information I. This change of interpretation of the symbols (note that the mathematical axioms of probability do not change under Bayesianism) introduces the notion of the posterior probability distribution for models and model parameters (as Laplace did for the mass of Saturn in one of the first applications of the Bayesian method[1]).

Let us return to our bus stop example and see how the prior helps. Eq. 5.1 corresponds to $p(D|M, \boldsymbol{\theta}, I)$ in eq. 5.2, with the vector of parameters $\boldsymbol{\theta}$ having only one component: τ. We take the prior $p(\tau|I)$ as proportional to τ^{-1} for reasons that will be explained in §5.2.1 (τ here is a scale parameter). The posterior probability density function for τ then becomes

$$p(\tau|t, I) = \begin{cases} t/\tau^2 & \text{if } \tau \geq t, \\ 0 & \text{otherwise.} \end{cases} \tag{5.5}$$

This posterior pdf is a result of multiplying the likelihood and prior, which are both proportional to τ^{-1}, and then normalizing the integral of $p(\tau|t, I)$ as

$$\int_t^\infty p(\tau|t, I)\,d\tau = \int_t^\infty \frac{C}{\tau^2} d\tau = 1, \tag{5.6}$$

which gives $C = t$ as the normalization constant. The divergent integral over τ encountered in likelihood analysis is mitigated by the extra τ^{-1} term from the prior. The median τ given by the posterior $p(\tau|t, I)$ is now equal to $2t$, in agreement with our expectations. An interesting side result is that the $p\%$ quantiles are

---

[1] Laplace said: "I find that it is a bet of 11,000 against one that the error of this result is not 1/100th of its value ..." (see p. 79 in [25]). Therefore, Laplace clearly interpreted measurements as giving a probability statement about the mass of Saturn, although there is only one Saturn and its true mass is what it is, and it is not a random variable according to classical statistics.

equal to $(1 - t/\tau_p)$; for example, the 95% confidence region for τ, known as the *credible region* in the Bayesian context, spans $1.03\,t < \tau < 40\,t$. If we waited for a bus for 1 minute, then, adopting the usual 95% confidence region, we cannot reject the possibility that τ is as large as 40 minutes. Equivalently, if we waited for a bus for 1 minute, we can paraphrase Laplace and say that "it is a bet of 20 against 1 that the bus will arrive in the interval between 0.03 minutes and 39 minutes from now."

We will now discuss the main ingredients of the Bayesian methodology in more detail, and then illustrate the main concepts using a few simple practical examples in §5.6–5.7.

5.1.3. Summary of the Bayesian Statistical Inference

To simplify the notation, we will shorten $M(\boldsymbol{\theta})$ to simply M whenever the absence of explicit dependence on $\boldsymbol{\theta}$ is not confusing. A completely Bayesian data analysis has the following conceptual steps:

1. The formulation of the data likelihood $p(D|M, I)$. Of course, if the adopted $p(D|M, I)$ is a poor description of this process, then the resulting posterior pdf will be inaccurate, too.
2. The choice of the prior $p(\boldsymbol{\theta}|M, I)$, which incorporates all other knowledge that might exist, but is *not* used when computing the likelihood (e.g., prior measurements of the same type, different measurements, or simply an uninformative prior, as discussed below).
3. Determination of the posterior pdf, $p(M|D, I)$, using Bayes' theorem. In practice, this step can be computationally intensive for complex multidimensional problems. Often $p(D|I)$ is not explicitly specified because $p(M|D, I)$ can be properly normalized by renormalizing the product $p(D|M, I)\, p(M|I)$.
4. The search for the best model parameters M, which maximize $p(M|D, I)$, yielding the *maximum a posteriori* (MAP) estimate. This *point estimate* is the natural analog to the maximum likelihood estimate (MLE) from classical statistics. Another natural Bayesian estimator is the *posterior mean*:

$$\bar{\theta} = \int \theta \, p(\theta|D) \, d\theta. \tag{5.7}$$

   In multidimensional cases, $p(\theta|D)$, where θ is one of many model parameters, is obtained from $p(M, \boldsymbol{\theta}|D, I)$ using *marginalization*, or integration of $p(M, \boldsymbol{\theta}|D, I)$ over all other model parameters and renormalization ($\int p(\theta|D) \, d\theta = 1$, and the model index is not explicitly acknowledged because it is implied by the context). Both the MAP and posterior mean are only convenient ways to summarize the information provided by the posterior pdf, and often do not capture its full information content.
5. Quantification of uncertainty in parameter estimates, via *credible regions* (the Bayesian counterpart to frequentist confidence regions). As in MLE, such an estimate can be obtained analytically by doing mathematical derivations specific to the chosen model. Also as in MLE, various numerical techniques can be used to simulate samples from the posterior. This can be viewed as an analogy to the frequentist bootstrap approach, which can simulate

draws of samples from the true underlying distribution of the data. In both cases, various descriptive statistics can then be computed on such samples to examine the uncertainties surrounding the data and estimators of model parameters based on that data.

6. *Hypothesis testing* as needed to make other conclusions about the model or parameter estimates. Unlike hypothesis tests in classical statistics, in Bayesian inference hypothesis tests incorporate the prior and thus may give different results.

The Bayesian approach can be thought of as formalizing the process of continually refining our state of knowledge about the world, beginning with no data (as encoded by the prior), then updating that by multiplying in the likelihood once the data D are observed to obtain the posterior. When more data are taken, then the posterior based on the first data set can be used as the prior for the second analysis. Indeed, the data sets can be fundamentally different: for example, when estimating cosmological parameters using observations of supernovas, the prior often comes from measurements of the cosmic microwave background, the distribution of large-scale structure, or both (e.g., [18]). This procedure is acceptable as long as the pdfs refer to the same quantity. For a pedagogical discussion of probability calculus in a Bayesian context, please see [14].

5.2. Bayesian Priors

How do we choose the prior[2] $p(\theta|I) \equiv p(\theta|M, I)$ in eq. 5.4? The prior incorporates all other knowledge that might exist, but is not used when computing the likelihood, $p(D|M, \theta, I)$. To reiterate, despite the name, the data may chronologically precede the information in the prior. The latter can include the knowledge extracted from prior measurements of the same type as the data at hand, or different measurements that constrain the same quantity whose posterior pdf we are trying to constrain with the new data. For example, we may know from older work that the mass of an elementary particle is m_A, with a Gaussian uncertainty parametrized by σ_A, and now we wish to utilize a new measuring apparatus or method. Hence, m_A and σ_A may represent a convenient summary of the posterior pdf from older work that is now used as a prior for the new measurements. Therefore, the terms prior and posterior do not have an absolute meaning. Such priors that incorporate information based on other measurements (or other sources of meaningful information) are called *informative priors*.

5.2.1. Priors Assigned by Formal Rules

When no other information, except for the data we are analyzing, is available, we can assign priors by formal rules. Sometimes these priors are called *uninformative priors* but this term is a misnomer because these priors can incorporate weak but objective information such as "the model parameter describing variance cannot be negative."

---

[2]While common in the physics literature, the adjective "Bayesian" in front of "prior" is rare in the statistics literature.

Note that even the most uninformative priors still affect the estimates, and the results are not generally equivalent to the frequentist or maximum likelihood estimates.

As an example, consider a *flat prior*,

$$p(\theta|I) \propto C, \tag{5.8}$$

where $C > 0$ is a constant. Since $\int p(\theta|I)\, d\theta = \infty$, this is not a pdf; this is an example of an *improper prior*. In general, improper priors are not a problem as long as the resulting posterior is a well-defined pdf (because the likelihood effectively controls the result of integration). Alternatively, we can adopt a lower and an upper limit on θ which will prevent the integral from diverging (e.g., it is reasonable to assume that the mass of a newly discovered elementary particle must be positive and smaller than the Earth's mass). Flat priors are sometimes considered ill defined because a flat prior on a parameter does not imply a flat prior on a transformed version of the parameter (e.g., if $p(\theta)$ is a flat prior, $\ln \theta$ does not have a flat prior).

Although uninformative priors do not contain specific information, they can be assigned according to several general principles. The main point here is that for the same prior information, these principles result in assignments of the same priors.

The oldest method is the *principle of indifference* which states that a set of basic, mutually exclusive possibilities need to be assigned equal probabilities (e.g., for a fair six-sided die, each of the outcomes has a prior probability of 1/6). The *principle of consistency*, based on transformation groups, demands that the prior for a location parameter should not change with translations of the coordinate system, and yields a flat prior. Similarly, the prior for a scale parameter should not depend on the choice of units. If the scale parameter is σ and we rescale our measurement units by a positive factor a, we get a constraint

$$p(\sigma|I)\, d\sigma = p(a\sigma|I)\, d(a\sigma). \tag{5.9}$$

The solution is $p(\sigma|I) \propto \sigma^{-1}$ (or a flat prior for $\ln \sigma$), called a scale-invariant prior.

When we have additional weak prior information about some parameter, such as a low-order statistic, we can use the *principle of maximum entropy* to construct priors consistent with that information.

5.2.2. The Principle of Maximum Entropy

Entropy measures the information content of a pdf. We shall use S as the symbol for entropy, although we have already used s for the sample standard deviation (eq. 3.32), because we never use both in the same context. Given a pdf defined by N discrete values p_i, with $\sum_{i=1}^{N} p_i = 1$, its entropy is defined as

$$S = -\sum_{i=1}^{N} p_i \ln(p_i) \tag{5.10}$$

(note that $\lim_{p \to 0} [p \ln p] = 0$). This particular functional form can be justified using arguments of logical consistency (see Siv06 for an illuminating introduction) and information theory (using the concept of minimum description length, see HTF09). It is also called Shannon's entropy because Shannon was the first one to derive it in the context of information in 1948. It resembles thermodynamic entropy: this observation is how it got its name (this similarity is not coincidental; see Jay03). The unit for entropy is the *nat* (from natural unit; when ln is replaced by the base 2 logarithm, then the unit is the more familiar *bit*; 1 nat = 1.44 bits).

Sivia (see Siv06) discusses the derivation of eq. 5.10 and its extension to the continuous case

$$S = -\int_{-\infty}^{\infty} p(x) \ln\left(\frac{p(x)}{m(x)}\right) dx, \tag{5.11}$$

where the "measure" $m(x)$ ensures that entropy is invariant under a change of variables.

The idea behind the principle of maximum entropy for assigning uninformative priors is that by maximizing the entropy over a suitable set of pdfs, we find the distribution that is least informative (given the constraints). The power of the principle comes from a straightforward ability to add additional information about the prior distribution, such as the mean value and variance. Computational details are well exposed in Siv06 and Greg05, and here we only review the main results.

Let us start with Sivia's example of a six-faced die, where we need to assign six prior probabilities. When no specific information is available, the principle of indifference states that each of the outcomes has a prior probability of 1/6. If additional information is available (with its source unspecified), such as the mean value of a large number of rolls, μ, (for a fair die the expected mean value is 3.5), then we need to adjust prior probabilities to be consistent with this information. Given the six probabilities p_i, the expected mean value is

$$\sum_{i=1}^{6} i\, p_i = \mu, \tag{5.12}$$

and of course

$$\sum_{i=1}^{6} p_i = 1. \tag{5.13}$$

We have two constraints for the six unknown values p_i. The problem of assigning individual p_i can be solved using the principle of maximum entropy and the method of Lagrangian multipliers. We need to maximize the following quantity with respect to six individual p_i:

$$Q = S + \lambda_0 \left(1 - \sum_{i=1}^{6} p_i\right) + \lambda_1 \left(\mu - \sum_{i=1}^{6} i p_i\right), \tag{5.14}$$

where the first term is entropy:

$$S = -\sum_{i=1}^{6} p_i \ln\left(\frac{p_i}{m_i}\right), \tag{5.15}$$

and the second and third term come from additional constraints (λ_0 and λ_1 are called Lagrangian multipliers). In the expression for entropy, m_i are the values that would be assigned to p_i in the case when no additional information is known (i.e., without constraint on the mean value; in this problem $m_i = 1/6$). By differentiating Q with respect to p_i, we get conditions

$$-\left[\ln\left(\frac{p_i}{m_i}\right) + 1\right] - \lambda_0 - i\lambda_1 = 0, \tag{5.16}$$

and solutions

$$p_i = m_i \exp(-1 - \lambda_0)\exp(i\lambda_1). \tag{5.17}$$

The two remaining unknown values of λ_0 and λ_1 can be determined numerically using constraints given by eqs. 5.12 and 5.13. Therefore, *although our knowledge about p_i is incomplete and based on only two constraints, we can assign all six p_i!* When the number of possible discrete events is infinite (as opposed to six here), the maximum entropy solution for assigning p_i is the Poisson distribution parametrized by the expectation value μ.

In the corresponding continuous case, the maximum entropy solution for the prior is

$$p(\theta|\mu) = \frac{1}{\mu}\exp\left(\frac{-\theta}{\mu}\right). \tag{5.18}$$

This result is based on the constraint that we only know the expectation value for θ ($\mu = \int \theta p(\theta)\,d\theta$), and assuming a flat distribution $m(\theta)$ (the prior for θ when the additional constraint given by μ is not imposed). Another useful result is that when only the mean and the variance are known in advance, with the distribution defined over the whole real line, the maximum entropy solution is a Gaussian distribution with those values of mean and variance.

A quantity closely related to entropy is the Kullback–Leibler (KL) divergence from $p(x)$ to $m(x)$,

$$\text{KL} = \sum_i p_i \ln\left(\frac{p_i}{m_i}\right), \tag{5.19}$$

and analogously for the continuous case (i.e., KL is equal to S from eq. 5.11 except for the minus sign).

Sometimes, the KL divergence is called the KL distance between two pdfs. However, the KL distance is not a true distance metric because its value is not the same when $p(x)$ and $m(x)$ are switched. In Bayesian statistics the KL divergence can be used to measure the information gain when moving from a prior distribution to a posterior distribution. In information theory, the KL divergence can be interpreted as the additional message-length per datum if the code that is optimal for $m(x)$ is used to transmit information about $p(x)$. The KL distance will be discussed in a later chapter (see §9.7.1).

5.2.3. Conjugate Priors

In special combinations of priors and likelihood functions, the posterior probability has the same functional form as the prior probability. These priors are called *conjugate priors* and represent a convenient way for generalizing computations.

When the likelihood function is a Gaussian, then the conjugate prior is also a Gaussian. If the prior is parametrized as $\mathcal{N}(\mu_p, \sigma_p)$, and the data can be summarized as $\mathcal{N}(\overline{x}, s)$ (see eqs. 3.31 and 3.32), then the posterior[3] is $\mathcal{N}(\mu^0, \sigma^0)$, with

$$\mu^0 = \frac{\mu_p/\sigma_p^2 + \overline{x}/s^2}{1/\sigma_p^2 + 1/s^2} \quad \text{and} \quad \sigma^0 = \left(1/\sigma_p^2 + 1/s^2\right)^{-1/2}. \tag{5.20}$$

If the data have a smaller scatter (s) than the width of the prior (σ_p), then the resulting posterior (i.e., μ^0) is closer to $\overline{x}$ than to μ_p. Since μ^0 is obviously *different* from $\overline{x}$, this Bayesian estimator is biased! On the other hand, if we choose a very informative prior with $\sigma_p \ll s$, then the data will have little impact on the resulting posterior and μ^0 will be much closer to μ_p than to $\overline{x}$.

In the discrete case, the most frequently encountered conjugate priors are the beta distribution for binomial likelihood, and the gamma distribution for Poissonian likelihood (refer to §3.3 for descriptions of these distributions). For a more detailed discussion, see Greg05. We limit discussion here to the first example.

The beta distribution (see §3.3.10) allows for more flexibility when additional information about discrete measurements, such as the results of prior measurements, is available. A flat prior corresponds to $\alpha = 1$ and $\beta = 1$. When the likelihood function is based on a binomial distribution described by parameters N and k (see §3.3.3), and the prior is the beta distribution, then the posterior is also a beta distribution. It can be shown that parameters describing the posterior are given by $\alpha^0 = \alpha_p + k$ and $\beta^0 = \beta_p + N - k$, where α_p and β_p describe the prior. Evidently, as both k and $N - k$ become much larger than α_p and β_p, the "memory" of the prior information is, by and large, gone. This behavior is analogous to the case with $s \ll \sigma_p$ for the Gaussian conjugate prior discussed above.

5.2.4. Empirical and Hierarchical Bayes Methods

Empirical Bayes refers to an approximation of the Bayesian inference procedure where the parameters of priors (or *hyperparameters*) are estimated from the data. It differs from the standard Bayesian approach, in which the parameters of priors are chosen before any data are observed. Rather than integrate out the hyperparameters as in the standard approach, they are set to their most likely values. Empirical Bayes is also sometimes known as *maximum marginal likelihood*; for more details, see [1].

The empirical Bayes method represents an approximation to a fully Bayesian treatment of a *hierarchical Bayes* model. In hierarchical, or multilevel, Bayesian analysis a prior distribution depends on unknown variables, the hyperparameters, that describe the group (population) level probabilistic model. Their priors, called

---

[3]The posterior pdf is by definition normalized to 1. However, the product of two Gaussian functions, before renormalization, has an extra multiplicative term compared to $\mathcal{N}(\mu^0, \sigma^0)$. Strictly speaking, the product of two Gaussian pdfs is not a Gaussian pdf because it is not properly normalized.

hyperpriors, resemble the priors in simple (single-level) Bayesian models. This approach is useful for quantifying uncertainty in various aspects of the pdf for the prior, and for using overall population properties when estimating parameters of a single population member (for a detailed discussion, see [9]). Hierarchical Bayes models are especially useful for complex data analysis problems because they can effectively handle multiple sources of uncertainty at all stages of the data anaysis. For a recent application of hierarchical Bayes modeling in astronomy, see [17].

5.3. Bayesian Parameter Uncertainty Quantification

The posterior pdf in the Bayesian framework is treated as any other probabilistic pdf. In practice, it is often summarized in terms of various point estimates (e.g., MAP) or in terms of parameter intervals defined by certain properties of cumulative probability. Although very useful in practice, these summaries rarely capture the full information content of a posterior pdf.

5.3.1. Posterior Intervals

To obtain a Bayesian *credible region* estimate, we find a and b such that $\int_{-\infty}^{a} f(\theta) \, d\theta = \int_{b}^{\infty} f(\theta) \, d\theta = \alpha/2$. Then the probability that the true value of parameter θ is in the interval (a, b) is equal to $1 - \alpha$, in analogy with the classical confidence intervals, and the interval (a, b) is called a $1 - \alpha$ *posterior interval*.

In practice, the posterior pdf, $p(\theta)$, is often not an analytic function (e.g., it can only be evaluated numerically). In such cases, we can compute statistics such as the posterior mean and the $1 - \alpha$ posterior interval, using *simulation* (sampling). If we know how to compute $p(\theta)$, then we can use the techniques for random number generation described in §3.7 to draw N values θ_j. Then we can approximate the posterior mean as the sample mean, and approximate the $1 - \alpha$ posterior interval by finding $\alpha/2$ and $(1-\alpha/2)$ sample quantiles. Several examples of this type of procedure in one and two dimensions are found in §5.6.

5.3.2. Marginalization of Parameters

Consider a problem where the model in the posterior pdf, $p(M, \boldsymbol{\theta}|D, I)$, is parametrized by a vector of k free parameters, $\boldsymbol{\theta}$. A subset of these parameters are of direct interest, while the remaining parameters are used to describe certain aspects of data collection that are not of primary interest. For example, we might be measuring the properties of a spectral line (position, width, strength) detected in the presence of an unknown and variable background. We need to account for the background when describing the statistical behavior of our data, but the main quantities of interest are the spectral line properties. In order to obtain the posterior pdf for each interesting parameter, we can integrate the multidimensional posterior pdf over all other parameters. Alternatively, if we want to understand covariances between interesting parameters, we can integrate the posterior pdf only over uninteresting parameters called *nuisance parameters*. This integration procedure is known as *marginalization* and the resulting pdf is called the *marginal posterior pdf*. An analog of marginalization of parameters does not exist in classical statistics.

We have already discussed multidimensional pdfs and marginalization in the context of conditional probability (in §3.1.3). An example of integrating a two-dimensional pdf to obtain one-dimensional marginal distributions is shown in figure 3.2. Let us assume that x in that figure corresponds to an interesting parameter, and y is a nuisance parameter. The right panels show the posterior pdfs for x if somehow we knew the value of the nuisance parameter, for three different values of the latter. When we do not know the value of the nuisance parameter, we integrate over all plausible values and obtain the marginalized posterior pdf for x, shown at the bottom of the left panel. Note that the marginalized pdf spans a wider range of x than the three pdfs in the right panel. This difference is a general result.

Several practical examples of Bayesian analysis discussed in §5.6 use and illustrate the concept of marginalization.

5.4. Bayesian Model Selection

Bayes' theorem as introduced by eq. 5.3 quantifies the posterior pdf of parameters describing a single model, with that model *assumed to be true*. In model selection and hypothesis testing, we formulate alternative scenarios and ask which ones are best supported by the available data. For example, we can ask whether a set of measurements $\{x_i\}$ is better described by a Gaussian or by a Cauchy distribution, or whether a set of points is better fit by a straight line or a parabola.

To find out which of two models, say M_1 and M_2, is better supported by data, we compare their posterior probabilities via the *odds ratio* in favor of model M_2 over model M_1 as

$$O_{21} \equiv \frac{p(M_2|D, I)}{p(M_1|D, I)}. \tag{5.21}$$

The posterior probability for model M (M_1 or M_2) given data D, $p(M|D, I)$ in this expression, can be obtained from the posterior pdf $p(M, \boldsymbol{\theta}|D, I)$ in eq. 5.3 using marginalization (integration) over the model parameter space spanned by $\boldsymbol{\theta}$. The posterior probability that the model M is correct given data D (a number between 0 and 1) can be derived using eqs. 5.3 and 5.4 as

$$p(M|D, I) = \frac{p(D|M, I)\, p(M|I)}{p(D|I)}, \tag{5.22}$$

where

$$E(M) \equiv p(D|M, I) = \int p(D|M, \boldsymbol{\theta}, I)\, p(\boldsymbol{\theta}|M, I)\, d\boldsymbol{\theta} \tag{5.23}$$

is called the *marginal likelihood* for model M and it quantifies the probability that the data D would be observed *if* the model M were the correct model. In the physics literature, the marginal likelihood is often called *evidence* (despite the fact that to scientists, evidence and data mean essentially the same thing) and we adopt this term hereafter. Since the evidence $E(M)$ involves integration of the data likelihood $p(D|M, \boldsymbol{\theta}, I)$, it is also called the *global likelihood* for model M. The global likelihood, or evidence, is a *weighted average* of the likelihood function, with the prior for model parameters acting as the weighting function.

The hardest term to compute is $p(D|I)$, but it cancels out when the odds ratio is considered:

$$O_{21} = \frac{E(M_2)\, p(M_2|I)}{E(M_1)\, p(M_1|I)} = B_{21} \frac{p(M_2|I)}{p(M_1|I)}. \tag{5.24}$$

The ratio of global likelihoods, $B_{21} \equiv E(M_2)/E(M_1)$, is called the *Bayes factor*, and is equal to

$$B_{21} = \frac{\int p(D|M_2, \boldsymbol{\theta}_2, I)\, p(\boldsymbol{\theta}_2|M_2, I)\, d\boldsymbol{\theta}_2}{\int p(D|M_1, \boldsymbol{\theta}_1, I)\, p(\boldsymbol{\theta}_1|M_1, I)\, d\boldsymbol{\theta}_1}. \tag{5.25}$$

The vectors of parameters, $\boldsymbol{\theta}_1$ and $\boldsymbol{\theta}_2$, are explicitly indexed to emphasize that the two models may span vastly different parameter spaces (including the number of parameters per model).

How do we interpret the values of the odds ratio in practice? Jeffreys proposed a five-step scale for interpreting the odds ratio, where $O_{21} > 10$ represents "strong" evidence in favor of M_2 (M_2 is ten times more probable than M_1), and $O_{21} > 100$ is "decisive" evidence (M_2 is one hundred times more probable than M_1). When $O_{21} < 3$, the evidence is "not worth more than a bare mention."

As a practical example, let us consider coin flipping (this problem is revisited in detail in §5.6.2). We will compare two hypotheses; M_1: the coin has a known heads probability b_*, and M_2: the heads probability b is unknown, with a uniform prior in the range 0–1. Note that the prior for model M_1 is a delta function, $\delta(b - b_*)$. Let us assume that we flipped the coin N times, and obtained k heads. Using eq. 3.50 for the data likelihood, and assuming equal prior probabilities for the two models, it is easy to show that the odds ratio is

$$O_{21} = \int_0^1 \left(\frac{b}{b_*}\right)^k \left(\frac{1-b}{1-b_*}\right)^{N-k} db. \tag{5.26}$$

Figure 5.1 illustrates the behavior of O_{21} as a function of k for two different values of N and for two different values of b_*: $b_* = 0.5$ (M_1: the coin is fair) and $b_* = 0.1$. As this example shows, the ability to distinguish the two hypothesis improves with the sample size. For example, when $b_* = 0.5$ and $k/N = 0.1$, the odds ratio in favor of M_2 increases from ~ 9 for $N = 10$ to ~ 263 for $N = 20$. When $k = b_* N$, the odds ratio is 0.37 for $N = 10$ and 0.27 for $N = 20$. In other words, the simpler model is favored by the data, and the support strengthens with the sample size. It is easy to show by integrating eq. 5.26 that $O_{21} = \sqrt{\pi/(2N)}$ when $k = b_* N$ and $b_* = 0.5$. For example, to build strong evidence that a coin is fair, $O_{21} < 0.1$, it takes as many as $N > 157$ tosses. With $N = 10,000$, the heads probability of a fair coin is measured with a precision of 1% (see the discussion after eq. 3.51); the corresponding odds ratio is $O_{21} \approx 1/80$, approaching Jeffreys' decisive evidence level. Three more examples of Bayesian model comparison are discussed in §5.7.1–5.7.3.

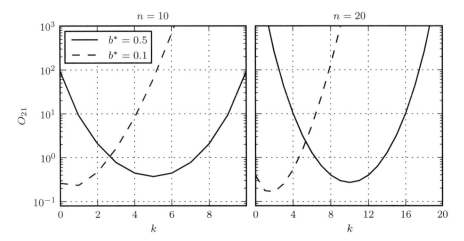

Figure 5.1. Odds ratio for two models, O_{21}, describing coin tosses (eq. 5.26). Out of N tosses (left: $N = 10$; right: $N = 20$), k tosses are heads. Model 2 is a one-parameter model with the heads probability determined from data ($b^0 = k/N$), and model 1 claims an a priori known heads probability equal to b_*. The results are shown for two values of b_*, as indicated in the legend). Note that the odds ratio is minimized and below 1 (model 1 wins) when $k = b_* N$.

5.4.1. Bayesian Hypothesis Testing

A special case of model comparison is Bayesian hypothesis testing. In this case, $M_2 = \overline{M_1}$ is a complementary hypothesis to M_1 (i.e., $p(M_1) + p(M_2) = 1$). Taking M_1 to be the "null" hypothesis, we can ask whether the data supports the alternative hypothesis M_2, i.e., whether we can reject the null hypothesis. Taking equal priors $p(M_1|I) = p(M_2|I)$, the odds ratio is

$$O_{21} = B_{21} = \frac{p(D|M_1)}{p(D|M_2)}. \tag{5.27}$$

Given that M_2 is simply a complementary hypothesis to M_1, it is not possible to compute $p(D|M_2)$ (recall that we had a well-defined alternative to M_1 in our coin example above). This inability to reject M_1 in the absence of an alternative hypothesis is very different from the hypothesis testing procedure in classical statistics (see §4.6). The latter procedure rejects the null hypothesis if it does not provide a good description of the data, that is, when it is very unlikely that the given data could have been generated as prescribed by the null hypothesis. In contrast, the Bayesian approach is based on the posterior rather than on the data likelihood, and cannot reject a hypothesis if there are no alternative explanations for observed data.

Going back to our coin example, assume we flipped the coin $N = 20$ times and obtained $k = 16$ heads. In the classical formulation, we would ask whether we can reject the null hypothesis that our coin is fair. In other words, we would ask whether $k = 16$ is a very unusual outcome (at some significance level α, say 0.05; recall §4.6) for a fair coin with $b_* = 0.5$ when $N = 20$. Using the results from §3.3.3, we find that the scatter around the expected value $k^0 = b_* N = 10$ is $\sigma_k = 2.24$. Therefore, $k = 16$ is about $2.7\sigma_k$ away from k^0, and at the adopted significance level $\alpha = 0.05$

we reject the null hypothesis (i.e., it is unlikely that $k = 16$ would have arisen by chance). Of course, $k = 16$ does not imply that it is *impossible* that the coin is fair (infrequent events happen, too!).

In the Bayesian approach, we offer an alternative hypothesis that the coin has an unknown heads probability. While this probability can be estimated from provided data (b^0), we consider *all the possible values* of b^0 when comparing the two proposed hypotheses. As shown in figure 5.1, the chosen parameters ($N = 20$, $k = 16$) correspond to the Bayesian odds ratio of ~ 10 in favor of the unfair coin hypothesis.

5.4.2. Occam's Razor

The principle of selecting the simplest model that is in fair agreement with the data is known as Occam's razor. This principle was already known to Ptolemy who said, "We consider it a good principle to explain the phenomena by the simplest hypothesis possible"; see [8]. Hidden in the above expression for the odds ratio is its ability to penalize complex models with many free parameters; that is, Occam's razor is naturally included into the Bayesian model comparison.

To reveal this fact explicitly, let us consider a model $M(\boldsymbol{\theta})$, and examine just one of the model parameters, say $\mu = \theta_1$. For simplicity, let us assume that its prior pdf, $p(\mu|I)$, is flat in the range $-\Delta_\mu/2 < \mu < \Delta_\mu/2$, and thus $p(\mu|I) = 1/\Delta_\mu$. In addition, let us assume that the data likelihood can be well described by a Gaussian centered on the value of μ that maximizes the likelihood, μ^0 (see eq. 4.2), and with the width σ_μ (see eq. 4.7). When the data are much more informative than the prior, $\sigma_\mu \ll \Delta$. The integral of this approximate data likelihood is proportional to the product of σ_μ and the maximum value of the data likelihood, say $L^0(M) \equiv \max[p(D|M)]$. The global likelihood for the model M is thus approximately

$$E(M) \approx \sqrt{2\pi}\, L^0(M) \frac{\sigma_\mu}{\Delta_\mu}. \tag{5.28}$$

Therefore, $E(M) \ll L^0(M)$ when $\sigma_\mu \ll \Delta_\mu$. Each model parameter constrained by the model carries a similar multiplicative penalty, $\propto \sigma/\Delta$, when computing the Bayes factor. If a parameter, or a degenerate parameter combination, is unconstrained by the data (i.e., $\sigma_\mu \approx \Delta_\mu$), there is no penalty. The odds ratio can justify an additional model parameter only if this penalty is offset by either an increase of the maximum value of the data likelihood, $L^0(M)$, or by the ratio of prior model probabilities, $p(M_2|I)/p(M_1|I)$. If both of these quantities are similar for the two models, the one with fewer parameters typically wins.

Going back to our practical example based on coin flipping, we can illustrate how model 2 gets penalized for its free parameter. The data likelihood for model M_2 is (details are discussed in §5.6.2)

$$L(b|M_2) = C_{Nk}\, b^k\, (1 - b)^{N-k}, \tag{5.29}$$

where $C_{Nk} = N!/[k!(N - k)!]$ is the binomial coefficient. The likelihood can be approximated as

$$L(b|M_2) \approx C_{Nk}\, \sqrt{2\pi}\, \sigma_b\, (b^0)^k\, (1 - b^0)^{N-k}\, \mathcal{N}(b^0, \sigma_b) \tag{5.30}$$

with $b^0 = k/N$ and $\sigma_b = \sqrt{b^0(1 - b^0)/N}$ (see §3.3.3). Its maximum is at $b = b^0$ and has the value

$$L^0(M_2) = C_{Nk}(b^0)^k(1 - b^0)^{N-k}. \tag{5.31}$$

Assuming a flat prior in the range $0 \leq b \leq 1$, it follows from eq. 5.28 that the evidence for model M_2 is

$$E(M_2) \approx \sqrt{2\pi}\, L^0(M_2)\, \sigma_b. \tag{5.32}$$

Of course, we would get the same result by directly integrating $L(b|M_2)$ from eq. 5.29.

For model M_1, the approximation given by eq. 5.28 cannot be used because the prior is not flat but rather $p(b|M_1) = \delta(b - b_*)$ (the data likelihood is analogous to eq. 5.29). Instead, we can use the exact result

$$E(M_1) = C_{Nk}(b_*)^k(1 - b_*)^{N-k}. \tag{5.33}$$

Hence,

$$O_{21} = \frac{E(M_2)}{E(M_1)} \approx \sqrt{2\pi}\, \sigma_b \left(\frac{b^0}{b_*}\right)^k \left(\frac{1 - b^0}{1 - b_*}\right)^{N-k}, \tag{5.34}$$

which is an approximation to eq. 5.26. Now we can explicitly see that the evidence in favor of model M_2 decreases (the model is "penalized") proportionally to the posterior pdf width of its free parameter. If indeed $b^0 \approx b_*$, model M_1 wins because it explained the data without any free parameter. On the other hand, the evidence in favor of M_2 increases as the data-based value b^0 becomes very different from the prior claim b_* by model M_1 (as illustrated in figure 5.1). Model M_1 becomes disfavored because it is unable to explain the observed data.

5.4.3. Information Criteria

The Bayesian information criterion (BIC, also known as the Schwarz criterion) is a concept closely related to the odds ratio, and to the Aikake information criterion (AIC; see §4.3.2 and eq. 4.17). The BIC attempts to simplify the computation of the odds ratio by making certain assumptions about the likelihood, such as Gaussianity of the posterior pdf; for details and references, see [21]. The BIC is easier to compute and, similarly to the AIC, it is based on the maximum value of the data likelihood, $L^0(M)$, rather than on its integration over the full parameter space (evidence $E(M)$ in eq. 5.23). The BIC for a given model M is computed as

$$\mathrm{BIC} \equiv -2\ln\left[L^0(M)\right] + k\ln N, \tag{5.35}$$

where k is the number of model parameters and N is the number of data points. The BIC corresponds to $-2\ln[E(M)]$ (to make it consistent with the AIC), and can be derived using the approximation for $E(M)$ given by eq. 5.28 and assuming $\sigma_\mu \propto 1/\sqrt{N}$.

When two models are compared, their BIC (or AIC) are compared analogously to the odds ratio, that is, the model with the smaller value wins (sometimes BIC and AIC are defined with an opposite sign, in which case the model with the largest value wins). If they are equally successful in describing the data (the first term above), then the model with fewer free parameters wins. Note that the BIC penalty for additional model parameters is stronger than for the AIC when N is sufficiently large (10–20, depending on k); very complex models are penalized more severely by the BIC than by the AIC.

Both the BIC and AIC are approximations and might not be valid if the underlying assumptions are not met (e.g., for model M_1 in our coin example above). Furthermore, unlike the odds ratio, both of them penalize unconstrained parameters. In general, it is better to compute the odds ratio when computationally feasible. We will see an example of this below in §5.8.4.

5.5. Nonuniform Priors: Eddington, Malmquist, and Lutz–Kelker Biases

In many cases, Bayesian analysis is equivalent to maximum likelihood analysis even when priors are taken into account (as we will see shortly using concrete examples discussed in §5.6). In this section, we address several important cases from astronomy where prior information *greatly affects* data interpretation. However, the same principles apply to any data set where measurement errors are not negligible and the population distribution of the measured quantity is strongly skewed, and can be easily generalized to nonastronomical contexts.

The effects of the selection function on the resulting sample are discussed in §4.9. The difference between the true distribution function $h(x)$ and its data-based estimate $f(x)$ (see eqs. 4.83 and 4.84), when caused by sample truncation, is known as the selection bias or Malmquist bias. Unfortunately, a fundamentally different effect is also often called Malmquist bias, as well as Eddington–Malmquist bias. The latter effect, addressed here, is usually encountered in brightness (magnitude) measurements and is due to the combined effects of measurement errors and nonuniform true distribution $h(x)$. Furthermore, a bias with an identical mathematical description in the context of trigonometric parallax measurements is known as Lutz–Kelker bias;[4] see [24]. The main distinguishing characteristic between the truncation (selection) bias discussed in §4.9 and the Eddington–Malmquist and Lutz–Kelker biases discussed here is that the latter two biases disappear when the measurement error for x vanishes, but the former does not.

Consider the situation illustrated in figure 5.2: an observable quantity with true values $x_{\rm true}$ is measured for a sample with true distribution $h(x_{\rm true})$. The measurements are affected by a known error distribution $e(x_{\rm obs}|x_{\rm true})$, where $x_{\rm obs}$ are the actual measured (observed) values. When $h(x_{\rm true})$ is a nonuniform distribution and measurement errors are not negligible, the distribution of $x_{\rm obs}$ for a subsample

---

[4]This taxonomic confusion in the literature (for an excellent summary, see [34]) apparently stems from the fact that Malmquist published the two relevant papers only two years apart (1920 and 1922) in journals that are not readily accessible today, and that those papers partially overlap with other works such as Eddington's. It may also stem from the early reluctance by astronomers to embrace Bayesian statistics (for an illuminating discussion, see [33]).

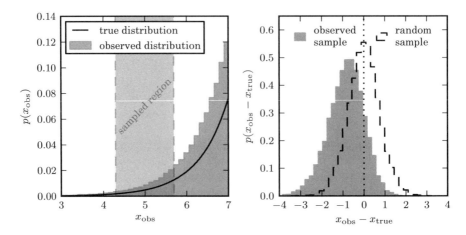

Figure 5.2. An illustration of the bias in a subsample selected using measurements with finite errors, when the population distribution is a steep function. The sample is drawn from the distribution $p(x) \propto 10^{0.6x}$, shown by the solid line in the left panel, and convolved with heteroscedastic errors with widths in the range $0.5 < \sigma < 1.5$. When a subsample is selected using "measured" values, as illustrated in the left panel, the distribution of differences between the "observed" and true values is biased, as shown by the histogram in the right panel. The distribution is biased because more objects with larger true x are scattered into the subsample from the right side, than from the left side where the true x are smaller.

selected using x_{obs} can be substantially different from the distribution of x_{true}. For example, all localized features from $h(x)$, such as peaks, are "blurred" in $f(x)$ due to errors. More formally, the true and observed distributions are related via convolution (see eq. 3.44),

$$f(x) = h(x) \star e(x) = \int_{-\infty}^{\infty} h(x')\, e(x - x')\, dx'. \tag{5.36}$$

Obviously, if $e(x) = \delta(x)$ then $f(x) = h(x)$, and if $h(x) = $ constant then $f(x) = $ constant, too. This expression can be easily derived using Bayesian priors, as follows.

The prior pdf, the probability of x in the absence of measurements, is $h(x)$, the population distribution of the measured quantity. The posterior pdf, the probability of measuring value x_{obs}, or the distribution of measured values, is

$$f(x_{\mathrm{obs}}|x_{\mathrm{true}}) \propto h(x_{\mathrm{true}})\, e(x_{\mathrm{obs}}|x_{\mathrm{true}}). \tag{5.37}$$

A given value of x_{obs} samples the range of x_{true} "allowed" by $e(x_{\mathrm{obs}}|x_{\mathrm{true}})$. When $h(x)$ is a steep rising function of x, it is more probable that x_{obs} corresponds to $x_{\mathrm{true}} > x_{\mathrm{obs}}$ than to $x_{\mathrm{true}} < x_{\mathrm{obs}}$, even when $e(x_{\mathrm{obs}}|x_{\mathrm{true}})$ is symmetric. Given an observed value of x_{obs}, we do not know its corresponding x_{true}. Nevertheless, we can still estimate $f(x_{\mathrm{obs}})$ by marginalizing over an unknown x_{true}; that is, we integrate eq. 5.37 over x_{true}, which yields eq. 5.36.

Figure 5.2 illustrates the bias (a systematic difference between the true and observed values) in the distribution of x_{obs} in samples selected by x_{obs} (or the full sample, if finite). It is the act of *selecting a subsample using measured values that produces the bias*; if instead we selected a finite subsample by other means (e.g., we ask what are optical magnitudes for a subsample of radio sources detected in our optical image), the difference between measured and true magnitudes would simply follow the corresponding error distribution and not be biased, as shown in the right panel for a random subsample; for a more detailed discussion, see [32].

The example shown in figure 5.2 assumes heteroscedastic Gaussian errors. When errors are homoscedastic and Gaussian, the bias in x_{obs} can be computed analytically for a given $h(x)$. In this case the relationship between $f(x)$ and $h(x)$ is

$$f(x) = \frac{1}{\sigma\sqrt{2\pi}} \int_{-\infty}^{\infty} h(x') \exp\left(\frac{-(x-x')^2}{2\sigma^2}\right) dx'. \tag{5.38}$$

Eddington was the first to show that to first order, the net effect is simply an offset between x_{obs} and x_{true}, $\Delta x = x_{obs} - x_{true}$, and $f(x)$ is obtained by "sliding" $h(x)$ by Δx (toward smaller values of x if $h(x)$ increases with x). The offset Δx can be expressed in terms of *measured* $f(x)$ as

$$\Delta x = -\sigma^2 \frac{1}{f(x)} \frac{df(x)}{dx}, \tag{5.39}$$

evaluated at $x = x_{obs}$. Note that Δx vanishes for $\sigma = 0$: Eddington–Malmquist and Lutz–Kelker biases become negligible for vanishing measurement errors (e.g., for a photometric error of $\sigma = 0.01$ mag, the bias becomes less than 0.0001 mag). A flat $h(x)$ (i.e., $dh(x)/dx = 0$) yields a flat $f(x)$ and thus there is no bias in this case ($\Delta x = 0$).

Special cases of much interest in astronomy correspond to very steep $h(x)$ where Δx might be nonnegligible even if σ appears small. There are two contexts, magnitude (flux) measurements and trigonometric parallax measurements, that have been frequently addressed in astronomical literature; see [33, 34]. For a spatially uniform distribution of sources, the magnitude distribution follows[5]

$$h(x) = h_o \, 10^{kx}, \tag{5.40}$$

with $k = 0.6$ (the so-called Euclidean counts), and the trigonometric parallax distribution follows[6]

$$h(x) = h_o \, x^{-p}, \tag{5.41}$$

with $p = 4$. Both of these distributions are very steep and may result in significant biases.

---

[5]This expression can be derived in a straightforward way using the definition of magnitude in terms of flux, and the dependence of flux on the inverse squared distance.

[6]This expression can be derived in a straightforward way using the fact that the trigonometric parallax is proportional to inverse distance.

For the case of the magnitude distribution given by eq. 5.40, the magnitude offset is

$$\Delta x = -\sigma^2 k \ln 10, \tag{5.42}$$

with the classical result for the Malmquist bias correction, $\Delta x = -1.38\sigma^2$, valid when $k = 0.6$ (measured magnitudes are brighter than true magnitudes, and inferred distances, if luminosity is known, are underestimated). As long as $\sigma \ll 1$, the offset Δx is not large compared to σ. Notably, $f(x)$ retains the same shape as $h(x)$, which justifies the use of eq. 5.40 when determining Δx using eq. 5.39. Given eq. 5.40 and the offset between $x_{\rm obs}$ and $x_{\rm true}$ given by eq. 5.42, $f(x_{\rm obs})$ is *always larger* than the corresponding $h(x_{\rm true})$: more sources with $x_{\rm obs}$ were scattered from the $x_{\rm true} > x_{\rm obs}$ range than from $x_{\rm true} < x_{\rm obs}$ (for $k > 0$).

Similarly, in the case of parallax measurements, the parallax offset is

$$\Delta x = \sigma^2 \frac{p}{x}. \tag{5.43}$$

The classical result for the Lutz–Kelker bias is expressed as $(x_{\rm obs}/x_{\rm true}) = 1 + p(\sigma/x)^2$; see [24]. For a constant *fractional* error (σ/x), the ratio $(x_{\rm obs}/x_{\rm true})$ is constant and thus $h(x)$ and $f(x)$ must have the same shape. For this reason, the use of $h(x)$ instead of $f(x)$ in eq. 5.39 is justified a posteriori.

If a sample is selected using parallax measurements, measured parallaxes are biased high, and implied distances are underestimated. If used for luminosity calibration, the resulting luminosity scale will be biased low. The resulting bias is as large as $\sim$26% even when parallax measurements have relative error as small as 20%. The fractional random luminosity error for a single measurement is twice as large as the fractional parallax error; 40% in this example. When the number of calibration sources is as large as, say, 10^4, the sample calibration error will not be 0.4% as naively expected from $\sqrt{N}$, but rather 26%! An interesting side result in the context of parallax measurements is that the parallax distribution given by eq. 5.41 is sufficiently steep that parallax measurements with relative errors exceeding about 15% $(\sigma/x > 0.15)$ are practically useless in the case of individual stars: in this case the posterior pdf has a peak at 0 without another local maximum.

These classical results for bias corrections rest on assumptions that are too simplistic to be used with modern astronomical survey data sets (e.g., homoscedastic measurement errors and uniform source distribution in Euclidean geometry). For example, the magnitude measurement error is rarely constant in practice and thus the usefulness of eq. 5.42 is limited. To realistically estimate these biases, the population distribution of the measured quantity, and the dependence of measurement error on it, need to be modeled. Two practical examples illustrated in figure 5.3 simulate upcoming measurements from Gaia and LSST.

To illustrate a bias in photometric calibration, we assume that we have pairs of measurements of the same stars (e.g., from two different nights) and we compare their magnitudes. The true magnitude distribution, $h(x)$, is generated using eq. 5.40 in the range $20 < m < 25$. Gaussian heteroscedastic photometric errors are simulated using a relation expected for LSST (see [15]),

$$\sigma^2 = (0.04 - \gamma) x + \gamma x^2 \; (\text{mag}^2), \tag{5.44}$$

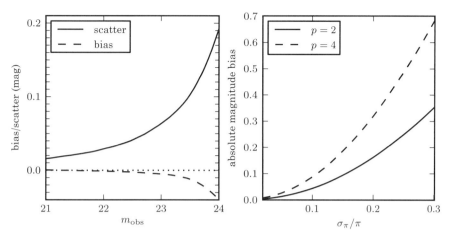

Figure 5.3. An illustration of the Eddington–Malmquist (left) and Lutz–Kelker (right) biases for mock data sets that simulate upcoming LSST and Gaia surveys (see text). The left panel shows a bias in photometric calibration when using pairs of measurements of the same stars with realistic photometric error distributions. Depending on the adopted faint limit (x-axis), the median difference between two measurements (dashed line) is biased when this limit is too close to the 5σ data limit (corresponding to errors of 0.2 mag); in this example the 5σ magnitude limit is set to 24. The solid line shows the assumed random measurement errors—if the number of stars in the sample is large, the random error for magnitude difference may become much smaller than the bias. The right panel shows the bias in absolute magnitude for samples calibrated using trigonometric parallax measurements with relative errors σ_π/π, and two hypothetical parallax distributions given by eq. 5.41 and $p = 2, 4$.

where $x = 10^{0.4\,(m-m_5)}$ and $\gamma = 0.039$ for optical sky noise-limited photometry. Sources are typically extracted from astronomical images down to the "5σ" limiting magnitude, m_5, corresponding to $\sigma = 0.2$ mag, with σ decreasing fast toward brighter magnitudes, $m < m_5$. Since σ varies with magnitude, $f(x)$ does not retain the same shape as $h(x)$. We generate two hypothetical observations, m_1 and m_2, and measure the median and the scatter for magnitude difference $\Delta m = m_1 - m_2$ as a function of magnitude limits enforced using m_1 (see the left panel in figure 5.3). Depending on the adopted faint limit, the median magnitude difference is biased when m_1 is too close to m_5.

In order to minimize the impact of the Eddington–Malmquist bias, the faint limit of the sample should be at least a magnitude brighter than m_5. Alternatively, if one of the two sets of measurements has a much fainter m_5 than the other one, then the sample should be selected using that set.

To illustrate a bias in calibration of absolute magnitudes using parallax measurements, we simulate an LSST–Gaia sample. The magnitude distribution is generated using eq. 5.40 in the r magnitude range $17 < r < 20$ that roughly corresponds to the brightness overlap between Gaia (faint limit at $r \sim 20$) and LSST (saturates at $r \sim 17$). Trigonometric parallax errors are simulated using a relation similar to that expected for Gaia,

$$\sigma_\pi = 0.3 \times 10^{0.2\,(r-20)} \quad \text{milliarcsec}, \tag{5.45}$$

and we consider a hypothetical sample of stars whose absolute magnitudes are about 10. We compute a bias in absolute magnitude measurement as

$$\Delta M = 5 \log_{10} \left[1 + 4 \left(\frac{\sigma_\pi}{\pi} \right)^2 \right], \tag{5.46}$$

where π is the parallax corresponding to a star with given r and absolute magnitude $M_r = 10$, and we analyze two different parallax distributions described by $p = 2$ and $p = 4$ (see eq. 5.41). As illustrated in the right panel in figure 5.3, in order to minimize this bias below, say, 0.05 mag, the sample should be selected by $\sigma_\pi/\pi < 0.1$.

5.6. Simple Examples of Bayesian Analysis: Parameter Estimation

In this section we illustrate the important aspects of Bayesian parameter estimation using specific practical examples. In the following section (§5.7), we will discuss several examples of Bayesian model selection. The main steps for Bayesian parameter estimation and model selection are summarized in §5.1.3.

5.6.1. Parameter Estimation for a Gaussian Distribution

First, we will solve a simple problem where we have a set of N measurements, $\{x_i\}$, of, say, the length of a rod. The measurement errors are Gaussian, and the measurement error for each measurement is known and given as σ_i (heteroscedastic errors). We seek the posterior pdf for the length of the rod, μ: $p(\mu|\{x_i\}, \{\sigma_i\})$.

Given that the likelihood function for a single measurement, x_i, is *assumed* to follow a Gaussian distribution (see below for a generalization), the likelihood for obtaining data $D = \{x_i\}$ given μ (and $\{\sigma_i\}$) is simply the product of likelihoods for individual data points,

$$p(\{x_i\}|\mu, I) = \prod_{i=1}^{N} \frac{1}{\sqrt{2\pi}\sigma_i} \exp \left(\frac{-(x_i - \mu)^2}{2\sigma_i^2} \right). \tag{5.47}$$

For the prior for μ, we shall adopt the least informative choice: a uniform distribution over some very wide interval ranging from $\mu_{\min}$ to $\mu_{\max}$:

$$p(\mu|I) = C, \quad \text{for } \mu_{\min} < \mu < \mu_{\max}, \tag{5.48}$$

where $C = (\mu_{\max} - \mu_{\min})^{-1}$, and 0 otherwise (the choice of $\mu_{\min}$ and $\mu_{\max}$ will not have a major impact on the solution: for example, in this case we could assume $\mu_{\min} = 0$ and $\mu_{\max}$ is the radius of Earth). The logarithm of the posterior pdf for μ is then

$$L_p = \ln \left[p(\mu|\{x_i\}, \{\sigma_i\}, I) \right] = \text{constant} - \sum_{i=1}^{N} \frac{(x_i - \mu)^2}{2\sigma_i^2} \tag{5.49}$$

(remember that we use L and $\ln L$ as notation for the data likelihood and its logarithm; L_p is reserved for the logarithm of the posterior pdf). This result is analogous to eq. 4.8 and the only difference is in the value of the constant term (due to the prior for μ and ignoring the $p(D|I)$ term), which is unimportant in this analysis.

Again, as a result of the Gaussian error distribution we can derive an analytic solution for the maximum likelihood estimator of μ by setting $\left(dL_p/d\mu \right) |_{(\mu=\mu_0)} = 0$,

$$\mu_0 = \frac{\sum_{i=1}^{N} w_i x_i}{\sum_{i=1}^{N} w_i}, \tag{5.50}$$

with weights $w_i = \sigma_i^{-2}$. That is, μ_0 is simply a weighted arithmetic mean of all measurements. When all σ_i are equal, we obtain the standard result given by eq. 3.31 and eq. 4.5. The posterior pdf for μ is a Gaussian centered on μ_0, and with the width given by (in analogy with eq. 4.6)

$$\sigma_\mu = \left(-\frac{d^2 L_p}{d\mu^2} \Big|_{\mu=\mu_0} \right)^{-1/2} = \left(\sum_{i=1}^{N} \frac{1}{\sigma_i^2} \right)^{-1/2} = \left(\sum_i w_i \right)^{-1/2}. \tag{5.51}$$

Because we used a flat prior for $p(M|I)$, these results are identical to those that follow from the maximum likelihood method. Note that although eq. 5.51 is only an approximation based on quadratic Taylor expansion of the logarithm of the posterior pdf, it is exact in this case because there are no terms higher than μ^2 in eq. 5.49. Again, when all σ_i are equal to σ, we obtain the standard result given by eqs. 3.34 and 4.7. The key conclusion is that the posterior pdf for μ is *Gaussian in cases when σ_i are known*, regardless of the data set size N. This is not true when σ is unknown and also determined from data, as follows.

Let us now solve a similar but more complicated problem when a set of N values (measurements), $\{x_i\}$, is drawn from an unspecified Gaussian distribution, $\mathcal{N}(\mu, \sigma)$. That is, here σ *also needs to be determined from data;* for example, it could be that individual measurement errors are always negligible compared to the intrinsic spread σ of the measured quantity (e.g., when measuring the weight of a sample of students with a microgram precision), or that all measurements of a rod do have the same unknown precision.

We seek the two-dimensional posterior pdf $p(\mu, \sigma | \{x_i\})$. This problem is frequently encountered in practice and the most common solution for estimators of μ and σ is given by eqs. 3.31 and 3.32. Much less common is the realization that the assumption of the Gaussian uncertainty of μ, with its width given by eq. 3.34, is valid *only in the large N limit* when σ is not known a priori. When N is not large, the posterior pdf for μ follows Student's t distribution. Here is a how to derive this result using the Bayesian framework.

Given that the likelihood function for a single measurement, x_i, is assumed to follow a Gaussian distribution $\mathcal{N}(\mu, \sigma)$, the likelihood for all measurements is given by

$$p(\{x_i\}|\mu, \sigma, I) = \prod_{i=1}^{N} \frac{1}{\sqrt{2\pi}\sigma} \exp\left(\frac{-(x_i - \mu)^2}{2\sigma^2} \right). \tag{5.52}$$

This equation is identical to eq. 4.2. However, the main and fundamental difference is that σ in eq. 4.2 was assumed to be known, while σ in eq. 5.52 is to be estimated, thus making the posterior pdf a function of two parameters (similarly, σ_i in eq. 5.47 were also assumed to be known).

We shall again adopt a uniform prior distribution for the location parameter μ and, following the discussion in §5.2, a uniform prior distribution for $\ln \sigma$, which leads to

$$p(\mu, \sigma | I) \propto \frac{1}{\sigma}, \quad \text{for } \mu_{\min} \leq \mu \leq \mu_{\max} \text{ and } \sigma_{\min} \leq \sigma \leq \sigma_{\max}. \tag{5.53}$$

The exact values of the distribution limits are not important as long as they do not significantly truncate the likelihood. However, because we will need a properly normalized pdf in the context of the model comparison discussed in §5.7.1, we explicitly write the full normalization here:

$$p(\{x_i\} | \mu, \sigma, I)\, p(\mu, \sigma | I) = C\, \frac{1}{\sigma^{(N+1)}} \prod_{i=1}^{N} \exp \left(\frac{-(x_i - \mu)^2}{2\sigma^2} \right), \tag{5.54}$$

where

$$C = (2\pi)^{-N/2} \left(\mu_{\max} - \mu_{\min} \right)^{-1} \left[\ln \left(\frac{\sigma_{\max}}{\sigma_{\min}} \right) \right]^{-1}. \tag{5.55}$$

The value of C can be a very small number, especially when N is large. For example, with ($\mu_{\min} = -10$, $\mu_{\max} = 10$, $\sigma_{\min} = 0.01$, $\sigma_{\max} = 100$, $N = 10$), then $C = 5.6 \times 10^{-7}$.

The logarithm of the posterior pdf now becomes (cf. eq. 5.49)

$$L_p \equiv \ln \left[p(\mu, \sigma | \{x_i\}, I) \right] = \text{constant} - (N+1) \ln \sigma - \sum_{i=1}^{N} \frac{(x_i - \mu)^2}{2\sigma^2}. \tag{5.56}$$

Had we assumed a uniform distribution of σ, instead of $\ln \sigma$, then the factor multiplying $\ln \sigma$ would change from $(N+1)$ to N. Using the equality

$$\sum_{i=1}^{N} (x_i - \mu)^2 = N (\bar{x} - \mu)^2 + \sum_{i=1}^{N} (x_i - \bar{x})^2 \tag{5.57}$$

and with the substitution $V = N^{-1} \sum_{i=1}^{N} (x_i - \bar{x})^2$ (note that $V = (N-1)s^2/N$, where s is the sample standard deviation given by eq. 3.32), we can rewrite eq. 5.56 in terms of three data-based quantities, $N, \bar{x}$ and V:

$$L_p = \text{constant} - (N+1) \ln \sigma - \frac{N}{2\sigma^2} \left((\bar{x} - \mu)^2 + V \right). \tag{5.58}$$

Note that irrespective of the size of our data set we only need these three numbers ($N, \bar{x}$, and V) to *fully capture its entire information content* (because we assumed

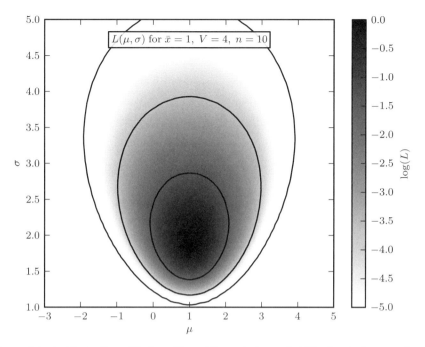

Figure 5.4. An illustration of the logarithm of the posterior probability density function for μ and σ, $L_p(\mu, \sigma)$ (see eq. 5.58) for data drawn from a Gaussian distribution and $N = 10, \bar{x} = 1$, and $V = 4$. The maximum of L_p is renormalized to 0, and color coded as shown in the legend. The maximum value of L_p is at $\mu_0 = 1.0$ and $\sigma_0 = 1.8$. The contours enclose the regions that contain 0.683, 0.955, and 0.997 of the cumulative (integrated) posterior probability.

a Gaussian likelihood function). They are called *sufficient statistics* because they summarize all the information in the data that is relevant to the problem (for formal definitions see Wass10).

An illustration of $L_p(\mu, \sigma)$ for $N = 10, \bar{x} = 1$, and $V = 4$ is shown in figure 5.4. The position of its maximum (μ_o, σ_o) can be found using eqs. 4.3 and 5.58: $\mu_0 = \bar{x}$ and $\sigma_0^2 = VN/(N+1)$ (i.e., σ_0 is equal to the sample standard deviation; see eq. 3.32).

The region of the (μ, σ) plane which encloses a given cumulative probability for the posterior pdf (or regions in the case of multimodal posterior pdfs) can be found by the following simple numerical procedure. The posterior pdf up to a normalization constant is simply $\exp(L_p)$ (e.g., see eqs. 5.56 or 5.58). The product of $\exp(L_p)$ and the pixel area (assuming sufficiently small pixels so that no interpolation is necessary) can be summed up to determine the normalization constant (the integral of the posterior pdf over all model parameters must be unity). The renormalized posterior pdf can be sorted, while keeping track of the corresponding pixel for each value, and the corresponding cumulative distribution computed. Given a threshold of $p\%$, all the pixels for which the cumulative probability is larger than $(1 - p/100)$ will outline the required region.

For a given $\sigma = \sigma'$, the maximum value of L_p is found along the $\mu = \mu_0 = \bar{x}$ line, and the posterior probability $p(\mu|\sigma = \sigma')$ is a Gaussian (same result as given by eqs. 4.5 and 4.7). However, now we do not know the true value of σ. When deriving the posterior probability $p(\mu)$, we need to marginalize (integrate) over all possible values of σ (in figure 5.4, at each value of μ, we "sum all the pixels" in the corresponding vertical slice through the image and renormalize the result),

$$p(\mu|\{x_i\}, I) = \int_0^\infty p(\mu, \sigma|\{x_i\}, I)\, d\sigma, \qquad (5.59)$$

yielding (starting with eq. 5.58 and using the substitution $t = 1/\sigma$ and integration by parts)

$$p(\mu|\{x_i\}, I) \propto \left[1 + \frac{(\overline{x} - \mu)^2}{V}\right]^{-N/2}. \qquad (5.60)$$

It is easy to show that this result corresponds to Student's t distribution (see eq. 3.60) with $k = N - 1$ degrees of freedom for the variable $t = (\overline{x} - \mu)/(s/\sqrt{N})$, where the sample standard deviation s is given by eq. 3.32. Had we assumed a uniform prior for σ, we would have obtained Student's t distribution with $k = (N - 2)$ degrees of freedom (the difference between these two solutions becomes negligible for large N).

The posterior marginal pdf $p(\mu|\{x_i\}, I)$ for $N = 10, \overline{x} = 1$, and $V = 4$ is shown in figure 5.5, for both σ priors (note that the uniform prior for σ gives a slightly wider posterior pdf). While the core of the distribution is similar to a Gaussian with parameters given by eqs. 3.31 and 3.34, its tails are much more extended. As N increases, Student's t distribution becomes more similar to a Gaussian distribution and thus $p(\mu|\{x_i\}, I)$ eventually becomes Gaussian, as expected from the central limit theorem.

Analogously to determining $p(\mu|\{x_i\}, I)$, the posterior pdf for σ is derived using marginalization,

$$p(\sigma|\{x_i\}, I) = \int_0^\infty p(\mu, \sigma|\{x_i\}, I)\, d\mu, \qquad (5.61)$$

yielding (starting with eq. 5.58 again)

$$p(\sigma|\{x_i\}, I) \propto \frac{1}{\sigma^N} \exp\left(\frac{-NV}{2\sigma^2}\right). \qquad (5.62)$$

Had we assumed a uniform prior for σ, the first term would have been $1/\sigma^{(N-1)}$. Eq. 5.62 is equivalent to the χ^2 distribution with $k = N - 1$ degrees of freedom for variable $Q = NV/\sigma^2$ (see eq. 3.58). An analogous result is known as Cochran's theorem in classical statistics. For a uniform prior for σ, the number of degrees of freedom is $k = N + 1$.

The posterior marginal pdf $p(\sigma|\{x_i\}, I)$ for our example is shown in figure 5.5. As is easily discernible, the posterior pdf for σ given by eq. 5.62 is skewed and not Gaussian, although the standard result given by eq. 3.35 implies the latter. The result from eq. 3.35 can be easily derived from eq. 5.62 using the approximation given by eq. 4.6, and is also shown in figure 5.5 (eq. 3.35 corresponds to a uniform σ prior; for a prior proportional to σ^{-1}, there is an additional $(N - 1)/(N + 1)$ multiplicative term).

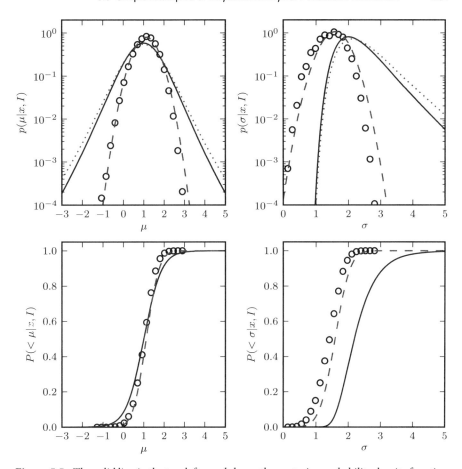

Figure 5.5. The solid line in the top-left panel shows the posterior probability density function $p(\mu|\{x_i\}, I)$ described by eq. 5.60, for $N = 10$, $\bar{x} = 1$ and $V = 4$ (integral over σ for the two-dimensional distribution shown in figure 5.4). The dotted line shows an equivalent result when the prior for σ is uniform instead of proportional to σ^{-1}. The dashed line shows the Gaussian distribution with parameters given by eqs. 3.31 and 3.34. For comparison, the circles illustrate the distribution of the bootstrap estimates for the mean given by eq. 3.31. The solid line in the top-right panel shows the posterior probability density function $p(\sigma|\{x_i\}, I)$ described by eq. 5.62 (integral over μ for the two-dimensional distribution shown in figure 5.4). The dotted line shows an equivalent result when the prior for σ is uniform. The dashed line shows a Gaussian distribution with parameters given by eqs. 3.32 and 3.35. The circles illustrate the distribution of the bootstrap estimates for σ given by eq. 3.32. The bottom two panels show the corresponding cumulative distributions for solid and dashed lines, and for bootstrap estimates, from the top panel.

When N is small, the posterior probability for large σ is much larger than that given by the Gaussian approximation. For example, the cumulative distributions shown in the bottom-right panel in figure 5.5 indicate that the probability of $\sigma > 3$ is in excess of 0.1, while the Gaussian approximation gives $\sim$0.01 (with the discrepancy increasing fast for larger σ). Sometimes, this inaccuracy of the Gaussian approximation can have a direct impact on derived scientific conclusions. For example, assume that we measured velocity dispersion (i.e., the sample standard deviation) of a subsample of 10 stars from the Milky Way's halo and obtained a value of 50 km s^{-1}. The Gaussian approximation tells us that (within the classical framework) we can reject the hypothesis that this subsample came from a population with velocity dispersion greater than 85 km s^{-1} (representative of the halo population) at a highly significant level (p value $\sim$ 0.001) and thus we might be tempted to argue that we discovered a stellar stream (i.e., a population with much smaller velocity dispersion). However, eq. 5.62 tells us that a value of 85 km s^{-1} or larger cannot be rejected even at a generous $\alpha = 0.05$ significance level! In addition, the Gaussian approximation and classical framework formally allow $\sigma \leq 0$, an impossible conclusion which is easily avoided by adopting a proper prior in the Bayesian approach. That is, the problem with negative s that we mentioned in the context of eq. 3.35 is resolved when using eq. 5.62. Therefore, when N is small (less than 10, though $N < 100$ is a safe bet for most applications), the confidence interval (i.e., credible region in the Bayesian framework) for σ should be evaluated using eq. 5.62 instead of eq. 3.35.

For a comparison of classical and Bayesian results, figure 5.5 also shows bootstrap confidence estimates for μ and σ (circles). As is evident, when the sample size is small, they have unreliable (narrower) tails and are more similar to Gaussian approximations with the widths given by eqs. 3.34 and 3.35. Similar widths are obtained using the jackknife method, but in this case we would use Student's t distribution with $N - 1$ degrees of freedom (see §4.5). The agreement with the above posterior marginal probability distributions would be good in the case of μ, but the asymmetric behavior for σ would not be reproduced. Therefore, as discussed in §4.5, the bootstrap and jackknife methods should be used with care.

Gaussian distribution with Gaussian errors

The posterior pdfs for μ and σ given by eqs. 5.60 and 5.62 correspond to a case where $\{x_i\}$ are drawn from an unspecified Gaussian distribution, $\mathcal{N}(\mu, \sigma)$. The width σ can be interpreted in two ways: it could correspond to the *intrinsic* spread σ of the measured quantity when measurement errors are always negligible, or it could simply be the unknown homoscedastic measurement error when measuring a single-valued quantity (such as the length of a rod in the above examples). A more general case is when the measured quantity is drawn from some distribution whose parameters we are trying to estimate, and the known measurement errors are heteroscedastic. For example, we might be measuring the radial velocity dispersion of a stellar cluster using noisy estimates of radial velocity for individual stars.

If the errors are homoscedastic, the resulting distribution of measurements is Gaussian: this is easily shown by recognizing the fact that the sum of two random variables has a distribution equal to the convolution of the input distributions, and that the convolution of two Gaussians is itself a Gaussian. However, when errors are heteroscedastic, the resulting distribution of measurements is not itself a Gaussian. As an example, figure 5.6 shows the pdf for the $\mathcal{N}(0, 1)$ distribution sampled with heteroscedastic Gaussian errors with widths uniformly distributed between 0 and 3. The Anderson–Darling statistic (see §4.7.4) for the resulting distribution is $A^2 = 3088$ (it is so large because N is large), strongly indicating that the data are not drawn from a normal distribution. The best-fit normal curves (based on both the sample variance and interquartile range) are shown for comparison.

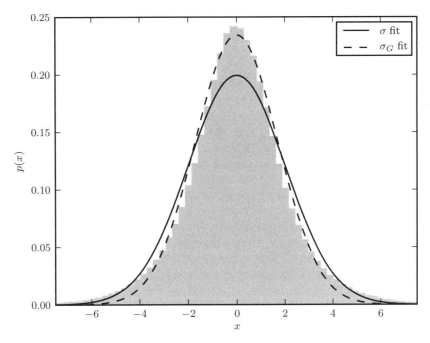

Figure 5.6. The distribution of 10^6 points drawn from $\mathcal{N}(0, 1)$ and sampled with heteroscedastic Gaussian errors with widths, e_i, uniformly distributed between 0 and 3. A linear superposition of these Gaussian distributions with widths equal to $\sqrt{1 + e_i^2}$ results in a non-Gaussian distribution. The best-fit Gaussians centered on the sample median with widths equal to sample standard deviation and quartile-based σ_G (eq. 3.36) are shown for comparison.

In order to proceed with parameter estimation in this situation, we shall assume that data were drawn from an intrinsic $\mathcal{N}(\mu, \sigma)$ distribution, and that measurement errors are also Gaussian and described by the known width e_i. Starting with an analog of eq. 5.52,

$$p(\{x_i\}|\mu, \sigma, I) = \prod_{i=1}^{N} \frac{1}{\sqrt{2\pi}(\sigma^2 + e_i^2)^{1/2}} \exp\left(\frac{-(x_i - \mu)^2}{2(\sigma^2 + e_i^2)}\right), \qquad (5.63)$$

and following the same steps as above (with uniform priors for μ and σ), we get an analog of eq. 5.56:

$$L_p = \text{constant} - \frac{1}{2} \sum_{i=1}^{N} \left(\ln(\sigma^2 + e_i^2) + \frac{(x_i - \mu)^2}{\sigma^2 + e_i^2}\right). \qquad (5.64)$$

However, in this case σ is "coupled" to e_i and must remain inside the sum, thus preventing us from capturing the entire information content of our data set using only 3 numbers (N, $\bar{x}$ and V), as we did when $e_i \ll \sigma$ (see eq. 5.58). This difficulty arises because the underlying distribution of $\{x_i\}$ is no longer Gaussian—instead it is a weighted sum of Gaussians with varying widths (recall figure 5.6; of course, the likelihood function for a single measurement is Gaussian). Compared to a Gaussian with the same σ_G (see eq. 3.36), the distribution of $\{x_i\}$ has more pronounced tails that reflect the distribution of e_i with finite width.

By setting the derivative of L_p with respect to μ to zero, we can derive an analog of eq. 5.50,

$$\mu_0 = \frac{\sum_i w_i x_i}{\sum_i w_i}, \tag{5.65}$$

except that weights are now

$$w_i = \frac{1}{\sigma_0^2 + e_i^2}. \tag{5.66}$$

These weights are fundamentally different from the case where σ (or σ_i) is known: σ_0 is now a quantity we are trying to estimate! By setting the derivative of L_p with respect to σ to zero, we get the constraint

$$\sum_{i=1}^{N} \frac{1}{\sigma_0^2 + e_i^2} = \sum_{i=1}^{N} \frac{(x_i - \mu_0)^2}{(\sigma_0^2 + e_i^2)^2}. \tag{5.67}$$

Therefore, we cannot obtain a closed-form expression for the MAP estimate of σ_0. In order to obtain MAP estimates for μ and σ, we need to solve the system of two complicated equations, eqs. 5.65 and 5.67. We have encountered a very similar problem when discussing the expectation maximization algorithm in §4.4.3. A straightforward way to obtain solutions is an iterative procedure: we start with a guess for μ_0 and σ_0 and obtain new estimates from eq. 5.65 (trivial) and eq. 5.67 (needs to be solved numerically).

Of course, there is nothing to stop us from simply evaluating L_p given by eq. 5.64 on a grid of μ and σ, as we did earlier in the example illustrated in figure 5.4. We generate a data set using $N = 10$, $\mu = 1$, $\sigma = 1$, and errors $0 < e_i < 3$ drawn from a uniform distribution. This e_i distribution is chosen to produce a similar sample variance as in the example from figure 5.4 ($V \approx 4$). Once e_i is generated, we draw a data value from a Gaussian distribution centered on μ and width equal to $(\sigma^2 + e_i^2)^{1/2}$. The resulting posterior pdf is shown in figure 5.7. Unlike the case with homoscedastic errors (see figure 5.4), the posterior pdf in this case is not symmetric with respect to the $\mu = 1$ line.

In practice, approximate estimates of μ_0 and σ_0 can be obtained without the explicit computation of the posterior pdf. Numerical simulations show that the sample median is an efficient and unbiased estimator of μ_0 (by symmetry), and its uncertainty can be estimated using eq. 3.34, with the standard deviation replaced by the quartile-based width estimator given by eq. 3.36, σ_G. With a data-based estimate of σ_G and the median error e_{50}, σ_0 can be estimated as

$$\sigma_0^2 = \zeta^2 \sigma_G^2 - e_{50}^2 \tag{5.68}$$

(to avoid solutions consistent with zero when N is small, this should be used only for large N, say $N > 100$). Here, ζ depends on the details of the distribution of e_i and

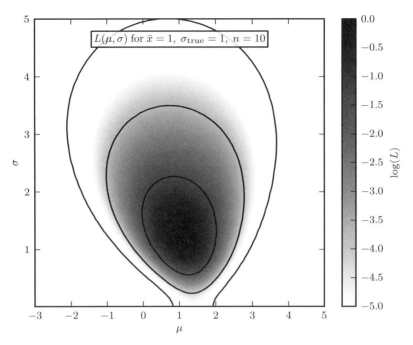

Figure 5.7. The logarithm of the posterior probability density function for μ and σ, $L_p(\mu, \sigma)$, for a Gaussian distribution with heteroscedastic Gaussian measurement errors (sampled uniformly from the 0–3 interval), given by eq. 5.64. The input values are $\mu = 1$ and $\sigma = 1$, and a randomly generated sample has 10 points. Note that the posterior pdf is not symmetric with respect to the $\mu = 1$ line, and that the outermost contour, which encloses the region that contains 0.997 of the cumulative (integrated) posterior probability, allows solutions with $\sigma = 0$.

can be estimated as

$$\zeta = \frac{\text{median}(\tilde{\sigma}_i)}{\text{mean}(\tilde{\sigma}_i)}, \qquad (5.69)$$

where

$$\tilde{\sigma}_i = (\sigma_G^2 + e_i^2 - e_{50}^2)^{1/2}. \qquad (5.70)$$

If all $e_i = e$, then $\zeta = 1$ and $\sigma_0^2 = \sigma_G^2 - e^2$, as expected from the convolution of two Gaussians. Of course, if $e_i \ll \sigma_G$ then $\zeta \to 1$ and $\sigma_0 \to \sigma_G$ (i.e., $\{x_i\}$ are drawn from $\mathcal{N}(\mu, \sigma_0)$).

These closed-form solutions follow from the result that σ_G for a weighted sum of Gaussians $\mathcal{N}(\mu, \sigma)$ with varying σ is *approximately* equal to the mean value of σ (i.e., $\text{mean}[(\sigma_o^2 + e_i^2)^{1/2}] = \sigma_G$), and the fact that the median of a random variable which is a function of another random variable is equal to the value of that function evaluated for the median of the latter variable (i.e., $\text{median}(\sigma_o^2 + e_i^2) = \sigma_o^2 + e_{50}^2$). For very large samples ($N > 1000$), the error for σ_0 given by eq. 5.68 becomes smaller than its bias (10–20%; that is, this estimator is not consistent). If this bias level is important in a specific application, a quick remedy is to compute the posterior pdf

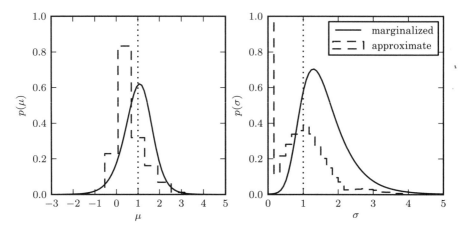

Figure 5.8. The solid lines show marginalized posterior pdfs for μ (left) and σ (right) for a Gaussian distribution with heteroscedastic Gaussian measurement errors (i.e., integrals over σ and μ for the two-dimensional distribution shown in figure 5.7). For comparison, the dashed histograms show the distributions of approximate estimates for μ and σ (the median and given by eq. 5.68, respectively) for 10,000 bootstrap resamples of the same data set. The true values of μ and σ are indicated by the vertical dotted lines.

given by eq. 5.64 in the neighborhood of approximate solutions and find the true maximum.

These "poor man's" estimators for the example discussed here ($\zeta = 0.94$) are also shown in figure 5.8. Because the sample size is fairly small ($N = 10$), in a few percent of cases $\sigma_0 < 0.1$. Although these estimators are only approximate, they are a *much better* solution than to completely ignore σ_0 and use, for example, the weighted mean formula (eq. 5.50) with $w_i = 1/e_i^2$. The main reason for this better performance is that here $\{x_i\}$ does not follow a Gaussian distribution (although both the intrinsic distribution of the measured quantity is Gaussian *and* measurement errors are Gaussian). Of course, the best option is to use the full Bayesian solution.

We have by now collected a number of different examples based on Gaussian distributions. They are summarized in table 5.1. The last row addresses the seemingly hopeless problem when both σ_0 and the heteroscedastic errors $\{e_i\}$ are unknown. Nevertheless, even in this case data contain information and can be used to place an upper limit $\sigma_0 \leq \sigma_G$ (in addition to using the median to estimate μ). Furthermore, $\{e_i\}$ can be considered model parameters, too, and marginalized over with some reasonable priors (e.g., a flat prior between 0 and the maximum value of $|x_i - \mu|$) to derive a better constraint for σ_0. This is hard to do analytically, but easy numerically, and we shall address this case in §5.8.5.

5.6.2. Parameter Estimation for the Binomial Distribution

We have already briefly discussed the coin flip example in §5.4. We revisit it here in the more general context of parameter estimation for the binomial distribution. Given a set of N measurements (or trials), $\{x_i\}$, drawn from a binomial distribution described with parameter b (see §3.3.3), we seek the posterior probability distribution $p(b|\{x_i\})$. Similarly to the Gaussian case discussed above, when N is large, b and its

TABLE 5.1.

Summary of the results for estimating parameters of $\mathcal{N}(\mu, \sigma_0)$ when data $\{x_i\}$ have a Gaussian error distribution given by $\{e_i\}$ (each data point is drawn from $\mathcal{N}(\mu, \sigma_i)$ with $\sigma_i^2 = \sigma_0^2 + e_i^2$). Weights w_i refer to eqs. 5.50 and 5.51, and s is the standard deviation given by eq. 3.32. If the error distribution is homoscedastic, but not necessarily Gaussian, use the median instead of the weighted mean, and the quartile-based width estimate σ_G (eq. 3.36) instead of the standard deviation.

σ_0	$\{e_i\}$	Weights	Description
σ_0	$e_i = e$	$w_i = 1$	homoscedastic, both σ_0 and e are known
σ_0	$e_i = 0$	$w_i = 1$	homoscedastic, errors negligible, $\sigma_0 = s$
0	$e_i = e$	$w_i = 1$	homoscedastic, single-valued quantity, $e = s$
σ_0	$e_i = e$	$w_i = 1$	homoscedastic, e known, $\sigma_0^2 = (s^2 - e^2)$
0	e_i known	$w_i = e_i^{-2}$	errors heteroscedastic but assumed known
σ_0	e_i known	no closed form	σ_0 unknown; see eq. 5.64 and related discussion
σ_0	unknown	no closed form	upper limit for σ_0; numerical modeling; see text (also §5.8.5).

(presumably Gaussian) uncertainty can be determined as discussed in §3.3.3. For small N, the proper procedure is as follows.

Here the data set $\{x_i\}$ is discrete: all outcomes are either 0 (heads) or 1 (tails, which we will consider "success"). An astronomical analog might be the computation of the fraction of galaxies which show evidence for a black hole in their center. Given a model parametrized by the probability of success b, the likelihood that the data set contains k outcomes equal to 1 is given by eq. 3.50. Assuming that the prior for b is flat in the range 0–1, the posterior probability for b is

$$p(b|k, N) = C\, b^k\, (1 - b)^{N-k}, \tag{5.71}$$

where k is now the actual observed number of successes in a data set of N values, and the normalization constant C can be determined from the condition $\int_0^1 p(b|k, N)\, db = 1$ (alternatively, we can make use of the fact that the beta distribution is a conjugate prior for binomial likelihood; see §5.2.3). The maximum posterior occurs at $b_0 = k/N$.

For a concrete numerical example, let us assume that we studied $N = 10$ galaxies and found a black hole in $k = 4$ of them. Our best estimate for the fraction of galaxies with black holes is $b_0 = k/N = 0.4$. An interesting question is, "What is the probability that, say, $b < 0.1$?" For example, your colleague's theory placed an upper limit of 10% for the fraction of galaxies with black holes and you want to test this theory using classical framework ("Can it be rejected at a confidence level $\alpha = 0.01$?").

Using the Gaussian approximation discussed in §3.3.3, we can compute the standard error for b_0 as

$$\sigma_b = \left[\frac{b_0\,(1 - b_0)}{N}\right]^{1/2} = 0.155, \tag{5.72}$$

and conclude that the probability for $b < 0.1$ is ~0.03 (the same result follows from eq. 4.6). Therefore, at a confidence level $\alpha = 0.01$ the theory is not rejected. However,

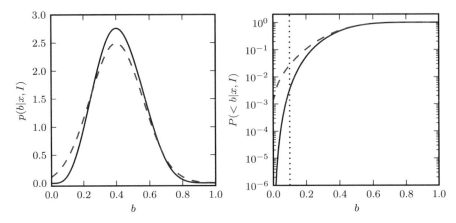

Figure 5.9. The solid line in the left panel shows the posterior pdf $p(b|k, N)$ described by eq. 5.71, for $k = 4$ and $N = 10$. The dashed line shows a Gaussian approximation described in §3.3.3. The right panel shows the corresponding cumulative distributions. A value of 0.1 is marginally likely according to the Gaussian approximation ($p_{approx}(< 0.1) \approx 0.03$) but strongly rejected by the true distribution ($p_{true}(< 0.1) \approx 0.003$).

the exact solution given by eq. 5.71 and shown in figure 5.9 is not a Gaussian! By integrating eq. 5.71, you can show that $p(b < 0.1|k = 4, N = 10) = 0.003$, and therefore your data do reject your colleague's theory[7] (note that in the Bayesian framework we need to specify an alternative hypothesis; see §5.7). When N is not large, or b_0 is close to 0 or 1, one should avoid using the Gaussian approximation when estimating the credible region (or the confidence interval) for b.

5.6.3. Parameter Estimation for the Cauchy (Lorentzian) Distribution

As already discussed in §3.3.5, the mean of a sample drawn from the Cauchy distribution is not a good estimator of the distribution's location parameter. In particular, the mean value for many independent samples will themselves follow the same Cauchy distribution, and will not benefit from the central limit theorem (because the variance does not exist). Instead, the location and scale parameters for a Cauchy distribution (μ and γ) can be simply estimated using the median value and interquartile range for $\{x_i\}$. We shall now see how we can estimate the parameters of a Cauchy distribution using a Bayesian approach.

As a practical example, we will use the lighthouse problem due to Gull and discussed in Siv06 (a mathematically identical problem is also discussed in Lup93; see problem 18). A lighthouse is positioned at $(x, y) = (\mu, \gamma)$ and it emits discrete light signals in random directions. The coastline is defined as the $y = 0$ line, and the lighthouse's distance from it is γ. Let us define the angle θ as the angle between the line that connects the lighthouse and the point $(x, y) = (\mu, 0)$, and the direction of a signal. The signals will be detected along the coastline with the positions

$$x = \mu + \gamma \tan(\theta), \tag{5.73}$$

---

[7]It is often said that it takes a 2σ result to convince a theorist that his theory is correct, a 5σ result to convince an observer that an effect is real, and a 10σ result to convince a theorist that his theory is wrong.

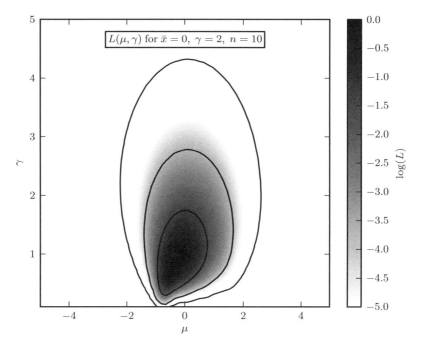

Figure 5.10. An illustration of the logarithm of posterior probability distribution for μ and γ, $L(\mu, \gamma)$ (see eq. 5.75) for $N = 10$ (the sample is generated using the Cauchy distribution with $\mu = 0$ and $\gamma = 2$). The maximum of L is renormalized to 0, and color coded as shown in the legend. The contours enclose the regions that contain 0.683, 0.955 and 0.997 of the cumulative (integrated) posterior probability.

with $-\pi/2 \leq \theta \leq \pi/2$. If the angle θ is distributed uniformly, it is easy to show that x follows the Cauchy distribution given by eq. 3.53 (use $p(x) = (\pi \, dx/d\theta)^{-1} p(\theta)$), and the data likelihood is

$$p(\{x_i\}|\mu, \gamma, I) = \prod_{i=1}^{N} \frac{1}{\pi} \left(\frac{\gamma}{\gamma^2 + (x_i - \mu)^2} \right). \tag{5.74}$$

Given a data set of measured positions $\{x_i\}$, we need to estimate μ and γ. Analogously to the Gaussian case discussed above, we shall adopt a uniform prior distribution for the location parameter μ, and a uniform prior distribution for $\ln \gamma$, for $\mu_{\min} \leq \mu \leq \mu_{\max}$ and $\gamma_{\min} < \gamma \leq \gamma_{\max}$. The logarithm of the posterior pdf is

$$L_p \equiv \ln[p(\mu, \gamma|\{x_i\}, I)] = \text{constant} + (N-1) \ln \gamma - \sum_{i=1}^{N} \ln\left[\gamma^2 + (x_i - \mu)^2\right]. \tag{5.75}$$

An example, based on $N = 10$ values of x_i, generated using the Cauchy distribution with $\mu = 0$ and $\gamma = 2$, is shown in figure 5.10. In this particular realization, the maximum of L is at $\mu_0 = -0.36$ and $\gamma_0 = 0.81$. This maximum can be found by setting the derivatives of L to 0 (using eqs. 4.3 and 5.75), but the resulting

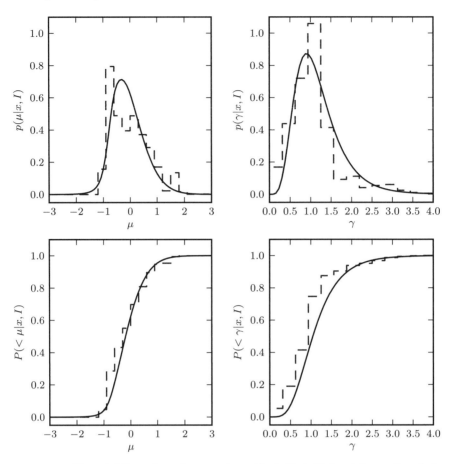

Figure 5.11. The solid lines show the posterior pdf $p(\mu|\{x_i\}, I)$ (top-left panel) and the posterior pdf $p(\gamma|\{x_i\}, I)$ (top-right panel) for the two-dimensional pdf from figure 5.10. The dashed lines show the distribution of approximate estimates of μ and γ based on the median and interquartile range. The bottom panels show the corresponding cumulative distributions.

equations still have to be solved numerically (compare eq. 5.75 to its Gaussian cousin, eq. 5.56; note that in this case we cannot form an analog to eq. 5.56).

The posterior marginal distributions for μ and γ are shown in figure 5.11. Note that the posterior distribution for γ is very asymmetric. For this reason, its peak is not as good an estimator of the true γ as the distribution's median (see the cumulative distribution in the bottom-right panel). As the sample size N increases, both posterior marginal distributions become asymptotically Gaussian and eq. 4.6 can be used to estimate the confidence intervals for μ and γ.

It turns out that for the Cauchy distribution the median and σ_G (see eq. 3.54) can be used as a good shortcut to determine the best-fit parameters (instead of computing marginal posterior pdfs). For example, the sample median and the median of posterior marginal distributions for μ have Kendall's correlation coefficient $\tau \sim 0.7$. For the particular example discussed in figure 5.10, the median and interquartile range imply $\mu = -0.26$ and $\gamma = 1.11$. When using this shortcut, the bootstrap method can be used to estimate parameter uncertainties (see figure 5.11 for comparison with Bayesian results; the agreement is good though not perfect).

In practice, it is often prohibitive to do an exhaustive search of the parameter space to find the maximum of the posterior pdf (as we did in figures 5.4 and 5.10). Computation becomes especially hard in the case of a high-dimensional parameter space. Instead, various numerical minimization techniques may be used to find the best parameters (discussed in §5.8).

5.6.4. Beating $\sqrt{N}$ for Uniform Distribution

We have discussed how to estimate the location parameter for a uniform distribution in §3.4. Due to the absence of tails, the extreme values of x_i provide a more efficient estimator of the location parameter than the mean value, with errors that improve with the sample size as $1/N$, rather than as $1/\sqrt{N}$ when using the mean. Here we derive this result using the Bayesian method.

Given the uniform distribution described by eq. 3.39, the likelihood of observing x_i is given by

$$p(x_i|\mu, W) = \frac{1}{W} \text{ for } |x_i - \mu| \leq \frac{W}{2}, \tag{5.76}$$

and 0 otherwise (i.e., x_i can be at most $W/2$ away from μ). We assume that both μ and W are unknown and need to be estimated from data. We shall assume a uniform prior for μ between $-\Delta$ and Δ, where Δ greatly exceeds the plausible range for $|\mu + W|$, and a scale-invariant prior for W. The likelihood of observing the observed data set $\{x_i\}$ is

$$p(\mu, W|\{x_i\}, I) \propto \frac{1}{W^{N+1}} \prod_{i=1}^{N} g(\mu, W|x_i), \tag{5.77}$$

where

$$g(\mu, W|x_i) = 1 \text{ for } W \geq 2|x_i - \mu|, \tag{5.78}$$

and 0 otherwise. In the (μ, W) plane, the region allowed by x_i (where $p(\mu, W|\{x_i\}, I) > 0$) is the wedge defined by $W \geq 2(x_i - \mu)$ and $W \geq 2(\mu - x_i)$. For the whole data set, the allowed region becomes a wedge defined by $W \geq 2(x_{max} - \mu)$ and $W \geq 2(\mu - x_{min})$, where x_{min} and x_{max} are the minimum and maximum values in the data set $\{x_i\}$ (see figure 5.12). The minimum allowed W is $W_{min} = (x_{max} - x_{min})$, and the posterior is symmetric with respect to $\tilde{\mu} = (x_{min} + x_{max})/2$. The highest value of the posterior probability is at the point ($\mu = \tilde{\mu}$, $W = W_{min}$). Within the allowed range, the posterior is proportional to $1/W^{N+1}$ without a dependence on μ, and is 0 otherwise. The logarithm of the posterior probability $p(\mu, W|\{x_i\}, I)$ based on $N = 100$ values of x_i, generated using $\mu = 0$, $W = 1$ (for this particular realization, $x_{min} = 0.047$ and $x_{max} = 9.884$), is shown in figure 5.12. We see that in this case, the likelihood contours are not well described by ellipses!

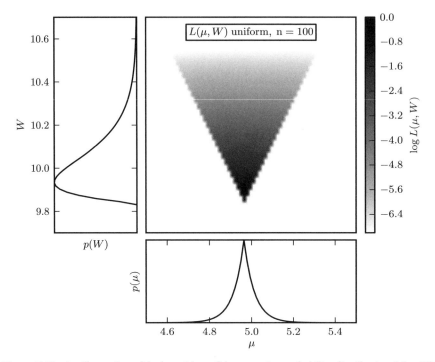

Figure 5.12. An illustration of the logarithm of the posterior probability distribution $L(\mu, W)$ (see eq. 5.77) for $N = 100$, $\mu = 5$, and $W = 10$. The maximum of L is renormalized to 0, and color coded on a scale from -5 to 0, as shown in the legend. The bottom panel shows the marginal posterior for μ (see eq. 5.79), and the left panel shows the marginal posterior for W (see eq. 5.80).

It is straightforward to show that the marginal posterior pdf for μ is

$$p(\mu) = \int_0^\infty p(\mu, W|\{x_i\}, I)\, dW \propto \frac{1}{(|\mu - \tilde{\mu}| + W_{\min}/2)^N}, \tag{5.79}$$

and for W,

$$p(W) = \int_{-\Delta}^{\Delta} p(\mu, W|\{x_i\}, I)\, d\mu \propto \frac{W - W_{\min}}{W^{N+1}} \quad \text{for } W \geq W_{\min}. \tag{5.80}$$

Taking expectation values for μ and W using these marginal distributions reproduces eqs. 3.68 and 3.69. As shown in figure 5.12, the shape of $p(\mu)$ is more peaked than for a Gaussian distribution.

We can understand how the width of $p(\mu)$ scales with N using eq. 5.79, by considering the scale of the half-max width $d\mu$ satisfying $p(\tilde{\mu} + d\mu) = p(\tilde{\mu})/2$. Putting these values into eq. 5.79 and simplifying leads to

$$d\mu = \frac{W_{\min}}{2}\left(2^{1/N} - 1\right). \tag{5.81}$$

Expanding this in an asymptotic series about $N = \infty$ yields the approximation

$$d\mu = \frac{W_{\min} \ln 2}{2N} + \mathcal{O}\left[\left(\frac{1}{N}\right)^2\right], \tag{5.82}$$

which scales as $1/N$, and is in close agreement with the Gaussian approximation to the standard deviation, $(2W_{\min})/(N\sqrt{12})$, derived in §3.4. This $1/N$ scaling is demonstrated using a numerical example shown in figure 3.21.

5.6.5. Parameter Estimation for a Gaussian and a Uniform Background

We shall now solve a slightly more complicated parameter estimation problem involving a Gaussian distribution than those from §5.6.1. Here the underlying model is a mixture of a Gaussian distribution and a uniform distribution in some interval W. This example illustrates parameter covariance and marginalization over a nuisance parameter.

The likelihood of obtaining a measurement x_i is given by

$$p(x_i | A, \mu, \sigma) = \frac{A}{\sqrt{2\pi}\sigma} \exp\left(\frac{-(x_i - \mu)^2}{2\sigma^2}\right) + \frac{1-A}{W} \tag{5.83}$$

for $0 < x < W$, and 0 otherwise. It is implied that the measurement error for x_i is negligible compared to σ (or alternatively that σ is an unknown measurement error when we measure a single-valued quantity μ). We will assume that the location parameter μ for the Gaussian component is known, and that W is sufficiently large so that the Gaussian tails are not truncated. Note that there is a "trade-off" between the first and second component: a larger A means a weaker background component with the strength $B = (1 - A)/W$ (we cannot simply substitute an unknown B for the second term because of the normalization constraint). Since an increase of σ will widen the Gaussian so that its tails partially compensate for the strength of the background component, we expect a covariance between A and σ (because the background strength is not a priori known).

We shall adopt uniform prior distributions for A and σ and require that they are both nonnegative,

$$p(A, \sigma | I) = \text{constant}, \quad \text{for } 0 \leq A < A_{\max} \text{ and } 0 \leq \sigma < \sigma_{\max}. \tag{5.84}$$

This model might correspond to a spectral line of known central wavelength but with an unknown width (σ) and strength (amplitude A), measured in the presence of a background with strength B. It could also correspond to a profile of a source in an image with an unknown background B; we know the source position and are trying to estimate its flux (A) and width (σ; and perhaps compare it later to the point spread function to check if the source is resolved). Using the numerical methods developed in §5.8, one might even imagine solving a model with all three parameters unknown: location, width, and background level. This extension is addressed in §5.8.6.

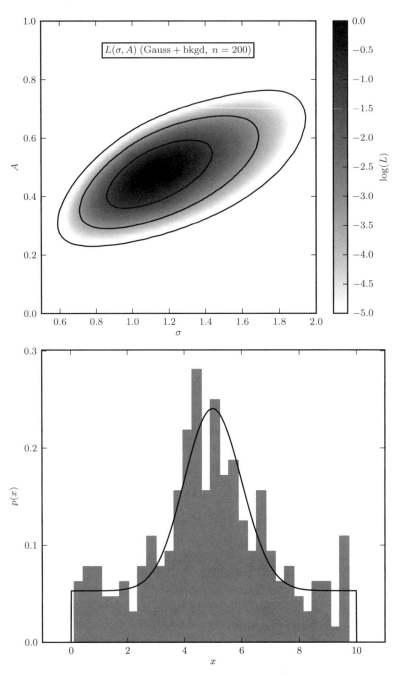

Figure 5.13. An illustration of the logarithm of the posterior probability density function $L(\sigma, A)$ (see eq. 5.85) for data generated using $N = 200$, $\mu = 5$, $\sigma = 1$, and $A = 0.5$, with the background strength $(1 - A)/W = 0.05$ in the interval $0 < x < W$, $W = 10$. The maximum of $L(\sigma, A)$ is renormalized to 0, and color coded on a scale -5 to 0, as shown in the legend. The contours enclose the regions that contain 0.683, 0.955, and 0.997 of the cumulative (integrated) posterior probability. Note the covariance between A and σ. The histogram in the bottom panel shows the distribution of data values used to construct the posterior pdf in the top panel, and the probability density function from which the data were drawn as the solid line.

In the case of a known location μ, the logarithm of the posterior pdf is

$$L_p \equiv \ln \left[p(A, \sigma | \{x_i\}, \mu, W) \right] = \sum_{i=1}^{n} \ln \left[A \frac{\exp\left(\frac{-(x_i - \mu)^2}{2\sigma^2} \right)}{\sqrt{2\pi}\sigma} + \frac{1 - A}{W} \right]. \quad (5.85)$$

It is not possible to analytically find A and σ corresponding to the maximum of this posterior pdf. A numerical example, based on $N = 200$ values of x_i, generated using $A = 0.5$, $\sigma = 1$, $\mu = 5$, and $W = 10$ is shown in figure 5.13.

As expected, there is a trade-off between A and σ: an error in σ is compensated by a proportional error in A. The posterior pdfs for A and σ can be determined by marginalizing the posterior pdf, as we did in §5.6.3 (see figure 5.11). On the other hand, if additional information were available about either parameter, the other one would be better constrained. In this example, if W were significantly increased, the knowledge of the background strength, and thus of A too, would be much improved. In the limit of a perfectly known A, the posterior pdf for σ would simply correspond to a slice through $L_p(A, \sigma)$ (on a linear scale and renormalized!), and would be narrower than the marginal posterior pdf for σ (cf. §3.5.2).

Note that our data $\{x_i\}$ were never binned when computing the posterior probability. We only used the binning and histogram in figure 5.13 to visualize the simulated sample. Nevertheless, the measuring process often results in binned data (e.g., pixels in an image, or spectral measurements). If we had to solve the same problem of a Gaussian line on top of a flat background with binned data, the computation would be similar but not identical. The data would include counts y_i for each bin position x_i. The model prediction for y_i, or $p(D|M, I)$, would still be based on eq. 5.83, and we would need to use counting (Poisson) statistics to describe the variation around the expected count values. This example would be solved similarly to the example discussed next.

5.6.6. Regression Preview: Gaussian vs. Poissonian Likelihood

In examples presented so far, we dealt with distributions of a single random variable. Let us now consider a problem where the data are pairs of random variables, $(x_1, y_1), \ldots, (x_M, y_M)$. Let y correspond to counts in bins centered on x, with a constant and very small bin width. We would like to estimate a and b for a model $y = ax + b$. This is an example of *regression*; this topic is discussed in detail in chapter 8. Here we only want to illustrate how assumptions about data likelihood can affect inferences about model parameters, and discuss the effects of data binning on inferred parameters.

We start the discussion with the unbinned data case, that is, first we will assume that we know the exact x position for each count described by y. We have a data set $\{x_i\}$, of size $N = \sum_{j=1}^{M} y_j$, drawn from a pdf given by $p(x) = ax + b$, with $p(x) = 0$ outside the known range $x_{\min} \leq x \leq x_{\max}$, and we want to estimate a and b. For notational simplicity, we introduce $W = (x_{\max} - x_{\min})$ and $x_{1/2} = (x_{\min} + x_{\max})/2$. It is assumed that the uncertainty for each x_i is negligible compared to W.

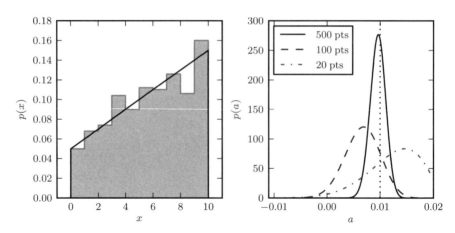

Figure 5.14. Regression of unbinned data. The distribution of $N = 500$ data points is shown in the left panel; the true pdf is shown by the solid curve. Note that although the data are binned in the left panel for visualization purposes, the analysis is performed on the *unbinned* data. The right panel shows the likelihood for the slope a (eq. 5.88) for three different sample sizes. The input value is indicated by the vertical dotted line.

Since $p(x)$ must be a properly normalized pdf, this parameter estimation problem is one-dimensional: a and b are related via the normalization constraint

$$b = \frac{1}{W} - a\, x_{1/2} \tag{5.86}$$

(recall the analysis of truncated data discussed in §4.2.7). The probability of observing a value x_i given a is thus

$$p(x_i|a) = a\,(x_i - x_{1/2}) + \frac{1}{W}. \tag{5.87}$$

Because $p(x) \geq 0$, each data point defines an allowed range for a, $a > -W^{-1}/(x_i - x_{1/2})$. Taking the entire data set, a is confined to the range $a_{\min} \leq a \leq a_{\max}$, with $a_{\min} = -W^{-1}/\max(x_i - x_{1/2})$ and $a_{\max} = -W^{-1}/\min(x_i - x_{1/2})$. For $a > 0$, values of x_i larger than $x_{1/2}$ are more probable than smaller ones. Consequently, data values $x_i > x_{1/2}$ favor $a > 0$ and $x_i < x_{1/2}$ favor $a < 0$. Assuming a uniform prior for a, the logarithm of the posterior pdf is

$$L_p(a|\{x_i\}, x_{\min}, x_{\max}) = \sum_{i=1}^{N} \ln\left[a\,(x_i - x_{1/2}) + \frac{1}{W}\right]. \tag{5.88}$$

For illustration, we generate three data sets with $x_{\min} = 0$, $x_{\max} = 10$, $a = 0.01$, and, from eq. 5.86, $b = 0.05$. Figure 5.14 shows the distribution of $N = 500$ generated values, and the posterior pdf for a obtained using 20, 100, and all 500 data values.

We now return to the binned data case (recall that the number of data points in the unbinned case is $N = \sum_{i=1}^{M} y_i$). The model for predicting the expected counts value in a given bin, $y(x) = a^*x + b^*$, can be thought of as

$$y(x) = Y \int_{x-\Delta/2}^{x+\Delta/2} p(x) \, dx, \qquad (5.89)$$

where Δ is the bin width and $p(x) = ax + b$ is a properly normalized pdf (i.e., b is related to a via eq. 5.86). The proportionality constant Y provides a scale for the counts and is approximately equal to N (the sum of all counts; we will return to this point below). Here we will assume that Δ is sufficiently small that $p(x)$ is constant within a bin, and thus $a^* = Y \Delta a$ and $b^* = Y \Delta b$. When estimating a^* and b^*, Δ need not be explicitly known, as long as it has the same value for all bins and the above assumption that $p(x)$ is constant within a bin holds.

The actual values y_i are "scattered around" their true values $\mu_i = a^*x_i + b^*$, in a manner described by the data likelihood. First we will assume that all μ_i are sufficiently large that the data likelihood can be approximated by a Gaussian, $p(y_i|x_i, a^*, b^*, I) = \mathcal{N}(\mu_i, \sigma_i)$, with $\sigma_i = \mu_i^{1/2}$. With uniform priors for a^* and b^*, the logarithm of the posterior pdf is

$$L_p^G(a^*, b^*) \equiv \ln\left[p(a^*, b^*|\{x_i, y_i\})\right] = \text{constant}$$
$$- \frac{1}{2} \sum_{i=1}^{M} \left[\ln(a^*x_i + b^*) + \frac{(y_i - a^*x_i - b^*)^2}{a^*x_i + b^*}\right]. \qquad (5.90)$$

Compared to eq. 5.56, which is also based on a Gaussian likelihood, the term inside the sum is now more complex. Aside from Poisson fluctuations, no source of error is assumed when measuring y_i. In practice, y_i (more precisely, μ_i) can be so large that these counting errors are negligible, and that instead each measurement has an uncertainty σ_i completely unrelated to a^* and b^*. In such a case, the first term in the sum would disappear, and the denominator in the second term would be replaced by σ_i^2.

When the count values are not large, the data likelihood for each y_i must be modeled using the Poisson distribution, $p(k|\mu_i) = \mu_i^k \exp(-\mu_i)/k!$, where $k = y_i$ and as before $\mu_i = a^*x_i + b^*$. With uniform priors for a^* and b^*,

$$L_p^P(a^*, b^*) = \text{constant} + \sum_{i=1}^{M} \left[y_i \ln(a^*x_i + b^*) - a^*x_i - b^*\right]. \qquad (5.91)$$

Figure 5.15 compares L_p^G and L_p^P. When the counts in each bin are high, the Gaussian expression L_p^G is a good approximation of the true likelihood L_p^P. When there are fewer counts per bin, the Gaussian approximation becomes biased toward smaller a and b. The difference is not large however: the expectation value and standard deviation for y_i is the same for the Poisson and Gaussian case. It is only the shape of the distribution that changes.

Note also in figure 5.15 that there is a correlation between the best-fit parameters: i.e., the posterior probability contours in the (a^*, b^*) plane are not aligned with the coordinate axes. The orientation of the semimajor axis reflects the normalization

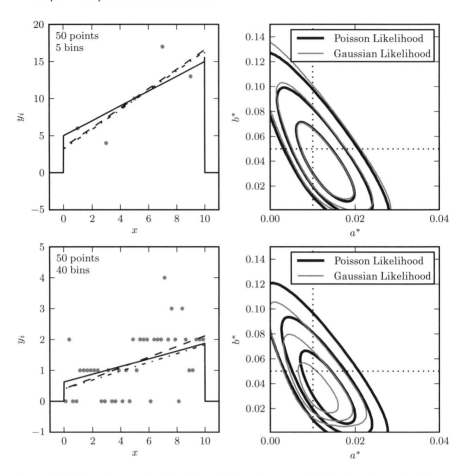

Figure 5.15. Binned regression. The left panels show data sets with 50 points, binned in 5 bins (upper panels) and 40 bins (lower panels). The curves show the input distribution (solid), the Poisson solution (dashed), and the Gaussian solution (dotted). The right panels show 1σ, 2σ, and 3σ likelihood contours for eqs. 5.91 (dark lines) and 5.90 (light lines). With 5 bins (top row) there are enough counts in each bin so that the Gaussian and Poisson predictions are very similar. As the number of bins is increased, the counts decrease and the Gaussian approximation becomes biased.

constraint for the underlying pdf. Analogously to eq. 5.86,

$$b^* = \frac{Y\Delta}{W} - a^* x_{1/2}. \tag{5.92}$$

This equation corresponds to a straight line in the (a^*, b^*) plane that coincides with the major axis of the probability contour for the Poissonian likelihood in figure 5.15. Given this normalization constraint, it may be puzzling that the posterior pdf at a given value of a^* has a finite (nonzero) width in the b^* direction. The reason is that for a given a^* and b^*, the implied Y corresponds to the *predicted* sum of y_i, and the actual sum can fluctuate around the predicted value.

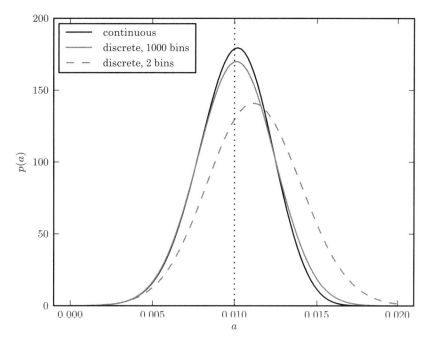

Figure 5.16. The comparison of the continuous method (figure 5.14) and the binned method (figure 5.15) on the same data set. In the limit of a large number of bins, most bins register only zero or one count, and the binned Poisson statistic gives nearly the same marginalized distribution for a as the continuous statistic. For as few as two bins, the constraint on the slope is only slightly biased.

Finally, we would expect that as the number of bins becomes very large, the solution would approach that of the unbinned case. In figure 5.16, we compare the marginalized distribution for a in the unbinned case and the binned case with 1000 bins. The results are nearly identical, as expected.

5.6.7. A Mixture Model: How to "Throw Out" Bad Data Points

We have seen in §5.6.1 and §5.6.3 that standard results for estimating parameters that are based on the assumption of Gaussianity, such as eq. 5.50, do not work well when the sampled distribution is not Gaussian. For example, a few outliers in a data set can significantly affect the weighted mean given by eq. 5.50.

When the model for the underlying distribution is known (e.g., a Cauchy distribution, or a Gaussian convolved with known Gaussian errors), we can maximize the posterior pdf as in any other case. When the model is not known a priori, we can use a Bayesian framework to construct a model in terms of unknown nuisance model parameters, and then marginalize over them to estimate the quantities of interest.

We shall consider here a one-dimensional case and revisit the problem of outliers in more detail in the context of regression (see §8.9). Let us assume as in §5.6.1 that we have a set of N measurements, $\{x_i\}$, of a single-valued quantity μ, and the measurement errors are Gaussian, heteroscedastic, known and given as σ_i. We seek the posterior pdf for μ: $p(\mu|\{x_i\}, \{\sigma_i\})$. The difference compared to §5.6.1 is that here some data points have unreliable σ_i; that is, some points were drawn from a different distribution with a much larger width. We shall assume no bias for simplicity; that is, these outliers are also drawn from a distribution centered on μ.

If we knew which points were outliers, then we would simply exclude them and apply standard Gaussian results to the remaining points (assuming that outliers represent a small fraction of the data set). We will assume that we do not have this information—this is a case of hidden variables similar to that discussed in §4.4.2, and indeed the analysis will be very similar.

Before proceeding, let us reiterate that an easy and powerful method to estimate μ when we suspect non-Gaussian errors is to simply use the median instead of the mean, and the σ_G instead of the standard deviation in eq. 3.38 when estimating the uncertainty of μ. Nevertheless, the full Bayesian analysis enables a more formal treatment, as well as the ability to estimate which points are likely outliers using an objective framework.

First, given $\{x_i\}$ and $\{\sigma_i\}$, how do we assess whether non-Gaussianity is important? Following the discussion in §4.3.1, we expect that

$$\chi^2_{\text{dof}} = \frac{1}{N-1} \sum_{i=1}^{N} \frac{(x_i - \bar{x})^2}{\sigma_i} \approx 1. \tag{5.93}$$

If $\chi^2_{\text{dof}} - 1$ is a few times larger than $\sqrt{2/(N-1)}$, then it is unlikely (as given by the cumulative pdf for χ^2_{dof} distribution) that our data set $\{x_i\}$ was drawn from a Gaussian error distribution with the claimed $\{\sigma_i\}$.

We start by formulating the data likelihood as

$$p(x_i|\mu, \sigma_i, g_i, I) = g_i \, \mathcal{N}(\mu, \sigma_i) + (1 - g_i) \, p_{\text{bad}}(x_i|\mu, I). \tag{5.94}$$

Here g_i is 1 if the data point is "good" and 0 if it came from the distribution of outliers, $p_{\text{bad}}(x_i|\mu, I)$. In this model $p_{\text{bad}}(x_i|\mu, I)$ applies to all outliers. Again, if we knew g_i this would be an easy problem to solve. Since $\{g_i\}$ represent hidden variables, we shall treat them as model parameters and then marginalize over them to get $p(\mu|\{x_i\}, \{\sigma_i\}, I)$. With a separable prior, which implies that the reliability of the measurements is decoupled from the true value of the quantity we are measuring,

$$p(\mu, \{g_i\}|I) = p(\mu|I) \, p(\{g_i\}|I), \tag{5.95}$$

we get

$$p(\mu, \{g_i\}|\{x_i\}, I) \propto p(\mu|I) \prod_{i=1}^{N} [g_i \, \mathcal{N}(\mu, \sigma_i) + (1 - g_i) \, p_{\text{bad}}(x_i|\mu, I)] \, p(\{g_i\}|I), \tag{5.96}$$

and finally, marginalizing over g_i gives

$$p(\mu|\{x_i\}, \{\sigma_i\}) \propto \int p(\mu, \{g_i\}|\{x_i\}, I) \, d^N g_i. \tag{5.97}$$

To proceed further, we need to choose specific models for the priors and $p_{\text{bad}}(x_i|\mu, I)$. We shall follow an example from [27], which discusses this problem in the context of incompatible measurements of the Hubble constant available at the end of the twentieth century.

For the distribution of outliers, we adopt an appropriately wide Gaussian distribution

$$p_{\text{bad}}(x_i|\mu, I) = \mathcal{N}(\mu, \sigma^*), \tag{5.98}$$

where a good choice for σ^* is a few times σ_G determined from data, although details vary depending on the actual $\{x_i\}$. If $\sigma^* \gg \sigma_G$, $p_{\text{bad}}(x_i|\mu, I)$ effectively acts as a uniform distribution. Alternatively, we could treat σ^* as yet another model parameter and marginalize over it.

The prior for μ can be taken as uniform for simplicity, and there are various possible choices for $p(\{g_i\}|I)$ depending on our beliefs, or additional information, about the quality of measurements. In the example discussing the Hubble constant from [27], each data point is the result of a given method and team, and the prior for g_i might reflect past performance for both. In order to enable an analytic solution here, we shall adopt uniform priors for all g_i. In this case, marginalization over all g_i is effectively replacing every g_i by 1/2, and leads to

$$p(\mu|\{x_i\}, \{\sigma_i\}) \propto \prod_{i=1}^{N} \left[\mathcal{N}(\mu, \sigma_i) + \mathcal{N}(\mu, \sigma^*) \right]. \tag{5.99}$$

Of course, other choices are possible, and they need not be the same for all data points. For example, we could combine uniform priors (no information) with Gaussians, skewed distributions toward $g_i = 0$ or $g_i = 1$, or even delta functions, representing that we are certain that some data points are trustworthy, or more generally, that we know their probability of being correct with certainty.

A distinctive feature of Bayesian analysis is that we can marginalize eq. 5.96 over μ and all g_i but one—we can obtain a posterior probability that a given data point is good (or bad, of course). Assuming uniform priors, let us integrate over all g_i except the first point (without a loss of generality) and μ to get (note that the product starts from $i = 2$)

$$p(\mu, g_1|\{x_i\}, \{\sigma_i\}) = \left[g_1 \mathcal{N}(\mu, \sigma_1) + (1 - g_1)\mathcal{N}(\mu, \sigma^*) \right] \prod_{i=2}^{N} \left[\mathcal{N}(\mu, \sigma_i) + \mathcal{N}(\mu, \sigma^*) \right]. \tag{5.100}$$

This expression cannot be integrated analytically over μ, but we can easily treat it numerically since it is only two-dimensional. In figure 5.17, we assume a sample of 10 points, with 8 points drawn from $\mathcal{N}(0, 1)$ and 2 points (outliers) drawn from $\mathcal{N}(0, 3)$. We solve eq. 5.100 and plot the likelihood as a function of μ and g_i for two points: a "bad" point and a "good" point. The "bad" point has a maximum at $g_i = 0$: our model identifies it as an outlier. Furthermore, it is clear that as the weight g_i increases, this has a noticeable effect on the value of μ, skewing it toward positive values. The "good" point has a maximum at $g_i = 1$: our model correctly identifies it as coming from the distribution.

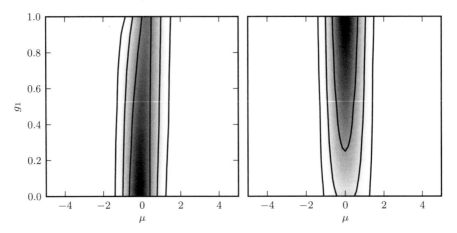

Figure 5.17. The marginal joint distribution between μ and g_i, as given by eq. 5.100. The left panel shows a point identified as bad ($\hat{g}_i = 0$), while the right panel shows a point identified as good ($\hat{g}_i = 1$).

We can also ask about the probability of a point being good or bad, independent of any other parameters. In this case, we would marginalize over μ in eq. 5.100,

$$p(g_1|\{x_i\}, \{\sigma_i\}) = \int p(\mu, g_1|\{x_i\}, \{\sigma_i\}) \, d\mu. \tag{5.101}$$

The result is linear in g_i for our uniform priors, and gives the posterior probability of the believability of each point.

We can take an approximate shortcut and evaluate $p(\mu, g_1|\{x_i\}, \{\sigma_i\})$ at the MAP estimate of μ, μ^0, implied by eq. 5.99,

$$p(g_1|\mu^0, \{x_i\}, \{\sigma_i\}) = 2 \frac{g_1 \mathcal{N}(\mu^0, \sigma_1) + (1 - g_1)\mathcal{N}(\mu^0, \sigma^*)}{\left[\mathcal{N}(\mu^0, \sigma_1) + \mathcal{N}(\mu^0, \sigma^*)\right]}, \tag{5.102}$$

where the term in the denominator comes from normalization. With $\sigma^*/\sigma_1 = 3$, it is easy to show that for a likely good point with $x_1 = \mu^0$, $p(g_1) = (1/2) + g_1$, and for a likely bad point a few σ_1 away from μ^0, $p(g_1) = 2(1 - g_1)$. That is, for good data points, the posterior for g_1 is skewed toward 1, and for bad data points it is skewed toward 0. The posterior for g_1 is flat if x_1 satisfies

$$|x_1 - \mu| = \sigma^* \left[\frac{2\ln(\sigma^*/\sigma_1)}{(\sigma^*/\sigma_1)^2 - 1}\right]^{1/2}. \tag{5.103}$$

Although insightful, these results are only approximate. In general, $p(g_1|\mu^0, \{x_i\}, \{\sigma_i\})$ and $p(g_1|\{x_i\}, \{\sigma_i\})$ are not equal, as illustrated in figure 5.18. Note that in the case of the bad point, this approximation leads to a higher likelihood of $g_i = 0$, because moving g_i toward 1 is no longer accompanied by a compensating shift in μ.

We conclude by noting that the data likelihood given by eq. 5.94 is very similar to that encountered in the context of the expectation maximization algorithm

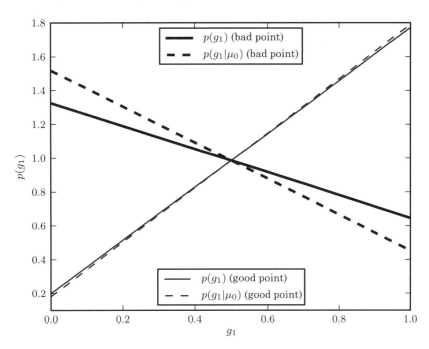

Figure 5.18. The marginal probability for g_i for the "good" and "bad" points shown in figure 5.17. The solid curves show the marginalized probability: that is, eq. 5.100 is integrated over μ. The dashed curves show the probability conditioned on $\mu = \mu_0$, the MAP estimate of μ (eq. 5.102).

(see §4.4.3 and eq. 4.18)—in both cases it is assumed that the data are drawn from a mixture of Gaussians with unknown class labels. If uniform priors are acceptable in a given problem, as we assumed above, then the ideas behind the EM algorithm could be used to efficiently find MAP estimates for both μ and all g_i, without the need to marginalize over other parameters.

5.7. Simple Examples of Bayesian Analysis: Model Selection

5.7.1. Gaussian or Lorentzian Likelihood?

Let us now revisit the examples discussed in §5.6.1 and 5.6.3. In the first example we *assumed* that the data $\{x_i\}$ were drawn from a Gaussian distribution and computed the two-dimensional posterior pdf for its parameters μ and σ. In the second example we did a similar computation, except that we *assumed* a Cauchy (Lorentzian) distribution and estimated the posterior pdf for its parameters μ and γ. What if we do not know what pdf our data were drawn from, and want to find out which of these two possibilities is better supported by our data?

We will assume that the data are identical to those used to compute the posterior pdf for the Cauchy distribution shown in figure 5.10 ($N = 10$, with $\mu = 0$ and $\gamma = 2$). We can integrate the product of the data likelihood and the prior pdf for the

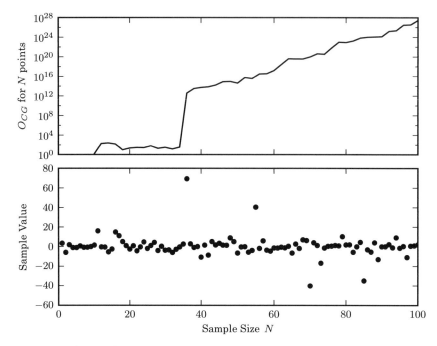

Figure 5.19. The Cauchy vs. Gaussian model odds ratio for a data set drawn from a Cauchy distribution ($\mu = 0$, $\gamma = 2$) as a function of the number of points used to perform the calculation. Note the sharp increase in the odds ratio when points falling far from the mean are added.

model parameters (see eqs. 5.74 and 5.75) to obtain model evidence (see eq. 5.23)

$$E(M = \text{Cauchy}) = \int p(\{x_i\}|\mu, \gamma, I)\, p(\mu, \gamma|I)\, d\mu\, d\gamma = 1.18 \times 10^{-12}. \quad (5.104)$$

When using the pdf illustrated in figure 5.10, we first compute exp(pixel value) for each pixel since the logarithm of the posterior is shown in the figure, then we multiply the result by the pixel area, and then sum all the values. In addition, we need to explicitly evaluate the constant of proportionality (see eq. 5.55). Since we assumed the same priors for both the Gaussian and the Cauchy case, they happen to be irrelevant in this example of a model comparison (but are nevertheless explicitly computed in the code accompanying figure 5.19).

We can construct the posterior pdf for the same data set using the Gaussian posterior pdf given by eq. 5.56 (and explicitly accounting for the proportionality constant) and obtain

$$E(M = \text{Gaussian}) = \int p(\{x_i\}|\mu, \sigma, I)\, p(\mu, \sigma|I)\, d\mu\, d\sigma = 8.09 \times 10^{-13}. \quad (5.105)$$

As no other information is available to prefer one model over the other one, we can assume that the ratio of model priors $p(M_C|I)/p(M_G|I) = 1$, and thus the odds ratio for the Cauchy vs. Gaussian model is the same as the Bayes factor,

$$O_{CG} = \frac{1.18 \times 10^{-12}}{9.09 \times 10^{-13}} = 1.45. \tag{5.106}$$

The odds ratio is very close to unity and is therefore inconclusive.

Why do we get an inconclusive odds ratio? Recall that this example used a sample of only 10 points; the probability of drawing at least one point far away from the mean, which would strongly argue against the Gaussian model, is fairly small. As the number of data values is increased, the ability to discriminate between the models will increase, too. Figure 5.19 shows the odds ratio for this problem as a function of the number of data points. As expected, when we increase the size of the observed sample, the odds ratio quickly favors the Cauchy over the Gaussian model.

Note the particularly striking feature that the addition of the 36th point causes the odds ratio to jump by many orders of magnitude: this point is extremely far from the mean, and thus is very unlikely under the assumption of a Gaussian model. The effect of this single point on the odds ratio illustrates another important caveat: the presence of even a single outlier may have a large effect on the computed likelihood, and as a result affect the conclusions. If your data has potential outliers, it is very important that these be accounted for within the distribution used for modeling the data likelihood (as was done in §5.6.7).

5.7.2. Understanding Knuth's Histograms

With the material covered in this chapter, we can now return to the discussion of histograms (see §4.8.1) and revisit them from the Bayesian perspective. We pointed out that Scott's rule and the Freedman–Diaconis rule for estimating optimal bin width produce the same answer for multimodal and unimodal distributions as long as their data set size and scale parameter are the same. This undesired result is avoided when using a method developed by Knuth [19]; an earlier discussion of essentially the same method is given in [13].

Knuth shows that the best piecewise constant model has the number of bins, M, which maximizes the following function (up to an additive constant, this is the logarithm of the posterior probability):

$$
\begin{aligned}
F(M|\{x_i\}, I)) = {}& N \log M + \log \left[\Gamma \left(\frac{M}{2} \right) \right] - M \log \left[\Gamma \left(\frac{1}{2} \right) \right] \\
& - \log \left[\Gamma \left(N + \frac{M}{2} \right) \right] + \sum_{k=1}^{M} \log \left[\Gamma \left(n_k + \frac{1}{2} \right) \right],
\end{aligned} \tag{5.107}
$$

where Γ is the gamma function, and n_k is the number of measurements x_i, $i = 1, \ldots, N$, which are found in bin k, $k = 1, \ldots, M$. Although this expression is more involved than the "rules of thumb" listed in §4.8.1, it can be easily evaluated for an *arbitrary* data set.

Knuth derived eq. 5.107 using Bayesian model selection and treating the histogram as a piecewise constant model of the underlying density function. By assumption, the bin width is constant and the number of bins is the result of model selection. Given the number of bins, M, the model for the underlying pdf is

$$h(x) = \sum_{k=1}^{M} h_k \Pi(x|x_{k-1}, x_k),$$ (5.108)

where the boxcar function $\Pi = 1$ if $x_{k-1} < x \le x_k$, and 0 otherwise. The M model parameters, $h_k, k = 1, \ldots, M$, are subject to normalization constraints, so that there are only $M-1$ free parameters. The uninformative prior distribution for $\{h_k\}$ is given by

$$p(\{h_k\}|M, I) = \frac{\Gamma(\frac{M}{2})}{\Gamma(\frac{1}{2})^M} \left[h_1 h_2 \ldots h_{M-1} \left(1 - \sum_{k=1}^{M-1} h_k \right) \right]^{-1/2},$$ (5.109)

which is known as the Jeffreys prior for the multinomial likelihood. The joint data likelihood is a multinomial distribution (see §3.3.3)

$$p(\{x_i\}|\{h_k\}, M, I) \propto h_1^{n_1} h_2^{n_2} \ldots h_M^{n_M}.$$ (5.110)

The posterior pdf for model parameters h_k is obtained by multiplying the prior and data likelihood. The posterior probability for the number of bins M is obtained by marginalizing the posterior pdf over all h_k. This marginalization includes a series of nested integrals over the $(M - 1)$-dimensional parameter space, and yields eq. 5.107; details can be found in Knuth's paper.

Knuth also derived the posterior pdf for h_k, and summarized it by deriving its expectation value and variance. The expectation value is

$$h_k = \frac{n_k + \frac{1}{2}}{N + \frac{M}{2}},$$ (5.111)

which is an interesting result (the naive expectation is $h_k = n_k/N$): even when there are no counts in a given bin, $n_k = 0$, we still get a nonvanishing estimate $h_k = 1/(2N + M)$. The reason is that the assumed prior distribution effectively places one half of a datum in each bin.

Comparison of different rules for optimal histogram bin width

The number of bins in Knuth's expression (eq. 5.107) is defined over the observed data range (i.e., the difference between the maximum and minimum value). Since the observed range generally increases with the sample size, it is not obvious how the optimal bin width varies with it. The variation depends on the actual underlying distribution from which data are drawn, and for a Gaussian distribution numerical simulations with N up to 10^6 show that

$$\Delta_b = \frac{2.7\sigma_G}{N^{1/4}}.$$ (5.112)

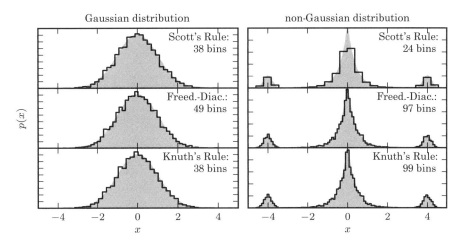

Figure 5.20. The results of Scott's rule, the Freedman–Diaconis rule, and Knuth's rule for selecting the optimal bin width for a histogram. These histograms are based on 5000 points drawn from the shown pdfs. On the left is a simple normal distribution. On the right is a Laplacian distribution at the center, with two small Gaussian peaks added in the wings.

We have deliberately replaced σ by σ_G (see eq. 3.36) to emphasize that the result is applicable to non-Gaussian distributions if they do not show complex structure, such as multiple modes, or extended tails. Of course, for a multimodal distribution the optimal bin width is smaller than given by eq. 5.112 (so that it can "resolve" the substructure in $f(x)$), and can be evaluated using eq. 5.107. Compared to the Freedman–Diaconis rule, the "rule" given by eq. 5.112 has a slower decrease of Δ_b with N; for example, for $N = 10^6$ the Freedman–Diaconis Δ_b is 3 times smaller than that given by eq. 5.112.

Despite the attractive simplicity of eq. 5.112, to utilize the full power of Knuth's method, eq. 5.107 should be used, as done in the following example. Figure 5.20 compares the optimal histogram bins for two different distributions, as selected by Scott's rule, the Freedman–Diaconis rule, and Knuth's method. For the non-Gaussian distribution, Scott's rule greatly underestimates the optimal number of histogram bins, resulting in a histogram that does not give as much intuition as to the shape of the underlying distribution.

The usefulness of Knuth's analysis and the result summarized by eq. 5.107 goes beyond finding the optimal bin size. The method is capable of recognizing substructure in data and, for example, it results in $M = 1$ when the data are consistent with a uniform distribution, and suggests more bins for a multimodal distribution than for a unimodal distribution even when both samples have the same size and σ_G (again, eq. 5.112 is an approximation valid only for unimodal centrally concentrated distributions; if in doubt, use eq. 5.107; for the latter, see the Python code used to generate figure 5.20).

Lastly, remember that Knuth's derivation assumed that the uncertainty of each x_i is negligible. When this is not the case, including the case of heteroscedastic errors, techniques introduced in this chapter can be used for general model selection, including the case of a piecewise constant model, as well as varying bin size.

Bayesian blocks

Though Knuth's Bayesian method is an improvement over the rules of thumb from §4.8.1, it still has a distinct weakness: it assumes a uniform width for the optimal histogram bins. The Bayesian model used to derive Knuth's rule suggests that this limitation could be lifted, by maximizing a well-designed likelihood function over bins of varying width. This approach has been explored in [30, 31], and dubbed *Bayesian blocks*. The method was first developed in the field of time-domain analysis (see §10.3.5), but is readily applicable to histogram data as well; the same ideas are also discussed in [13].

In the Bayesian blocks formalism, the data are segmented into *blocks*, with the borders between two blocks being set by *changepoints*. Using a Bayesian analysis based on Poissonian statistics within each block, an objective function, called the log-likelihood fitness function, can be defined for each block:

$$F(N_i, T_i) = N_i(\log N_i - \log T_i), \tag{5.113}$$

where N_i is the number of points in block i, and T_i is the width of block i (or the duration, in time-series analysis). Because of the additive nature of log-likelihoods, the fitness function for any set of blocks is simply the sum of the fitness functions for each individual block. This feature allows for the configuration space to be explored quickly using dynamic programming concepts: for more information see [31] or the Bayesian blocks implementation in AstroML.

In figure 5.21, we compare a Bayesian blocks segmentation of a data set to a segmentation using Knuth's rule. The adaptive bin width of the Bayesian blocks histogram leads to a better representation of the underlying data, especially when there are fewer points in the data set. An important feature of this method is that the bins are optimal in a quantitative sense, meaning that statistical significance can be attached to the bin configuration. This has led to applications in the field of time-domain astronomy, especially in signal detection.

Finally, we should mention that the fitness function in eq. 5.113 is just one of many possible fitness functions that can be used in the Bayesian blocks method. For more information, see [31] and references therein.

AstroML includes tools for easy computation of the optimal bins derived using Bayesian blocks. The interface is similar to that described in §4.8.1:

```
In [1]: %pylab
In [2]: from astroML.plotting import hist
In [3]: x = np.random.normal(size=1000)
In [4]: hist(x, bins='blocks')  # can also choose
        # bins='knuth'
```

This will internally call the `bayesian_blocks` function in the `astroML.density_estimation` module, and display the resulting histogram. The `hist` function in AstroML operates analogously to the `hist` function in Matplotlib, but can optionally use Bayesian blocks or Knuth's method to choose the binning. For more details see the source code associated with figure 5.21.

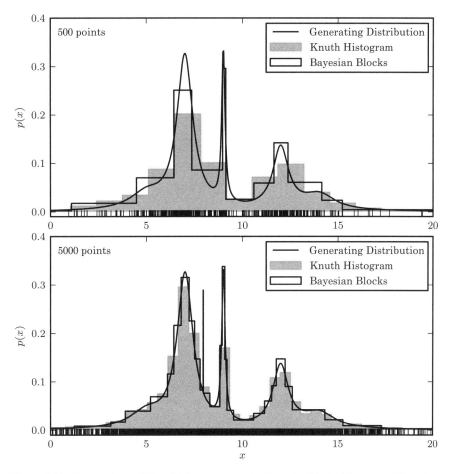

Figure 5.21. Comparison of Knuth's histogram and a Bayesian blocks histogram. The adaptive bin widths of the Bayesian blocks histogram yield a better representation of the underlying data, especially with fewer points.

5.7.3. One Gaussian or Two Gaussians?

In analogy with the example discussed in §5.7.1, we can ask whether our data were drawn from a Gaussian distribution, or from a distribution that can be described as the sum of two Gaussian distributions. In this case, the number of parameters for the two competing models is different: two for a single Gaussian, and five for the sum of two Gaussians. This five-dimensional pdf is hard to treat analytically, and we need to resort to numerical techniques as described in the next section. After introducing these techniques, we will return to this model comparison problem (see §5.8.4).

5.8. Numerical Methods for Complex Problems (MCMC)

When the number of parameters, k, in a model, $M(\theta)$, with the vector of parameters θ specified by θ_p, $p = 1, \ldots, k$, is large, direct exploration of the posterior pdf by exhaustive search becomes impractical, and often impossible. For example, if the grid for computing the posterior pdf, such as those illustrated in figures 5.4 and 5.10, includes only 100 points per coordinate, the five-dimensional model from the previous example (§5.7.3) will require on order 10^{10} computations of the posterior pdf. Fortunately, a number of numerical methods exist that utilize more efficient approaches than an exhaustive grid search.

Let us assume that we know how to compute the posterior pdf (we suppress the vector notation for $\boldsymbol{\theta}$ for notational clarity since in the rest of this section we always discuss multidimensional cases)

$$p(\theta) \equiv p(M(\boldsymbol{\theta})|D, I) \propto p(D|M(\boldsymbol{\theta}), I)p(\boldsymbol{\theta}|I). \tag{5.114}$$

In general, we wish to evaluate the multidimensional integral

$$I(\theta) = \int g(\theta)p(\theta)\, d\theta. \tag{5.115}$$

There are two classes of frequently encountered problems:

1. *Marginalization and parameter estimation,* where we seek the posterior pdf for parameters $\theta_i, i = 1, \ldots, P$, and the integral is performed over the space spanned by nuisance parameters $\theta_j, j = (P + 1), \ldots, k$ (for notational simplicity we assume that the last $k - P$ parameters are nuisance parameters). In this case, $g(\theta) = 1$. As a special case, we can seek the posterior mean (see eq. 5.7) for parameter θ_m, where $g(\theta) = \theta_m$, and the integral is performed over all other parameters. Analogously, we can also compute the credible region, defined as the interval that encloses $1 - \alpha$ of the posterior probability. In all of these computations, it is sufficient to evaluate the integral in eq. 5.115 up to an unknown normalization constant because the posterior pdf can be renormalized to integrate to unity.
2. *Model comparison,* where $g(\theta) = 1$ and the integral is performed over all parameters (see eq. 5.23). Unlike the first class of problems, here the proper normalization is mandatory.

One of the simplest numerical integration methods is generic Monte Carlo. We generate a random set of M values $\theta, \theta_j, j = 1, \ldots, M$, uniformly sampled within the integration volume V_θ, and estimate the integral from eq. 5.115 as

$$I \approx \frac{V_\theta}{M} \sum_{j=1}^{M} g(\theta_j)\, p(\theta_j). \tag{5.116}$$

This method is very inefficient when the integrated function greatly varies within the integration volume, as is the case for the posterior pdf. This problem is especially acute with high-dimensional integrals.

A number of methods exist that are much more efficient than generic Monte Carlo integration. The most popular group of techniques is known as Markov chain Monte Carlo (MCMC) methods. They return a sample of points, or chain, from the k-dimensional parameter space, with a distribution that is asymptotically proportional to $p(\theta)$. The constant of proportionality is not important in the first class of problems listed above. In model comparison problems, the proportionality constant from eq. 5.117 must be known; we return to this point in §5.8.4.

Given such a chain of length M, the integral from eq. 5.115 can be estimated as

$$I = \frac{1}{M} \sum_{j=1}^{M} g(\theta_j). \tag{5.117}$$

As a simple example, to estimate the expectation value for θ_1 (i.e., $g(\theta) = \theta_1$), we simply take the mean value of all θ_1 in the chain.

Given a Markov chain, quantitative description of the posterior pdf becomes a density estimation problem (density estimation methods are discussed in Chapter 6). To visualize the posterior pdf for parameter θ_1, marginalized over all other parameters, $\theta_2, \ldots, \theta_k$, we can construct a histogram of all θ_1 values in the chain, and normalize its integral to 1. To get a MAP estimate for θ_1, we find the maximum of this marginalized pdf. A generalization of this approach to multidimensional projections of the parameter space is illustrated in figure 5.22.

5.8.1. Markov Chain Monte Carlo

A Markov chain is a sequence of random variables where a given value nontrivially depends *only on its preceding value*. That is, given the *present* value, past and future values are independent. In this sense, a Markov chain is "memoryless." The process generating such a chain is called the Markov process and can be described as

$$p(\theta_{i+1}|\{\theta_i\}) = p(\theta_{i+1}|\theta_i), \tag{5.118}$$

that is, the next value depends only on the current value.

In our context, θ can be thought of as a vector in multidimensional space, and a realization of the chain represents a path through this space. To reach an equilibrium, or stationary, distribution of positions, it is necessary that the transition probability is symmetric:

$$p(\theta_{i+1}|\theta_i) = p(\theta_i|\theta_{i+1}). \tag{5.119}$$

This condition is called the detailed balance or reversibility condition. It shows that the probability of a jump between two points does not depend on the direction of the jump.

There are various algorithms for producing Markov chains that reach some prescribed equilibrium distribution, $p(\theta)$. The use of resulting chains to perform Monte Carlo integration of eq. 5.115 is called *Markov chain Monte Carlo* (MCMC).

5.8.2. MCMC Algorithms

Algorithms for generating Markov chains are numerous and greatly vary in complexity and applicability. Many of the most important ideas were generated in physics, especially in the context of statistical mechanics, thermodynamics, and quantum field theory [23]. We will only discuss in detail the most famous Metropolis–Hastings algorithm, and refer the reader to Greg05 and BayesCosmo, and references therein, for a detailed discussion of other algorithms.

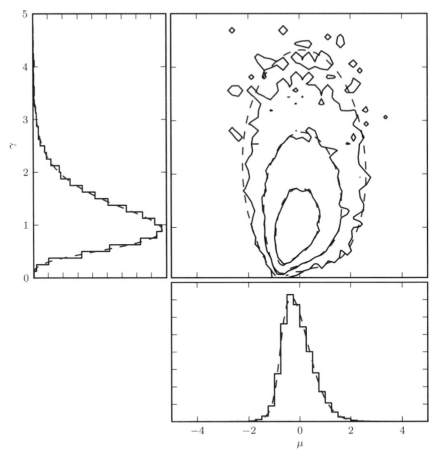

Figure 5.22. Markov chain Monte Carlo (MCMC) estimates of the posterior pdf for parameters describing the Cauchy distribution. The data are the same as those used in figure 5.10: the dashed curves in the top-right panel show the results of direct computation on a regular grid from that diagram. The solid curves are the corresponding MCMC estimates using 10,000 sample points. The left and the bottom panels show marginalized distributions.

In order for a Markov chain to reach a stationary distribution proportional to $p(\theta)$, the probability of arriving at a point θ_{i+1} must be proportional to $p(\theta_{i+1})$,

$$p(\theta_{i+1}) = \int T(\theta_{i+1}|\theta_i)\, p(\theta_i)\, d\theta_i, \qquad (5.120)$$

where the transition probability $T(\theta_{i+1}|\theta_i)$ is called the jump kernel or transition kernel (and it is assumed that we know how to compute $p(\theta_i)$). This requirement will be satisfied when the transition probability satisfies the detailed balance condition

$$T(\theta_{i+1}|\theta_i)\, p(\theta_i) = T(\theta_i|\theta_{i+1})\, p(\theta_{i+1}). \qquad (5.121)$$

Various MCMC algorithms differ in their choice of transition kernel (see Greg05 for a detailed discussion).

The Metropolis–Hastings algorithm adopts the kernel

$$T(\theta_{i+1}|\theta_i) = p_{\mathrm{acc}}(\theta_i, \theta_{i+1})\, K(\theta_{i+1}|\theta_i), \qquad (5.122)$$

where the proposed density distribution $K(\theta_{i+1}|\theta_i,)$ is an *arbitrary* function. The proposed point θ_{i+1} is randomly accepted with the acceptance probability

$$p_{\mathrm{acc}}(\theta_i, \theta_{i+1}) = \frac{K(\theta_i|\theta_{i+1})\, p(\theta_{i+1})}{K(\theta_{i+1}|\theta_i)\, p(\theta_i)} \qquad (5.123)$$

(when exceeding 1, the proposed point θ_{i+1} is always accepted). When θ_{i+1} is rejected, θ_i is added to the chain instead. A Gaussian distribution centered on θ_i is often used for $K(\theta_{i+1}|\theta_i)$.

The original Metropolis algorithm is based on a symmetric proposal distribution, $K(\theta_{i+1}|\theta_i) = K(\theta_i|\theta_{i+1})$, which then cancels out from the acceptance probability. In this case, θ_{i+1} is always accepted if $p(\theta_{i+1}) > p(\theta_i)$, and if not, then it is accepted with a probability $p(\theta_{i+1})/p(\theta_i)$.

Although $K(\theta_{i+1}|\theta_i)$ satisfies a Markov chain requirement that it must be a function of only the current position θ_i, it takes a number of steps to reach a stationary distribution from an initial arbitrary position θ_0. These early steps are called the "burn-in" and need to be discarded in analysis. There is no general theory for finding transition from the burn-in phase to the stationary phase; several methods are used in practice. Gelman and Rubin proposed to generate a number of chains and then compare the ratio of the variance between the chains to the mean variance within the chains (this ratio is known as the R statistic). For stationary chains, this ratio will be close to 1. The autocorrelation function (see §10.5) for the chain can be used to determine the required number of evaluations of the posterior pdf to get estimates of posterior quantities with the desired precision; for a detailed practical discussion see [7]. The autocorrelation function can also be used to estimate the increase in Monte Carlo integration error due to the fact that the sequence is correlated (see eq. 10.93).

When the posterior pdf is multimodal, the simple Metropolis–Hastings algorithm can become stuck in a local mode and not find the globally best mode within a reasonable running time. There are a number of better algorithms, such as Gibbs sampling, parallel tempering, various genetic algorithms, and nested sampling. For a good overview, see [3].

5.8.3. PyMC: MCMC in Python

For the MCMC examples in this book, we use the Python package PyMC.[8] PyMC comprises a set of flexible tools for performing MCMC using the Metropolis–Hastings algorithm, as well as maximum a priori estimates, normal approximations, and other sampling techniques. It includes built-in models for common distributions and priors (e.g. Gaussian distribution, Cauchy distribution, etc.) as well as an easy framework to define arbitrarily complicated distributions. For examples of the use of PyMC in practice, see the code accompanying MCMC figures throughout this text.

---

[8]https://github.com/pymc-devs/pymc

While PyMC offers some powerful tools for fine-tuning of MCMC chains, such as varying step methods, fitting algorithms, and convergence diagnostics, for simplicity we use only the basic features for the examples in this book. In particular, the burn-in for each chain is accomplished by simply setting the burn-in size high enough that we can assume the chain has become stationary. For more rigorous approaches to this, as well as details on the wealth of diagnostic tools available, refer to the PyMC documentation.

A simple fit with PyMC can be accomplished as follows. Here we will fit the mean of a distribution—perhaps an overly simplistic example for MCMC, but useful as an introductory example:

```python
import numpy as np
import pymc

# generate random Gaussian data with mu=0, sigma=1
N = 100
x = np.random.normal(size=N)

# define the MCMC model: uniform prior on mu,
  # fixed (known) sigma
mu = pymc.Uniform('mu', -5, 5)
sigma = 1
M = pymc.Normal('M', mu, sigma, observed=True,
                value=x)
model = dict(M=M, mu=mu)

# run the model, and get the trace of mu
S = pymc.MCMC(model)
S.sample(10000, burn=1000)
mu_sample = S.trace('mu')[:]

# print the MCMC estimate
print("Bayesian (MCMC): %.3f +/- %.3f"
      % (np.mean(mu_sample), np.std(mu_sample)))

# compare to the frequentist estimate
print("Frequentist: %.3f +/- %.3f"
      % (np.mean(x), np.std(x, ddof=1) / np.sqrt(N)))
```

The resulting output for one particular random seed:

```
Bayesian (MCMC): -0.054 +/- 0.103
Frequentist: -0.050 +/- 0.096
```

As expected for a uniform prior on μ, the Bayesian and frequentist estimates (via eqs. 3.31 and 3.34) are consistent. For examples of higher-dimensional MCMC problems, see the online source code associated with the MCMC figures throughout the text.

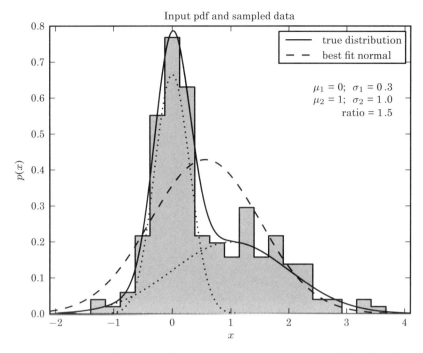

Figure 5.23. A sample of 200 points drawn from a Gaussian mixture model used to illustrate model selection with MCMC.

PyMC is far from the only option for MCMC computation in Python. One other tool that deserves mention is emcee,[9] a package developed by astronomers, which implements a variant of MCMC where the sampling is invariant to affine transforms (see [7, 11]). Affine-invariant MCMC is a powerful algorithm and offers improved runtimes for some common classes of problems.

5.8.4. Example: Model Selection with MCMC

Here we return to the problem of model selection from a Bayesian perspective. We have previously mentioned the odds ratio (§5.4), which takes into account the entire posterior distribution, and the Aikake and Bayesian information criteria (AIC and BIC—see §5.4.3), which are based on normality assumptions of the posterior.

Here we will examine an example of distinguishing between unimodal and bimodal models of a distribution in a Bayesian framework. Consider the data sample shown in figure 5.23. The sample is drawn from a bimodal distribution: the sum of two Gaussians, with the parameter values indicated in the figure. The best-fit normal distribution is shown as a dashed line. The question is, can we use a Bayesian framework to determine whether a single-peak or double-peak Gaussian is a better fit to the data?

A double Gaussian model is a five-parameter model: the first four parameters include the mean and width for each distribution, and the fifth parameter is the

---

[9]Cleverly dubbed "MCMC Hammer," http://danfm.ca/emcee/

TABLE 5.2.
Comparison of the odds ratios for a single and double Gaussian model using maximum a posteriori log-likelihood, AIC, and BIC.

	M1: single Gaussian	M2: double Gaussian	M1 − M2
$-2 \ln L^0$	465.4	406.0	59.4
BIC	476.0	432.4	43.6
AIC	469.4	415.9	53.5

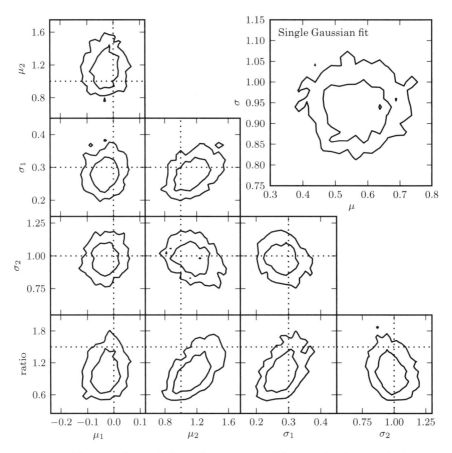

Figure 5.24. The top-right panel shows the posterior pdf for μ and σ for a single Gaussian fit to the data shown in figure 5.23. The remaining panels show the projections of the five-dimensional pdf for a Gaussian mixture model with two components. Contours are based on a 10,000 point MCMC chain.

relative normalization (weight) of the two components. Computing the AIC and BIC for the two models is relatively straightforward: the results are given in table 5.2, along with the maximum a posteriori log-likelihood $\ln L^0$ (the code for maximization of the likelihood and computation of the AIC/BIC can be found in the source of figure 5.24).

It is clear that by all three measures, the double Gaussian model is preferred. But these measures are only accurate if the posterior distribution is approximately Gaussian. For non-Gaussian posteriors, the best statistic to use is the odds ratio (§5.4). While odds ratios involving two-dimensional posteriors can be computed relatively easily (see §5.7.1), integrating five-dimensional posteriors is computationally difficult. This is one manifestation of the curse of dimensionality (see §7.1). So how do we proceed? One way to estimate an odds ratio is based on MCMC sampling.

Computing the odds ratio involves integrating the unnormalized posterior for a model (see §5.7.1):

$$L(M) = \int p(\boldsymbol{\theta}|\{x_i\}, I)\, d^k\theta, \tag{5.124}$$

where the integration is over all k model parameters. How can we compute this based on an MCMC sample? Recall that the set of points derived by MCMC is designed to be distributed according to the posterior distribution $p(\boldsymbol{\theta}|\{x_i\}, I)$, which we abbreviate to simply $p(\boldsymbol{\theta})$. This means that the local density of points $\rho(\boldsymbol{\theta})$ is proportional to this posterior distribution: for a well-behaved MCMC chain with N points,

$$\rho(\boldsymbol{\theta}) = C\, N\, p(\boldsymbol{\theta}), \tag{5.125}$$

where C is an unknown constant of proportionality. Integrating both sides of this equation and using $\int \rho(\boldsymbol{\theta})\, d^k\theta = N$, we find

$$L(M) = 1/C. \tag{5.126}$$

This means that at each point $\boldsymbol{\theta}$ in parameter space, we can estimate the integrated posterior using

$$L(M) = \frac{Np(\boldsymbol{\theta})}{\rho(\boldsymbol{\theta})}. \tag{5.127}$$

We see that the result can be computed from quantities that can be estimated from the MCMC chain: $p(\boldsymbol{\theta}_i)$ is the posterior evaluation at each point, and the local density $\rho(\boldsymbol{\theta}_i)$ can be estimated from the local distribution of points in the chain. The odds ratio problem has now been expressed as a density estimation problem, which can be approached in a variety of ways; see [3, 12]. Several relevant tools and techniques can be found in chapter 6. Because we can estimate the density at the location of each of the N points in the MCMC chain, we have N separate estimators of $L(M)$.

Using this approach, we can evaluate the odds ratio for model 1 (a single Gaussian: 2 parameters) vs. model 2 (two Gaussians: 5 parameters) for our example data set. Figure 5.24 shows the MCMC-derived likelihood contours (using 10,000 points) for each parameter in the two models. For model 1, the contours appear to be nearly Gaussian. For model 2, they are further from Gaussian, so the AIC and BIC values become suspect.

Using the density estimation procedure above,[10] we compute the odds ratio $O_{21} \equiv L(M_2)/L(M_1)$ and find that $O_{21} \approx 10^{11}$, strongly in favor of the

---

[10]We use a *kernel density estimator* here, with a top-hat kernel for computational simplicity; see §6.1.1 for details.

two-peak solution. For comparison, the implied difference in BIC is $-2\ln(O_{21}) =$ 50.7, compared to the approximate value of 43.6 from table 5.2. The Python code that implements this estimation can be found in the source of figure 5.24.

5.8.5. Example: Gaussian Distribution with Unknown Gaussian Errors

In §5.6.1, we explored several methods to estimate parameters for a Gaussian distribution from data with heteroscedastic errors e_i. Here we take this to the extreme, and allow each of the errors e_i to vary as part of the model. Thus our model has $N+2$ parameters: the mean μ, the width σ, and the data errors $e_i, i = 1, \ldots, N$. To be explicit, our model here (cf. eq. 5.63) is given by

$$p(\{x_i\}|\mu, \sigma, \{e_i\}, I) = \prod_{i=1}^{N} \frac{1}{\sqrt{2\pi}(\sigma^2 + e_i^2)^{1/2}} \exp\left(\frac{-(x_i - \mu)^2}{2(\sigma^2 + e_i^2)}\right). \qquad (5.128)$$

Though this pdf cannot be maximized analytically, it is relatively straightforward to compute via MCMC, by setting appropriate priors and marginalizing over the e_i as nuisance parameters. Because the e_i are scale factors like σ, we give them scale-invariant priors.

There is one interesting detail about this choice. Note that because σ and e_i appear together as a sum, the likelihood in eq. 5.128 has a distinct degeneracy. For any point in the model space, an identical likelihood can be found by scaling $\sigma^2 \rightarrow \sigma^2 + K, e_i^2 \rightarrow e_i^2 - K$ for all i (subject to positivity constraints on each term). Moreover, this degeneracy exists at the maximum just as it does elsewhere. Because of this, using priors of different forms on σ and e_i can lead to suboptimal results. If we chose, for example, a scale-invariant prior on σ and a flat prior on e_i, then our posterior would strongly favor $\sigma \rightarrow 0$, with the e_i absorbing its effect. This highlights the importance of carefully choosing priors on model parameters, even when those priors are flat or uninformative!

The result of an MCMC analysis on all $N+2$ parameters, marginalized over e_i, is shown in figure 5.25. For comparison, we also show the contours from figure 5.7. The input distribution is within 1σ of the most likely marginalized result, and this is with *no prior knowledge* about the error in each point!

5.8.6. Example: Unknown Signal with an Unknown Background

In §5.6.5 we explored Bayesian parameter estimation for the width of a Gaussian in the presence of a uniform background. Here we consider a more general model and find the width σ and location μ of a Gaussian signal within a uniform background. The likelihood is given by eq. 5.83, where σ, μ, and A are unknown. The results are shown in figure 5.26. The procedure for fitting this, which can be seen in the online source code for figure 5.26, is very general. If the signal shape were not Gaussian, it would be easy to modify this procedure to include another model. We could also evaluate a range of possible signal shapes and compare the models using the model odds ratio, as we did above.

Note that here the data are unbinned; if the data were binned (i.e., if we were trying to fit the number of counts in a data histogram), then this would be very similar to the matched filter analysis discussed in §10.4.

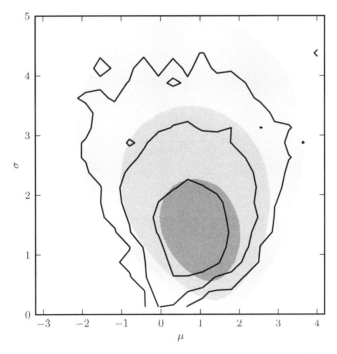

Figure 5.25. The posterior pdf for μ and σ for a Gaussian distribution with heteroscedastic errors. This is the same data set as used in figure 5.7, but here each measurement error is assumed unknown, treated as a model parameter with a scale-invariant prior, and marginalized over to obtain the distribution of μ and σ shown by contours. For comparison, the posterior pdf from figure 5.7 is shown by shaded contours.

5.9. Summary of Pros and Cons for Classical and Bayesian Methods

With the material covered in the last two chapters, we can now compare and contrast the frequentist and Bayesian approaches. It is possible to live completely in either paradigm as a data analyst, performing all of the needed types of analysis tasks that come up in real problems in some manner. So, what should an aspiring young—or not so young—scientist do?

Technical differences We will first discuss differences of a technical nature, then turn to subjective issues. Volumes of highly opinionated text have been written arguing for each of the two sides over many decades (whose main points we can only summarize), with no clear victory yet as of the writing of this book. Given this, the eagerly bellicose partisan looking to eliminate the other side should not be surprised to find that, upon deeper examination, the issues are complex and subtle, and any win is mixed and partial at best. A brief visit to some common battlegrounds, in the form of various basic inference tasks, reveals this:

- **Point estimates**: We can quantify the comparison through an appeal to decision theory, which is the study, using asymptotic theory, of the relative quality of estimators as the amount of data increases. What do we find when

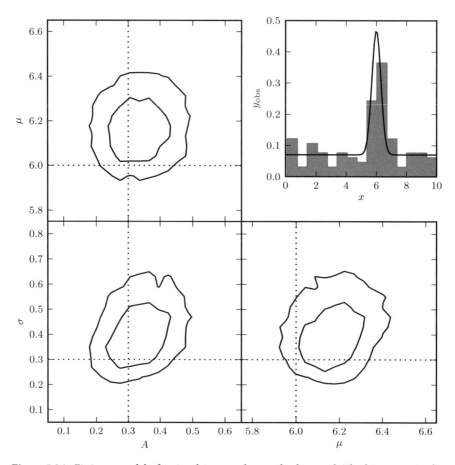

Figure 5.26. Fitting a model of a signal in an unknown background. The histogram in the top-right panel visualizes a sample drawn from a Gaussian signal plus a uniform background model given by eq. 5.83 and shown by the line. The remaining panels show projections of the three-dimensional posterior pdf, based on a 20,000 point MCMC chain.

we use its tools to examine, say, MLE[11] and Bayes estimators? A minimax estimator minimizes the maximum risk, and is considered one notion of a "gold-standard" desirable estimator. It can be shown for typical parametric models under weak conditions that the maximum likelihood estimator is approximately minimax, that the Bayes estimator with constant risk is also minimax, and that the MLE is approximately the Bayes estimator. Thus with enough samples each paradigm provides a good estimate, and furthermore the results are typically not very different. This is just an example of the results in the literature surrounding such analyses, which include analysis of Bayesian esimators under frequentist criteria and vice versa—see [20, 35] for some entryways into it. In the small-sample case, the prior has a larger effect, differentiating the frequentist and Bayes estimators. Whether this is good or

---

[11]We will sometimes use maximum likelihood as a representative of frequentist estimation, as it is the most studied, though MLE is only one option among many within frequentism.

bad leads directly to one's subjective assessment of the reasonableness of using priors, which we will come to below.

• **Uncertainty estimates**: Both MLE and Bayes estimators yield analytic confidence bands based on the asymptotic convergence of distributions to a normal distribution, for typical parametric models. For general complex models typical of machine learning, each paradigm has a way to obtain uncertainty estimates around point predictions. In Bayesian estimation, this comes in the form of posterior intervals, obtained via MCMC, and in frequentist estimation, confidence sets via resampling techniques such as the jackknife and bootstrap. Both arguably work under fairly generic circumstances, are easy to implement, and are computationally expensive. Achieving the exact guarantees of the bootstrap can require care to avoid certain model settings [6]; this is counterbalanced by the comparative difficulty in ensuring reliable results from MCMC [10], especially models with complicated likelihood landscapes. Thus in practice, neither procedure can truly be advertised as a black box that can be used in a hands-off manner.

• **Hypothesis testing**: Hypothesis testing is an area where the frequentist and Bayesian approaches differ substantially. Bayesian hypothesis testing is done by either placing a prior on the null and alternative hypotheses and computing the Bayes factor, or placing an overarching prior on both hypotheses and calculating the posterior probability of the null. Both of these rely heavily on one's confidence in the prior. In Bayesian hypothesis testing, unlike in point estimation, the influence of the prior is not attenuated with an increasing number of samples: this forces the subjective issue of the analyst's comfort level with relying on priors.

• **Goodness-of-fit testing**: Goodness-of-fit tests can be framed as hypothesis tests with no alternative hypothesis, which either makes them ill posed or useful, depending on one's point of view. Either way, they must be treated with care—if we reject the hypothesis we can conclude that we should not use the model; however if we do not reject it, we cannot conclude that the model is correct. Unlike in frequentism, in the Bayesian formulation this feature is made clear through the requirement of an alternative hypothesis. Nevertheless, the distinction is muddled by work straddling the boundaries, such as the ability to interpret frequentist goodness-of-fit tests in Bayesian terms (see [29]), the existence of Bayesian goodness-of-fit tests, and the ability to formalize goodness-of-fit tests as positing a parametric null against a nonparametric alternative; see [16, 36].

• **Nuisance parameters**: Nuisance parameters are common: we may not be interested in their values, but their values modify the distribution of our observations, and must thus be accounted for. In some cases within both frameworks, they can be eliminated analytically, and this should always be done if possible. In the Bayesian framework we can work with the joint posterior distribution using random samples drawn from it via MCMC. We can obtain the joint distribution of only the parameters of interest by simply marginalizing out the nuisance parameters. This is possible because of the availability of priors for each parameter. A frequentist approach can utilize likelihood ratio tests (see [2, 28]) to provide confidence intervals and significance tests for the parameters of interest, which account for the presence of nuisance parameters. The Bayesian approach is more general and straightforward. On the other hand the approach relies on the quality of MCMC samples, which can be uncertain in complex models.

Subjective differences The tiebreaker, then, is to a large degree a matter of taste. These are the most common arguments for Bayesianism, over frequentism:

- **Symmetry and grand unification**: Bayesianism allows you to make probability statements about parameters. It takes the likelihood function to its logical conclusion, removes the distinction between data and parameters, thus creating a beautiful symmetry that warms the heart of many a natural scientist.
- **Extra information**: The Bayesian method gives a convenient and natural way to put in prior information. It is particularly sensible for small-sample situations where the data might be insufficient to make useful conclusions on their own. Another natural and important use arises when wanting to model the measurement error for each data point.
- **Honesty/disclosure**: The fact that we are *forced* to put in subjective priors is actually good: we always have prior beliefs, and this forces us to make them explicit. There is no such thing as an objective statistical inference procedure.
- **More elegant in practice**: Just by generating samples from the posterior, one can obtain the key results of the Bayesian method (see §5.8).

Indeed, many of the examples in this chapter illustrate the wonderful elegance possible with Bayesian methods.

These are the most common arguments against Bayesianism, for frequentism:

- **Are we really being scientific?** The posterior interval is not a true confidence interval in the sense of long-run frequencies: we cannot make statements about the true parameter with these intervals, only about our beliefs. Bayesian estimates are based on information beyond what is verifiable. Are we really going to trust, say, a major scientific conclusion based on anything but the observed data?
- **The effect of the prior is always there**: The Bayesian estimate is always biased due to the prior, even when no actual prior information (i.e., the least informative prior) is used. Bayesian hypothesis testing, in particular, is very sensitive to the choice of prior. While the effect of the prior goes away asymptotically, for any finite sample size it is still there in some unquantified way.
- **Unnecessarily complicated and computationally intensive**: The Bayesian approach requires specifying prior functions even when there is no true prior information. In practice, it often requires computationally intractable or unreliable integrals and approximations, even for models that are computationally relatively simple in the frequentist case.
- **Unnecessarily brittle and limiting**: The Bayesian method is crucially dependent on the likelihood function. As a cost function, the likelihood, while enjoying nice optimality properties when the assumed model is correct, is often highly brittle when the assumptions are wrong or there are outliers in the data. Similarly, the likelihood is often not a good choice in a nonparametric setting.

Due to the various pragmatic obstacles, it is rare for a mission-critical analysis to be done in the "fully Bayesian" manner, i.e., without the use of tried-and-true frequentist tools at the various stages. Philosophy and beauty aside, the reliability and efficiency of the underlying computations required by the Bayesian framework are the main practical issues. A central technical issue at the heart of this is that it is much easier to do optimization (reliably and efficiently) in high dimensions than it is to do integration in high dimensions. Thus the workhorse machine learning methods, while there are ongoing efforts to adapt them to Bayesian framework, are almost all rooted in frequentist methods. A work-around is to perform MAP inference, which is optimization based.

Most users of Bayesian estimation methods, in practice, are likely to use a mix of Bayesian and frequentist tools. The reverse is also true—frequentist data analysts, even if they stay formally within the frequentist framework, are often influenced by "Bayesian thinking," referring to "priors" and "posteriors." The most advisable position is probably to know both paradigms well, in order to make informed judgments about which tools to apply in which situations. Indeed, the remaining chapters in this book discuss both classical and Bayesian analysis methods. Examples of arguments for each of the respective approaches are [4] and [22], and a more recent attempt at synthesis can be found in [5]. For more about the relative merits of the frequentist and Bayesian frameworks, please see Greg05 and Wass10.

References

[1] Casella, G. (1985). An introduction to empirical Bayes data analysis. *The American Statistician 39*(2), 83–87.

[2] Cash, W. (1979). Parameter estimation in astronomy through application of the likelihood ratio. *ApJ 228*, 939–947.

[3] Clyde, M. A., J. O. Berger, F. Bullard, and others (2007). Current challenges in Bayesian model choice. In G. J. Babu and E. D. Feigelson (Eds.), *Statistical Challenges in Modern Astronomy IV*, Volume 371 of *Astronomical Society of the Pacific Conference Series*, pp. 224.

[4] Efron, B. (1986). Why isn't everyone a Bayesian? *The American Statistician 40*(1), 1–5.

[5] Efron, B. (2004). Bayesians, frequentists, and scientists. Text of the 164th ASA presidential address, delivered at the awards ceremony in Toronto on August 10.

[6] Efron, B. and R. J. Tibshirani (1993). *An Introduction to the Bootstrap*. Chapman and Hall.

[7] Foreman-Mackey, D., D. W. Hogg, D. Lang, and J. Goodman (2012). emcee: The MCMC Hammer. *ArXiv:astro-ph/1202.3665*.

[8] Franklin, J. (2001). *The Science of Conjecture: Evidence and Probability Before Pascal*. Johns Hopkins University Press.

[9] Gelman, A., J. B. Carlin, H. S. Stern, and D. B. Rubin (2003). *Bayesian Data Analysis*. Chapman and Hall.

[10] Gilks, W., S. Richardson, and D. Spiegelhalter (1995). *Markov Chain Monte Carlo in Practice*. Chapman and Hall/CRC Interdisciplinary Statistics. Chapman and Hall/CRC.

[11] Goodman, J. and J. Weare (2010). Ensemble samplers with affine invariance. *Commun. Appl. Math. Comput. Sci. 5*, 65–80.

[12] Gregory, P. C. (2007). A Bayesian Kepler periodogram detects a second planet in HD208487. *MNRAS 374*, 1321–1333.

[13] Gregory, P. C. and T. J. Loredo (1992). A new method for the detection of a periodic signal of unknown shape and period. *ApJ 398*, 146–168.

[14] Hogg, D. W. (2012). Data analysis recipes: Probability calculus for inference. *ArXiv:astro-ph/1205.4446*.

[15] Ivezić, Ž., J. A. Tyson, E. Acosta, and others (2008). LSST: From science drivers to reference design and anticipated data products. *ArXiv:astro-ph/0805.2366*.

[16] Johnson, V. E. (2004). A Bayesian chi-squared test for goodness-of-fit. *Annals of Statistics 32*(6), 2361–2384.

[17] Kelly, B. C., R. Shetty, A. M. Stutz, and others (2012). Dust spectral energy distributions in the era of Herschel and Planck: A hierarchical Bayesian-fitting technique. *ApJ 752*, 55.

[18] Kessler, R., A. C. Becker, D. Cinabro, and others (2009). First-year Sloan Digital Sky Survey-II supernova results: Hubble diagram and cosmological parameters. *ApJS 185*, 32–84.

[19] Knuth, K. H. (2006). Optimal data-based binning for histograms. *ArXiv:physics/0605197*.

[20] Lehmann, E. L. and G. Casella (1998). *Theory of Point Estimation* (2nd ed.). Springer Texts in Statistics. Springer.

[21] Liddle, A. R. (2007). Information criteria for astrophysical model selection. *MNRAS 377*, L74–L78.

[22] Loredo, T. J. (1992). Promise of Bayesian inference for astrophysics. In E. D. Feigelson and G. J. Babu (Eds.), *Statistical Challenges in Modern Astronomy*, pp. 275–306.

[23] Loredo, T. J. (1999). Computational technology for Bayesian inference. In D. M. Mehringer, R. L. Plante, and D. A. Roberts (Eds.), *Astronomical Data Analysis Software and Systems VIII*, Volume 172 of *Astronomical Society of the Pacific Conference Series*, pp. 297.

[24] Lutz, T. E. and D. H. Kelker (1973). On the use of trigonometric parallaxes for the calibration of luminosity systems: Theory. *PASP 85*, 573.

[25] Marquis de Laplace (1995). *A Philosophical Essay on Probabilities*. Dover.

[26] McGrayne, S. B. (2011). *The Theory That Would Not Die: How Bayes' Rule Cracked the Enigma Code, Hunted Down Russian Submarines, and Emerged Triumphant from Two Centuries of Controversy*. Yale University Press.

[27] Press, W. H. (1997). Understanding data better with Bayesian and global statistical methods. In J. N. Bahcall, and J. P. Ostriker (Ed.), *Unsolved Problems in Astrophysics*. Princeton University Press.

[28] Protassov, R., D. A. van Dyk, A. Connors, V. L. Kashyap, and A. Siemiginowska (2002). Statistics, handle with care: Detecting multiple model components with the likelihood ratio test. *ApJ 571*, 545–559.

[29] Rubin, H. and J. Sethuraman (1965). Bayes risk efficiency. *Sankhyā: The Indian Journal of Statistics, Series A (1961–2002) 27*, 347–356.

[30] Scargle, J. D. (1998). Studies in astronomical time series analysis. V. Bayesian blocks, a new method to analyze structure in photon counting data. *ApJ 504*, 405.

[31] Scargle, J. D., J. P. Norris, B. Jackson, and J. Chiang (2012). Studies in astronomical time series analysis. VI. Bayesian block representations. *ArXiv:astro-ph/1207.5578*.

[32] Smith, H. (2003). Is there really a Lutz-Kelker bias? Reconsidering calibration with trigonometric parallaxes. *MNRAS 338*, 891–902.

[33] Smith, Jr., H. (1987). The calibration problem. I - Estimation of mean absolute magnitude using trigonometric parallaxes. II - Trigonometric parallaxes selected according to proper motion and the problem of statistical parallaxes. *A&A 171*, 336–347.

[34] Teerikorpi, P. (1997). Observational selection bias affecting the determination of the extragalactic distance scale. *ARAA 35*, 101–136.

[35] van der Vaart, A. W. (2000). *Asymptotic Statistics*. Cambridge Series in Statistical and Probabilistic Mathematics. Cambridge University Press.

[36] Verdinelli, I. and L. Wasserman (1998). Bayesian goodness-of-fit testing using infinite-dimensional exponential families. *Annals of Statistics 26*, 1215–1241.

PART III
Data Mining and Machine Learning

6 Searching for Structure in Point Data

"Through space the universe encompasses and swallows me up like an atom; through thought I comprehend the world." (Blaise Pascal)

We begin the third part of this book by addressing methods for exploring and quantifying structure in a multivariate distribution of points. One name for this kind of activity is *exploratory data analysis* (EDA). Given a sample of N points in D-dimensional space, there are three classes of problems that are frequently encountered in practice: density estimation, cluster finding, and statistical description of the observed structure. The space populated by points in the sample can be real physical space, or a space spanned by the measured quantities (attributes). For example, we can consider the distribution of sources in a multidimensional color space, or in a six-dimensional space spanned by three-dimensional positions and three-dimensional velocities.

To infer the pdf from a sample of data is known as density estimation. The same methodology is often called data smoothing. We have already encountered density estimation in the one-dimensional case when discussing histograms in §4.8 and §5.7.2, and in this chapter we extend it to multidimensional cases. Density estimation is one of the most critical components of extracting knowledge from data. For example, given a pdf estimated from point data, we can generate simulated distributions of data and compare them against observations. If we can identify regions of low probability within the pdf, we have a mechanism for the detection of unusual or anomalous sources. If our point data can be separated into subsamples using provided class labels, we can estimate the pdf for each subsample and use the resulting set of pdfs to classify new points: the probability that a new point belongs to each subsample/class is proportional to the pdf of each class evaluated at the position of the point (see §9.3.5). Density estimation relates directly to regression discussed in chapter 8 (where we simplify the problem to the prediction of a single variable from the pdf), and is at the heart of many of the classification procedures described in chapter 9. We discuss nonparametric and parametric methods for density estimation in §6.1–6.3.

Given a point data set, we can further ask whether it displays any structure (as opposed to a random distribution of points). Finding concentrations of multivariate points (or groups of sources) is known in astronomy as "clustering" (when a density estimate is available, clusters correspond to "overdensities"). Clusters can be defined to be distinct objects (e.g., gravitationally bound clusters of galaxies), or loose groups of sources with common properties (e.g., the identification of quasars based on their color properties). Unsupervised clustering refers to cases where there is no prior information about the number and properties of clusters in data. Unsupervised classification, discussed in chapter 9, assigns to each cluster found by unsupervised clustering, a class based on additional information (e.g., clusters identified in color space might be assigned the labels "quasar" and "star" based on supplemental spectral data). Finding clusters in data is discussed in §6.4.

In some cases, such as when considering the distribution of sources in multi-dimensional color space, clusters can have specific physical meaning (e.g., hot stars, quasars, cold stars). On the other hand, in some applications, such as large-scale clustering of galaxies, clusters carry information only in a statistical sense. For example, we can test cosmological models of structure formation by comparing clustering statistics in observed and simulated data. Correlation functions are commonly used in astronomy for the statistical description of clustering, and are discussed in §6.5.

Data sets used in this chapter

In this chapter we use four data sets: a subset of the SDSS spectroscopic galaxy sample (§1.5.5), a set of SDSS stellar spectra with stellar parameter estimates (§1.5.7), SDSS single-epoch stellar photometry (§1.5.3), and the SDSS Standard Star Catalog from Stripe 82 (§1.5.8). The galaxy sample contains 8014 galaxies selected to be centered on the SDSS "Great Wall" (a filament of galaxies that is over 100 Mpc in extent; see [14]). These data comprise measures of the positions, luminosities and colors of the galaxies, and are used to illustrate density estimation and the spatial clustering of galaxies. The stellar spectra are used to derive measures of effective temperature, surface gravity, and two quantities that summarize chemical composition: metallicity (parametrized as [Fe/H]) and α-element abundance (parametrized as [α/Fe]). These measurements are used to illustrate clustering in multidimensional parameter space. The precise multiepoch averaged photometry from the Standard Star Catalog is used to demonstrate the performance of algorithms that account for measurement errors when estimating the underlying density (pdf) from the less precise single-epoch photometric data.

6.1. Nonparametric Density Estimation

In some sense chapters 3, 4, and 5 were about estimating the underlying density of the data, using parametric models. Chapter 3 discussed parametric models of probability density functions and chapters 4 and 5 discussed estimation of their parameters from frequentist and Bayesian perspectives. We now look at how to estimate a density *nonparametrically*, that is, without specifying a specific functional model. Real data rarely follow simple distributions—nonparametric methods are meant to capture every aspect of the density's shape. What we lose by taking this route is the convenience, relative computational simplicity (usually), and easy interpretability of parametric models.

The go-to method for nonparametric density estimation, that is, modeling of the underlying distribution, is the method of *kernel density estimation* (KDE).

While a very simple method in principle, it also comes with impressive theoretical properties.

6.1.1. Kernel Density Estimation

As a motivation for kernel density estimation, let us first reconsider the one-dimensional histograms introduced in §4.8. One problem with a standard histogram is the fact that the exact locations of the bins can make a difference, and yet it is not clear how to choose in advance where the bins should be placed (see §4.8.1). We illustrate this problem in the two top panels in figure 6.1. They show histograms constructed with an identical data set, and with identical bin widths, but with bins offset in x by 0.25. This offset leads to very different histograms and possible interpretations of the data: the difference between seeing it as a bimodal distribution vs. an extended flat distribution.

How can we improve on a basic histogram and avoid this problem? Each point within a histogram contributes one unit to the height of the histogram at the position of its bin. One possibility is to allow each point to have its own bin, rather than arranging the bins in a regular grid, and furthermore allow the bins to overlap. In essence, each point is replaced by a box of unit height and some predefined width. The result is the distribution shown in the middle-left panel of figure 6.1. This distribution does not require a specific choice of bin boundaries—the data drive the bin positioning—and does a much better job of showing the bimodal character of the underlying distribution than the histogram shown in the top-right panel.

The above simple recipe for producing a histogram is an example of kernel density estimation. Here the kernel is a top-hat distribution centered on each individual point. It can be shown theoretically that this kernel density estimator (KDE) is a better estimator of the density than the ordinary histogram (see Wass10). However, as is discernible from the middle-left panel of figure 6.1, the rectangular kernel does not lead to a very smooth distribution and can even display suspicious spikes. For this reason, other kernels, for example Gaussians, are often used. The remaining panels of figure 6.1 show the kernel density estimate, described below, of the same data but now using Gaussian kernels of different widths. Using too narrow a kernel (middle-right panel) leads to a noisy distribution, while using too wide a kernel (bottom-left panel) leads to excessive smoothing and washing out of information. A well-tuned kernel (bottom-right panel) can lead to accurate estimation of the underlying distribution; the choice of kernel width is discussed below.

Given a set of measurements $\{x_i\}$, the kernel density estimator (i.e., an estimator of the underlying pdf) at an arbitrary position x is defined as

$$\widehat{f}_N(x) = \frac{1}{Nh^D} \sum_{i=1}^{N} K\left(\frac{d(x, x_i)}{h}\right), \tag{6.1}$$

where $K(u)$ is the *kernel function* and h is known as the bandwidth (which defines the size of the kernel). The local density is estimated as a weighted mean of all points, where the weights are specified via $K(u)$ and typically decrease with distance $d(x, x_i)$. Alternatively, KDE can be viewed as replacing each point with a "cloud" described by $K(u)$. The kernel function $K(u)$ can be any smooth function that is positive at all points ($K(u) \geq 0$), normalizes to unity ($\int K(u)\, du = 1$), has a mean of zero

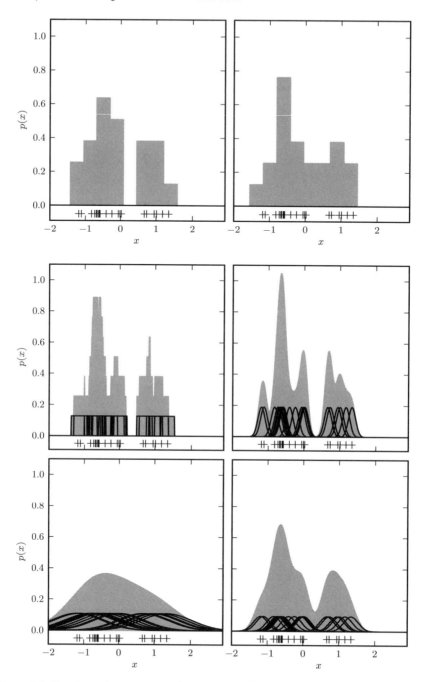

Figure 6.1. Density estimation using histograms and kernels. The top panels show two histogram representations of the same data (shown by plus signs in the bottom of each panel) using the same bin width, but with the bin centers of the histograms offset by 0.25. The middle-left panel shows an adaptive histogram where each bin is centered on an individual point and these bins can overlap. This adaptive representation preserves the bimodality of the data. The remaining panels show kernel density estimation using Gaussian kernels with different bandwidths, increasing from the middle-right panel to the bottom-right, and with the largest bandwidth in the bottom-left panel. The trade-off of variance for bias becomes apparent as the bandwidth of the kernels increases.

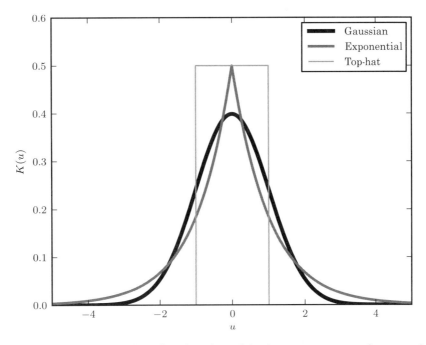

Figure 6.2. A comparison of the three kernels used for density estimation in figure 6.3: the Gaussian kernel (eq. 6.2), the top-hat kernel (eq. 6.3), and the exponential kernel (eq. 6.4).

($\int u K(u)\,du = 0$), and has a variance ($\sigma_K^2 = \int u^2 K(u)\,du$) greater than zero. An often-used kernel is the Gaussian kernel,

$$K(u) = \frac{1}{(2\pi)^{D/2}} e^{-u^2/2}, \tag{6.2}$$

where D is the number of dimensions of the parameter space and $u = d(x, x_i)/h$. Other kernels that can be useful are the top-hat (box) kernel,

$$K(u) = \begin{cases} \frac{1}{V_D(1)} & \text{if } u \le 1, \\ 0 & \text{if } u > 1, \end{cases} \tag{6.3}$$

and the exponential kernel,

$$K(u) = \frac{1}{D!\, V_D(1)} e^{-|u|}, \tag{6.4}$$

where $V_D(r)$ is the volume of a D-dimensional hypersphere of radius r (see eq. 7.3). A comparison of the Gaussian, exponential, and top-hat kernels is shown in figure 6.2.

Selecting the KDE bandwidth using cross-validation

Both histograms and KDE do, in fact, have a parameter: the kernel or bin width. The proper choice of this parameter is critical, much more so than the choice of a specific kernel, particularly when the data set is large; see [41]. We will now show a rigorous procedure for choosing the optimal kernel width in KDE (which can also be applied to finding the optimal bin width for a histogram).

Cross-validation can be used for any cost function (see §8.11); we just have to be able to evaluate the cost on *out-of-sample* data (i.e., points not in the training set). If we consider the likelihood cost for KDE, for which we have the leave-one-out *likelihood cross-validation*, then the cost is simply the sum over all points in the data set (i.e., $i = 1, \ldots, N$) of the log of the likelihood of the density, where the density, $\widehat{f}_{h,-i}(x_i)$, is estimated leaving out the ith data point. This can be written as

$$\mathrm{CV}_l(h) = \frac{1}{N} \sum_{i=1}^{N} \log \widehat{f}_{h,-i}(x_i), \tag{6.5}$$

and, by minimizing $C V_l(h)$ as a function of bandwidth, we can optimize for the width of the kernel h.

An alternative to likelihood cross-validation is to use the mean integrated square error (MISE), introduced in eq. 4.14, as the cost function. To determine the value of h that minimizes the MISE we can write

$$\int (\widehat{f}_h - f)^2 = \int \widehat{f}_h^2 - 2 \int \widehat{f}_h f + \int f^2. \tag{6.6}$$

As before, the first term can be obtained analytically, and the last term does not depend on h. For the second term we have expectation value

$$\mathbb{E}\left[\int \widehat{f}_h(x) f(x)\, dx \right] = \mathbb{E}\left[\frac{1}{N} \sum_{i=1}^{N} \widehat{f}_{h,-1}(x) \right]. \tag{6.7}$$

This motivates the L_2 *cross-validation* score:

$$\mathrm{CV}_{L_2}(h) = \int \widehat{f}_h^2 - 2\frac{1}{N} \sum_{i=1}^{N} \widehat{f}_{h,-i}(x_i) \tag{6.8}$$

since $\mathbb{E}[\mathrm{CV}_{L_2}(h) + \int f^2] = \mathbb{E}[\mathrm{MISE}(\widehat{f}_h)]$.

The optimal KDE bandwidth decreases at the rate $\mathcal{O}(N^{-1/5})$ (in a one-dimensional problem), and the error of the KDE using the optimal bandwidth converges at the rate $\mathcal{O}(N^{-4/5})$; it can be shown that histograms converge at a rate $\mathcal{O}(N^{-2/3})$; see [35]. KDE is, therefore, theoretically superior to the histogram as an estimator of the density. It can also be shown that there does not exist a density estimator that converges faster than $\mathcal{O}(N^{-4/5})$ (see Wass10).

Ideally we would select a kernel that has h as small as possible. If h becomes too small we increase the variance of the density estimation. If h is too large then the variance decreases but at the expense of the bias in the derived density. The optimal kernel function, in terms of minimum variance, turns out to be

$$K(x) = \frac{3}{4}\left(1 - x^2\right) \tag{6.9}$$

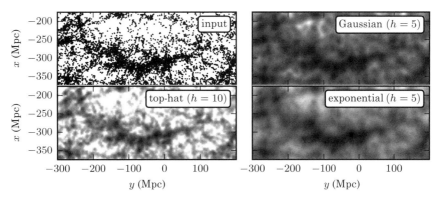

Figure 6.3. Kernel density estimation for galaxies within the SDSS "Great Wall." The top-left panel shows points that are galaxies, projected by their spatial locations (right ascension and distance determined from redshift measurement) onto the equatorial plane (declination $\sim 0°$). The remaining panels show estimates of the density of these points using kernel density estimation with a Gaussian kernel (upper right), a top-hat kernel (lower left), and an exponential kernel (lower right). Compare also to figure 6.4.

for $|x| \leq 1$ and 0 otherwise; see [37]. This function is called the *Epanechnikov kernel.*

AstroML contains an implementation of kernel density estimation in D dimensions using the above kernels:

```
import numpy as np
from astroML.density_estimation import KDE

X = np.random.normal(size=(1000, 2)) # 1000 points
    # in 2 dims
kde = KDE('gaussian', h=0.1) # select the gaussian
    # kernel
kde.fit(X) # fit the model to the data
dens = kde.eval(X) # evaluate the model at the data
```

There are several choices for the kernel. For more information and examples, see the AstroML documentation or the code associated with the figures in this chapter.

Figure 6.3 shows an example of KDE applied to a two-dimensional data set: a sample of galaxies centered on the SDSS "Great Wall." The distribution of points (galaxies) shown in the top-left panel is used to estimate the "smooth" underlying distribution using three types of kernels. The top-hat kernel (bottom-left panel) is the most "spread out" of the kernels, and its imprint on the resulting distribution is apparent, especially in underdense regions. Between the Gaussian and exponential kernels, the exponential is more sharply peaked and has wider tails, but both recover similar features in the distribution. For a comparison of other density estimation methods for essentially the same data set, see [12].

Computation of kernel density estimates To obtain the height of the density estimate at a single query point (position) x, we must sum over N kernel functions.
For many (say, $\mathcal{O}(N)$) queries, when N grows very large this brute-force approach
can lead to very long ($\mathcal{O}(N^2)$) computation time. Because the kernels usually have a
limited bandwidth h, points $\mathbf{x}_i$ with $|\mathbf{x_i}-\mathbf{x}| \gg h$ contribute a negligible amount to the
density at a point $\mathbf{x}$, and the bulk of the contribution comes from neighboring points.
However, a simplistic cutoff approach in an attempt to directly copy the nearest-
neighbor algorithm we discussed in §2.5.2 leads to potentially large, unquantified
errors in the estimate, defeating the purpose of an accurate nonparametric density
estimator. More principled approaches were introduced in [15] but improved upon
in subsequent research: for the highest accuracy and speed in low to moderate
dimensionalities see the dual-tree fast Gauss transforms in [22, 24], and in higher
dimensionalities see [23]. Ram et al. [33] showed rigorously that such algorithms
reduce the runtime of a single query of KDE from the naive $\mathcal{O}(N)$ to $\mathcal{O}(\log N)$, and
for $\mathcal{O}(N)$ queries from the naive $\mathcal{O}(N^2)$ to $\mathcal{O}(N)$. An example of the application of
such algorithms in astronomy was shown in [3]. A parallelization of such algorithms
is shown in [25]. A overview of tree algorithms for KDE and other problems can be
found in chapter 21 of WSAS.

6.1.2. KDE with Measurement Errors

Suppose now that the points (i.e., their coordinates) are measured with some error σ.
We begin with the simple one-dimensional case with homoscedastic errors. Assume
that the data is drawn from the true pdf $h(x)$, and the error is described by the
distribution $g(x|\sigma)$. Then the observed distribution $f(x)$ is given by the convolution
(see §3.44)

$$f(x) = (h \star g)(x) = \int_{-\infty}^{\infty} h(x')g(x - x')\, dx'. \tag{6.10}$$

This suggests that in order to obtain the underlying noise-free density $h(x)$, we
can obtain an estimate $f(x)$ from the noisy data first, and then "deconvolve" the
noise pdf. The nonparametric method of *deconvolution KDE* does precisely this;
see [10, 38]. According to the convolution theorem, a convolution in real space
corresponds to a product in Fourier space (see §10.2.2 for details). Because of this,
deconvolution KDE can be computed using the following steps:

1. Find the kernel density estimate of the observed data, $f(x)$, and compute the
   Fourier transform $F(k)$.
2. Compute the Fourier transform $G(k)$ of the noise distribution $g(x)$.
3. From eq. 6.10 and the convolution theorem, the Fourier transform of the true
   distribution $h(x)$ is given by $H(k) = F(k)/G(k)$. The underlying noise-free
   pdf $h(x)$ can be computed via the inverse Fourier transform of $H(k)$.

For certain kernels $K(x)$ and certain noise distributions $g(x)$, this deconvolution can
be performed analytically and the result becomes another modified kernel, called the
deconvolved kernel. Examples of kernel and noise forms which have these properties
can be found in [10, 38]. Here we will describe one example of a D-dimensional
version of this method, where the noise scale is assumed to be heteroscedastic and

dependent on the dimension. The noise model leads to an analytic treatment of the deconvolved kernel. We will assume that the noise is distributed according to the multivariate exponential

$$g(\mathbf{x}) = \frac{1}{\sqrt{2^D}\sigma_1\sigma_2\ldots\sigma_D} \exp\left[-\sqrt{2}\left(\frac{|x_1|}{\sigma_1} + \frac{|x_2|}{\sigma_2} + \cdots + \frac{|x_D|}{\sigma_D}\right)\right], \qquad (6.11)$$

where the σ_i represent the standard deviations in each dimension. We will assume that $(\sigma_1, \ldots, \sigma_D)$ are known for each data point. For the case of a Gaussian kernel function, the deconvolution kernel is then

$$K_{h,\sigma}(\mathbf{x}) = \frac{1}{\sqrt{(2\pi)^D}} \exp(-|\mathbf{x}|^2/2) \prod_i \left(1 - \frac{\sigma_i^2}{2h^2}(x_i^2 - 1)\right). \qquad (6.12)$$

This deconvolution kernel can then be used in place of the kernels discussed in §6.1.1 above, noting the additional dependence on the error σ of each point.

6.1.3. Extensions and Related Methods

The idea of kernel density estimation can be extended to other tasks, including classification (kernel discriminant analysis, §9.3.5), regression (kernel regression, §8.5), and conditional density estimation (kernel conditional density estimation, §3.1.3). Some of the ideas that have been developed to make kernel regression highly accurate, discussed in 8.5, can be brought back to density estimation, including the idea of using variable bandwidths, in which each data point can have its own kernel width.

6.2. Nearest-Neighbor Density Estimation

Another often used and simple density estimation technique is based on the distribution of nearest neighbors. For each point (e.g., a pixel location on the two-dimensional grid) we can find the distance to the Kth-nearest neighbor, d_K. In this method, originally proposed in an astronomical context by Dressler et al. [11], the implied point density at an arbitrary position x is estimated as

$$\widehat{f}_K(x) = \frac{K}{V_D(d_K)}, \qquad (6.13)$$

where the volume V_D is evaluated according to the problem dimensionality, D (e.g., for $D = 2$, $V_2 = \pi d^2$; for $D = 3$, $V_3 = 4\pi d^3/3$; for higher dimensions, see eq. 7.3). The simplicity of this estimator is a consequence of the assumption that the underlying density field is locally constant. In practice, the method is even simpler because one can compute

$$\widehat{f}_K(x) = \frac{C}{d_K^D}, \qquad (6.14)$$

and evaluate the scaling factor C at the end by requiring that the sum of the product of $\widehat{f}_K(x)$ and pixel volume is equal to the total number of data points. The error in $\widehat{f}_K(x)$ is $\sigma_f = K^{1/2}/V_D(d_K)$, and the fractional (relative) error is $\sigma_f/\widehat{f} = 1/K^{1/2}$. Therefore, the fractional accuracy increases with K at the expense of the spatial resolution (the effective resolution scales with $K^{1/D}$). In practice, K should be at least 5 because the estimator is biased and has a large variance for smaller K; see [7].

This general method can be improved (the error in $\widehat{f}$ can be decreased without a degradation in the spatial resolution, or alternatively the resolution can be increased without increasing the error in $\widehat{f}$) by considering distances to *all* K nearest neighbors instead of only the distance to the Kth-nearest neighbor; see [18]. Given distances to all K neighbors, $d_i, i = 1, \ldots, K$,

$$\widehat{f}_K(x) = \frac{C}{\sum_{i=1}^{K} d_i^D}. \tag{6.15}$$

Derivation of eq. 6.15 is based on Bayesian analysis, as described in [18]. The proper normalization when computing local density without regard to overall mean density is

$$C = \frac{K(K+1)}{2V_D(1)}. \tag{6.16}$$

When searching for local overdensities in the case of sparse data, eq. 6.15 is superior to eq. 6.14; in the constant density case, both methods have similar statistical power. For an application in an astronomical setting, see [36].

> AstroML implements nearest-neighbor density estimation using a fast ball-tree algorithm. This can be accomplished as follows:
>
> ```
> import numpy as np
> from astroML.density_estimation import
>     KNeighborsDensity
>
> X = np.random.normal(size=(1000, 2)) # 1000 points
>     # in 2 dims
> knd = KNeighborsDensity("bayesian", 10) # bayesian
>     # method, 10 nbrs
> knd.fit(X) # fit the model to the data
> dens = knd.eval(X) # evaluate the model at the data
> ```
>
> The method can be either "simple" to use eq. 6.14, or "bayesian" to use eq. 6.15. See the AstroML documentation or the code associated with figure 6.4 for further examples.

Figure 6.4 compares density estimation using the Gaussian kernel with a bandwidth of 5 Mpc and using the nearest-neighbor method (eq. 6.15) with $K = 5$ and $K = 40$ for the same sample of galaxies as shown in figure 6.3. For small K the

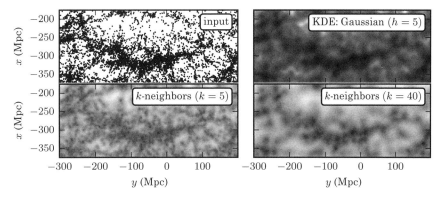

Figure 6.4. Density estimation for galaxies within the SDSS "Great Wall." The upper-left panel shows points that are galaxies, projected by their spatial locations onto the equatorial plane (declination $\sim 0°$). The remaining panels show estimates of the density of these points using kernel density estimation (with a Gaussian kernel with width 5 Mpc), a K-nearest-neighbor estimator (eq. 6.15) optimized for a small-scale structure (with $K = 5$), and a K-nearest-neighbor estimator optimized for a large-scale structure (with $K = 40$).

fine structure in the galaxy distribution is preserved but at the cost of a larger variance in the density estimation. As K increases the density distribution becomes smoother, at the cost of additional bias in the other estimates.

Figure 6.5 compares Bayesian blocks, KDE, and nearest-neighbor density estimation for two one-dimensional data sets drawn from the same (relatively complicated) generating distribution (this is the same generated data set used previously in figure 5.21). The generating distribution includes several "peaks" that are described by the Cauchy distribution (§3.3.5). KDE and nearest-neighbor methods are much noisier than the Bayesian blocks method in the case of the smaller sample; for the larger sample all three methods produce similar results.

6.3. Parametric Density Estimation

KDE estimates the density of a set of points by affixing a kernel to each point in the data set. An alternative is to use fewer kernels, and fit for the kernel locations as well as the widths. This is known as a *mixture model*, and can be viewed in two ways: at one extreme, it is a density estimation model similar to KDE. In this case one is not concerned with the locations of individual clusters, but the contribution of the full set of clusters at any given point. At the other extreme, it is a clustering algorithm, where the location and size of each component is assumed to reflect some underlying property of the data.

6.3.1. Gaussian Mixture Model

The most common mixture model uses Gaussian components, and is called a *Gaussian mixture model* (GMM). A GMM models the underlying density (pdf) of points as a sum of Gaussians. We have already encountered one-dimensional mixtures of Gaussians in §4.4; in this section we extend those results to multiple

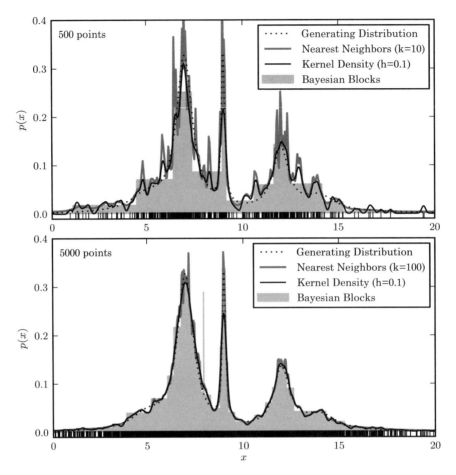

Figure 6.5. A comparison of different density estimation methods for two simulated one-dimensional data sets (cf. figure 5.21). The generating distribution is same in both cases and shown as the dotted line; the samples include 500 (top panel) and 5000 (bottom panel) data points (illustrated by vertical bars at the bottom of each panel). Density estimators are Bayesian blocks (§5.7.2), KDE (§6.1.1) and the nearest-neighbor method (eq. 6.15).

dimensions. Here the density of the points is given by (cf. eq. 4.18)

$$\rho(\mathbf{x}) = Np(\mathbf{x}) = N \sum_{j=1}^{M} \alpha_j \mathcal{N}(\mu_j, \Sigma_j), \tag{6.17}$$

where the model consists of M Gaussians with locations μ_j and covariances Σ_j. The likelihood of the data can be evaluated analogously to eq. 4.20. Thus there is not only a clear score that is being optimized, the log-likelihood, but this is a special case where that function is a generative model, that is, it is a full description of the data.

The optimization of this likelihood is more complicated in multiple dimensions than in one dimension, but the expectation maximization methods discussed in §4.4.3 can be readily applied in this situation; see [34]. We have already shown a simple example in one dimension for a toy data set (see figure 4.2). Here we will show

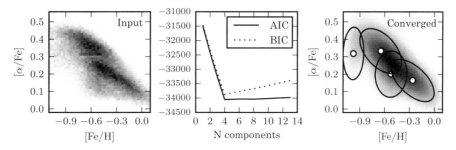

Figure 6.6. A two-dimensional mixture of Gaussians for the stellar metallicity data. The left panel shows the number density of stars as a function of two measures of their chemical composition: metallicity ([Fe/H]) and α-element abundance ([α/Fe]). The right panel shows the density estimated using mixtures of Gaussians together with the positions and covariances (2σ levels) of those Gaussians. The center panel compares the information criteria AIC and BIC (see §4.3.2 and §5.4.3).

an implementation of Gaussian mixture models for data sets in two dimensions, taken from real observations. In a later chapter, we will also apply this method to data in up to seven dimensions (see §10.3.4).

Scikit-learn includes an implementation of Gaussian mixture models in D dimensions:

```
import numpy as np
from sklearn.mixture import GMM

X = np.random.normal(size=(1000, 2)) # 1000 points
    # in 2 dims
gmm = GMM(3) # three component mixture
gmm.fit(X) # fit the model to the data
log_dens = gmm.score(X) # evaluate the log density
BIC = gmm.bic(X) # evaluate the BIC
```

For more involved examples, see the Scikit-learn documentation or the source code for figures in this chapter.

The left panel of figure 6.6 shows a Hess diagram (essentially a two-dimensional histogram) of the [Fe/H] vs. [α/Fe] metallicity for a subset of the SEGUE Stellar Parameters data (see §1.5.7). This diagram shows two distinct clusters in metallicity. For this reason, one may expect (or hope!) that the best-fit mixture model would contain two Gaussians, each containing one of those peaks. As the middle panel shows, this is not the case: the AIC and BIC (see §4.3.2) both favor models with four or more components. This is due to the fact that the components exist within a background, and the background level is such that a two-component model is insufficient to fully describe the data.

Following the BIC, we select $N = 4$ components, and plot the result in the rightmost panel. The reconstructed density is shown in grayscale and the positions of

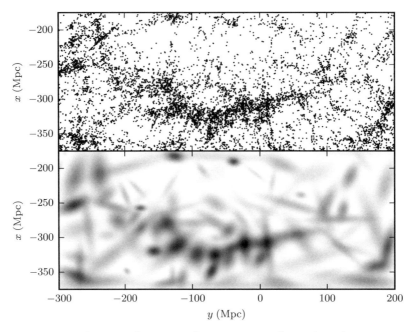

Figure 6.7. A two-dimensional mixture of 100 Gaussians (bottom) used to estimate the number density distribution of galaxies within the SDSS Great Wall (top). Compare to figures 6.3 and 6.4, where the density for the same distribution is computed using both kernel density and nearest-neighbor-based estimates.

the Gaussians in the model as solid ellipses. The two strongest components do indeed fall on the two peaks, where we expected them to lie. Even so, these two Gaussians do not completely separate the two clusters.

This is one of the common misunderstandings of Gaussian mixture models: the fact that the information criteria, such as BIC/AIC, prefer an N-component peak does not necessarily mean that there are N components. If the clusters in the input data are not near Gaussian, or if there is a strong background, the number of Gaussian components in the mixture will not generally correspond to the number of clusters in the data. On the other hand, if the goal is to simply describe the underlying pdf, many more components than suggested by BIC can be (and should be) used.

Figure 6.7 illustrates this point with the SDSS "Great Wall" data where we fit 100 Gaussians to the point distribution. While the underlying density representation is consistent with the distribution of galaxies and the positions of the Gaussians themselves correlate with the structure, there is not a one-to-one mapping between the Gaussians and the positions of clusters within the data. For these reasons, mixture models are often more appropriate when used as a density estimator as opposed to cluster identification (see, however, §10.3.4 for a higher-dimensional example of using GMM for clustering).

Figure 6.8 compares one-dimensional density estimation using Bayesian blocks, KDE, and a Gaussian mixture model using the same data sets as in figure 6.5. When the sample is small, the GMM solution with three components is favored by the BIC criterion. However, one of the components has a very large width ($\mu = 8$, $\sigma = 26$) and effectively acts as a nearly flat background. The reason for such a bad GMM

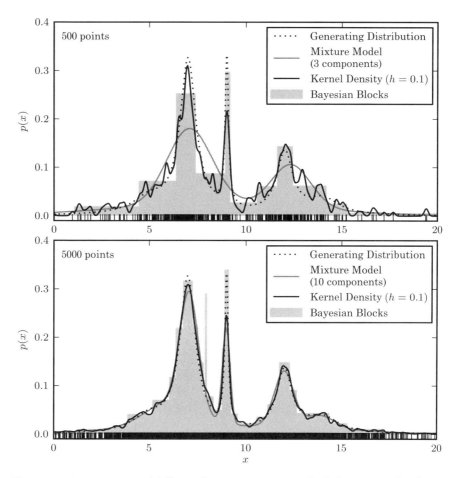

Figure 6.8. A comparison of different density estimation methods for two simulated one-dimensional data sets (same as in figure 6.5). Density estimators are Bayesian blocks (§5.7.2), KDE (§6.1.1), and a Gaussian mixture model. In the latter, the optimal number of Gaussian components is chosen using the BIC (eq. 5.35). In the top panel, GMM solution has three components but one of the components has a very large width and effectively acts as a nearly flat background.

performance (compared to Bayesian blocks and KDE which correctly identify the peak at $x \sim 9$) is the fact that individual "peaks" are generated using the Cauchy distribution: the wide third component is trying (hard!) to explain the wide tails. In the case of the larger sample, the BIC favors ten components and they obtain a similar level of performance to the other two methods.

BIC is a good tool to find how many statistically significant clusters are supported by the data. However, when density estimation is the only goal of the analysis (i.e., when individual components or clusters are not assigned any specific meaning) we can use any number of mixture components (e.g., when underlying density is very complex and hard to describe using a small number of Gaussian components). With a sufficiently large number of components, mixture models approach the flexibility of nonparametric density estimation methods.

Determining the number of components

Most mixture methods require that we specify the number of components as an input to the method. For those methods which are based on a score or error, determination of the number of components can be treated as a model selection problem like any other (see chapter 5), and thus be performed via cross-validation (as we did when finding optimal kernel bandwidth; see also §8.11), or using BIC/AIC criteria (§5.4.3). The hierarchical clustering method (§6.4.5) addresses this problem by finding clusterings at all possible scales.

It should be noted, however, that specifying the number of components (or clusters) is a relatively poorly posed question in astronomy. It is rare, despite the examples given in many machine learning texts, to find distinct, isolated and Gaussian clusters of data in an astronomical distribution. Almost all distributions are continuous. The number of clusters (and their positions) relates more to how well we can characterize the underlying density distribution. For clustering studies, it may be useful to fit a mixture model with many components and to divide components into "clusters" and "background" by setting a density threshold; for an example of this approach see figures 10.20 and 10.21.

An additional important factor that influences the number of mixture components supported by data is the sample size. Figure 6.9 illustrates how the best-fit GMM changes dramatically as the sample size is increased from 100 to 1000. Furthermore, even when the sample includes as many as 10,000 points, the underlying model is not fully recovered (only one of the two background components is recognized).

6.3.2. Cloning Data in $D > 1$ Dimensions

Here we return briefly to a subject we discussed in §3.7: cloning a distribution of data. The rank-based approach illustrated in figure 3.25 works well in one dimension, but cloning an arbitrary higher-dimensional distribution requires an estimate of the local density at each point. Gaussian mixtures are a natural choice for this, because they can flexibly model density fields in any number of dimensions, and easily generate new points within the model.

Figure 6.10 shows the procedure: from 1000 observed points, we fit a ten-component Gaussian mixture model to the density. A sample of 5000 points drawn from this density model mimics the input to the extent that the density model is accurate. This idea can be very useful when simulating large multidimensional data sets based on small observed samples. This idea will also become important in the following section, in which we explore a variant of Gaussian mixtures in order to create denoised samples from density models based on noisy observed data sets.

6.3.3. GMM with Errors: Extreme Deconvolution

Bayesian estimation of multivariate densities modeled as mixtures of Gaussians, with data that have measurement error, is known in astronomy as "extreme deconvolution" (XD); see [6]. As with the Gaussian mixtures above, we have already encountered this situation in one dimension in §4.4. Recall the original mixture of Gaussians, where each data point $\mathbf{x}$ is sampled from one of M different Gaussians with given means and variances, (μ_i, Σ_i), with the weight for each Gaussian being

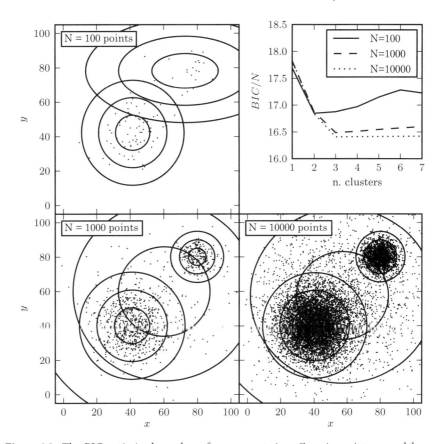

Figure 6.9. The BIC-optimized number of components in a Gaussian mixture model as a function of the sample size. All three samples (with 100, 1000, and 10,000 points) are drawn from the same distribution: two narrow foreground Gaussians and two wide background Gaussians. The top-right panel shows the BIC as a function of the number of components in the mixture. The remaining panels show the distribution of points in the sample and the 1, 2, and 3 standard deviation contours of the best-fit mixture model.

α_i. Thus, the pdf of $\mathbf{x}$ is given as

$$p(\mathbf{x}) = \sum_j \alpha_j \mathcal{N}(\mathbf{x}|\mu_j, \boldsymbol{\Sigma}_j), \qquad (6.18)$$

where, recalling eq. 3.97,

$$\mathcal{N}(\mathbf{x}|\mu_j, \boldsymbol{\Sigma}_j) = \frac{1}{\sqrt{(2\pi)^D \det(\boldsymbol{\Sigma}_j)}} \exp\left(-\frac{1}{2}(\mathbf{x}-\mu)^T \boldsymbol{\Sigma}_j^{-1}(\mathbf{x}-\mu)\right). \qquad (6.19)$$

Extreme deconvolution generalizes the EM approach to a case with measurement errors. More explicitly, one assumes that the noisy observations $\mathbf{x}_i$ and the true values $\mathbf{v}_i$ are related through

$$\mathbf{x}_i = \mathbf{R}_i \mathbf{v}_i + \epsilon_i, \qquad (6.20)$$

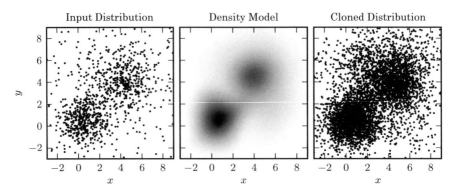

Figure 6.10. Cloning a two-dimensional distribution. The left panel shows 1000 observed points. The center panel shows a ten-component Gaussian mixture model fit to the data (two components dominate over other eight). The third panel shows 5000 points drawn from the model in the second panel.

where $\mathbf{R}_i$ is the so-called projection matrix, which may or may not be invertible. The noise ϵ_i is assumed to be drawn from a Gaussian with zero mean and variance $\mathbf{S}_i$. Given the matrices $\mathbf{R}_i$ and $\mathbf{S}_i$, the aim of XD is to find the parameters μ_i, $\boldsymbol{\Sigma}_i$ of the underlying Gaussians, and the weights α_i, as defined in §6.18, in a way that would maximize the likelihood of the observed data. The EM approach to this problem results in an iterative procedure that converges to (at least) a local maximum of the likelihood. The generalization of the EM procedure (see [6]) in §4.4.3 becomes the following:

- The expectation (E) step:

$$q_{ij} \leftarrow \frac{\alpha_j \mathcal{N}(\mathbf{w}_i | \mathbf{R}_i \mu_j, \mathbf{T}_{ij})}{\sum_j \alpha_k \mathcal{N}(\mathbf{w}_i | \mathbf{R}_i \mu_k, \mathbf{T}_{ij})}, \tag{6.21}$$

$$\mathbf{b}_{ij} \leftarrow \mu_j + \boldsymbol{\Sigma}_j \mathbf{R}_i^T \mathbf{T}_{ij}^{-1} (\mathbf{w}_i - \mathbf{R}_i \mu_j), \tag{6.22}$$

$$\mathbf{B}_{ij} \leftarrow \boldsymbol{\Sigma}_j - \boldsymbol{\Sigma}_j \mathbf{R}_i^T \mathbf{T}_{ij}^{-1} \mathbf{R}_i \boldsymbol{\Sigma}_j, \tag{6.23}$$

where $\mathbf{T}_{ij} = \mathbf{R}_i \boldsymbol{\Sigma}_j \mathbf{R}_i^T + \mathbf{S}_i$.
- The maximization (M) step:

$$\alpha_i \leftarrow \frac{1}{N} \sum_i q_{ij}, \tag{6.24}$$

$$\mu_j \leftarrow \frac{1}{q_j} \sum_i q_{ij} \mathbf{b}_{ij}, \tag{6.25}$$

$$\boldsymbol{\Sigma}_j \leftarrow \frac{1}{q_j} \sum_i q_{ij} [(\mu_j - \mathbf{b}_{ij})(\mu_j - \mathbf{b}_{ij}^T) + \mathbf{B}_{ij}], \tag{6.26}$$

where $q_j = \sum_i q_{ij}$.

The iteration of these steps increases the likelihood of the observations $\mathbf{w}_i$, given the model parameters. Thus, iterating until convergence, one obtains a solution that is a local maximum of the likelihood. This method has been used with success in quasar classification, by estimating the densities of quasar and nonquasar objects from flux measurements; see [5]. Details of the use of XD, including methods to avoid local maxima in the likelihood surface, can be found in [6].

AstroML contains an implementation of XD which has a similar interface to GMM in Scikit-learn:

```
import numpy as np
from astroML.density_estimation import XDGMM

X = np.random.normal(size=(1000, 1)) # 1000 pts in
    # 1 dim
Xerr = np.random.random((1000, 1, 1)) # 1000 1x1
        # covariance matrices
xdgmm = XDGMM(n_components=2)
xdgmm.fit(X, Xerr) # fit the model
logp=xdgmm.logprob_a(X, Xerr) # evaluate probability
X_new = xdgmm.sample(1000) # sample new points from
        # distribution
```

For further examples, see the source code of figures 6.11 and 6.12.

Figure 6.11 shows the performance of XD on a simulated data set. The top panels show the true data set (2000 points) and the data set with noise added. The bottom panels show the XD results: on the left is a new data set drawn from the mixture (as expected, it has the same characteristics as the noiseless sample). On the right are the 2σ limits of the ten Gaussians used in the fit. The important feature of this figure is that from the noisy data, we are able to recover a distribution that closely matches the true underlying data: we have deconvolved the data and the noise in a similar vein to deconvolution KDE in §6.1.2.

This deconvolution of measurement errors can also be demonstrated using a real data set. Figure 6.12 shows the results of XD when applied to photometric data from the Sloan Digital Sky Survey. The high signal-to-noise data (i.e., small color errors; top-left panel) come from the Stripe 82 Standard Star Catalog, where multiple observations are averaged to arrive at magnitudes with a smaller scatter (via the central limit theorem; see §3.4). The lower signal-to-noise data (top-right panel) are derived from single epoch observations. Though only two dimensions are plotted, the XD fit is performed on a five-dimensional data set, consisting of the g-band magnitude along with the $u - g, g - r, r - i$, and $i - z$ colors.

The results of the XD fit to the noisy data are shown in the two middle panels: the background distribution is fit by a single wide Gaussian, while the remaining clusters trace the main locus of points. The points drawn from the resulting distribution have a much tighter scatter than the input data. This decreased scatter can be

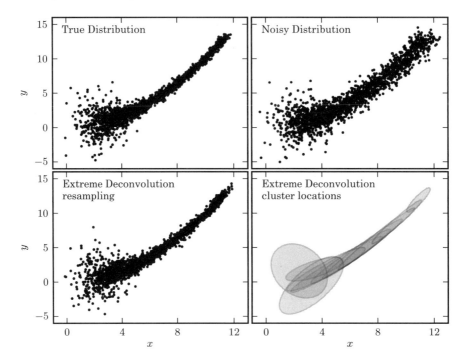

Figure 6.11. An example of extreme deconvolution showing a simulated two-dimensional distribution of points, where the positions are subject to errors. The top two panels show the distributions with small (left) and large (right) errors. The bottom panels show the densities derived from the noisy sample (top-right panel) using extreme deconvolution; the resulting distribution closely matches that shown in the top-left panel.

quantitatively demonstrated by analyzing the width of the locus perpendicular to its long direction using the so-called w color; see [17].

The w color is defined as

$$w = -0.227g + 0.792r - 0.567i + 0.05, \qquad (6.27)$$

and has a zero mean by definition. The lower panel of figure 6.12 shows a histogram of the width of the w color in the range $0.3 < g - r < 1.0$ (i.e., along the "blue" part of the locus where w has a small standard deviation). The noisy data show a spread in w of 0.016 (magnitude), while the extreme deconvolution model reduces this to 0.008, better reflective of the true underlying distribution. Note that the intrinsic width of the w color obtained by XD is actually a bit smaller than the corresponding width for the Standard Star Catalog (0.010) because even the averaged data have residual random errors. By subtracting 0.008 from 0.010 in quadrature, we can estimate these errors to be 0.006, in agreement with independent estimates; see [17].

Last but not least, XD can gracefully treat cases of missing data: in this case the corresponding measurement error can be simply set to a very large value (much larger than the dynamic range spanned by available data).

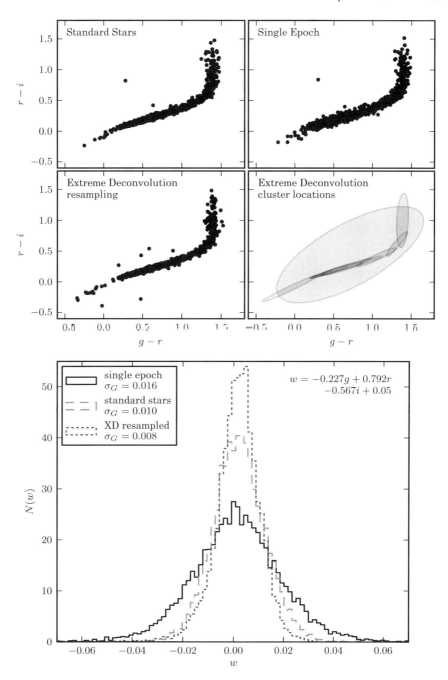

Figure 6.12. Extreme deconvolution applied to stellar data from SDSS Stripe 82. The top panels compare the color distributions for a high signal-to-noise sample of standard stars (left) with lower signal-to-noise, single epoch, data (right). The middle panels show the results of applying extreme deconvolution to the single epoch data. The bottom panel compares the distributions of a color measured perpendicularly to the locus (the so-called w color is defined following [16]). The distribution of colors from the extreme deconvolution of the noisy data recovers the tight distribution of the high signal-to-noise data.

6.4. Finding Clusters in Data

"Clustering" in astronomy refers to a number of different aspects of data analysis. Given a multivariate point data set, we can ask whether it displays any structure, that is, concentrations of points. Alternatively, when a density estimate is available we can search for "overdensities." Another way to interpret clustering is to seek a partitioning or segmentation of data into smaller parts according to some criteria. In the following section we describe the techniques used for the unsupervised identification of clusters within point data sets. Again, here "unsupervised" means that there is no prior information about the number and properties of clusters.

6.4.1. General Aspects of Clustering and Unsupervised Learning

Finding clusters is sometimes thought of as "black art" since the objective criteria for it seems more elusive than, say, for a prediction task such as classification (where we know the true underlying function for at least some subset of the sample). When we can speak of a *true underlying function* (as we do in most density estimation, classification, and regression methods) we mean that we have a score or error function with which to evaluate the effectiveness of our analysis. Under this model we can discuss optimization, error bounds, generalization (i.e., minimizing error on future data), what happens to the error as we get more data, etc. In other words we can leverage all the powerful tools of statistics we have discussed previously.

6.4.2. Clustering by Sum-of-Squares Minimization: K-Means

One of the simplest methods for partitioning data into a small number of clusters is K-means. K-means seeks a partitioning of the points into K disjoint subsets C_k with each subset containing N_k points such that the following sum-of-squares objective function is minimized:

$$\sum_{k=1}^{K} \sum_{i \in C_k} ||x_i - \mu_k||^2, \qquad (6.28)$$

where $\mu_k = \frac{1}{N_k} \sum_{i \in C_k} x_i$ is the mean of the points in set C_k, and $C(x_i) = C_k$ denotes that the class of x_i is C_k.

The procedure for K-means is to initially choose the centroid, μ_k, of each of the K clusters. We then assign each point to the cluster that it is closest to (i.e., according to $C(x_i) = \arg\min_k ||x_i - \mu_k||$). At this point we update the centroid of each cluster by recomputing μ_k according to the new assignments. The process continues until there are no new assignments.

While a globally optimal minimum cannot be guaranteed, the process can be shown to never increase the sum-of-squares error. In practice K-means is run multiple times with different starting values for the centroids of C_k and the result with the lowest sum-of-squares error is used. K-means can be interpreted as a "hard" version of the EM algorithm for a mixture of spherical Gaussians (i.e., we are assuming with K-means that the data can be described by spherical clusters with each cluster containing approximately the same number of points).

Scikit-learn implements K-means using expectation maximization:

```
import numpy as np
from sklearn.cluster import KMeans

X = np.random.normal(size=(1000, 2)) # 1000 pts in 2
    #dims
clf = KMeans(n_clusters=3)
clf.fit(X)
centers=clf.cluster_centers_ # locations of the
        # clusters
labels=clf.predict(X) # labels for each of the
        # points
```

For more information, see the Scikit-learn documentation.

In figure 6.13 we show the application of K-means to the stellar metallicity data used for the Gaussian mixture model example in figure 6.6. For the $K = 4$ clusters (consistent with figure 6.6) we find that the background distribution "pulls" the centroid of two of the clusters such that they are offset from the peak of the density distributions. This contrasts with the results found for the GMM described in §6.3 where we model the two density peaks (with the additional Gaussians capturing the structure in the distribution of background points).

6.4.3. Clustering by Max-Radius Minimization: the Gonzalez Algorithm

An alternative to minimizing the sum of square errors is to minimize the maximum radius of a cluster,

$$\min_k \max_{x_i \in C_k} ||x_i - \mu_k||, \tag{6.29}$$

where we assign one of the points within the data set to be the center of each cluster, μ_k.

An effective algorithm for finding the cluster centroids is known as the *Gonzalez algorithm*. Starting with no clusters we progressively add one cluster at a time (by arbitrarily selecting a point within the data set to be the center of the cluster). We then find the point x_i which maximizes the distance from the centers of existing clusters and set that as the next cluster center. This procedure is repeated until we achieve K clusters. At this stage each point in the data set is assigned the label of its nearest cluster center.

6.4.4. Clustering by Nonparametric Density Estimation: Mean Shift

Another way to find arbitrarily shaped clusters is to define clusters in terms of the modes or peaks of the nonparametric density estimate, associating each data point with its closest peak. This so-called *mean-shift* algorithm is a technique to find local modes (bumps) in a kernel density estimate of the data. The concept behind mean

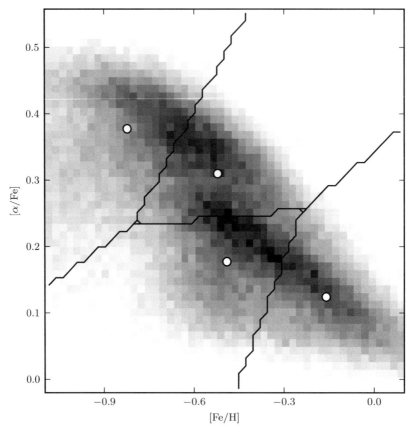

Figure 6.13. The K-means analysis of the stellar metallicity data used in figure 6.6. Note how the background distribution "pulls" the cluster centers away from the locus where one would place them by eye. This is why more sophisticated models like GMM are often better in practice.

shift is that we move the data points in the direction of the log of the gradient of the density of the data, until they finally converge to each other at the peaks of the bumps. The number of modes, K, is found implicitly by the method.

Suppose x_i^m is the position of the ith data point during iteration m of the procedure. A kernel density estimate $\widehat{f}^m$ is constructed from the points $\{x_i^m\}$. We obtain the next round of points according to an update procedure:

$$x_i^{m+1} = x_i^m + a\nabla \log \widehat{f}^m(x_i^m) \tag{6.30}$$

$$= x_i^m + \frac{a}{\widehat{f}^m(x_i^m)}\nabla \widehat{f}^m(x_i^m), \tag{6.31}$$

where $\widehat{f}^m(x_i^m)$ is found by kernel density estimation and $\nabla \widehat{f}^m(x_i^m)$ is found by kernel density estimation using the gradient of the original kernel.

The convergence of this procedure is defined by the bandwidth, h, of the kernel and the parametrization of a. For example, points drawn from a spherical Gaussian will jump to the centroid in one step when a is set to the variance of the Gaussian. The log of the gradient of the density ensures that the method converges in a few iterations, with points in regions of low density moving a considerable distance toward regions of high density in each iteration.

For the Epanechnikov kernel (see §6.1.1) and the value

$$a = \frac{h^2}{D+2},$$

(6.32)

the update rule reduces to the form

$$x_i^{m+1} = \text{mean position of points } x_i^m \text{ within distance } h \text{ of } x_i^m.$$

(6.33)

This is called the *mean-shift algorithm*.

Mean shift is implemented in Scikit-learn:

```
import numpy as np
from sklearn.cluster import MeanShift

X = np.random.normal(size=(1000, 2)) # 1000 pts in 2
    # dims
ms = MeanShift(bandwidth=1.0)
    # if no bandwidth is specified,
    # it will be learned from data
ms.fit(X) # fit the data
centers = ms.cluster_centers_ # centers of clusters
labels = ms.labels_ # labels of each point X
```

For more examples and information, see the source code for figure 6.14 and the Scikit-learn documentation.

An example of the mean-shift algorithm is shown in figure 6.14 using the same metallicity data set used in figures 6.6 and 6.13. The algorithm identifies the modes (or bumps) within the density distributions without attempting to model the correlation of the data within the clusters (i.e., the resulting clusters are axis aligned).

6.4.5. Clustering Procedurally: Hierarchical Clustering

A *procedural* method is a method which has not been formally related to some function of the underlying density. Such methods are more common in clustering

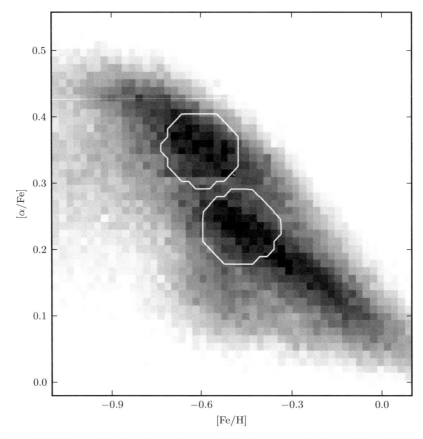

Figure 6.14. Mean-shift clustering on the metallicity data set used in figures 6.6 and 6.13. The method finds two clusters associated with local maxima of the distribution (interior of the circles). Points outside the circles have been determined to lie in the background. The mean shift does not attempt to model correlation in the clusters: that is, the resulting clusters are axis aligned.

and dimension reduction than in other tasks. Although this makes it hard, if not impossible, to say much about these methods analytically, they are nonetheless often still useful in practice.

Hierarchical clustering relaxes the need to specify the number of clusters K by finding all clusters at all scales. We start by partitioning the data into N clusters, one for each point in the data set. We can then join two of the clusters resulting in $N - 1$ clusters. This procedure is repeated until the Nth partition contains one cluster. If two points are in the same cluster at level m, and remain together at all subsequent levels, this is known as *hierarchical clustering* and is visualized using a tree diagram or *dendrogram*.

Hierarchical clustering can be approached as a top-down (*divisive*) procedure, where we progressively subdivide the data, or as a bottom-up (*agglomerative*) procedure, where we merge the nearest pairs of clusters. For our examples below we will consider the agglomerative approach.

At each step in the clustering process we merge the "nearest" pair of clusters. Options for defining the distance between two clusters, C_k and $C_{k'}$, include

$$d_{\min}(C_k, C_{k'}) = \min_{x \in C_k, x' \in C_{k'}} ||x - x'||, \tag{6.34}$$

$$d_{\max}(C_k, C_{k'}) = \max_{x \in C_k, x' \in C_{k'}} ||x - x'||, \tag{6.35}$$

$$d_{\text{avg}}(C_k, C_{k'}) = \frac{1}{N_k N_{k'}} \sum_{x \in C_k} \sum_{x' \in C_{k'}} ||x - x'||, \tag{6.36}$$

$$d_{\text{cen}}(C_k, C_{k'}) = ||\mu_k - \mu_{k'}||, \tag{6.37}$$

where x and x' are the points in cluster C_k and $C_{k'}$ respectively, N_k and $N_{k'}$ are the number of points in each cluster, and μ_k and $\mu_{k'}$ the centroid of the clusters.

Using the distance $d_{\min}$ results in a hierarchical clustering known as a minimum spanning tree (see [1, 4, 20], for some astronomical applications) and will commonly produce clusters with extended chains of points. Using $d_{\max}$ tends to produce hierarchical clustering with compact clusters. The other two distance examples have behavior somewhere between these two extremes.

A hierarchical clustering model, using $d_{\min}$, for the SDSS "Great Wall" data is shown in figure 6.15. The extended chains of points expressed by this minimum spanning tree trace the large-scale structure present within the data. Individual clusters can be isolated by sorting the links (or edges as they are known in graph theory) by increasing length, then deleting those edges longer than some threshold. The remaining components form the clusters. For a *single-linkage hierarchical clustering* this is also known as "friends-of-friends" clustering [31], and in astronomy is often used in cluster analysis for N-body simulations (e.g., [2, 9]).

Unfortunately a minimum spanning tree is naively $\mathcal{O}(N^3)$ to compute, using straightforward algorithms. Well-known graph-based algorithms, such as Kruskal's [19] and Prim's [32] algorithms, must consider the entire set of edges (see WSAS for details). In our Euclidean setting, there are $\mathcal{O}(N^2)$ possible edges, rendering these algorithms too slow for large data sets. However it has recently been shown how to perform the computation in approximately $\mathcal{O}(N \log N)$ time; see [27].

Figure 6.15 shows an *approximate* Euclidean minimum spanning tree, which finds the minimum spanning tree of the graph built using the k nearest neighbors of each point. The calculation is enabled by utility functions in SciPy and Scikit-learn:

```
from scipy.sparse.csgraph import\
    minimum_spanning_tree
from sklearn.neighbors import kneighbors_graph

X = np.random.random((1000, 2)) # 1000 pts in 2 dims
G = kneighbors_graph(X, n_neighbors=10,
                     mode='distance')
T = minimum_spanning_tree(G)
```

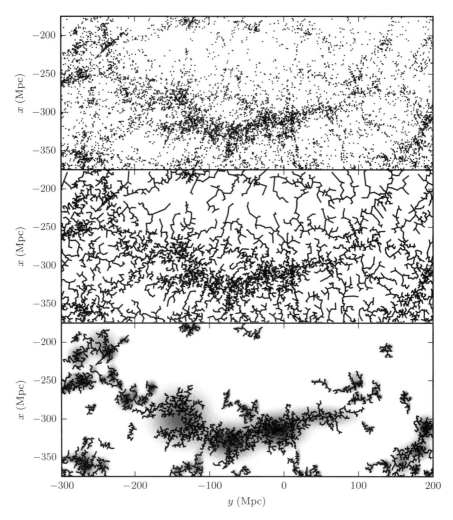

Figure 6.15. An approximate Euclidean minimum spanning tree over the two-dimensional projection of the SDSS Great Wall. The upper panel shows the input points, and the middle panel shows the dendrogram connecting them. The lower panel shows clustering based on this dendrogram, created by removing the largest 10% of the graph edges, and keeping the remaining connected clusters with 30 or more members. See color plate 4.

The result is that T is a 1000×1000 sparse matrix, with T[i, j] = 0 for nonconnected points, and T[i, j] is the distance between points i and j for connected points. This algorithm will be efficient for small n_neighbors. If we set n_neighbors=1000 in this case, then the approximation to the Euclidean minimum spanning tree will be exact. For well-behaved data sets, this approximation is often exact even for $k \ll N$. For more details, see the source code of figure 6.15.

6.5. Correlation Functions

In earlier sections we described the search for structure within point data using density estimation (§6.1) and cluster identification (§6.4). For point processes, a popular extension to these ideas is the use of correlation functions to characterize how far (and on what scales) the distribution of points differs from a random distribution; see [28, 30]. Correlation functions, and in particular autocorrelation functions, have been used extensively throughout astrophysics with examples of their use including the characterization of the fluctuations in the densities of galaxies and quasars as a function of luminosity, galaxy type and age of the universe. The key aspect of these statistics is that they can be used as metrics for testing models of structure formation and evolution directly against data.

We can define the correlation function by noting that the probability of finding a point in a volume element, dV, is directly proportional to the density of points, ρ. The probability of finding a pair of points in two volume elements, dV_1 and dV_2, separated by a distance, r, (see the left panel of figure 6.16) is then given by

$$dP_{12} = \rho^2 dV_1 dV_2 (1 + \xi(r)), \tag{6.38}$$

where $\xi(r)$ is known as the *two-point correlation function*.

From this definition, we see that the two-point correlation function describes the excess probability of finding a pair of points, as a function of separation, compared to a random distribution. Positive, negative, or zero amplitudes in $\xi(r)$ correspond to distributions that are respectively correlated, anticorrelated or random. The two-point correlation function relates directly to the power spectrum, $P(k)$, through the Fourier transform (see §10.2.2),

$$\xi(r) = \frac{1}{2\pi^2} \int dk \, k^2 \, P(k) \, \frac{\sin(kr)}{kr} \tag{6.39}$$

with the scale or wavelength of a fluctuation, λ is related to the wave number k by $k = 2\pi/\lambda$. As such, the correlation function can be used to describe the density fluctuations of sources by

$$\xi(r) = \left\langle \frac{\delta\rho(x)}{\rho} \frac{\delta\rho(x+r)}{\rho} \right\rangle, \tag{6.40}$$

where $\delta\rho(x)/\rho = (\rho - \bar{\rho})/\rho$ is the density contrast, relative to the mean value $\bar{\rho}$, at position x.

In studies of galaxy distributions, $\xi(r)$ is often parametrized in terms of a power law,

$$\xi(r) = \left(\frac{r}{r_0}\right)^{-\gamma}, \tag{6.41}$$

where r_0 is the clustering scale length and γ the power law exponent (with $r_0 \sim$ 6 Mpc and $\gamma \sim 1.8$ for galaxies in the local universe). Rather than considering the full three-dimensional correlation function given by eq. 6.38, we often desire instead to look at the angular correlation function of the apparent positions of objects on the

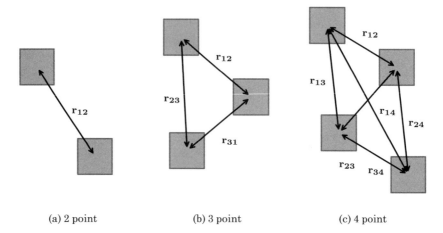

(a) 2 point   (b) 3 point   (c) 4 point

Figure 6.16. An example of n-tuple configurations for the two-point, three-point, and four-point correlation functions (reproduced from WSAS).

sky (e.g., [8]). In this case, the approximate form of the relation is given by

$$w(\theta) = \left(\frac{\theta}{\theta_0}\right)^\delta \tag{6.42}$$

with $\delta = 1 - \gamma$ (see [26]).

Correlation functions can be extended to orders higher than the two-point function by considering configurations of points that comprise triplets (three-point function), quadruplets (four-point function), and higher multiplicities (n-point functions). Figure 6.16 shows examples of these configurations for the three- and four-point correlation function. Analogously to the definition of the two-point function, we can express these higher-order correlation functions in terms of the probability of finding a given configuration of points. For example, for the three-point correlation function we define the probability dP_{123} of finding three points in volume elements dV_1, dV_2, and dV_3 that are defined by a triangle with sides r_{12}, r_{13}, r_{23}. We write the three-point correlation function as

$$dP_{123} = \rho^3 dV_1 dV_2 dV_3 (1 + \xi(r_{12}) + \xi(r_{23}) + \xi(r_{13}) + \zeta(r_{12}, r_{23}, r_{13})) \tag{6.43}$$

with ζ known as the *reduced* or *connected three-point correlation function* (i.e., it does not depend on the lower-order correlation functions). The additional two-point correlation function terms in eq. 6.43 simply reflect triplets that arise from the excess of pairs of galaxies due to the nonrandom nature of the data.

6.5.1. Computing the n-point Correlation Function

n-point correlation functions are an example of the general n-point problems discussed in chapter 2. For simplicity, we start with the two-point correlation function, $\xi(r)$, which can be estimated by calculating the excess or deficit of pairs of points within a distance r and $r + dr$ compared to a random distribution. These random points are generated with the same selection function as the data (i.e., within the same volume and with identical masked regions)—the random data represent a Monte Carlo integration of the window function (see §10.2.2) of the data.

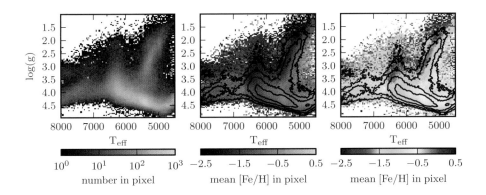

Plate 1. See figure 1.11.

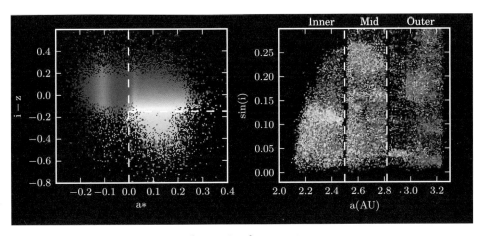

Plate 2. See figure 1.12.

HEALPix Pixels (Mollweide)

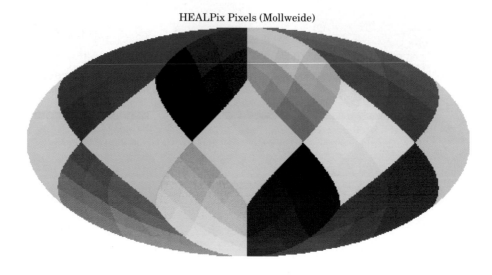

Raw WMAP data

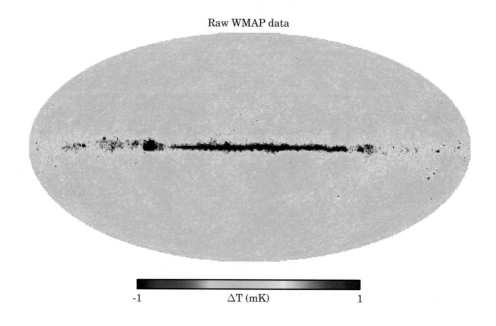

-1 ΔT (mK)   1

Plate 3. See figure 1.15.

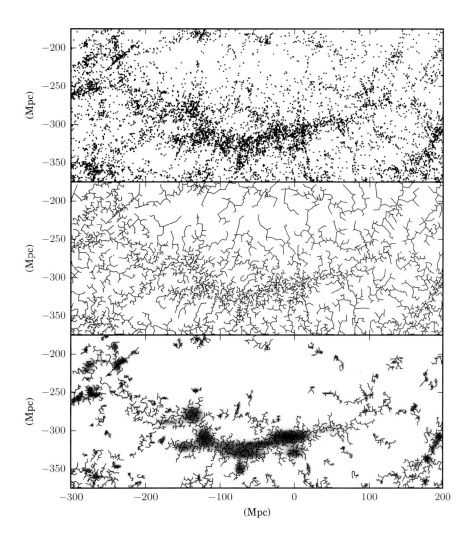

Plate 4. See figure 6.15.

PCA projection

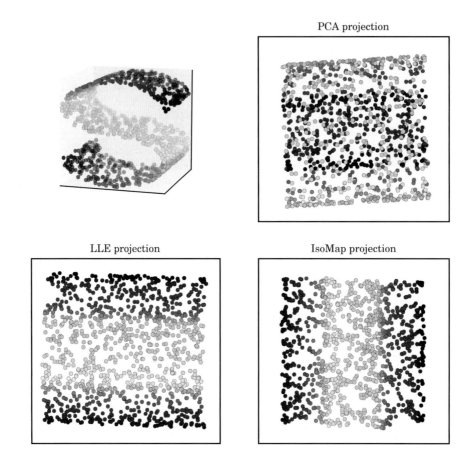

LLE projection

IsoMap projection

Plate 5. See figure 7.8.

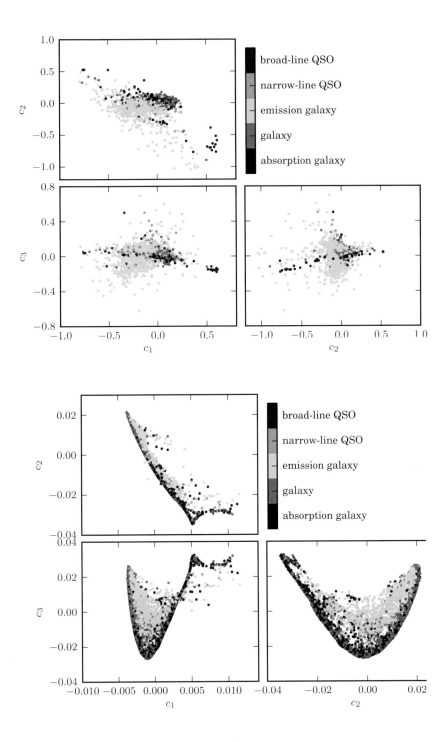

Plate 6. See figure 7.9.

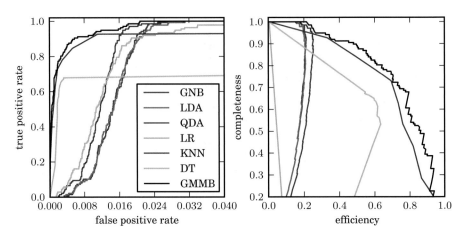

Plate 7. See figure 9.17.

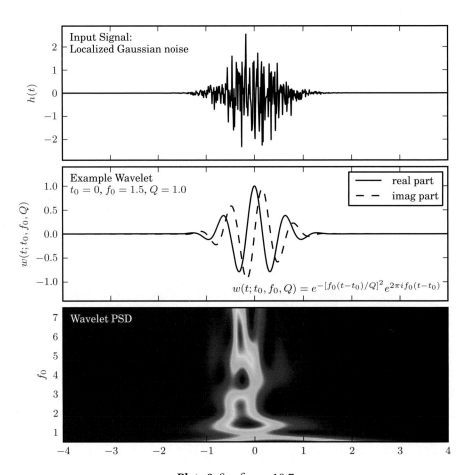

Plate 9. See figure 10.7.

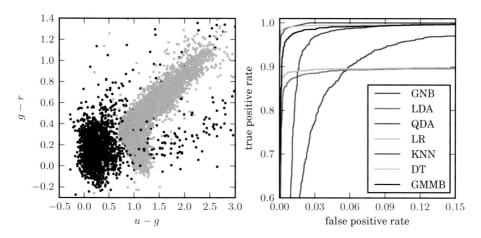

Plate 8. See figure 9.18.

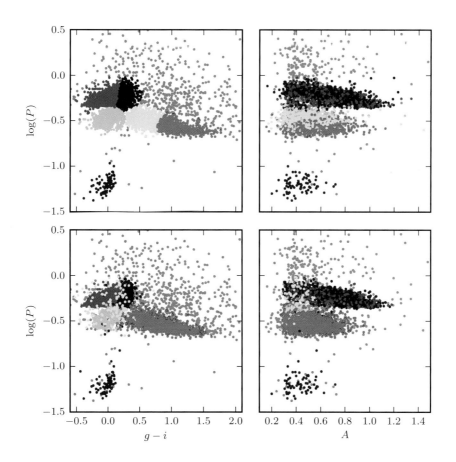

Plate 10. See figure 10.20.

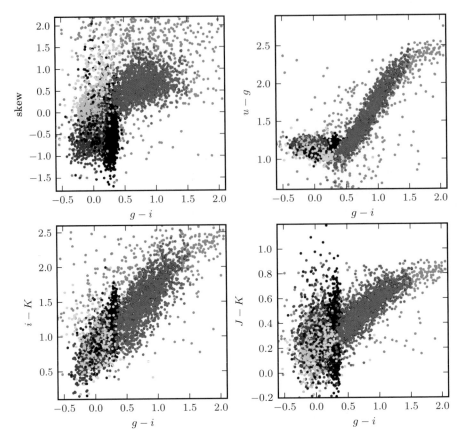

Plate 11. See figure 10.21.

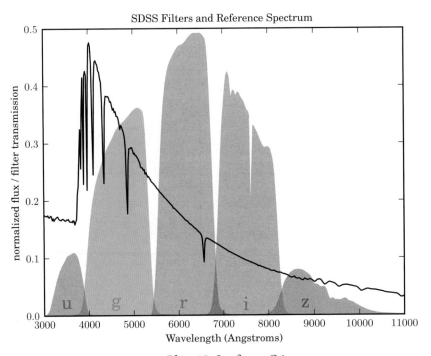

Plate 12. See figure C.1.

Typically the random data have a density $\sim$20 times higher than that of the data (to ensure that the shot noise of the randoms does not contribute to the variance of the estimator). This means that the computational cost of estimating the correlation function is dominated by the size of the random data set.

If we write the number of pairs of data points as $DD(r)$, the number of pairs of random points as $RR(r)$ and the number of data-random pairs as $DR(r)$, then we can write an estimator of the two-point correlation function as

$$\hat{\xi}(r) = \frac{DD(r)}{RR(r)} - 1. \qquad (6.44)$$

Edge effects, due to the interaction between the distribution of sources and the irregular survey geometry in which the data reside, bias estimates of the correlation function. Other estimators have been proposed which have better variance and are less sensitive to edge effects than the classic estimator of eq. 6.44. One example is the Landy–Szalay estimator (see [21]),

$$\hat{\xi}(r) = \frac{DD(r) - 2DR(r) + RR(r)}{RR(r)}. \qquad (6.45)$$

The Landy–Szalay estimator can be extended to higher-order correlation functions (see [39]). For the three-point function this results in

$$\hat{\xi}(r) = \frac{DDD(r) - 3DDR(r) + DRR(r) - RRR(r)}{RRR(r)}, \qquad (6.46)$$

where $DDD(r)$ represents the number of data triplets as defined by the triangular configuration shown in the central panel of figure 6.16 and $DDR(r)$, $DRR(r)$, and $RRR(r)$ are the associated configurations for the data-data-random, data-random-random, and random-random-random triplets, respectively. We note that eq. 6.46 is specified for an equilateral triangle configuration (i.e., all internal angles are held constant and the triangle configuration depends only on r). For more general triangular configurations, $DDD(r)$ and other terms depend on the lengths of all three sides of the triangle, or on the lengths of two sides and the angle between them.

AstroML implements a two-point correlation function estimator based on the Scikit-learn ball-tree:

```
import numpy as np
from astroML.correlation import two_point_angular

RA = 40 * np.random.random(1000) # RA and DEC in
     # degrees
DEC = 10 * np.random.random(1000)
bins = np.linspace(0.1, 10, 11) # evaluate in 10 bins
     # with these edges
corr = two_point_angular(RA, DEC, bins,
                  method='landy-szalay')
```

For more information, refer to the source code of figure 6.17.

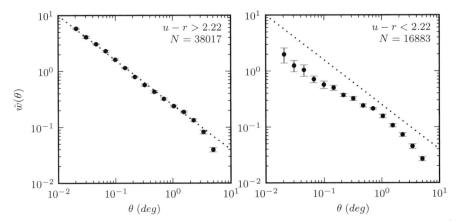

Figure 6.17. The two-point correlation function of SDSS spectroscopic galaxies in the range $0.08 < z < 0.12$, with $m < 17.7$. This is the same sample for which the luminosity function is computed in figure 4.10. Errors are estimated using ten bootstrap samples. Dotted lines are added to guide the eye and correspond to a power law proportional to $\theta^{-0.8}$. Note that the red galaxies (left panel) are clustered more strongly than the blue galaxies (right panel).

In figure 6.17, we illustrate the angular correlation function for a subset of the SDSS spectroscopic galaxy sample, in the range $0.08 < z < 0.12$. The left panel shows the correlation function for red galaxies with $u - r > 2.22$ and the right panel for blue galaxies with $u - r < 2.22$. The error bars on these plots are derived from ten bootstrap samples (i.e., independent volumes; see §4.5). Note that the clustering on small scales is much stronger for red than for blue galaxies. With 38,017 and 16,883 galaxies in the samples, a ball-tree-based implementation offers significant improvement in the computation time for the two-point correlation function over a brute-force method.

The naive computational scaling of the n-point correlation function, where we evaluate all permutations of points, is $\mathcal{O}(N^n)$ (with N the size of the data and n the order of the correlation function). For large samples of points, the computational expense of this operation can become prohibitive. Space-partitioning trees (as introduced in §2.5.2) can reduce this computational burden to $\mathcal{O}(N^{\log n})$. The underlying concept behind these efficient tree-based correlation function algorithms is the exclusion or pruning of regions of data that do not match the configuration of the correlation function (e.g., for the two-point function, pairs of points that lie outside of the range r to $r + dr$). By comparing the minimum and maximum pairwise distances between the bounding boxes of two nodes of a tree we can rapidly identify and exclude those nodes (and all child nodes) that do not match the distance constraints. This dramatically reduces the number of pairwise calculations.

A number of public implementations of tree-based correlation functions are available that utilize these techniques. These include applications optimized for single processors [29] and for parallel systems [13]. For a more thorough algorithmic discussion of computing n-point statistics, we refer the reader to WSAS.

6.6. Which Density Estimation and Clustering Algorithms Should I Use?

In this section, and the analogous sections at the end of each chapter in part III of the book, we will touch upon many of the broad issues in the "art" (as opposed to "science") of machine learning (ML), and therefore our discussions are necessarily subjective, incomplete, and cursory. However we hope this summary of unsupervised methods introduced in this chapter can serve as a springboard into the practice of ML.

We can define four axes of practical "goodness" along which one can place each machine learning method that we have described in this chapter, as well as each method in subsequent chapters:

- **Accuracy**: How well can it make accurate predictions or model data?
- **Interpretability**: How easy is it to understand why the model is making the predictions that it does, or reason about its behavior?
- **Simplicity**: Is the model "fiddly," that is, does it have numerous parameters that must be tuned and tweaked with manual effort? Is it difficult to program?
- **Speed**: Is the method fast, or is it possible via sophisticated algorithms to make it fast without altering its accuracy or other properties?

What are the most* accurate *unsupervised methods? Generally speaking, the more parameters a model has, the more flexibility it has to fit complex functions, thus reducing error. Consider first the task of *density estimation*, where the error in question is some notion of the difference between the true underlying probability density function and the estimated function. One broad division that can be made is between parametric and nonparametric methods. Parametric methods find the best model within a parametric model class, that is, one with a fixed number of model parameters. A Gaussian, or a mixture of k Gaussians, and the augmented mixture of Gaussians represented by the extreme deconvolution method are examples of a parametric model for density estimation. If the true underlying function is not a member of the chosen model class, even the model with the best setting of its parameters simply cannot fit the data. Nonparametric methods conceptually have a number of parameters that grows in some way as the number of data points N grows. Kernel density estimation is the standard nonparametric approach for density estimation in higher dimensions. Such methods can be shown to be able to fit virtually any underlying function (under very mild assumptions on the function). In general, achieving the highest accuracies requires nonparametric methods.

Note that while a mixture model (such as a mixture of Gaussians) is parametric for a fixed number of mixture components, in principle if that number is chosen in a way that can grow with N, such a model becomes nonparametric (e.g., when modeling the underlying density we can use many more components than suggested by criteria such as BIC). In practice, due to the complication and computational cost of trying every possible number of parameters, ad hoc approaches are typically used to decide the number of parameters for such models, making their status along the parametric–nonparametric scale murkier.

For the task of clustering, parametric-style options include K-means and max-radius minimization. In practice, the various heuristic procedures that are used to determine the number of clusters K often place them arguably in the aforementioned murky regime. Mean shift, being based on kernel density estimation, is a nonparametric method in spirit, though its final step is typically similar to K-means. Hierarchical clustering is also nonparametric in spirit, in that it finds clusters at all scales, possibly $\mathcal{O}(N)$ clusters. While density estimation methods have been studied in great mathematical detail, the analogous formal statements one can make for clustering methods (e.g., regarding whether the true underlying cluster structure is recovered by a method, and if so, at what convergence rate, etc.) are much scarcer, making it difficult to speak precisely about the accuracy of clustering methods in general.

In this chapter we have also touched upon other types of unsupervised tasks such as two-sample testing. Two-sample testing using n-point correlation functions increases in accuracy as n goes up, where accuracy in this setting is related to the subtlety of the distributional differences that are detectable and the power of the test. We have not contrasted these statistics with other available statistics, but generally speaking other statistics do not extend to higher orders as the n-point statistics do.

What are the most* interpretable *methods? All of the methods we have described have the concept of a "distance" underneath them, which arguably makes them all fairly intuitive. For density estimation, mixtures of Gaussians and kernel density estimation are based on functions of Euclidean distance. For clustering, K-means, max-radius minimization, and mean shift are all based on the Euclidean distance, while hierarchical clustering uses Euclidean distance as the base notion of similarity between points and thereafter uses some distance-like notion of similarity between clusters.

The question of interpretability, in the context of many unsupervised methods, ventures into the question of validity. In clustering, as we have already touched upon, this issue is particularly acute. Most of the clustering methods, at least, can be understood as the solutions to optimization problems, that is, they find the clusterings that are optimal under some score function for clusterings. Then the clusters are valid or invalid depending on one's assessment of whether the particular score function is sensible. Hierarchical clustering is more opaque in this regard—we do not have the same sense in which the clusters it finds are the "right" ones because it is simply a procedure that begins and ends, but does not necessarily optimize any clear score function.

Among density estimation methods, the nonparametric one, kernel estimation, is perhaps the least straightforward in the sense of understanding why it works. It is in a certain sense easy to understand as a smoothing of the data by adding up "blurred" versions of the points, in the form of miniature pdfs, rather than delta functions. However, understanding why this simple method yields a legitimate nonparametric estimator of the true underlying density requires a leap into the mathematics of asymptotic statistics (see, e.g., [40]).

Parametric models, generally speaking, are more interpretable than nonparametric models because the meanings of each of the parts of the model are usually clear. A Bayesian parametric model takes this to the ultimate level by making a further level of parametric assumptions explicit. While being parametric requires assumptions which may in fact be wrong, leading to worse error, they are at least clear and understood. A nonparametric model is often effectively a collection of small model bits, and so the effect or meaning of each part of the overall model is not clear, making nonparametric models more like black boxes.

What are the most* scalable *unsupervised methods? Max-radius minimization is generally the most efficient of the methods discussed in this chapter, requiring only K passes over the N points. K-means, for a single run (corresponding to one setting of the initial parameters and K), is quite efficient, requiring some number of iterations over the N points which is typically in the tens or hundreds. Computing Gaussian mixtures is similar to computing K-means, but each iteration is more expensive as it involves, among other things, inverting $D \times D$ covariance matrices in the evaluations of Gaussians. All of the aforementioned methods are fairly efficient because each of the N points is somehow compared to K centroids, and K is generally small, making them scale as $\mathcal{O}(N)$ per iteration with some unknown number of iterations. Kernel density estimation is similar except that the K is now replaced by N, making the computational cost quadratic in N. Mean shift, being based on KDE, is quadratic as well, and hierarchical clustering is even worse, becoming as bad as $\mathcal{O}(N^3)$ depending on the variant. Depending on the order of the statistic, n-point correlation functions are quadratic, cubic, etc. in N.

All of these methods can be sped up by fast exact (no approximation) tree-based algorithms, which generally reduce what was originally an $\mathcal{O}(N)$ cost to $\mathcal{O}(\log N)$, or an $\mathcal{O}(N^2)$ cost to $\mathcal{O}(N)$. Even $\mathcal{O}(N^3)$ costs can be reduced, in some cases to $\mathcal{O}(N \log N)$. Most of these runtimes have been proven mathematically, though some are only empirical and remain to be rigorously analyzed. The dimension of the points will affect the runtimes of such algorithms in the form of a constant exponential factor (in the dimension) in the worst case, that is, in high dimensions the algorithms will be slower. However, the quantity that matters is some notion of *intrinsic* dimension (roughly, the dimension of the submanifold upon which the data lie; see the discussion in §2.5.2), which is generally not known in advance. It is rare for a real data set not to have an intrinsic dimension which is substantially lower than its extrinsic dimension, due to the fact that dimensions tend to have some correlation with each other. In astronomical problems, even the extrinsic dimensionality is often low. When the dimensionality is too high for exact tree-based algorithms to be efficient, one can turn to approximate versions of such algorithms.

What are the* simplest *unsupervised methods to use? Max-radius minimization, not a usual staple of machine learning books, is interesting in part because of its simplicity. K-means is also one of the simplest machine learning methods to implement. Gaussian mixtures represent a step up in complexity, introducing probabilities, matrix inversions, and the resulting numerical issues. We will characterize all of these methods as somewhat "fiddly," meaning that they are not easily automatable—one must typically fiddle with the critical parameter(s), in this case K, and look at the results of many runs for each K tried, due to the fact that any particular run from an initial random starting point of the parameters may not achieve the global optimum in general. In principle, model selection for any ML method can be automated by performing cross-validation over the entire set of plausible values of the critical parameter(s) (see §6.1.1). It is more difficult in practice for these types of unsupervised problems because the best K typically does not jump out clearly from the resulting error curve.

Extreme deconvolution is an example of a Bayesian formulation of a standard ML method, in this case Gaussian mixtures. Standard ML methods, by and large, are frequentist, though for most common ones there have been Bayesianization attempts with various degrees of practical success. Broadly speaking, performing ML in a strictly Bayesian manner involves the highest amount of fiddliness for any problem, because the notion of priors adds extra distributional choices, at a minimum, generally with parameters that are explicitly not chosen in an automated data-driven manner, but by definition manually. Though simple to implement, at least at a baseline level, MCMC as a general computational workhorse adds significantly to the amount of fiddliness due to the ever-present difficulty in being sure that the Markov chain has converged, leading to many runs with different starting points and MCMC parameter settings, and frequently manual examination of the samples produced. Bayesian formulations of ML methods, extreme deconvolution being an example, which pose learning as optimization problems rather than using MCMC, can reduce this aspect of fiddliness.

Kernel density estimation, though representing an increase in statistical expressiveness over Gaussian mixtures, is simpler in that there is no parameter fitting at all, and only requires varying its critical parameter, the bandwidth h. Mean shift requires KDE-like steps in its optimization iterations, followed typically by K-means, putting it in the fiddly category. Hierarchical clustering is arguably both more powerful and simpler than methods like K-means that require selection of K because it finds clustering at all scales, with no parameter fitting or random initializations, finding the same clusters each time it is run.

Other considerations, and taste Other considerations include problem-specific circumstances which some methods may be able to handle more naturally or effectively than others. For example, probabilistic models that can be optimized via the EM algorithm, such as Gaussian mixtures, can be naturally augmented to perform missing value imputation, that is, using the best model so far to fill in guesses for any missing entries in the data matrix. Extreme deconvolution is a specific augmentation of Gaussian mixtures to incorporate measurement errors, while standard formulations of ML methods do not come with a way to incorporate errors.

Often in astronomy, physical insight may also play a role in the choice of analysis method. For example, the two-point correlation function is a mathematical transformation of the power spectrum, the expected form of which can be computed (at least in the simplest linear regime) analytically from first principles.

In reality the final consideration, though often not admitted or emphasized as much as practice reflects, is *taste*. A strong branching in taste often occurs at the decision of whether to join either the Bayesian or frequentist camp. The more common decision is to treat the set of available paradigms and methods as a smorgasbord from which to choose, according to the trade-offs they represent and the characteristics of one's problem. But in practice several different methods may not be strongly distinguishable, at least not initially, from a pure performance point of view. At that point, one may simply be attracted to the "style" or connotations of a method, such as the MLE roots of Gaussian mixtures, or the nonparametric theory roots of KDE.

TABLE 6.1.
Summary of the practical properties of different unsupervised techniques.

Method	Accuracy	Interpretability	Simplicity	Speed
K-nearest-neighbor	H	H	H	M
Kernel density estimation	H	H	H	H
Gaussian mixture models	H	M	M	M
Extreme deconvolution	H	H	M	M
K-means	L	M	H	M
Max-radius minimization	L	M	M	M
Mean shift	M	H	H	M
Hierarchical clustering	H	L	L	L
Correlation functions	H	M	M	M

Simple summary Table 6.1 is a quick summary of our assessment of each of the methods considered in this chapter, in terms of high (H), medium (M), and low (L). Note that the table mixes methods for different tasks (density estimation, clustering, correlation functions), so they are not all directly comparable to each other.

References

[1] Allison, R. J., S. P. Goodwin, R. J. Parker, and others (2009). Using the minimum spanning tree to trace mass segregation. *MNRAS 395*, 1449–1454.

[2] Audit, E., R. Teyssier, and J.-M. Alimi (1998). Non-linear dynamics and mass function of cosmic structures. II. Numerical results. *A&A 333*, 779–789.

[3] Balogh, M., V. Eke, C. Miller, and others (2004). Galaxy ecology: Groups and low-density environments in the SDSS and 2dFGRS. *MNRAS 348*, 1355–1372.

[4] Barrow, J. D., S. P. Bhavsar, and D. H. Sonoda (1985). Minimal spanning trees, filaments and galaxy clustering. *MNRAS 216*, 17–35.

[5] Bovy, J., J. Hennawi, D. Hogg, and others (2011). Think outside the color box: Probabilistic target selection and the SDSS-XDQSO Quasar Targeting Catalog. *ApJ 729*, 141.

[6] Bovy, J., D. Hogg, and S. Roweis (2011). Extreme deconvolution: Inferring complete distribution functions from noisy, heterogeneous and incomplete observations. *The Annals of Applied Statistics 5*(2B), 1657–1677.

[7] Casertano, S. and P. Hut (1985). Core radius and density measurements in N-body experiments. Connections with theoretical and observational definitions. *ApJ 298*, 80–94.

[8] Connolly, A. J., R. Scranton, D. Johnston, and others (2002). The angular correlation function of galaxies from early Sloan Digital Sky Survey data. *ApJ 579*, 42–47.

[9] Davis, M., G. Efstathiou, C. S. Frenk, and S. White (1985). The evolution of large-scale structure in a universe dominated by cold dark matter. *ApJ 292*, 371–394.

[10] Delaigle, A. and A. Meister (2008). Density estimation with heteroscedastic error. *Bernoulli 14*(2), 562–579.

[11] Dressler, A. (1980). Galaxy morphology in rich clusters—implications for the formation and evolution of galaxies. *ApJ 236*, 351–365.

[12] Ferdosi, B. J., H. Buddelmeijer, S. C. Trager, M. Wilkinson, and J. Roerdink (2011). Comparison of density estimation methods for astronomical datasets. *A&A 531*, A114.

[13] Gardner, J. P., A. Connolly, and C. McBride (2007). Enabling rapid development of parallel tree search applications. In *Proceedings of the 5th IEEE Workshop on Challenges of Large Applications in Distributed Environments*, CLADE '07, New York, NY, USA, pp. 1–10. ACM.

[14] Gott, III, J. R., M. Jurić, D. Schlegel, and others (2005). A map of the universe. *ApJ 624*, 463–484.

[15] Gray, A. G. and A. W. Moore (2003). Nonparametric density estimation: Toward computational tractability. In *SIAM International Conference on Data Mining (SDM)*.

[16] Ivezić, Ž., R. H. Lupton, D. Schlegel, and others (2004). SDSS data management and photometric quality assessment. *Astronomische Nachrichten 325*, 583–589.

[17] Ivezić, Ž., J. A. Smith, G. Miknaitis, and others (2007). Sloan Digital Sky Survey Standard Star Catalog for Stripe 82: The dawn of industrial 1% optical photometry. *AJ 134*, 973–998.

[18] Ivezić, Ž., A. K. Vivas, R. H. Lupton, and R. Zinn (2005). The selection of RR Lyrae stars using single-epoch data. *AJ 129*, 1096–1108.

[19] Kruskal, J. B. (1956). On the shortest spanning subtree of a graph and the traveling salesman problem. *Proceedings of the American Mathematical Society 7*(1), 48–50.

[20] Krzewina, L. G. and W. C. Saslaw (1996). Minimal spanning tree statistics for the analysis of large-scale structure. *MNRAS 278*, 869–876.

[21] Landy, S. D. and A. S. Szalay (1993). Bias and variance of angular correlation functions. *ApJ 412*, 64–71.

[22] Lee, D. and A. G. Gray (2006). Faster Gaussian summation: Theory and experiment. In *Conference on Uncertainty in Artificial Intelligence (UAI)*.

[23] Lee, D. and A. G. Gray (2009). Fast high-dimensional kernel summations using the Monte Carlo multipole method. In *Advances in Neural Information Processing Systems (NIPS) (Dec 2008)*. MIT Press.

[24] Lee, D., A. G. Gray, and A. W. Moore (2006). Dual-tree fast gauss transforms. In Y. Weiss, B. Schölkopf, and J. Platt (Eds.), *Advances in Neural Information Processing Systems (NIPS) (Dec 2005)*. MIT Press.

[25] Lee, D., R. Vuduc, and A. G. Gray (2012). A distributed kernel summation framework for general-dimension machine learning. In *SIAM International Conference on Data Mining (SDM)*.

[26] Limber, D. N. (1953). The analysis of counts of the extragalactic nebulae in terms of a fluctuating density field. *ApJ 117*, 134.

[27] March, W. B., P. Ram, and A. G. Gray (2010). Fast Euclidean minimum spanning tree: algorithm, analysis, and applications. In *Proceedings of the 16th ACM SIGKDD international conference on Knowledge discovery and data mining*, KDD '10, pp. 603–612. ACM.

[28] Martínez, V. and E. Saar (2002). *Statistics of the Galaxy Distribution*. Chapman and Hall/CRC.

[29] Moore, A., A. Connolly, C. Genovese, and others (2000). Fast algorithms and efficient statistics: N-point correlation functions. *Mining the Sky 370*(2), 71–+.

[30] Peebles, P. (1980). *The Large-Scale Structure of the Universe*. Princeton Series in Physics. Princeton University Press.

[31] Press, W. H. and M. Davis (1982). How to identify and weigh virialized clusters of galaxies in a complete redshift catalog. *ApJ 259*, 449–473.

[32] Prim, R. C. (1957). Shortest connection networks and some generalizations. *Bell System Technology Journal 36*, 1389–1401.

[33] Ram, P., D. Lee, W. March, and A. G. Gray (2010). Linear-time algorithms for pairwise statistical problems. In *Advances in Neural Information Processing Systems (NIPS) (Dec 2009)*. MIT Press.

[34] Roche, A. (2011). EM algorithm and variants: An informal tutorial. *ArXiv:statistics/1105.1476*.

[35] Scott, D. W. (2008). *Multivariate Density Estimation: Theory, Practice, and Visualization*. John Wiley & Sons, Inc., Hoboken, NJ, USA.

[36] Sesar, B., Ž. Ivezić, S. H. Grammer, and others (2010). Light curve templates and galactic distribution of RR Lyrae stars from Sloan Digital Sky Survey Stripe 82. *ApJ 708*, 717–741.

[37] Silverman, B. (1986). *Density Estimation for Statistics and Data Analysis*. Monographs on Statistics and Applied Probability. Chapman and Hall.

[38] Stefanski, L. and J. Raymond (1990). Deconvolving kernel density estimators. *Statistics: A Journal of Theoretical and Applied Statistics 21*(2), 169–184.

[39] Szapudi, S. and A. S. Szalay (1998). A new class of estimators for the N-point correlations. *ApJL 494*, L41.

[40] van der Vaart, A. W. (2000). *Asymptotic Statistics*. Cambridge Series in Statistical and Probabilistic Mathematics. Cambridge University Press.

[41] Wand, M. P. and M. C. Jones (1995). Kernel smoothing. Chapman and Hall.

7 Dimensionality and Its Reduction

"A mind that is stretched by a new idea can never go back to its original dimensions."
(Oliver Wendell Holmes)

With the dramatic increase in data available from a new generation of astronomical telescopes and instruments, many analyses must address the question of the complexity as well as size of the data set. For example, with the SDSS imaging data we could measure arbitrary numbers of properties or features for any source detected on an image (e.g., we could measure a series of progressively higher moments of the distribution of fluxes in the pixels that make up the source). From the perspective of efficiency we would clearly rather measure only those properties that are directly correlated with the science we want to achieve. In reality we do not know the correct measurement to use or even the optimal set of functions or bases from which to construct these measurements. This chapter deals with how we can learn which measurements, properties, or combinations thereof carry the most information within a data set. The techniques we will describe here are related to concepts we have discussed when describing Gaussian distributions (§3.5.2), density estimation (§6.1), and the concepts of information content (§5.2.2).

We will start in §7.1 with an exploration of the problems posed by high-dimensional data. In §7.2 we will describe the data sets used in this chapter, and in §7.3, we will introduce perhaps the most important and widely used dimensionality reduction technique, principal component analysis (PCA). In the remainder of the chapter, we will introduce several alternative techniques which address some of the weaknesses of PCA.

7.1. The Curse of Dimensionality

Imagine that you have decided to purchase a car and that your initial thought is to purchase a "fast" car. Whatever the exact implementation of your search strategy, let us assume that your selection results in a fraction $r < 1$ of potential matches. If you expand the requirements for your perfect car such that it is "red," has "8 cylinders," and a "leather interior" (with similar selection probabilities), then your selection probability would be r^4. If you also throw "classic car" into the mix, the selection probability becomes r^5. The more selection conditions you adopt, the tinier is the chance of finding your ideal car! These selection conditions are akin to dimensions in your data set, and this effect is at the core of the phenomenon known as the "curse of dimensionality"; see [2].

From a data analysis point of view, the curse of dimensionality impacts the size of data required to constrain a model, the complexity of the model itself, and the search times required to optimize the model. Considering just the first of these issues, we can quantitatively address the question of how much data is needed to estimate the probability of finding points within a high-dimensional space. Imagine that you have drawn N points from a D-dimensional uniform distribution, described by a hypercube (centered at the origin) with edge length 2 (i.e., each coordinate lies within the range $[-1, 1]$). Under a Euclidean distance metric, what proportion of points fall within a unit distance of the origin?

If the points truly are uniformly distributed throughout the volume, then a good estimate of this is the ratio of the volume of a unit hypersphere centered at the origin, to the volume of the side-length=2 hypercube centered at the origin. For two dimensions, this gives

$$f_2 = \frac{\pi r^2}{(2r)^2} = \pi/4 \approx 78.5\%. \tag{7.1}$$

For three dimensions, this becomes

$$f_3 = \frac{(4/3)\pi r^3}{(2r)^3} = \pi/6 \approx 52.3\%. \tag{7.2}$$

Generalizing to higher dimensions requires some less familiar formulas for the hypervolumes of D-spheres. It can be shown that the hypervolume of a D-dimensional hypersphere with radius r is given by

$$V_D(r) = \frac{2r^D \pi^{D/2}}{D\,\Gamma(D/2)}, \tag{7.3}$$

where $\Gamma(z)$ is the complete gamma function. The reader can check that evaluating this for $D = 2$ and $D = 3$ yield the familiar formulas for the area of a circle and the volume of a sphere. With this formula we can compute

$$f_D = \frac{V_D(r)}{(2r)^D} = \frac{\pi^{D/2}}{D2^{D-1}\Gamma(D/2)}. \tag{7.4}$$

One can quite easily show that

$$\lim_{D\to\infty} f_D = 0. \tag{7.5}$$

That is, in the limit of many dimensions, *not a single point* will be within a unit radius of the origin! In other words, the fraction of points within a search radius r (relative to the full space) will tend to zero as the dimensionality grows and, therefore, the number of points in a data set required to evenly sample this hypervolume will grow exponentially with dimension. In the context of astronomy, the SDSS [1] comprises

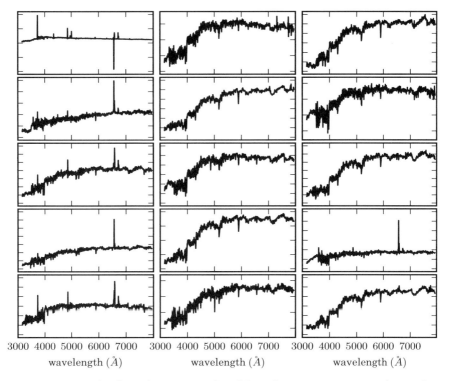

Figure 7.1. A sample of 15 galaxy spectra selected from the SDSS spectroscopic data set (see §1.5.5). These spectra span a range of galaxy types, from star-forming to passive galaxies. Each spectrum has been shifted to its rest frame and covers the wavelength interval 3000–8000 Å. The specific fluxes, $F_\lambda(\lambda)$, on the ordinate axes have an arbitrary scaling.

a sample of 357 million sources. Each source has 448 measured attributes (e.g., measures of flux, size, shape, and position). If we used our physical intuition to select just 30 of those attributes from the database (e.g., a subset of the magnitude, size, and ellipticity measures) and normalized the data such that each dimension spanned the range -1 to 1, the probability of having one of the 357 million sources reside within the unit hypersphere would be only one part in 1.4×10^5.

Given the dimensionality of current astronomical data sets, how can we ever hope to find or characterize any structure that might be present? The underlying assumption behind our earlier discussions has been that all dimensions or attributes are created equal. We know that is, however, not true. There exist projections within the data that capture the principal physical and statistical correlations between measured quantities (this idea lies behind the *intrinsic dimensionality* discussed in §2.5.2). Finding these dimensions or axes efficiently and thereby reducing the dimensionality of the underlying data is the subject of this chapter.

7.2. The Data Sets Used in This Chapter

Throughout this chapter we use the SDSS galaxy spectra described in §1.5.4 and §1.5.5, as a proxy for high-dimensional data. Figure 7.1 shows a representative sample of these spectra covering the interval 3200–7800 Å in 1000 wavelength bins. While a spectrum defined as $x(\lambda)$ may not immediately be seen as a point in a high-dimensional space, it can be represented as such. The function $x(\lambda)$ is in practice sampled at D discrete flux values, and written as a D-dimensional vector. And just as a three-dimensional vector is often visualized as a point in a three-dimensional space, this spectrum (represented by a D-dimensional vector) can be thought of as a single point in D-dimensional space. Analogously, a $D = N \times K$ image may also be expressed as a vector with D elements, and therefore a point in a D-dimensional space. So, while we use spectra as our proxy for high-dimensional space, the algorithms and techniques described in this chapter are applicable data as diverse as catalogs of multivariate data, two-dimensional images, and spectral hypercubes.

7.3. Principal Component Analysis

Figure 7.2 shows a two-dimensional distribution of points drawn from a Gaussian centered on the origin of the x- and y-axes. While the points are strongly correlated along a particular direction, it is clear that this correlation does not align with the initial choice of axes. If we wish to reduce the number of features (i.e., the number of axes) that are used to describe these data (providing a more compact representation) then it is clear that we should rotate our axes to align with this correlation (we have already encountered this rotation in eq. 3.82). Any rotation preserves the relative ordering or configuration of the data so we choose our rotation to maximize the ability to discriminate between the data points. This is accomplished if the rotation maximizes the variance along the resulting axes (i.e., defining the first axis, or principal component, to be the direction with maximal variance, the second principal component to be orthogonal to the first component that maximizes the residual variance, and so on). As indicated in figure 7.2, this is mathematically equivalent to a regression that minimizes the square of the orthogonal distances from the points to the principal axes.

This dimensionality reduction technique is known as a principal component analysis (PCA). It is also referred to in the literature as a Karhunen–Loéve [21, 25] or Hotelling transform. PCA is a linear transform, applied to multivariate data, that defines a set of uncorrelated axes (the principal components) ordered by the variance captured by each new axis. It is one of the most widely applied dimensionality reduction techniques used in astrophysics today and dates back to Pearson who, in 1901, developed a procedure for fitting lines and planes to multivariate data; see [28].

There exist a number of excellent texts on PCA that review its use across a broad range of fields and applications (e.g., [19] and references therein). We will, therefore, focus our discussion of PCA on a brief description of its mathematical formalism then concentrate on its application to astronomical data and its use as a tool for classification, data compression, regression, and signal-to-noise filtering of high-dimensional data sets.

Before progressing further with the application of PCA, it is worth noting that many of the applications of PCA to astronomical data describe the importance of the orthogonal nature of PCA (i.e., the ability to project a data set onto a set of uncorrelated axes). It is often forgotten that the observations themselves are already a

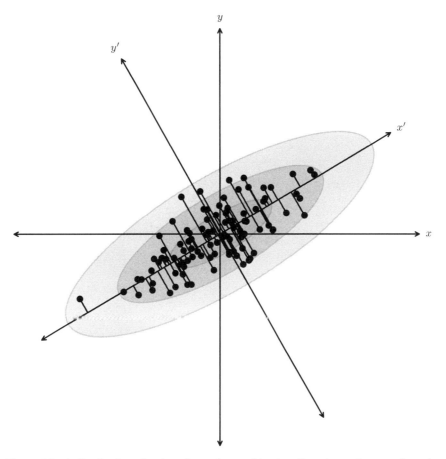

Figure 7.2. A distribution of points drawn from a bivariate Gaussian and centered on the origin of x and y. PCA defines a rotation such that the new axes (x' and y') are aligned along the directions of maximal variance (the principal components) with zero covariance. This is equivalent to minimizing the square of the perpendicular distances between the points and the principal components.

representation of an orthogonal basis (e.g., the axes $\{1,0,0,0,\dots\}$, $\{0,1,0,0,0,\dots\}$, etc.). As we will show, the importance of PCA is that *the new axes are aligned with the direction of maximum variance within the data* (i.e., the direction with the maximum signal).

7.3.1. The Derivation of Principal Component Analyses

Consider a set of data, $\{x_i\}$, comprising a series of N observations with each observation made up of K measured features (e.g., size, color, and luminosity, or the wavelength bins in a spectrum). We initially center the data by subtracting the mean of each feature in $\{x_i\}$ and then write this $N \times K$ matrix as X.[1] The covariance

---

[1] Often the opposite convention is used: that is, N points in K dimensions are stored in a $K \times N$ matrix rather than an $N \times K$ matrix. We choose the latter to align with the convention used in Scikit-learn and AstroML.

of the centered data, C_X, is given by

$$C_X = \frac{1}{N-1} X^T X, \tag{7.6}$$

where the $N - 1$ term comes from the fact that we are working with the sample covariance matrix (i.e., the covariances are derived from the data themselves).

Nonzero off-diagonal components within the covariance matrix arise because there exist correlations between the measured features (as we saw in figure 7.2; recall also the discussion of bivariate and multivariate distributions in §3.5). PCA wishes to identify a projection of $\{x_i\}$, say, R, that is aligned with the directions of maximal variance. We write this projection as $Y = XR$ and its covariance as

$$C_Y = R^T X^T X R = R^T C_X R \tag{7.7}$$

with C_X the covariance of X as defined above.

The first principal component, r_1, of R is defined as the projection with the maximal variance (subject to the constraint that $r_1^T r_1 = 1$). We can derive this principal component by using Lagrange multipliers and defining the cost function, $\phi(r_1, \lambda)$, as

$$\phi(r_1, \lambda_1) = r_1^T C_X r_1 - \lambda_1 (r_1^T r_1 - 1). \tag{7.8}$$

Setting the derivative of $\phi(r_1, \lambda)$ (with respect to r_1) to zero gives

$$C_X r_1 - \lambda_1 r_1 = 0. \tag{7.9}$$

λ_1 is, therefore, the root of the equation $\det(C_X - \lambda_1 \mathbf{I}) = 0$ and is an eigenvalue of the covariance matrix. The variance for the first principal component is maximized when

$$\lambda_1 = r_1^T C_X r_1 \tag{7.10}$$

is the largest eigenvalue of the covariance matrix. The second (and further) principal components can be derived in an analogous manner by applying the additional constraint to the cost function that the principal components are uncorrelated (e.g., $r_2^T C_X r_1 = 0$).

The columns of R are then the eigenvectors or principal components, and the diagonal values of C_Y define the amount of variance contained within each component. With

$$C_X = R C_Y R^T \tag{7.11}$$

and ordering the eigenvectors by their eigenvalue we can define the set principal components for X.

Efficient computation of principal components

One of the most direct methods for computing the PCA is through the eigenvalue decomposition of the covariance or correlation matrix, or equivalently through the

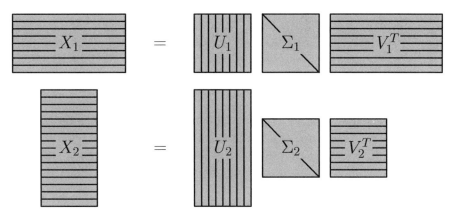

Figure 7.3. Singular value decomposition (SVD) can factorize an $N \times K$ matrix into $U\Sigma V^T$. There are different conventions for computing the SVD in the literature, and this figure illustrates the convention used in this text. The matrix of singular values Σ is always a square matrix of size $[R \times R]$ where $R = \min(N, K)$. The shape of the resulting U and V matrices depends on whether N or K is larger. The columns of the matrix U are called the left-singular vectors, and the columns of the matrix V are called the right-singular vectors. The columns are orthonormal bases, and satisfy $U^T U = V^T V = I$,

singular value decomposition (SVD) of the data matrix itself. The scaled SVD can be written

$$U \Sigma V^T = \frac{1}{\sqrt{N-1}} X, \tag{7.12}$$

where the columns of U are the *left-singular vectors*, and the columns of V are the *right-singular vectors*. There are many different conventions for the SVD in the literature; we will assume the convention that the matrix of singular values Σ is always a square, diagonal matrix, of shape $[R \times R]$ where $R = \min(N, K)$ is the rank of the matrix X (assuming all rows and columns of X are independent). U is then an $[N \times R]$ matrix, and V^T is an $[R \times K]$ matrix (see figure 7.3 for a visualization of this SVD convention). The columns of U and V form orthonormal bases, such that $U^T U = V^T V = I$.

Using the expression for the covariance matrix (eq. 7.6) along with the scaled SVD (eq. 7.12) gives

$$\begin{aligned} C_X &= \left[\frac{1}{\sqrt{N-1}} X \right]^T \left[\frac{1}{\sqrt{N-1}} X \right] \\ &= V \Sigma U^T U \Sigma V^T \\ &= V \Sigma^2 V^T. \end{aligned} \tag{7.13}$$

Comparing to eq 7.11, we see that the right singular vectors V correspond to the principal components R, and the diagonal matrix of eigenvalues C_Y is equivalent to the square of the singular values,

$$\Sigma^2 = C_Y. \tag{7.14}$$

Thus the eigenvalue decomposition of C_X, and therefore the principal components, can be computed from the SVD of X, without explicitly constructing the matrix C_X.

NumPy and SciPy contain powerful suites of linear algebra tools. For example, we can confirm the above relationship using svd for computing the SVD, and eigh for computing the symmetric (or in general Hermitian) eigenvalue decomposition:

```
>>> import numpy as np
>>> X = np.random.random((100, 3))
>>> CX = np.dot(X.T, X)
>>> U, Sdiag, VT = np.linalg.svd(X, full_matrices=
    False)
>>> CYdiag, R = np.linalg.eigh(CX)
```

The full_matrices keyword assures that the convention shown in figure 7.3 is used, and for both Σ and C_Y, only the diagonal elements are returned. We can compare the results, being careful of the different ordering conventions: svd puts the largest singular values first, while eigh puts the smallest eigenvalues first:

```
>>> np.allclose(CYdiag, Sdiag[::-1] ** 2)
 # [::-1] reverses the array
True
>>> np.set_printoptions(suppress=True)
 # clean output for below
>>> VT[::-1].T / R
array([[-1., -1.,  1.],
       [-1., -1.,  1.],
       [-1., -1.,  1.]])
```

The eigenvectors of C_X and the right singular vectors of X agree up to a sign, as expected. For more information, see appendix A or the documentation of numpy.linalg and scipy.linalg.

The SVD formalism can also be used to quickly see the relationship between the covariance matrix C_X, and the correlation matrix,

$$\begin{aligned}
M_X &= \frac{1}{N-1} X X^T \\
&= U \Sigma V^T V \Sigma U^T \\
&= U \Sigma^2 U^T
\end{aligned} \tag{7.15}$$

in analogy with above. The left singular vectors, U, turn out to be the eigenvectors of the correlation matrix, which has eigenvalues identical to those of the covariance matrix. Furthermore, the orthonormality of the matrices U and V means that if U is known, V (and therefore R) can be quickly determined using the linear algebraic manipulation of eq. 7.12:

$$R = V = \frac{1}{\sqrt{N-1}} X^T U \Sigma^{-1}. \tag{7.16}$$

Thus we have three equivalent ways of computing the principal components R and the eigenvalues C_X: the SVD of X, the eigenvalue decomposition of C_X, or the eigenvalue decomposition of M_X. The optimal procedure will depend on the relationship between the data size N and the dimensionality K. If $N \gg K$, then using the eigenvalue decomposition of the $K \times K$ covariance matrix C_X will in general be more efficient. If $K \gg N$, then using the $N \times N$ correlation matrix M_X will be more efficient. In the intermediate case, direct computation of the SVD of X will be the most efficient route.

7.3.2. The Application of PCA

PCA can be performed easily using Scikit-learn:

```
import numpy as np
from sklearn.decomposition import PCA

X = np.random.normal(size=(100, 3))
    # 100 points in 3 dimensions
R = np.random.random((3, 10))  # projection matrix
X = np.dot(X, R)  # X is now 10-dim, with 5 intrinsic
    # dims
pca = PCA(n_components=4)  # n_components can be
    # optionally set
pca.fit(X)
comp = pca.transform(X)  # compute the subspace
    # projection of X

mean = pca.mean_  # length 10 mean of the data
components = pca.components_  # 4 x 10 matrix of
            # components
var = pca.explained_variance_  # the length 4 array
    # of eigenvalues
```

In this case, the last element of var will be zero, because the data is inherently three-dimensional. For larger problems, RandomizedPCA is also useful. For more information, see the Scikit-learn documentation.

To form the data matrix X, the data vectors are centered by subtracting the mean of each dimension. Before this takes place, however, the data are often preprocessed to ensure that the PCA is maximally informative. In the case of heterogeneous data (e.g., galaxy shape and flux), the columns are often preprocessed by dividing by

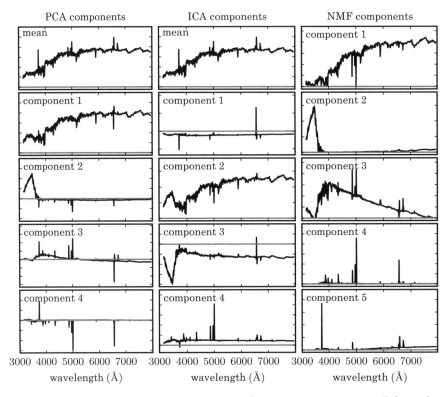

Figure 7.4. A comparison of the decomposition of SDSS spectra using PCA (left panel—see §7.3.1), ICA (middle panel—see §7.6) and NMF (right panel—see §7.4). The rank of the component increases from top to bottom. For the ICA and PCA the first component is the mean spectrum (NMF does not require mean subtraction). All of these techniques isolate a common set of spectral features (identifying features associated with the continuum and line emission). The ordering of the spectral components is technique dependent.

their variance. This so-called whitening of the data ensures that the variance of each feature is comparable, and can lead to a more physically meaningful set of principal components. In the case of spectra or images, a common preprocessing step is to normalize each row, such that the integrated flux of each object is one. This helps to remove uninteresting correlations based on the overall brightness of the spectrum or image.

For the case of the galaxy spectra in figure 7.1, each spectrum has been normalized to a constant total flux, before being centered such that the spectrum has zero mean (this subtracted mean spectrum is shown in the upper-left panel of figure 7.4). The principal directions found in the high-dimensional data set are often referred to as the "eigenspectra," and just as a vector can be represented by the sum of its components, a spectrum can be represented by the sum of its eigenspectra. The left panel of figure 7.4 shows, from top to bottom, the mean spectrum and the first four eigenspectra. The eigenspectra are ordered by their associated eigenvalues shown in figure 7.5. Figure 7.5 is often referred to as a scree plot (related to the shape of rock debris after it has fallen down a slope; see [6]) with the eigenvalues reflecting

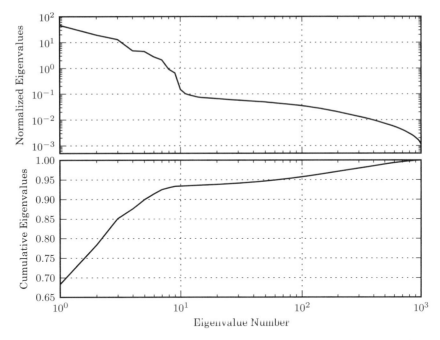

Figure 7.5. The eigenvalues for the PCA decomposition of the SDSS spectra described in §7.3.2. The top panel shows the decrease in eigenvalue as a function of the number of eigenvectors, with a break in the distribution at ten eigenvectors. The lower panel shows the cumulative sum of eigenvalues normalized to unity. 94% of the variance in the SDSS spectra can be captured using the first ten eigenvectors.

the amount of variance contained within each of the associated eigenspectra (with the constraint that the sum of the eigenvalues equals the total variance of the system).

The cumulative variance associated with the eigenvectors measures the amount of variance of the *entire data set* which is encoded in the eigenvectors. From figure 7.5, we see that ten eigenvectors are responsible for 94% of the variance in the sample: this means that by projecting each spectrum onto these first ten eigenspectra, an average of 94% of the "information" in each spectrum is retained, where here we use the term "information" loosely as a proxy for variance. This amounts to a compression of the data by a factor of 100 (using ten of the 1000 eigencomponents) with a very small loss of information. This is the sense in which PCA allows for dimensionality reduction.

This concept of data compression is supported by the shape of the eigenvectors. Eigenvectors with large eigenvalues are predominantly low-order components (in the context of astronomical data they primarily reflect the continuum shape of the galaxies). Higher-order components (with smaller eigenvalues) are predominantly made up of sharp features such as emission lines. The combination of continuum and line emission within these eigenvectors can describe any of the input spectra. The remaining eigenvectors reflect the noise within the ensemble of spectra in the sample.

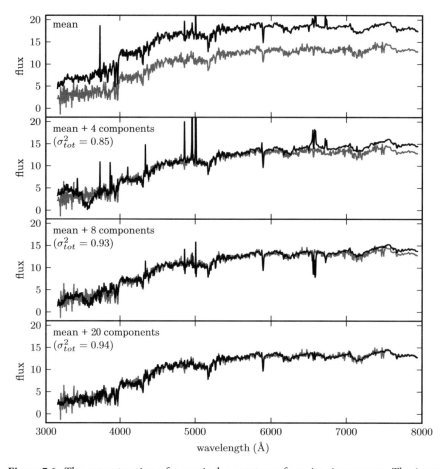

Figure 7.6. The reconstruction of a particular spectrum from its eigenvectors. The input spectrum is shown in gray, and the partial reconstruction for progressively more terms is shown in black. The top panel shows only the mean of the set of spectra. By the time 20 PCA components are added, the reconstruction is very close to the input, as indicated by the expected total variance of 94%.

The reconstruction of an example spectrum, $x(k)$, from the eigenbasis, $e_i(k)$ is shown in figure 7.6. Each spectrum $x_i(k)$ can be described by

$$x_i(k) = \mu(k) + \sum_{j}^{R} \theta_{ij} e_j(k), \qquad (7.17)$$

where i represents the number of the input spectrum, j represents the number of the eigenspectrum, and, for the case of a spectrum, k represents the wavelength. Here, $\mu(k)$ is the mean spectrum and θ_{ij} are the linear expansion coefficients derived from

$$\theta_{ij} = \sum_{k} e_j(k)(x_i(k) - \mu(k)). \qquad (7.18)$$

R is the total number of eigenvectors (given by the rank of X, $\min(N,K)$). If the summation is over all eigenvectors, the input spectrum is fully described with no loss of information. Truncating this expansion (i.e., $r < R$),

$$x_i(k) = \sum_{i}^{r < R} \theta_i \, e_i(k), \qquad (7.19)$$

will exclude those eigencomponents with smaller eigenvalues. These components will, predominantly, reflect the noise within the data set. This is reflected in figure 7.6: truncating the reconstruction at 20 components captures the overall shape and important features of the spectrum: the differences between the reconstruction and the input spectrum are mostly high-frequency spectral noise.

A number of aspects of PCA are worth noting. Comparisons between the eigenvectors derived from PCA and known spectral types of galaxies have shown that these statistically orthogonal components correlate strongly with specific physical properties (i.e., they relate to the star formation and the composition of the stellar types within a galaxy spectrum, e.g., [33, 34]). Second, given the cumulative nature of the sum of variances used in PCA, astrophysically interesting components within the spectra (e.g., sharp spectral lines or transient features for certain galaxy populations) may not be reflected in the largest PCA components. Because of this, care must be taken when truncating at a small number of components. Additionally, the assumption that a sum of linear components can efficiently reconstruct the features within the data does not always hold. An example of this is the variation in broad emission lines (such as those from quasars). The variation in line width is an inherently nonlinear process and can require a large number of components to fully characterize: for broad line quasars over 30 components are required to reproduce the underlying spectra compared to the 10 required for quiescent and star-forming galaxies. In these cases, dimensionality reduction techniques based on the local structure need to be considered (see §7.5). Finally, up to this point we have ignored errors and missing data when considering the application of PCA. We address this in §7.3.3.

Choosing the Level of Truncation in an Expansion

One of the critical issues when reconstructing a data set from a linear combination of eigenvectors is choosing the number of components, r, to keep. Too many components will introduce noise into the reconstruction. Too few may not capture the complete physical correlations within the data. While many attempts have been made to place the choice of r on a sound statistical footing, the techniques that are used today are typically either based on empirical relations derived from simplified experiments or derived from a series of somewhat ad hoc assumptions (see [19] for a detailed discussion).

The most common criterion for defining r is based on the total variance captured in the first r eigenvectors. If we specify a bound, α, on the fraction of the variance we wish to capture then we can define r from the summation of the eigenvalues, σ_i, that are the diagonals of the matrix Σ:

$$\frac{\sum_i^{i=r} \sigma_i}{\sum_i^{i=R} \sigma_i} < \alpha. \tag{7.20}$$

Typical values for α range from 0.70 to 0.95, though the choice of threshold is sensitive to the shape of the scree plot (figure 7.5); for a shallow slope in the scree plot, whether r or $r+1$ crosses the threshold is somewhat arbitrary.

The shape of the scree plot can be used to define the level of truncation. Cattell [6], using factor analysis, proposed that a sharp change in the gradient of the eigenvalues (i.e., a knee in the scree plot) could be used to define the cutoff value, r. The knee is defined by [6] as $\Sigma_r^2 - \Sigma_{r+1}^2$. If no clearly defined break in the scree plot exists, the definition of this cutoff becomes problematic. A modification of the technique in [6] is the LEV diagram, which plots the logarithm of the eigenvalue against the number of eigenvectors. The rationale for this modification is that if noise decays geometrically, variation in the eigenvalues should drop as a linear function.

For the correlation matrix PCA, Kaiser's rule [20] or the Guttman–Kaiser criterion can be applied. This states that, if all of the elements of x are independent then all principal components would have unit variance. In this case, r can be set to the limit where the eigenvalues in the scree plot fall below unity. In the context of the covariance matrix this can be reformulated as the setting of r to the number of components at which the eigenvalue falls to the average of all eigenvalues. Jolliffe [19] proposed a modification of this truncation to 70% of the average eigenvalues to allow for sample variance (which increases the number of components returned). Experiments with Kaiser's rule show that it tends to overpredict the number of components that remain after truncation.

Each of the criteria described above are sensitive to the shape of the scree plot and the choice of truncation rule remains application specific.

7.3.3. PCA with Missing Data

Until now we have assumed that the data we are working with are complete, without gaps or censored elements. In real-world applications, the presence of detector glitches, variable noise, or masking effects (e.g., sky lines in astronomical spectra) can make these assumptions invalid. Truncation of the expansion provides a signal-to-noise filtering of the data. The PCA bases should also be able to correct for missing elements within the data: because the PCA components encode the correlation of each flux with the other measured fluxes, these components should provide a natural way to determine these missing values.

One complication we must address is that eigenspectra are only defined to be orthogonal over the data range on which they are constructed. If a data vector does not fully cover that space then projecting the data onto the eigenbases will result in a biased set of expansion coefficients. Everson and Sirovich [12] have, however, shown that, when we know how the input data are masked or censored, we can correct for the nonorthogonality of the eigenbases. Following Connolly and Szalay [9], we consider an observed spectrum, x^o, as the combination of the true spectrum (i.e., without gaps), x, and a wavelength-dependent weight, w. This weight is zero where data are missing and $1/\sigma^2$ for the remaining spectral range (with σ^2 the variance of the spectral elements). Minimizing the quadratic deviation between the original spectrum, x^o, and its truncated reconstruction, $\sum_i \theta_i e_i$ and solving for θ_i gives

$$\sum_k \theta_i \boldsymbol{w}(k)\boldsymbol{e}_i(k)\boldsymbol{e}_j(k) = \sum_k \boldsymbol{w}(k)\boldsymbol{x}^o(k)\boldsymbol{e}_j(k), \tag{7.21}$$

where $\sum_k$ represents the sum over the length of the vector $\boldsymbol{x}(k)$ (i.e., over wavelength for the case of the spectra). If we define $M_{ij} = \sum_k \boldsymbol{w}(k)\boldsymbol{e}_i(k)\boldsymbol{e}_j(k)$ and $F_i = \sum_k \boldsymbol{w}(k)\boldsymbol{x}^o(k)\boldsymbol{e}_i(k)$ then this simplifies to

$$\theta_i = \sum_j M_{ij}^{-1} F_j, \tag{7.22}$$

where F_j represent the coefficients derived from the gappy data and M_{ij}^{-1} expresses how correlated the eigenvectors are over the missing regions.

Figure 7.7 shows the reconstruction of missing data using the PCA components. The gray regions represent intervals in wavelength space where the spectra are censored (the underlying data values within these censored regions are shown by the black line). The gray lines represent the reconstruction of these spectral regions using the eigenbases. An estimate of the uncertainty on the reconstruction coefficients is given by

$$\mathrm{Cov}(\theta_i, \theta_j) = M_{ij}^{-1}. \tag{7.23}$$

The accuracy of this reconstruction will depend on the distribution of the gaps within the data vector (though this is reflected in $\mathrm{Cov}(\theta_i, \theta_j)$). For an uncorrelated set of gaps, reconstruction with the number of sampled points, $N_{\mathrm{samples}} > r$, is possible. This observation is at the heart of the fields of lossy compression and compressed sensing.

7.3.4. Scaling to Large Data Sets

There are a number of limitations of PCA that can make it impractical for application to very large data sets. Principal to this are the computational and memory requirements of the SVD, which scale as $\mathcal{O}(D^3)$ and $\mathcal{O}(2 \times D \times D)$, respectively. In §7.3.1 we derived the PCA by applying an SVD to the covariance and correlation matrices of the data X. Thus, the computational requirements of the SVD are set by the rank of the data matrix, X, with the covariance matrix the preferred route if $K < N$ and the correlation matrix if $K > N$. Given the symmetric nature of both the covariance and correlation matrix, eigenvalue decompositions (EVD) are often more efficient than SVD approaches.

Even given these optimizations, with data sets exceeding the size of the memory available per core, applications of PCA can be very computationally challenging. This is particularly the case for real-world applications when the correction techniques for missing data are iterative in nature. One approach to address these limitations is to make use of online algorithms for an iterative calculation of the mean. As shown in

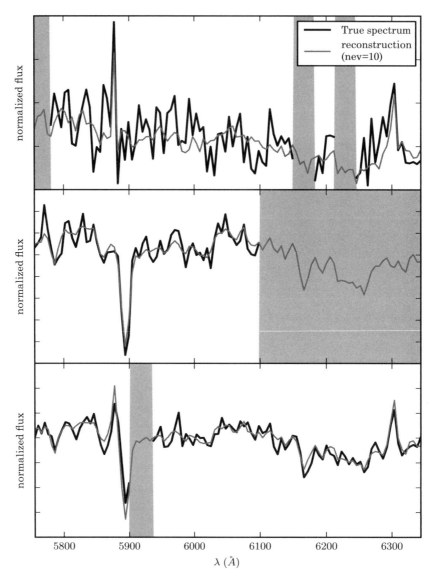

Figure 7.7. The principal component vectors defined for the SDSS spectra can be used to interpolate across or reconstruct missing data. Examples of three masked spectral regions are shown comparing the reconstruction of the input spectrum (black line) using the mean and the first ten eigenspectra (gray line) The gray bands represent the masked region of the spectrum.

[5], the sample covariance matrix can be defined as

$$C = \gamma C_{\text{prev}} + (1 - \gamma)x^T x \tag{7.24}$$

$$\sim \gamma Y_p D_p Y_p^T + (1 - \gamma)x^T x, \tag{7.25}$$

where C_{prev} is the covariance matrix derived from a previous iteration, x is the new observation, Y_p are the first p eigenvectors of the previous covariance matrix, D_p

are the associated eigenvalues, and γ is a weight parameter that can scale with step size.

For each new observation the covariance matrix is updated, and the PCA components based on this matrix enable the identification and filtering of outliers within a data set. Assuming that the number of required eigenvectors, p, is small, the cost of periodically reevaluating the SVD of the updated covariance matrix is a cheap operation.

7.4. Nonnegative Matrix Factorization

One of the challenges in interpreting PCA bases comes from the fact that the eigenvectors are defined relative to the mean data vector. This results in principal components that can be positive or negative. In contrast, for many physical systems we have a priori knowledge that a data vector can be represented by a linear sum of positive components. For example, in the case of the SDSS spectra, a galaxy spectrum can be assumed to be a linear sum of stellar components (consistent with the general form of the eigenspectra seen in figure 7.4).

Nonnegative matrix factorization (NMF) applies an additional constraint on the components that comprise the data matrix X; see [23]. It assumes that any data matrix can be factored into two matrices, W and Y, such that

$$X = WY, \tag{7.26}$$

where both W and Y are nonnegative (i.e., all elements in these matrices are nonnegative). WY is, therefore, an approximation of X. By minimizing the reconstruction error $||(X - WY)^2||$, it can be shown that nonnegative bases can be derived using a simple update rule,

$$W_{ki} = W_{ki} \frac{[XY^T]_{ki}}{[WYY^T]_{ki}}, \tag{7.27}$$

$$Y_{in} = Y_{in} \frac{[W^T X]_{in}}{[W^T WY]_{in}}, \tag{7.28}$$

where n, k, and i denote the wavelength, spectrum and template indices, respectively [23]. This iterative process does not guarantee nonlocal minima (as with many iterative machine learning approaches such as K-means and EM). With random initialization and cross-validation the solutions for the NMF bases are, however, often appropriate.

The central panel of figure 7.7 shows the results of NMF applied to the spectra used in the PCA analysis of §7.3.2. The components derived by NMF are broadly consistent with those from PCA but with a different ordering of the basis functions. Given that both applications assume a linear transform between X and Y this might not be that surprising. For the case of NMF, the assumption is that each spectrum can be represented by a smaller number of nonnegative components than the underlying dimensionality of the data. The number of components are, therefore, defined prior to the generation of the NMF and can be derived through cross-validation techniques described in §8.11.

Projecting onto the NMF bases is undertaken in a similar manner to eq. 7.27 except that, in this case, the individual components are held fixed; see [3].

Scikit-learn contains an implementation of NMF. The basic usage is as follows:

```python
import numpy as np
from sklearn.decomposition import NMF

X = np.random.random((100, 3))  # 100 points in 3
    # dims, all positive
nmf = NMF(n_components=3)  # setting n_components is
    # optional
nmf.fit(X)
proj = nmf.transform(X)  # project to 3 dimensions

comp = nmf.components_  # 3 x 10 array of components
err = nmf.reconstruction_err_  # how well 3
    # components captures data
```

There are many options to tune this procedure: for more information, refer to the Scikit-learn documentation.

7.5. Manifold Learning

PCA, NMF, and other linear dimensionality techniques are powerful ways to reduce the size of a data set for visualization, compression, or to aid in classification and regression. Real-world data sets, however, can have very nonlinear features which are hard to capture with a simple linear basis. For example, as we noted before, while quiescent galaxies can be well described by relatively few principal components, emission-line galaxies and quasars can require up to ∼ 30 linear components to completely characterize. These emission lines are nonlinear features of the spectra, and nonlinear methods are required to project that information onto fewer dimensions.

Manifold learning comprises a set of recent techniques which aim to accomplish this sort of nonlinear dimensionality reduction. A classic test case for this is the *S-curve* data set, shown in figure 7.8. This is a three-dimensional space, but the points are drawn from a two-dimensional manifold which is embedded in that space. Principal component analysis cannot capture this intrinsic information (see the upper-right panel of figure 7.8). There is no linear projection in which distant parts of the nonlinear manifold do not overlap. Manifold learning techniques, on the other hand, do allow this surface to be unwrapped or unfolded so that the underlying structure becomes clear.

In light of this simple example, one may wonder what can be gained from such an algorithm. While projecting from three to two dimensions is a neat trick, these

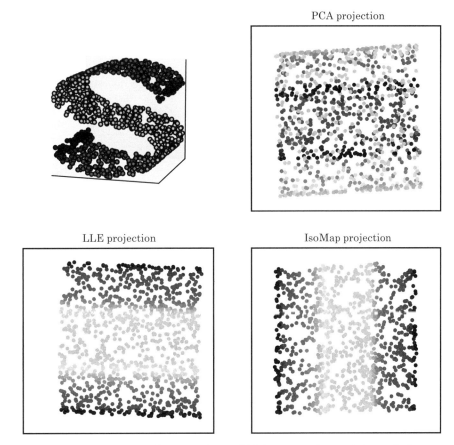

Figure 7.8. A comparison of PCA and manifold learning. The top-left panel shows an example S-shaped data set (a two-dimensional manifold in a three-dimensional space). PCA identifies three principal components within the data. Projection onto the first two PCA components results in a mixing of the colors along the manifold. Manifold learning (LLE and IsoMap) preserves the local structure when projecting the data, preventing the mixing of the colors. See color plate 5.

algorithms become very powerful when working with data like galaxy and quasar spectra, which lie in up to 4000 dimensions. Vanderplas and Connolly [32] first applied manifold learning techniques to galaxy spectra, and found that as few as two nonlinear components are sufficient to recover information which required dozens of components in a linear projection.

There are a variety of manifold learning techniques and variants available. Here we will discuss the two most popular: locally linear embedding (LLE) and IsoMap, short for isometric mapping.

7.5.1. Locally Linear Embedding

Locally linear embedding [29] is an unsupervised learning algorithm which attempts to embed high-dimensional data in a lower-dimensional space while preserving the geometry of local neighborhoods of each point. These local neighborhoods are determined by the relation of each point to its k nearest neighbors. The LLE algorithm consists of two steps: first, for each point, a set of weights is derived which best reconstruct the point from its k nearest neighbors. These weights encode the local geometry of each neighborhood. Second, with these weights held fixed, a new lower-dimensional data set is found which maintains the neighborhood relationships described by these weights.

Let us be more specific. Let X be an $N \times K$ matrix representing N points in K dimensions. We seek an $N \times N$ weight matrix W which minimizes the reconstruction error

$$\mathcal{E}_1(W) = |X - WX|^2 \tag{7.29}$$

subject to certain constraints on W which we will mention shortly.

Let us first examine this equation and think about what it means. With some added notation, we can write this in a way that is a bit more intuitive. Each point in the data set represented by X is a K-dimensional row vector. We will denote the ith row vector by $\mathbf{x_i}$. Each point also has a corresponding weight vector given by the ith row of the weight matrix W. The portion of the reconstruction error associated with this single point can be written

$$\mathcal{E}_1(W) = \sum_{i=1}^{N} \left| \mathbf{x_i} - \sum_{j=1}^{N} W_{ij} \mathbf{x_j} \right|^2 . \tag{7.30}$$

What does it mean to minimize this equation with respect to the weights W? What we are doing is finding the linear combination of points in the data set which best reconstructs each point from the others. This is, essentially, finding the hyperplane that best describes the local surface at each point within the data set. Each row of the weight matrix W gives a set of weights for the corresponding point. As written above, the expression can be trivially minimized by setting $W = I$, the identity matrix. In this case, $WX = X$ and $\mathcal{E}_1(W) = 0$. To prevent this simplistic solution, we can constrain the problem such that the diagonal $W_{ii} = 0$ for all i. This constraint leads to a much more interesting solution. In this case the matrix W would in some sense encode the *global* geometric properties of the data set: how each point relates to all the others.

The key insight of LLE is to take this one step further, and constrain all $W_{ij} = 0$ *except* when point j is one of the k nearest neighbors of point i. With this constraint in place, the resulting matrix W has some interesting properties. First, W becomes very sparse for $k \ll N$. Out of the N^2 entries in W, only Nk are nonzero. Second, the rows of W encode the *local* properties of the data set: how each point relates to its nearest neighbors. W as a whole encodes the aggregate of these local properties, and thus contains global information about the geometry of the data set, viewed through the lens of connected local neighborhoods.

The second step of LLE mirrors the first step, but instead seeks an $N \times d$ matrix Y, where $d < D$ is the dimension of the embedded manifold. Y is found by minimizing the quantity

$$\mathcal{E}_2(Y) = |Y - WY|^2 , \tag{7.31}$$

where this time W is kept fixed. The symmetry between eqs. 7.29 and 7.31 is clear. Because of this symmetry and the constraints put on W, local neighborhoods in the low-dimensional embedding, Y, will reflect the properties of corresponding local neighborhoods in X. This is the sense in which the embedding Y is a good nonlinear representation of X.

Algorithmically, the solutions to eqs. 7.29 and 7.31 can be obtained analytically using efficient linear algebra techniques. The details are available in the literature [29, 32], but we will summarize the results here. Step 1 requires a nearest-neighbor search (see §2.5.2), followed by a least-squares solution to the corresponding row of the weight matrix W. Step 2 requires an eigenvalue decomposition of the matrix $C_W \equiv (I - W)^T (I - W)$, which is an $N \times N$ sparse matrix, where N is the number of points in the data set. Algorithms for direct eigenvalue decomposition scale as $\mathcal{O}(N^3)$, so this calculation can become prohibitively expensive as N grows large. Iterative methods can improve on this: Arnoldi decomposition (related to the Lanczos method) allows a few extremal eigenvalues of a sparse matrix to be found relatively efficiently. A well-tested tool for Arnoldi decomposition is the Fortran package ARPACK [24]. A full Python wrapper for ARPACK is available in the functions `scipy.sparse.linalg.eigsh` (for symmetric matrices) and `scipy.sparse.linalg.eigs` (for asymmetric matrices) in SciPy version 0.10 and greater. These tools are used in the manifold learning routines available in Scikit-learn: see below.

In the astronomical literature there are cases where LLE has been applied to data as diverse as galaxy spectra [32], stellar spectra [10], and photometric light curves [27]. In the case of spectra, the authors showed that the LLE projection results in a low-dimensional representation of the spectral information, while maintaining physically important nonlinear properties of the sample (see figure 7.9). In the case of light curves, the LLE has been shown useful in aiding automated classification of observed objects via the projection of high-dimensional data onto a one-dimensional nonlinear sequence in the parameter space.

Scikit-learn has a routine to perform LLE, which uses a fast tree for neighbor search, and ARPACK for a fast solution of the global optimization in the second step of the algorithm. It can be used as follows:

```python
import numpy as np
from sklearn.manifold import LocallyLinearEmbedding

X = np.random.normal(size=(1000, 2))
    # 100 pts in 2 dims
R = np.random.random((2, 10))   # projection matrix
X = np.dot(X, R)   # now a 2D linear manifold in 10D
    # space
k = 5   # number of neighbors used in the fit
n = 2   # number of dimensions in the fit
lle = LocallyLinearEmbedding(k, n)
lle.fit(X)
proj = lle.transform(X)   # 100 x 2 projection of data
```

There are many options available for the LLE computation, including more robust variants of the algorithm. For details, see the Scikit-learn documentation, or the code associated with the LLE figures in this chapter.

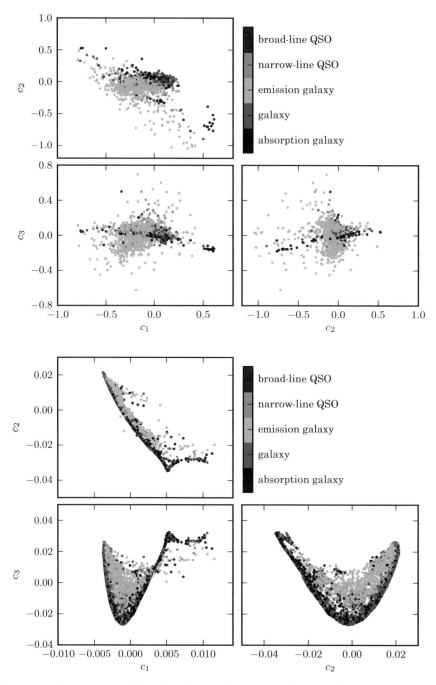

Figure 7.9. A comparison of the classification of quiescent galaxies and sources with strong line emission using LLE and PCA. The top panel shows the segregation of galaxy types as a function of the first three PCA components. The lower panel shows the segregation using the first three LLE dimensions. The preservation of locality in LLE enables nonlinear features within a spectrum (e.g., variation in the width of an emission line) to be captured with fewer components. This results in better segregation of spectral types with fewer dimensions. See color plate 6.

7.5.2. IsoMap

IsoMap [30], short for isometric mapping, is another manifold learning method which, interestingly, was introduced in the same issue of *Science* in 2000 as was LLE. IsoMap is based on a multidimensional scaling (MDS) framework. Classical MDS is a method to reconstruct a data set from a matrix of pairwise distances (for a detailed discussion of MDS see [4]).

If one has a data set represented by an $N \times K$ matrix X, then one can trivially compute an $N \times N$ distance matrix D_X such that $[D_X]_{ij}$ contains the distance between points i and j. Classical MDS seeks to reverse this operation: given a distance matrix D_X, MDS discovers a new data set Y which minimizes the error

$$\mathcal{E}_{XY} = |\tau(D_X) - \tau(D_Y)|^2, \tag{7.32}$$

where τ is an operator with a form chosen to simplify the analytic form of the solution. In metric MDS the operator τ is given by

$$\tau(D) = \frac{HSH}{2}, \tag{7.33}$$

where S is the matrix of square distances $S_{ij} = D_{ij}^2$, and H is the "centering matrix" $H_{ij} = \delta_{ij} - 1/N$. This choice of τ is convenient because it can then be shown that the optimal embedding Y is identical to the top D eigenvectors of the matrix $\tau(D_X)$ (for a derivation of this property see [26]).

The key insight of IsoMap is that we can use this metric MDS framework to derive a nonlinear embedding by constructing a suitable stand-in for the distance matrix D_X. IsoMap recovers nonlinear structure by approximating geodesic curves which lie within the embedded manifold, and computing the distances between each point in the data set along these geodesic curves. To accomplish this, the IsoMap algorithm creates a connected graph G representing the data, where G_{ij} is the distance between point i and point j if points i and j are neighbors, and $G_{ij} = 0$ otherwise. Next, the algorithm constructs a matrix D_X' such that $[D_X']_{ij}$ contains the length of the shortest path between point i and j traversing the graph G. Using this distance matrix, the optimal d-dimensional embedding is found using the MDS algorithm discussed above.

IsoMap has a computational cost similar to that of LLE if clever algorithms are used. The first step (nearest-neighbor search) and final step (eigendecomposition of an $N \times N$ matrix) are similar to those of LLE. IsoMap has one additional hurdle however: the computation of the pairwise shortest paths on an order-N sparse graph G. A brute-force approach to this sort of problem is prohibitively expensive: for each point, one would have to test every combination of paths, leading to a total computation time of $\mathcal{O}(N^2 k^N)$. There are known algorithms which improve on this: the Floyd–Warshall algorithm [13] accomplishes this in $\mathcal{O}(N^3)$, while the Dijkstra algorithm using Fibonacci heaps [14] accomplishes this in $\mathcal{O}(N^2(k + \log N))$: a significant improvement over brute force.

Scikit-learn has a fast implementation of the IsoMap algorithm, using either the Floyd–Warshall algorithm or Dijkstra's algorithm for shortest-path search. The neighbor search is implemented with a fast tree search, and the final eigenanalysis is implemented using the Scikit-learn ARPACK wrapper. It can be used as follows:

```python
import numpy as np
from sklearn.manifold import Isomap

X = np.random.normal(size=(1000, 2))
    # 1000 pts in 2 dims
R = np.random.random((2, 10))   # projection matrix
X = np.dot(X, R)   # X is now a 2D manifold in
    # 10D space
k = 5   # number of neighbors used in the fit
n = 2   # number of dimensions in the fit
iso = Isomap(k, n)
iso.fit(X)
proj = iso.transform(X)   # 1000 x 2 projection of
    # data
```

For more details, see the documentation of Scikit-learn or the source code of the IsoMap figures in this chapter.

7.5.3. Weaknesses of Manifold Learning

Manifold learning is a powerful tool to recover low-dimensional nonlinear projections of high-dimensional data. Nevertheless, there are a few weaknesses that prevent it from being used as widely as techniques like PCA:

Noisy and gappy data: Manifold learning techniques are in general not well suited to fitting data plagued by noise or gaps. To see why, imagine that a point in the data set shown in figure 7.8 is located at $(x, y) = (0, 0)$, but not well constrained in the z direction. In this case, there are three perfectly reasonable options for the missing z coordinate: the point could lie on the bottom of the "S", in the middle of the "S", or on the top of the "S". For this reason, manifold learning methods will be fundamentally limited in the case of missing data. One may imagine, however, an iterative approach which would construct a (perhaps multimodal) Bayesian constraint on the missing values. This would be an interesting direction for algorithmic research, but such a solution has not yet been demonstrated.

Tuning parameters: In general, the nonlinear projection obtained using these techniques depends highly on the set of nearest neighbors used for each point. One may select the k neighbors of each point, use all neighbors within a radius r of each point, or choose some more sophisticated technique. There is currently no solid recommendation in the literature for choosing the optimal set of neighbors for a given embedding: the optimal choice will depend highly on the local density of each point, as well as the curvature of the manifold at each point. Once again, one may imagine an iterative approach to optimizing the selection of neighbors based on these insights. This also could be an interesting direction for research.

Dimensionality: One nice feature of PCA is that the dimensionality of the data set can be estimated, to some extent, from the eigenvalues associated with the projected dimensions. In manifold learning, there is no such clean mapping. In fact, we have no guarantee that the embedded manifold is unidimensional! One can easily imagine a situation where a data set is drawn from a two-dimensional manifold in one region, and from a three-dimensional manifold in an adjacent region. Thus in a manifold learning setting, the choice of output dimensionality is a free parameter. In practice, either $d = 1$, $d = 2$, or $d = 3$ is often chosen, for the simple reason that it leads to a projection which is easy to visualize! Some attempts have been made to use local variance of a data set to arrive at an estimate of the dimensionality (see, e.g., [11]) but this works only marginally well in practice.

Sensitivity to outliers: Another weakness of manifold learning is its sensitivity to outliers. In particular, even a single outlier between different regions of the manifold can act to "short-circuit" the manifold so that the algorithm cannot find the correct underlying embedding.

Reconstruction from the manifold: Because manifold learning methods generally do not provide a set of basis functions, any mapping from the embedded space to the higher-dimensional input space must be accomplished through a reconstruction based on the location of nearest neighbors. This means that a projection derived from these methods cannot be used to compress data in a way that is analogous to PCA. The full input data set and the full projected data must be accessed in order to map new points between the two spaces.

With these weaknesses in mind, manifold learning techniques can still be used successfully to analyze and visualize large astronomical data sets. LLE has been applied successfully to several classes of data, both as a classification technique and an outlier detection technique [10, 27, 32]. These methods are ripe for exploration of the high-dimensional data sets available from future large astronomical surveys.

7.6. Independent Component Analysis and Projection Pursuit

Independent component analysis (ICA) [8] is a computational technique that has become popular in the biomedical signal processing community to solve what has often been referred to as the "cocktail party problem"; see [7]. In this problem, there are multiple microphones situated through out a room containing N people. Each microphone picks up a linear combination of the N voices. The goal of ICA is to use the concept of statistical independence to isolate (or unmix) the individual signals. In the context of astronomical problems we will consider the application of ICA to the series of galaxy spectra used for PCA (see §7.3.2). In this example, each galaxy spectrum is considered as the microphone picking up a linear combination of input signals from individual stars and HII regions.

Each spectrum, $x_i(k)$, can now be described by

$$x_1(k) = a_{11}s_1(k) + a_{12}s_2(k) + a_{13}s_3(k) + \cdots, \tag{7.34}$$

$$x_2(k) = a_{21}s_1(k) + a_{22}s_2(k) + a_{23}s_3(k) + \cdots, \tag{7.35}$$

$$x_3(k) = a_{31}s_1(k) + a_{32}s_2(k) + a_{33}s_3(k) + \cdots, \tag{7.36}$$

where $s_i(k)$ are the individual stellar spectra and a_{ij} the appropriate mixing amplitudes. In matrix format we can write this as

$$X = AS, \tag{7.37}$$

where X and S are matrices for the set of input spectra and stellar spectra, respectively. Extracting these signal spectra is equivalent to estimating the appropriate weight matrix, W, such that

$$S = WX. \tag{7.38}$$

The principle that underlies ICA comes from the observation that the input signals, $s_i(k)$, should be statistically independent. Two random variables are considered statistically independent if their joint probability distribution, $f(x, y)$, can be fully described by a combination of their marginalized probabilities, that is,

$$f(x^p, y^q) = f(x^p)f(y^q), \tag{7.39}$$

where p and q represent arbitrary higher-order moments of the probability distributions. For the case of PCA, $p = q = 1$ and the statement of independence simplifies to the weaker condition of uncorrelated data (see §7.3.1 on the derivation of PCA).

In most implementations of ICA algorithms the requirement for statistical independence is expressed in terms of the non-Gaussianity of the probability distributions. The rationale for this is that the sum of any two independent random variables will always be more Gaussian than either of the individual random variables (i.e., from the central limit theorem). This would mean that, for the case of the stellar components that make up a galaxy spectrum, if we identify an unmixing matrix, W, that maximizes the non-Gaussianity of the distributions, then we would be identifying the input signals. Definitions of non-Gaussianity range from the use of the kurtosis of a distribution (see §5.2.2), the negentropy (the negative of the entropy of a distribution), and mutual information.

Related to ICA is projection pursuit [15, 16]. Both techniques seek interesting directions within multivariate data sets. One difference is that in projection pursuit these directions are identified one at a time, while for ICA the search for these signals can be undertaken simultaneously. For each case, the definition of "interesting" is often expressed in terms of how non-Gaussian the distributions are after projections. Projection Pursuit can be considered as a subset of ICA.

Scikit-learn has an implementation of ICA based on the FastICA algorithm [18]. It can be used as follows:

```python
import numpy as np
from sklearn.decomposition import FastICA

X = np.random.normal(size=(100, 2))
    # 100 pts in 2 dims
R = np.random.random((2, 5))  # mixing matrix
X = np.dot(X, R)  # X is now 2D data in 5D space
ica = FastICA(2)  # fit two components
ica.fit(X)
proj = ica.transform(X) # 100 x 2 projection of data

comp = ica.components_  # the 2 x 5 matrix of indep.
    # components
sources = ica.sources_  # the 100 x 2 matrix of
    # sources
```

There are several options to fine-tune the algorithm; for details refer to the Scikit-learn documentation.

7.6.1. The Application of ICA to Astronomical Data

In figure 7.1 we showed the set of spectra used to test PCA. From the eigenspectra and their associated eigenvalues it was apparent that each spectrum comprises a linear combination of basis functions whose shapes are broadly consistent with the spectral properties of individual stellar types (O, A, and G stars). In the middle panel of figure 7.4 we apply ICA to these same spectra to define the independent components. As with PCA, preprocessing of the input data is an important component of any ICA application. For each data set the mean vector is removed to center the data. Whitening of the distributions (where the covariance matrix is diagonalized and normalized to reduce it to the identity matrix) is implemented through an eigenvalue decomposition of the covariance matrix.

The cost function employed is that of FastICA [18] which uses an analytic approximation to the negentropy of the distributions. The advantage of FastICA is that each of the independent components can be evaluated one at a time. Thus the analysis can be terminated once a sufficient number of components has been identified.

Comparison of the components derived from PCA, NMF and ICA in figure 7.4 shows that each of these decompositions produces a set of basis functions that are broadly similar (including both continuum and line emission). The ordering of the importance of each component is dependent on technique: in the case of ICA, finding a subset of ICA components is not the same as finding all ICA components (see [17]), which means that the a priori assumption of the number of underlying components will affect the form of the resulting components. This choice of dimensionality is a problem common to all dimensionality techniques (from LLE to the truncation of

PCA components). As with many multivariate applications, as the size of the mixing matrix grows, computational complexity often makes it impractical to calculate the weight matrix W directly. Reduction in the complexity of the input signals through the use of PCA (either to filter the data or to project the data onto these basis functions) is often applied to ICA applications.

7.7. Which Dimensionality Reduction Technique Should I Use?

In chapter 6 we introduced the concept of defining four axes against which we can compare the techniques described in each chapter. Using the axes of "accuracy," "interpretability," "simplicity," and "speed" (see §6.6 for a description of these terms) we provide a rough assessment of each of the methods considered in this chapter.

What are the most *accurate* dimensionality reduction methods? First we must think about how we want to define the notion of accuracy for the task of dimension reduction. In some sense the different directions one can take here are what distinguishes the various methods. One clear notion of this is in terms of reconstruction error—some function of the difference between the original data matrix and the data matrix reconstructed using the desired number of components. When all possible components are used, every dimensionality reduction method, in principle, is an exact reconstruction (zero error), so the question becomes some function of the number of components—for example, which method tends to yield the most faithful reconstruction after K components. PCA is designed to provide the best square reconstruction error for any given K. LLE minimizes square reconstruction error, but in a nonlinear fashion. NMF also minimizes a notion of reconstruction error, under nonnegativity constraints. Other approaches, such as the multidimensional scaling family of methods ([4, 22]), of which IsoMap is a modern descendant and variant, attempt to minimize the difference between each pairwise distance in the original space and its counterpart in the reconstructed space. In other words, once one has made a choice of which notion of accuracy is desired, there is generally a method which directly maximizes that notion of accuracy. Generally speaking, for a fixed notion of error, a nonlinear model should better minimize error than one which is constrained to be linear (i.e., manifold learning techniques should provide more compact and accurate representations of the data). Additional constraints, such as nonnegativity, only reduce the flexibility the method has to fit the data, making, for example, NMF generally worse in the sense of reconstruction error than PCA. ICA can also be seen in the light of adding additional constraints, making it less aggressive in terms of reconstruction error.

NMF-type approaches do, however, have a significant performance advantage over PCA, ICA, and manifold learning for the case of low signal-to-noise data due to the fact that they have been expanded to account for heteroscedastic uncertainties; see [31]. NMF and its variants can model the *measured* uncertainties within the data (rather than assuming that the variance of the ensemble of data is representative of these uncertainties). This feature, at least in principle, provides a more accurate set of derived basis functions.

What are the most *interpretable* dimensionality reduction methods? In the context of dimensionality reduction methods, interpretability generally means the extent to which we can identify the meaning of the directions found by the method. For data sets where all values are known to be nonnegative, as in the case of spectra or image pixel brightnesses, NMF tends to yield more sensible components than PCA. This makes sense because the constraints ensure that reconstructions obtained from NMF are themselves valid spectra or images, which might not be true otherwise. Interpretability in terms of understanding each component through the original dimensions it captures can be had to some extent with linear models like PCA, ICA, and NMF, but is lost in the nonlinear manifold learning approaches like LLE and IsoMap (which do not provide the principal directions, but rather the positions of objects within a lower-dimensional space).

A second, and significant, advantage of PCA over the other techniques is the ability to estimate the importance of each principal component (in terms of its contribution to the variance). This enables simple, though admittedly ad hoc (see §7.3.2), criteria to be applied when truncating the projection of data onto a set of basis functions.

What are the most *scalable* dimensionality reduction methods? Faster algorithms can be obtained for PCA in part by focusing on the fact that only the top K components are typically needed. Approximate algorithms for SVD-like computations (where only parts of the SVD are obtained) based on sampling are available, as well as algorithms using similar ideas that yield full SVDs. Online algorithms can be fairly effective for SVD-like computations. ICA can be approached a number of ways, for example using the FastICA algorithm [18]. Fast optimization methods for NMF are discussed in [23]. In general, linear methods for dimensionality reduction are usually relatively tractable, even for high N. For nonlinear methods such as LLE and IsoMap, the most expensive step is typically the first one, an $\mathcal{O}(N^2)$ all-nearest-neighbor computation to find the K neighbors for each point. Fast algorithms for such problems, including one that is provably $\mathcal{O}(N)$, are discussed in §2.4 (see also WSAS). The second step is typically a kind of SVD computation. Overall LLE is $\mathcal{O}(N^2)$ and IsoMap is $\mathcal{O}(N^3)$, making them intractable on large data sets without algorithmic intervention.

What are the *simplest* dimensionality reduction methods? All dimensionality reduction methods are fiddly in the sense that they all require, to some extent, the selection of the number of components/dimensions to which the data should be reduced. PCA's objective is convex, meaning that there is no need for multiple random restarts, making it usable essentially right out of the box. This is not the case for ICA and NMF, which have nonconvex objectives. Manifold learning methods like LLE and IsoMap have convex objectives, but require careful evaluation of the parameter k which defines the number of nearest neighbors.

As mentioned in §6.6, selecting parameters can be done automatically in principle for any machine learning method for which cross-validation (see §8.11) can be applied, including unsupervised methods as well as supervised ones. Cross-validation can be applied whenever the model can be used to make predictions (in this case "predictions" means coordinates in the new low-dimensional space) on test data, that is, data not included in the training data. While PCA, NMF, and ICA can

TABLE 7.1.
Summary of the practical properties of the main dimensionality reduction techniques.

Method	Accuracy	Interpretability	Simplicity	Speed
Principal component analysis	H	H	H	H
Locally linear embedding	H	M	H	M
Nonnegative matrix factorization	H	H	M	M
Independent component analysis	M	M	L	L

be applied to such "out-of-sample" data, LLE and IsoMap effectively require, in order to get predictions, that test data be added to the existing training set and the whole model retrained. This requirement precludes the use of cross-validation to select k automatically and makes the choice of k a subjective exercise.

Other considerations, and taste. As we noted earlier in §7.5.3, other considerations can include robustness to outliers and missing values. Many approaches have been taken to making robust versions of PCA, since PCA, as a maximum-likelihood-based model, is sensitive to outliers. Manifold learning methods can also be sensitive to outliers. PCA, ICA, and NMF can be made to work with missing values in a number of ways: we saw one example for PCA in §7.3.3. There is no clear way to handle missing values with manifold learning methods, as they are based on the idea of distances, and it is not clear how to approach the problem of computing distances with missing values. Regarding taste, if we consider these criteria as a whole the simplest and most useful technique is principal component analysis (as has been recognized by the astronomical community with over 300 refereed astronomical publications between 2000 and 2010 that mention the use of PCA). Beyond PCA, NMF has the advantage of mapping to many astronomical problems (i.e., for positive data and in the low signal-to-noise regime) and is particularly suitable for smaller data sets. ICA has interesting roots in information theory and signal processing, and manifold learning has roots in geometry; their adoption within the astronomical community has been limited to date.

Simple summary. Table 7.1 is a simple summary of the trade-offs along our axes of accuracy, interpretability, simplicity, and speed in dimension reduction methods, expressed in terms of high (H), medium (M), and low (L) categories.

References

[1]   Abazajian, K. N., J. K. Adelman-McCarthy, M. A. Agüeros, and others (2009). The Seventh Data Release of the Sloan Digital Sky Survey. *ApJS 182*, 543–558.

[2]   Bellman, R. E. (1961). *Adaptive Control Processes*. Princeton, NJ: Princeton University Press.

[3]   Blanton, M. R. and S. Roweis (2007). K-corrections and filter transformations in the ultraviolet, optical, and near-infrared. *AJ 133*, 734–754.

[4]   Borg, I. and P. Groenen (2005). *Modern Multidimensional Scaling: Theory and Applications*. Springer.

[5]   Budavári, T., V. Wild, A. S. Szalay, L. Dobos, and C.-W. Yip (2009). Reliable eigenspectra for new generation surveys. *MNRAS 394*, 1496–1502.

[6]   Cattell, R. B. (1966). The scree test for the number of factors. *Multivariate Behavioral Research 1*(2), 245–276.

[7]   Cherry, E. C. (1953). Some experiments on the recognition of speech, with one and with two ears. *Acoustical Society of America Journal 25*, 975.

[8]   Comon, P. (1994). Independent component analysis—a new concept? *Signal Processing 36*(3), 287–314.

[9]   Connolly, A. J. and A. S. Szalay (1999). A robust classification of galaxy spectra: Dealing with noisy and incomplete data. *AJ 117*, 2052–2062.

[10] Daniel, S. F., A. Connolly, J. Schneider, J. Vanderplas, and L. Xiong (2011). Classification of stellar spectra with local linear embedding. *AJ 142*, 203.

[11] de Ridder, D. and R. P. Duin (2002). Locally linear embedding for classification. *Pattern Recognition Group Technical Report Series PH-2002-01.*

[12] Everson, R. and L. Sirovich (1995). Karhunen-Loeve procedure for gappy data. *J. Opt. Soc. Am. A 12*, 1657–1664.

[13] Floyd, R. W. (1962). Algorithm 97: Shortest path. *Commun. ACM 5*, 345.

[14] Fredman, M. L. and R. E. Tarjan (1987). Fibonacci heaps and their uses in improved network optimization algorithms. *J. ACM 34*, 596–615.

[15] Friedman, J. H. (1987). Exploratory projection pursuit. *Journal of the American Statistical Association 82*, 249–266.

[16] Friedman, J. H. and J. W. Tukey (1974). A projection pursuit algorithm for exploratory data analysis. *IEEE Transactions on Computers 23*, 881–889.

[17] Girolami, M. and C. Fyfe (1997). Negentropy and kurtosis as projection pursuit indices provide generalised ICA algorithms. In A. Cichocki and A.Back (Eds.), *NIPS-96 Blind Signal Separation Workshop*, Volume 8.

[18] Hyvaerinen, A. (1999). Fast and robust fixed-point algorithms for independent component analysis. *IEEE-NN 10*(3), 626.

[19] Jolliffe, I. T. (1986). *Principal Component Analysis*. Springer.

[20] Kaiser, H. F. (1960). The application of electronic computers to factor analysis. *Educational and Psychological Measurement 20*(1), 141–151.

[21] Karhunen, H. (1947). Uber Lineare Methoden in der Wahrscheinlichkeitsrechnung. *Ann. Acad. Sci. Fenn. Ser. A.I. 37*. See translation by I. Selin, The Rand Corp., Doc. T-131, 1960.

[22] Kruskal, J. and M. Wish (1978). *Multidimensional Scaling*. Number 11 in Quantitative Applications in the Social Sciences. SAGE Publications.

[23] Lee, D. D. and H. S. Seung (2001). Algorithms for non-negative matrix factorization. In T. K. Leen, T. G. Dietterich, and V. Tresp (Eds.), *Advances in Neural Information Processing Systems 13: Proceedings of the 2000 Conference*, Cambridge, Massachusetts, pp. 556–562. MIT Press.

[24] Lehoucq, R. B., D. C. Sorensen, and C. Yang (1997). ARPACK users guide: Solution of large scale eigenvalue problems by implicitly restarted Arnoldi methods.

[25] Loéve, M. (1963). *Probability Theory*. New York: Van Nostrand.

[26] Mardia, K. V., J. T. Kent, and J. M. Bibby (1979). *Multivariate Analysis*. Academic Press.

[27] Matijevič, G., A. Prša, J. A. Orosz, and others (2012). Kepler eclipsing binary stars. III. Classification of Kepler eclipsing binary light curves with locally linear embedding. *AJ 143*, 123.

[28] Pearson, K. (1901). On lines and planes of closest fit to systems of points in space. *The London, Edinburgh and Dublin Philosophical Magazine and Journal of Science 2*, 559–572.

[29] Roweis, S. T. and L. K. Saul (2000). Nonlinear dimensionality reduction by locally linear embedding. *Science 290*, 2323–2326.

[30] Tenenbaum, J. B., V. Silva, and J. C. Langford (2000). A global geometric framework for nonlinear dimensionality reduction. *Science 290*(5500), 2319–2323.

[31] Tsalmantza, P. and D. W. Hogg (2012). A data-driven model for spectra: Finding double redshifts in the Sloan Digital Sky Survey. *ApJ 753*, 122.

[32] Vanderplas, J. and A. Connolly (2009). Reducing the dimensionality of data: Locally linear embedding of Sloan Galaxy Spectra. *AJ 138*, 1365–1379.

[33] Yip, C. W., A. J. Connolly, A. S. Szalay, and others (2004). Distributions of galaxy spectral types in the Sloan Digital Sky Survey. *AJ 128*, 585–609.

[34] Yip, C. W., A. J. Connolly, D. E. Vanden Berk, and others (2004). Spectral classification of quasars in the Sloan Digital Sky Survey: Eigenspectra, redshift, and luminosity effects. *AJ 128*, 2603–2630.

8 Regression and Model Fitting

"Why are you trying so hard to fit in when you were born to stand out?" (Ian Wallace)

Regression is a special case of the general model fitting and selection procedures discussed in chapters 4 and 5. It can be defined as the relation between a dependent variable, y, and a set of independent variables, x, that describes the expectation value of y given x: $E[y|x]$. The purpose of obtaining a "best-fit" model ranges from scientific interest in the values of model parameters (e.g., the properties of dark energy, or of a newly discovered planet) to the predictive power of the resulting model (e.g., predicting solar activity). The usage of the word regression for this relationship dates back to Francis Galton, who discovered that the difference between a child and its parents for some characteristic is proportional to its parents' deviation from typical people in the whole population,[1] or that children "regress" toward the population mean. Therefore, modern usage of the word in a statistical context is somewhat different.

As we will describe below, regression can be formulated in a way that is very general. The solution to this generalized problem of regression is, however, quite elusive. Techniques used in regression tend, therefore, to make a number of simplifying assumptions about the nature of the data, the uncertainties of the measurements, and the complexity of the models. In the following sections we start with a general formulation for regression, list various simplified cases, and then discuss methods that can be used to address them, such as regression for linear models, kernel regression, robust regression and nonlinear regression.

8.1. Formulation of the Regression Problem

Given a multidimensional data set drawn from some pdf and the full error covariance matrix for each data point, we can attempt to infer the underlying pdf using either parametric or nonparametric models. In its most general incarnation, this is a

---

[1] If your parents have very high IQs, you are more likely to have a lower IQ than them, than a higher one. The expected probability distribution for your IQ if you also have a sister whose IQ exceeds your parents' IQs is left as an exercise for the reader. Hint: This is related to regression toward the mean discussed in § 4.7.1.

very hard problem to solve. Even with a restrictive assumption that the errors are Gaussian, incorporating the error covariance matrix within the posterior distribution is not trivial (cf. § 5.6.1). Furthermore, accounting for any selection function applied to the data can increase the computational complexity significantly (e.g., recall § 4.2.7 for the one-dimensional case), and non-Gaussian error behavior, if not accounted for, can produce biased results.

Regression addresses a slightly simpler problem: instead of determining the multidimensional pdf, we wish to infer the expectation value of y given x (i.e., the conditional expectation value). If we have a model for the conditional distribution (described by parameters $\boldsymbol{\theta}$) we can write this function[2] as $y = f(x|\boldsymbol{\theta})$. We refer to y as a scalar dependent variable and x as an independent vector. Here x does not need to be a random variable (e.g., x could correspond to deterministic sampling times for a time series). For a given model class (i.e., the function f can be an analytic function such as a polynomial, or a nonparametric estimator), we have k model parameters θ_p, $p = 1, \ldots, k$.

Figure 8.1 illustrates how the constraints on the model parameters, $\boldsymbol{\theta}$, respond to the observations x_i and y_i. In this example, we assume a simple straight-line model with $y_i = \theta_0 + \theta_1 x_i$. Each point provides a joint constraint on θ_0 and θ_1. If there were no uncertainties on the variables then this constraint would be a straight line in the (θ_0, θ_1) plane ($\theta_0 = y_i - \theta_1 x_i$). As the number of points is increased the best estimate of the model parameters would then be the intersection of all lines. Uncertainties within the data will transform the constraint from a line to a distribution (represented by the region shown as a gray band in figure 8.1). The best estimate of the model parameters is now given by the posterior distribution. This is simply the multiplication of the probability distributions (constraints) for all points and is shown by the error ellipses in the lower panel of figure 8.1. Measurements with upper limits (e.g., point x_4) manifest as half planes within the parameter space. Priors are also accommodated naturally within this picture as additional multiplicative constraints applied to the likelihood distribution (see § 8.2).

Computationally, the cost of this general approach to regression can be prohibitive (particularly for large data sets). In order to make the analysis tractable, we will, therefore, define several types of regression using three "classification axes":

- *Linearity.* When a parametric model is linear in all model parameters, that is, $f(x|\theta) = \sum_{p=1}^{k} \theta_p g_p(x)$, where functions $g_p(x)$ do not depend on any free model parameters (but can be nonlinear functions of x), regression becomes a significantly simpler problem, called linear regression. Examples of this include polynomial regression, and radial basis function regression. Regression of models that include nonlinear dependence on θ_p, such as $f(x|\theta) = \theta_1 + \theta_2 \sin(\theta_3 x)$, is called nonlinear regression.
- *Problem complexity.* A large number of independent variables increases the complexity of the error covariance matrix, and can become a limiting factor in nonlinear regression. The most common regression case found in practice is the $M = 1$ case with only a single independent variable (i.e., fitting a straight line to data). For linear models and negligible errors on the independent variables, the problem of dimensionality is not (too) important.

---

[2]Sometimes $f(x; \boldsymbol{\theta})$ is used instead of $f(x|\boldsymbol{\theta})$ to emphasize that here f is a function rather than pdf.

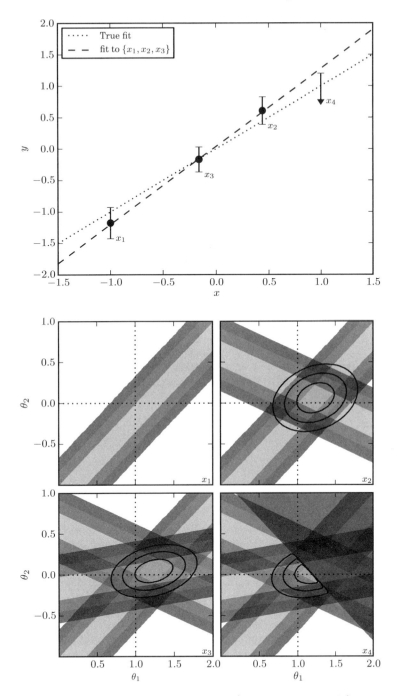

Figure 8.1. An example showing the online nature of Bayesian regression. The upper panel shows the four points used in regression, drawn from the line $y = \theta_1 x + \theta_0$ with $\theta_1 = 1$ and $\theta_0 = 0$. The lower panel shows the posterior pdf in the (θ_0, θ_1) plane as each point is added in sequence. For clarity, the implied dark regions for $\sigma > 3$ have been removed. The fourth point is an upper-limit measurement of y, and the resulting posterior cuts off half the parameter space.

- *Error behavior.* The uncertainties in the values of independent and dependent variables, and their correlations, are the primary factor that determines which regression method to use. The structure of the error covariance matrix, and deviations from Gaussian error behavior, can turn seemingly simple problems into complex computational undertakings. Here we will separately discuss the following cases:

  1. Both independent and dependent variables have negligible errors (compared to the intrinsic spread of data values); this is the simplest and most common "y vs. x" case, and can be relatively easily solved even for nonlinear models and multidimensional data.
  2. Only errors for the dependent variable (y) are important, and their distribution is Gaussian and homoscedastic (with σ either known or unknown).
  3. Errors for the dependent variable are Gaussian and known, but heteroscedastic.
  4. Errors for the dependent variable are non-Gaussian, and their behavior is known.
  5. Errors for the dependent variable are non-Gaussian, but their exact behavior is unknown.
  6. Errors for independent variables (x) are not negligible, but the full covariance matrix can be treated as Gaussian. This case is relatively straightforward when fitting a straight line, but can become cumbersome for more complex models.
  7. All variables have non-Gaussian errors. This is the hardest case and there is no ready-to-use general solution. In practice, the problem is solved on a case-by-case basis, typically using various approximations that depend on the problem specifics.

For the first four cases, when error behavior for the dependent variable is known, and errors for independent variables are negligible, we can easily use the Bayesian methodology developed in chapter 5 to write the posterior pdf for the model parameters,

$$p(\boldsymbol{\theta}|\{x_i, y_i\}, I) \propto p(\{x_i, y_i\}|\boldsymbol{\theta}, I)\, p(\boldsymbol{\theta}, I). \tag{8.1}$$

Here the information I describes the error behavior for the dependent variable. The data likelihood is the product of likelihoods for the individual points, and the latter can be expressed as

$$p(y_i|x_i, \boldsymbol{\theta}, I) = e(y_i|y), \tag{8.2}$$

where $y = f(x|\boldsymbol{\theta})$ is the adopted model class, and $e(y_i|y)$ is the probability of observing y_i given the true value (or the model prediction) y. For example, if the y error distribution is Gaussian, with the width for ith data point given by σ_i, and the errors on x are negligible, then

$$p(y_i|x_i, \boldsymbol{\theta}, I) = \frac{1}{\sigma_i\sqrt{2\pi}}\exp\left(\frac{-[y_i - f(x_i|\boldsymbol{\theta})]^2}{2\sigma_i^2}\right). \tag{8.3}$$

8.1.1. Data Sets Used in This Chapter

For regression and its application to astrophysics we focus on the relation between the redshifts of supernovas and their luminosity distance (i.e., a cosmological parametrization of the expansion of the universe [1]). To accomplish this we generate a set of synthetic supernova data assuming a cosmological model given by

$$\mu(z) = -5 \log_{10} \left((1+z) \frac{c}{H_0} \int \frac{dz}{(\Omega_m (1+z)^3 + \Omega_\Lambda)^{1/2}} \right), \tag{8.4}$$

where $\mu(z)$ is the distance modulus to the supernova, H_0 is the Hubble constant, Ω_m is the cosmological matter density and Ω_Λ is the energy density from a cosmological constant. For our fiducial cosmology we choose $\Omega_m = 0.3$, $\Omega_\Lambda = 0.7$ and $H_0 = 70\,\mathrm{km\,s^{-1}\,Mpc^{-1}}$, and add heteroscedastic Gaussian noise that increases linearly with redshift. The resulting $\mu(z)$ cannot be expressed as a sum of simple closed-form analytic functions, including low-order polynomials. This example addresses many of the challenges we face when working with observational data sets: we do not know the intrinsic complexity of the model (e.g., the form of dark energy), the dependent variables can have heteroscedastic uncertainties, there can be missing or incomplete data, and the dependent variables can be correlated. For the majority of techniques described in this chapter we will assume that uncertainties in the independent variables are small (relative to the range of data and relative to the dependent variables). In real-world applications we do not get to make this choice (the observations themselves define the distribution in uncertainties irrespective of the models we assume). For the supernova data, an example of such a case would be if we estimated the supernova redshifts using broadband photometry (i.e., photometric redshifts). Techniques for addressing such a case are described in § 8.8.1. We also note that this toy model data set is a simplification in that it does not account for the effect of K-corrections on the observed colors and magnitudes; see [7].

8.2. Regression for Linear Models

Given an independent variable x and a dependent variable y, we will start by considering the simplest case, a linear model with

$$y_i = \theta_0 + \theta_1 x_i + \epsilon_i. \tag{8.5}$$

Here θ_0 and θ_1 are the coefficients that describe the regression (or objective) function that we are trying to estimate (i.e., the slope and intercept for a straight line $f(x) = \theta_0 + \theta_1 x_i$), and ϵ_i represents an additive noise term.

The assumptions that underlie our linear regression model include the uncertainties on the independent variables that are considered to be negligible, and the dependent variables have known heteroscedastic uncertainties, $\epsilon_i = \mathcal{N}(0, \sigma_i)$. From eq. 8.3 we can write the data likelihood as

$$p(\{y_i\}|\{x_i\}, \boldsymbol{\theta}, I) = \prod_{i=1}^{N} \frac{1}{\sqrt{2\pi}\sigma_i} \exp \left(\frac{-(y_i - (\theta_0 + \theta_1 x_i))^2}{2\sigma_i^2} \right). \tag{8.6}$$

For a flat or uninformative prior pdf, $p(\theta|I)$, where we have no knowledge about the distribution of the parameters θ, the posterior will be directly proportional to the likelihood function (which is also known as the error function). If we take the logarithm of the posterior then we arrive at the classic definition of regression in terms of the log-likelihood:

$$\ln(L) \equiv \ln((\boldsymbol{\theta}|\{x_i, y_i\}, I)) \propto \sum_{i=1}^{N} \left(\frac{-(y_i - (\theta_0 + \theta_1 x_i))^2}{2\sigma_i^2} \right). \tag{8.7}$$

Maximizing the log-likelihood as a function of the model parameters, $\boldsymbol{\theta}$, is achieved by minimizing the sum of the square errors. This observation dates back to the earliest applications of regression with the work of Gauss [6] and Legendre [14], when the technique was introduced as the "method of least squares."

The form of the likelihood function and the "method of least squares" optimization arises from our assumption of Gaussianity for the distribution of uncertainties in the dependent variables. Other forms for the likelihoods can be assumed (e.g., using the L_1 norm, see § 4.2.8, which actually precedes the use of the L_2 norm [2, 13], but this is usually at the cost of increased computational complexity). If it is known that measurement errors follow an exponential distribution (see § 3.3.6) instead of a Gaussian distribution, then the L_1 norm should be used instead of the L_2 norm and eq. 8.7 should be replaced by

$$\ln(L) \propto \sum_{i=1}^{N} \left(\frac{-|y_i - (\theta_0 + \theta_1 x_i)|}{\Delta_i} \right). \tag{8.8}$$

For the case of Gaussian homoscedastic uncertainties, the minimization of eq. 8.7 simplifies to

$$\theta_1 = \frac{\sum_i^N x_i y_i - \bar{x}\bar{y}}{\sum_i^N (x_i - \bar{x})^2}, \tag{8.9}$$

$$\theta_0 = \bar{y} - \theta_1 \bar{x}, \tag{8.10}$$

where $\bar{x}$ is the mean value of x and $\bar{y}$ is the mean value of y. As an illustration, these estimates of θ_0 and θ_1 correspond to the center of the ellipse shown in the bottom-left panel in figure 8.1. An estimate of the variance associated with this regression and the standard errors on the estimated parameters are given by

$$\sigma^2 = \sum_{i=1}^{N} (y_i - \theta_0 + \theta_1 x_i)^2, \tag{8.11}$$

$$\sigma_{\theta_1}^2 = \sigma^2 \frac{1}{\sum_i^N (x_i - \bar{x})^2}, \tag{8.12}$$

$$\sigma_{\theta_0}^2 = \sigma^2 \left(\frac{1}{N} + \frac{\bar{x}^2}{\sum_i^N (x_i - \bar{x})^2} \right). \tag{8.13}$$

For heteroscedastic errors, and in general for more complex regression functions, it is easier and more compact to generalize regression in terms of matrix notation. We, therefore, define regression in terms of a design matrix, M, such that

$$Y = M\boldsymbol{\theta}, \tag{8.14}$$

where Y is an N-dimensional vector of values y_i,

$$Y = \begin{bmatrix} y_0 \\ y_1 \\ y_2 \\ \cdot \\ y_{N-1} \end{bmatrix}. \tag{8.15}$$

For our straight-line regression function, $\boldsymbol{\theta}$ is a two-dimensional vector of regression coefficients,

$$\boldsymbol{\theta} = \begin{bmatrix} \theta_0 \\ \theta_1 \end{bmatrix}, \tag{8.16}$$

and M is a $2 \times N$ matrix,

$$M = \begin{bmatrix} 1 & x_0 \\ 1 & x_1 \\ 1 & x_2 \\ \cdot & \cdot \\ 1 & x_{N-1} \end{bmatrix}, \tag{8.17}$$

where the constant value in the first column captures the θ_0 term in the regression.

For the case of heteroscedastic uncertainties, we define a covariance matrix, C, as an $N \times N$ matrix,

$$C = \begin{bmatrix} \sigma_0^2 & 0 & \cdot & 0 \\ 0 & \sigma_1^2 & \cdot & 0 \\ \cdot & \cdot & \cdot & \cdot \\ 0 & 0 & \cdot & \sigma_{N-1}^2 \end{bmatrix} \tag{8.18}$$

with the diagonals of this matrix containing the uncertainties, σ_i, on the dependent variable, Y.

The maximum likelihood solution for this regression is

$$\boldsymbol{\theta} = (M^T C^{-1} M)^{-1} (M^T C^{-1} Y), \tag{8.19}$$

which again minimizes the sum of the square errors, $(Y - \boldsymbol{\theta} M)^T C^{-1} (Y - \boldsymbol{\theta} M)$, as we did explicitly in eq. 8.9. The uncertainties on the regression coefficients, $\boldsymbol{\theta}$, can now be expressed as the symmetric matrix

$$\Sigma_\theta = \begin{bmatrix} \sigma_{\theta_0}^2 & \sigma_{\theta_0 \theta_1} \\ \sigma_{\theta_0 \theta_1} & \sigma_{\theta_1}^2 \end{bmatrix} = [M^T C^{-1} M]^{-1}. \tag{8.20}$$

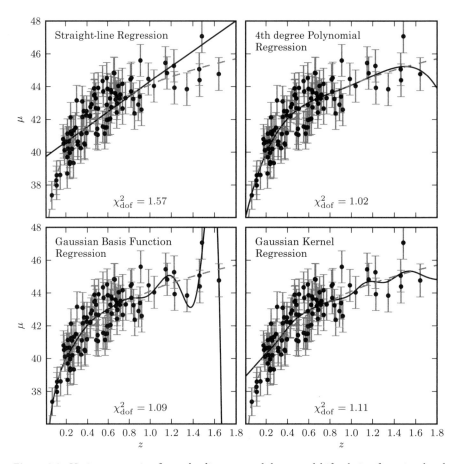

Figure 8.2. Various regression fits to the distance modulus vs. redshift relation for a simulated set of 100 supernovas, selected from a distribution $p(z) \propto (z/z_0)^2 \exp[(z/z_0)^{1.5}]$ with $z_0 = 0.3$. Gaussian basis functions have 15 Gaussians evenly spaced between $z = 0$ and 2, with widths of 0.14. Kernel regression uses a Gaussian kernel with width 0.1.

Whether we have sufficient data to constrain the regression (i.e., sufficient degrees of freedom) is defined by whether $M^T M$ is an invertible matrix.

The top-left panel of figure 8.2 illustrates a simple linear regression of redshift, z, against distance modulus, μ, for the set of 100 supernovas described in § 8.1.1. The solid line shows the regression function for the straight-line model and the dashed line the underlying cosmological model from which the data were drawn (which of course cannot be described by a straight line). It is immediately apparent that the chosen regression model does not capture the structure within the data at the high and low redshift limits—the model does not have sufficient flexibility to reproduce the correlation displayed by the data. This is reflected in the $\chi^2_{\rm dof}$ for this fit which is 1.54 (see § 4.3.1 for a discussion of the interpretation of $\chi^2_{\rm dof}$).

We now relax the assumptions we made at the start of this section, allowing not just for heteroscedastic uncertainties but also for correlations between the measures of the dependent variables. With no loss in generality, eq. 8.19 can be extended to allow for covariant data through the off-diagonal elements of the covariance matrix C.

8.2.1. Multivariate Regression

For multivariate data (where we fit a hyperplane rather than a straight line) we simply extend the description of the regression function to multiple dimensions, with $y = f(x|\boldsymbol{\theta})$ given by

$$y_i = \theta_0 + \theta_1 x_{i1} + \theta_2 x_{i2} + \cdots + \theta_k x_{ik} + \epsilon_i \qquad (8.21)$$

with θ_i the regression parameters and x_{ik} the kth component of the ith data entry within a multivariate data set. This multivariate regression follows naturally from the definition of the design matrix with

$$M = \begin{pmatrix} 1 & x_{01} & x_{02} & . & x_{0k} \\ 1 & x_{11} & x_{12} & . & x_{1k} \\ . & . & . & . & . \\ 1 & x_{N1} & x_{N2} & . & x_{Nk} \end{pmatrix}. \qquad (8.22)$$

The regression coefficients (which are estimates of $\boldsymbol{\theta}$ and are often differentiated from the true values by writing them as $\hat{\boldsymbol{\theta}}$) and their uncertainties are, as before,

$$\boldsymbol{\theta} = (M^T C^{-1} M)^{-1} (M^T C^{-1} Y) \qquad (8.23)$$

and

$$\Sigma_\theta = [M^T C^{-1} M]^{-1}. \qquad (8.24)$$

Multivariate linear regression with homoscedastic errors on dependent variables can be performed using the routine `sklearn.linear_model.LinearRegression`. For data with homoscedastic errors, AstroML implements a similar routine:

```
import numpy as np
from astroML.linear_model import LinearRegression
X = np.random.random((100, 2))   # 100 points in
    2 dimensions
dy = np.random.random(100)   # heteroscedastic errors
y = np.random.normal(X[:, 0] + X[:, 1], dy)

model = LinearRegression()
model.fit(X, y, dy)
y_pred = model.predict(X)
```

LinearRegression in Scikit-learn has a similar interface, but does not explicitly account for heteroscedastic errors. For a more realistic example, see the source code of figure 8.2.

8.2.2. Polynomial and Basis Function Regression

Due to its simplicity, the derivation of regression in most textbooks is undertaken using a straight-line fit to the data. However, the straight line can simply be interpreted as a first-order expansion of the regression function $y = f(x|\boldsymbol{\theta})$. In general we can express $f(x|\boldsymbol{\theta})$ as the sum of arbitrary (often nonlinear) functions as long as the model is linear in terms of the regression parameters, $\boldsymbol{\theta}$. Examples of these general linear models include a Taylor expansion of $f(x)$ as a series of polynomials where we solve for the amplitudes of the polynomials, or a linear sum of Gaussians with fixed positions and variances where we fit for the amplitudes of the Gaussians.

Let us initially consider polynomial regression and write $f(x|\boldsymbol{\theta})$ as

$$y_i = \theta_0 + \theta_1 x_i + \theta_2 x_i^2 + \theta_3 x_i^3 + \cdots . \tag{8.25}$$

The design matrix for this expansion becomes

$$M = \begin{pmatrix} 1 & x_0 & x_0^2 & x_0^3 \\ 1 & x_1 & x_1^2 & x_1^3 \\ . & . & . & . \\ 1 & x_N & x_N^2 & x_N^3 \end{pmatrix}, \tag{8.26}$$

where the terms in the design matrix are 1, x, x^2, and x^3, respectively. The solution for the regression coefficients and the associated uncertainties are again given by eqs. 8.19 and 8.20.

A fourth-degree polynomial fit to the supernova data is shown in the top-right panel of figure 8.2. The increase in flexibility of the model improves the fit (note that we have to be aware of overfitting the data if we just arbitrarily increase the degree of the polynomial; see § 8.11). The $\chi^2_{\rm dof}$ of the regression is 1.02, which indicates a much better fit than the straight-line case. At high redshift, however, there is a systematic deviation between the polynomial regression and the underlying generative model (shown by the dashed line), which illustrates the danger of extrapolating this model beyond the range probed by the data.

Polynomial regression with heteroscedastic errors can be performed using the `PolynomialRegression` function in AstroML:

```
import numpy as np
from astroML.linear_model import PolynomialRegression

X = np.random.random((100, 2)) # 100 points in 2 dims
y = X[:, 0] ** 2 + X[:, 1] ** 3
model = PolynomialRegression(3)
        # fit 3rd degree polynomial
model.fit(X, y)
y_pred = model.predict(X)
```

Here we have used homoscedastic errors for simplicity. Heteroscedastic errors in y can be used in a similar way to `LinearRegression`, above. For a more realistic example, see the source code of figure 8.2.

The number of terms in the polynomial regression grows exponentially with order. Given a data set with k dimensions to which we fit a p-dimensional polynomial, the number of parameters in the model we are fitting is given by

$$m = \frac{(p+k)!}{p!\,k!},$$
(8.27)

including the intercept or offset. The number of degrees of freedom for the regression model is then $\nu = N - m$ and the probability of that model is given by a χ^2 distribution with ν degrees of freedom.

We can generalize the polynomial model to a basis function representation by noting that each row of the design matrix can be replaced with any series of linear or nonlinear functions of the variables x_i. Despite the use of arbitrary basis functions, the resulting problem remains linear, because we are fitting only the coefficients multiplying these terms. Examples of commonly used basis functions include Gaussians, trigonometric functions, inverse quadratic functions, and splines.

Basis function regression can be performed using the routine BasisFunctionRegression in AstroML. For example, Gaussian basis function regression is as follows:

```
import numpy as np
from astroML.linear_model import
    BasisFunctionRegression

X = np.random.random((100, 1))  # 100 points in 1
    # dimension dy = 0.1
y = np.random.normal(X[:, 0], dy)
mu = np.linspace(0, 1, 10)[:, np.newaxis]
    # 10 x 1 array of mu
sigma = 0.1

model = BasisFunctionRegression('gaussian', mu=mu,
    sigma=sigma)
model.fit(X, y, dy)
y_pred = model.predict(X)
```

For a further example, see the source code of figure 8.2.

The application of Gaussian basis functions to our example regression problem is shown in figure 8.2. In the lower-left panel, 15 Gaussians, evenly spaced between redshifts $0 < z < 2$ with widths of $\sigma_z = 0.14$, are fit to the supernova data. The χ^2_{dof} for this fit is 1.09, comparable to that for polynomial regression.

8.3. Regularization and Penalizing the Likelihood

All regression examples so far have sought to minimize the mean square errors between a model and data with known uncertainties. The Gauss–Markov theorem states that this least-squares approach results in the minimum variance unbiased estimator (see § 3.2.2) for the linear model. In some cases, however, the regression problem may be ill posed and the best unbiased estimator is not the most appropriate regression. Instead, we can trade an increase in bias for a reduction in variance. Examples of such cases include data that are highly correlated (which results in ill-conditioned matrices), or when the number of terms in the regression model decreases the number of degrees of freedom such that we must worry about overfitting of the data.

One solution to these problems is to penalize or limit the complexity of the underlying regression model. This is often referred to as regularization, or shrinkage, and works by applying a penalty to the likelihood function. Regularization can come in many forms, but usually imposes smoothness on the model, or limits the numbers of, or the values of, the regression coefficients.

In § 8.2 we showed that regression minimizes the least-squares equation,

$$(Y - M\boldsymbol{\theta})^T (Y - M\boldsymbol{\theta}). \tag{8.28}$$

We can impose a penalty on this minimization if we include a regularization term,

$$(Y - M\boldsymbol{\theta})^T (Y - M\boldsymbol{\theta}) + \lambda |\boldsymbol{\theta}^T \boldsymbol{\theta}|, \tag{8.29}$$

where λ is the regularization or smoothing parameter and $|\boldsymbol{\theta}^T \boldsymbol{\theta}|$ is an example of the penalty function. In this example, we penalize the size of the regression coefficients (which is known as ridge regression as we will discuss in the next section). Solving for $\boldsymbol{\theta}$ we arrive at a modification of eq. 8.19,

$$\boldsymbol{\theta} = (M^T C^{-1} M + \lambda I)^{-1} (M^T C^{-1} Y), \tag{8.30}$$

where I is the identity matrix. One aspect worth noting about robustness through regularization is that, even if $M^T C^{-1} M$ is singular, solutions can still exist for $(M^T C^{-1} M + \lambda I)$.

A Bayesian implementation of regularization would use the prior to impose constraints on the probability distribution of the regression coefficients. If, for example, we assumed that the prior on the regression coefficients was Gaussian with the width of this Gaussian governed by the regularization parameter λ then we could write it as

$$p(\boldsymbol{\theta}|I) \propto \exp\left(\frac{-(\lambda \boldsymbol{\theta}^T \boldsymbol{\theta})}{2}\right). \tag{8.31}$$

Multiplying the likelihood function by this prior results in a posterior distribution with an exponent $(Y - M\boldsymbol{\theta})^T (Y - M\boldsymbol{\theta}) + \lambda |\boldsymbol{\theta}^T \boldsymbol{\theta}|$, equivalent to the MLE regularized regression described above. This Gaussian prior corresponds to ridge regression. For LASSO regression, described below, the corresponding prior would be an exponential (Laplace) distribution.

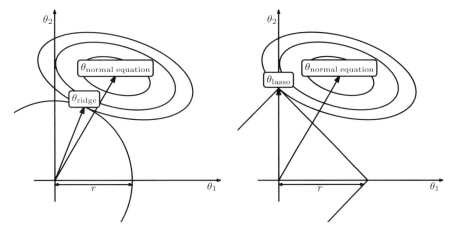

Figure 8.3. A geometric interpretation of regularization. The right panel shows L_1 regularization (LASSO regression) and the left panel L_2 regularization (ridge regularization). The ellipses indicate the posterior distribution for no prior or regularization. The solid lines show the constraints due to regularization (limiting $\boldsymbol{\theta}^2$ for ridge regression and $|\boldsymbol{\theta}|$ for LASSO regression). The corners of the L_1 regularization create more opportunities for the solution to have zeros for some of the weights.

8.3.1. Ridge Regression

The regularization example above is often referred to as ridge regression or Tikhonov regularization [22]. It provides a penalty on the sum of the squares of the regression coefficients such that

$$|\boldsymbol{\theta}|^2 < s, \tag{8.32}$$

where s controls the complexity of the model in the same way as the regularization parameter λ in eq. 8.29. By suppressing large regression coefficients this penalty limits the variance of the system at the expense of an increase in the bias of the derived coefficients.

A geometric interpretation of ridge regression is shown in figure 8.3. The solid elliptical contours are the likelihood surface for the regression with no regularization. The circle illustrates the constraint on the regression coefficients ($|\boldsymbol{\theta}|^2 < s$) imposed by the regularization. The penalty on the likelihood function, based on the squared norm of the regression coefficients, drives the solution to small values of $\boldsymbol{\theta}$. The smaller the value of s (or the larger the regularization parameter λ) the more the regression coefficients are driven toward zero.

The regularized regression coefficients can be derived through matrix inversion as before. Applying an SVD to the $N \times m$ design matrix (where m is the number of terms in the model; see § 8.2.2) we get $M = U\Sigma V^T$, with U an $N \times m$ matrix, V^T the $m \times m$ matrix of eigenvectors and Σ the $m \times m$ matrix of eigenvalues. We can now write the regularized regression coefficients as

$$\boldsymbol{\theta} = V\Sigma' U^T Y, \tag{8.33}$$

where Σ' is a diagonal matrix with elements $d_i/(d_i^2 + \lambda)$, with d_i the eigenvalues of MM^T.

As λ increases, the diagonal components are down weighted so that only those components with the highest eigenvalues will contribute to the regression. This relates directly to the PCA analysis we described in § 7.3. Projecting the variables onto the eigenvectors of MM^T such that

$$Z = MV, \tag{8.34}$$

with z_i the ith eigenvector of M, ridge regression shrinks the regression coefficients for any component for which its eigenvalues (and therefore the associated variance) are small.

The effective goodness of fit for a ridge regression can be derived from the response of the regression function,

$$\hat{y} = M(M^T M + \lambda I)^{-1} M^T y, \tag{8.35}$$

and the number of degrees of freedom,

$$\mathrm{DOF} = \mathrm{Trace}[M(M^T M + \lambda I)^{-1} M^T] = \sum_i \frac{d_i^2}{d_i^2 + \lambda}. \tag{8.36}$$

Ridge regression can be accomplished with the Ridge class in Scikit-learn:

```
import numpy as np
from sklearn.linear_model import Ridge

X = np.random.random((100, 10))
    # 100 points in 10 dims
y = np.dot(X, np.random.random(10))
    # random combination of X
model = Ridge(alpha = 0.05)  # alpha controls
        # regularization
model.fit(X, y)
y_pred = model.predict(X)
```

For more information, see the Scikit-learn documentation.

Figure 8.4 uses the Gaussian basis function regression of § 8.2.2 to illustrate how ridge regression will constrain the regression coefficients. The left panel shows the general linear regression for the supernovas (using 100 evenly spaced Gaussians with $\sigma = 0.2$). As we noted in § 8.2.2, an increase in the number of model parameters results in an overfitting of the data (the lower panel in figure 8.4 shows how the regression coefficients for this fit are on the order of 10^8). The central panel demonstrates how ridge regression (with $\lambda = 0.005$) suppresses the amplitudes of the regression coefficients and the resulting fluctuations in the modeled response.

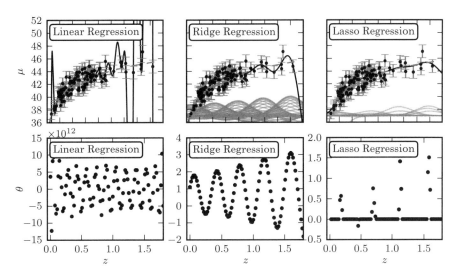

Figure 8.4. Regularized regression for the same sample as Fig 8.2. Here we use Gaussian basis function regression with a Gaussian of width $\sigma = 0.2$ centered at 100 regular intervals between $0 \leq z \leq 2$. The lower panels show the best-fit weights as a function of basis function position. The left column shows the results with no regularization: the basis function weights w are on the order of 10^8, and overfitting is evident. The middle column shows ridge regression (L_2 regularization) with $\lambda = 0.005$, and the right column shows LASSO regression (L_1 regularization) with $\lambda = 0.005$. All three methods are fit without the bias term (intercept).

8.3.2. LASSO Regression

Ridge regression uses the square of the regression coefficients to regularize the fits (i.e., the L_2 norm). A modification of this approach is to use the L_1 norm [2] to subset the variables within a model as well as applying shrinkage. This technique is known as LASSO (least absolute shrinkage and selection; see [21]). LASSO penalizes the likelihood as

$$(Y - M\boldsymbol{\theta})^T (Y - M\boldsymbol{\theta}) + \lambda |\boldsymbol{\theta}|, \tag{8.37}$$

where $|\boldsymbol{\theta}|$ penalizes the absolute value of $\boldsymbol{\theta}$. LASSO regularization is equivalent to least-squares regression with a penalty on the absolute value of the regression coefficients,

$$|\boldsymbol{\theta}| < s. \tag{8.38}$$

The most interesting aspect of LASSO is that it not only weights the regression coefficients, it also imposes sparsity on the regression model. Figure 8.3 illustrates the impact of the L_1 norm on the regression coefficients from a geometric perspective. The $\lambda |\boldsymbol{\theta}|$ penalty preferentially selects regions of likelihood space that coincide with one of the vertices within the region defined by the regularization. This corresponds to setting one (or more if we are working in higher dimensions) of the model attributes to zero. This subsetting of the model attributes reduces the underlying complexity of the model (i.e., we make zeroing of weights, or feature selection, more aggressive). As λ increases, the size of the region encompassed within the constraint decreases.

Ridge regression can be accomplished with the Lasso class in Scikit-learn:

```
import numpy as np
from sklearn.linear_model import Lasso

X = np.random.random((100, 10))
    # 100 points in 10 dims
y = np.dot(X, np.random.random(10))
    # random comb. of X
model = Lasso(alpha = 0.05)   # alpha controls
        # regularization
model.fit(X, y)
y_pred = model.predict(X)
```

For more information, see the Scikit-learn documentation.

Figure 8.4 shows this effect for the supernova data. Of the 100 Gaussians in the input model, with $\lambda = 0.005$, only 13 are selected by LASSO (note the regression coefficients in the lower panel). This reduction in model complexity suppresses the overfitting of the data.

A disadvantage of LASSO is that, unlike ridge regression, there is no closed-form solution. The optimization becomes a quadratic programming problem (though it is still a convex optimization). There are a number of numerical techniques that have been developed to address these issues including coordinate-gradient descent [12] and least angle regression [5].

8.3.3. How Do We Choose the Regularization Parameter λ?

In each of the regularization examples above we defined a "shrinkage parameter" that we refer to as the regularization parameter. The natural question then is how do we set λ? So far we have only noted that as we increase λ we increase the constraints on the range regression coefficients (with $\lambda = 0$ returning the standard least-squares regression). We can, however, evaluate its impact on the regression as a function of its amplitude.

Applying the k-fold cross-validation techniques described in § 8.11 we can define an error (for a specified value of λ) as

$$\text{Error}(\lambda) = k^{-1} \sum_k N_k^{-1} \sum_i^{N_k} \frac{[y_i - f(x_i|\boldsymbol{\theta})]^2}{\sigma_i^2}, \tag{8.39}$$

where N_k^{-1} is the number of data points in the kth cross-validation sample, and the summation over N_k represents the sum of the squares of the residuals of the fit. Estimating λ is then simply a case of finding the λ that minimizes the cross-validation error.

8.4. Principal Component Regression

For the case of high-dimensional data or data sets where the variables are collinear, the relation between ridge regression and principal component analysis can be exploited to define a regression based on the principal components. For centered data (i.e., zero mean) we recall from § 7.3 that we can define the principal components of a system from the data covariance matrix, $X^T X$, by applying an eigenvalue decomposition (EVD) or singular value decomposition (SVD),

$$X^T X = V \Sigma V^T, \tag{8.40}$$

with V^T the eigenvectors and Σ the eigenvalues.

Projecting the data matrix onto these eigenvectors we define a set of projected data points,

$$Z = X V^T, \tag{8.41}$$

and truncate this expansion to exclude those components with small eigenvalues. A standard linear regression can now be applied to the data transposed to this principal component space with

$$Y = M_z \boldsymbol{\theta} + \epsilon, \tag{8.42}$$

with M_z the design matrix for the projected components z_i. The PCA analysis (including truncation) and the regression are undertaken as separate steps in this procedure. The distinction between principal component regression (PCR) and ridge regression is that the number of model components in PCR is ordered by their eigenvalues and is absolute (i.e., we weight the regression coefficients by 1 or 0). For ridge regression the weighting of the regression coefficients is continuous.

The advantages of PCR over ridge regression arise primarily for data containing independent variables that are collinear (i.e., where the correlation between the variables is almost unity). For these cases, the regression coefficients have large variance and their solutions can become unstable. Excluding those principal components with small eigenvalues alleviates this issue. At what level to truncate the set of eigenvectors is, however, an open question (see § 7.3.2). A simple approach to take is to truncate based on the eigenvalue with common cutoffs ranging between 1% and 10% of the average eigenvalue (see § 8.2 of [10] for a detailed discussion of truncation levels for PCR). The disadvantage of such an approach is that an eigenvalue does not always correlate with the ability of a given principal component to predict the dependent variable. Other techniques, including cross-validation [17], have been proposed yet there is no well-adopted solution to this problem.

Finally, we note that in the case of ill-conditioned regression problems (e.g., those with collinear data), most implementations of linear regression will implicitly perform a form of principal component regression when inverting the singular matrix $M^T M$. This comes through the use of the robust pseudoinverse, which truncates small singular values to prevent numerical overflow in ill-conditioned problems.

8.5. Kernel Regression

The previous sections found the regression or objective function that "best fits" a set of data assuming a linear model. Before we address the question of nonlinear optimization we will describe a number of techniques that make use of locality within the data (i.e., local regression techniques).

Kernel or Nadaraya–Watson [18, 23] regression defines a kernel, $K(x_i, x)$, local to each data point, with the amplitude of the kernel depending only on the distance from the local point to all other points in the sample. The properties of the kernel are such that it is positive for all values and asymptotes to zero as the distance approaches infinity. The influence of the kernel (i.e., the region of parameter space over which we weight the data) is determined by its width or bandwidth, h. Common forms of the kernel include the top-hat function, and the Gaussian distribution.

The Nadaraya–Watson estimate of the regression function is given by

$$f(x|K) = \frac{\sum_{i=1}^{N} K\left(\frac{||x_i - x||}{h}\right) y_i}{\sum_{i=1}^{N} K\left(\frac{||x_i - x||}{h}\right)}, \tag{8.43}$$

which can be viewed as taking a weighted average of the dependent variable, y. This gives higher weight to points near x with a weighting function,

$$w_i(x) = \frac{K\left(\frac{||x_i - x||}{h}\right)}{\sum_{i=1}^{N} K\left(\frac{||x_i - x||}{h}\right)}. \tag{8.44}$$

Nadaraya–Watson kernel regression can be performed using AstroML in the following way:

```
import numpy as np
from astroML.linear_model import NadarayaWatson

X = np.random.random((100, 2))# 100 points in 2 dims
y = X[:, 0] + X[:, 1]
model = NadarayaWatson('gaussian', 0.05)
model.fit(X, y)
y_pred = model.predict(X)
```

Figure 8.2 shows the application of Gaussian kernel regression to the synthetic supernova data compared to the standard linear regression techniques introduced in § 8.2. For this example we use a Gaussian kernel with $h = 0.1$ that is constant across the redshift interval. At high redshift, where the data provide limited support for the model, we see that the weight function drives the predicted value of y to that of the nearest neighbor. This prevents the extrapolation problems common when fitting polynomials that are not constrained at the edges of the data (as we saw in the top panels of figure 8.2). The width of the kernel acts as a smoothing function.

At low redshift, the increase in the density of points at $z \approx 0.25$ biases the weighted estimate of $\hat{y}$ for $z < 0.25$. This is because, while the kernel amplitude is small for the higher redshift points, the increase in the sample size results in a higher than average weighting of points at $z \approx 0.25$. Varying the bandwidth as a function of the independent variable can correct for this. The rule of thumb for kernel-based regression (as with kernel density estimation described in § 6.3) is that the bandwidth is more important that the exact shape of the kernel.

Estimation of the optimal bandwidth of the kernel is straightforward using cross-validation (see § 8.11). We define the CV error as

$$
\mathrm{CV}_{L_2}(h) = \frac{1}{N} \sum_{i=1}^{N} \left(y_i - f\left(x_i | K\left(\frac{||x_i - x_j||}{h} \right) \right) \right)^2 . \tag{8.45}
$$

We actually do not need to compute separate estimates for each leave-one-out cross-validation subsample if we rewrite this as

$$
\mathrm{CV}_{L_2}(h) = \frac{1}{N} \frac{\sum_{i=1}^{N} (y_i - f(x_i | K))^2}{\left(1 - \frac{K(0)}{\sum_{j=1}^{N} K\left(\frac{||x_i - x_j||}{h} \right)} \right)^2} . \tag{8.46}
$$

It can be shown that, as for kernel density estimation, the optimal bandwidth decreases with sample size at a rate of $N^{-1/5}$.

8.6. Locally Linear Regression

Related to kernel regression is *locally linear regression*, where we solve a separate weighted least-squares problem at each point x, finding the $w(x)$ which minimizes

$$
\sum_{i=1}^{N} K\left(\frac{||x - x_i||}{h} \right) (y_i - w(x) x_i)^2 . \tag{8.47}
$$

The assumption for locally linear regression is that the regression function can be approximated by a Taylor series expansion about any local point. If we truncated this expansion at the first term (i.e., a locally constant solution) we recover kernel regression. For locally linear regression the function estimate is

$$
f(x|K) = \theta(x)x \tag{8.48}
$$
$$
= x^T \left(\mathbf{X}^T W(x) \mathbf{X} \right)^{-1} \mathbf{X}^T W(x) \mathbf{Y} \tag{8.49}
$$
$$
= \sum_{i=1}^{N} w_i(x) y_i , \tag{8.50}
$$

where $W(x)$ is an $N \times N$ diagonal matrix with the ith diagonal element given by $K ||x_i - x||/h$.

A common form for $K(x)$ is the tricubic kernel,

$$K(x_i, x) = \left(1 - \left(\frac{|x - x_i|}{h}\right)^3\right)^3 \tag{8.51}$$

for $|x_i - x| < h$, which is often referred to as lowess (or loess; locally weighted scatter plot smoothing); see [3].

There are further extensions possible for local linear regression:

- *Local polynomial regression.* We can consider any polynomial order. However there is a bias–variance (complexity) trade-off, as usual. The general consensus is that going past linear increases variance without decreasing bias much, since local linear regression captures most of the boundary bias.
- *Variable-bandwidth kernels.* Let the bandwidth for each training point be inversely proportional to its kth nearest neighbor's distance. Generally a good idea in practice, though there is less theoretical consensus on how to choose the parameters in this framework.

None of these modifications improves the convergence rate.

8.7. Nonlinear Regression

Forcing data to correspond to a linear model through the use of coordinate transformations is a well-used trick in astronomy (e.g., the extensive use of logarithms in the astronomical literature to linearize complex relations between attributes; fitting $y = A \exp(Bx)$ becomes a linear problem with $z = K + Bx$, where $z = \log y$ and $K = \log A$). These simplifications, while often effective, introduce other complications (including the non-Gaussian nature of the uncertainties for low signal-to-noise data). We must, eventually, consider the case of nonlinear regression and model fitting.

In the cosmological examples described previously we have fit a series of parametric and nonparametric models to the supernova data. Given that we know the theoretical form of the underlying cosmological model, these models are somewhat ad hoc (e.g., using a series of polynomials to parameterize the dependence of distance modulus on cosmological redshift). In the following we consider directly fitting the cosmological model described in eq. 8.4. Solving for Ω_m and Ω_Λ is a nonlinear optimization problem requiring that we maximize the posterior,

$$p(\Omega_m, \Omega_\Lambda | z, I) \propto \prod_{i=1}^{n} \frac{1}{\sqrt{2\pi}\sigma_i} \exp\left(\frac{-(\mu_i - \mu(z_i | \Omega_m, \Omega_\Lambda))^2}{2\sigma_i^2}\right) p(\Omega_m, \Omega_\Lambda) \tag{8.52}$$

with μ_i the distance modulus for the supernova and z_i the redshift.

In § 5.8 we introduced Markov chain Monte Carlo as a sampling technique that can be used for searching through parameter space. Figure 8.5 shows the resulting likelihood contours for our cosmological model after applying the Metropolis–Hastings algorithm to generate the MCMC chains and integrating the chains over the parameter space.

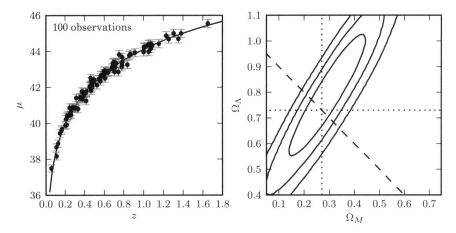

Figure 8.5. Cosmology fit to the standard cosmological integral. Errors in μ are a factor of ten smaller than for the sample used in figure 8.2. Contours are 1σ, 2σ, and 3σ for the posterior (uniform prior in Ω_M and Ω_Λ). The dashed line shows flat cosmology. The dotted lines show the input values.

An alternate approach is to use the Levenberg–Marquardt (LM) algorithm [15, 16] to optimize the maximum likelihood estimation. LM searches for the sum-of-squares minima of a multivariate distribution through a combination of gradient descent and Gauss–Newton optimization. Assuming that we can express our regression function as a Taylor series expansion then, to first order, we can write

$$f(x_i|\boldsymbol{\theta}) = f(x_i|\boldsymbol{\theta}_0) + J\,d\boldsymbol{\theta}, \tag{8.53}$$

where $\boldsymbol{\theta}_0$ is an initial guess for the regression parameters, J is the Jacobian about this point ($J = \partial f(x_i|\boldsymbol{\theta})/\partial\boldsymbol{\theta}$), and $d\boldsymbol{\theta}$ is a small change in the regression parameters. LM minimizes the sum of square errors,

$$\sum_i (y_i - f(x_i|\boldsymbol{\theta}_0) - J_i d\boldsymbol{\theta})^2 \tag{8.54}$$

for the perturbation $d\boldsymbol{\theta}$. This results in an update relation for $d\boldsymbol{\theta}$ of

$$(J^T C^{-1} J + \lambda \operatorname{diag}(J^T C^{-1} J))\,d\boldsymbol{\theta} = J^T C^{-1}(Y - f(X|\boldsymbol{\theta})), \tag{8.55}$$

with C the standard covariance matrix introduced in eq. 8.18. In this expression the λ term acts as a damping parameter (in a manner similar to ridge regression regularization discussed in § 8.3.1). For small λ the relation approximates a Gauss–Newton method (i.e., it minimizes the parameters assuming the function is quadratic). For large λ the perturbation $d\boldsymbol{\theta}$ follows the direction of steepest descent. The $\operatorname{diag}(J^T C^{-1} J)$ term, as opposed to the identity matrix used in ridge regression, ensures that the update of $d\boldsymbol{\theta}$ is largest along directions where the gradient is smallest (which improves convergence).

LM is an iterative process. At each iteration LM searches for the step $d\boldsymbol{\theta}$ that minimizes eq. 8.54 and then updates the regression model. The iterations cease when the step size, or change in likelihood values reaches a predefined value. Throughout the iteration process the damping parameter, λ, is adaptively varied (decreasing as the minimum is approached). A common approach involves decreasing (or increasing) λ by v^k, where k is the number of the iteration and v is a damping factor (with $v > 1$). Upon successful convergence the regression parameter covariances are given by $[J^T\theta J]^{-1}$. It should be noted that the success of the LM algorithm often relies on the initial guesses for the regression parameters being close to the maximum likelihood solution (in the presence of nonlocal minima), or the likelihood function surface being unimodal.

The submodule `scipy.optimize` includes several routines for optimizing linear and nonlinear equations. The Levenberg–Marquardt algorithm is used in `scipy.optimize.leastsq`. Below is a brief example of using the routine to estimate the first six terms of the Taylor series for the function $y = \sin x \approx \sum_n a_n x^n$:

```
import numpy as np
from scipy import optimize
x = np.linspace(-3, 3, 100) # 100 values between -3
                            # and 3
def taylor_err(a, x, f):
    p = np.arange(len(a))[:, np.newaxis]
        # column vector
    return f(x) - np.dot(a, x ** p)
a_start = np.zeros(6)  # starting guess
a_best, flag = optimize.leastsq(taylor_err, a_start,
                                args=(x, np.sin))
```

8.8. Uncertainties in the Data

In the opening section we introduced the problem of regression in its most general form. Computational complexity (particularly for the case of multivariate data) led us through a series of approximations that can be used in optimizing the likelihood and the prior (e.g., that the uncertainties have a Gaussian distribution, that the independent variables are error-free, that we can control complexity of the model, and that we can express the likelihood in terms of linear functions). Eventually we have to face the problem that many of these assumptions no longer hold for data in the wild. We have addressed the question of model complexity; working from linear to nonlinear regression. We now return to the question of *error behavior* and consider the uncertainty that is inherent in any analysis.

8.8.1. Uncertainties in the Dependent and Independent Axes

In almost all real-world applications, the assumption that one variable (the independent variable) is essentially free from any uncertainty is not valid. Both the dependent and independent variables will have measurement uncertainties. For our example data set we have assumed that the redshifts of the supernovas are known to a very high level of accuracy (i.e., that we measure these redshifts spectroscopically). If, for example, the redshifts of the supernovas were estimated based on the colors of the supernova (or host galaxy) then the errors on the redshift estimate can be significant. For our synthetic data we assume fractional uncertainties on the independent variable of 10%.

The impact of errors on the "independent" variables is a bias in the derived regression coefficients. This is straightforward to show if we consider a linear model with a dependent and independent variable, y^* and x^*. We can write the objective function as before,

$$y_i^* = \theta_0 + \theta_1 x_i^*. \tag{8.56}$$

Now let us assume that we observe y and x, which are noisy representations of y^* and x^*, i.e.,

$$x_i = x_i^* + \delta_i, \tag{8.57}$$

$$y_i = y^* + \epsilon_i, \tag{8.58}$$

with δ and ϵ centered normal distributions.

Solving for y we get

$$y = \theta_0 + \theta_1(x_i - \delta_i) + \epsilon_i. \tag{8.59}$$

The uncertainty in x is now part of the regression equation and scales with the regression coefficients (biasing the regression coefficient). This problem is known in the statistics literature as *total least squares* and belongs to the class of "errors-in-variables" problems; see [4]. A very detailed discussion of regression and the problem of uncertainties from an astronomical perspective can be found in [8, 11].

How can we account for the measurement uncertainties in both the independent and dependent variables? Let us start with a simple example where we assume the errors are Gaussian so we can write the covariance matrix as

$$\Sigma_i = \begin{bmatrix} \sigma_{x_i}^2 & \sigma_{xy_i} \\ \sigma_{xy_i} & \sigma_{y_i}^2 \end{bmatrix}. \tag{8.60}$$

For a straight-line regression we express the slope of the line, θ_1, in terms of its normal vector,

$$\mathbf{n} = \begin{bmatrix} -\sin\alpha \\ \cos\alpha \end{bmatrix}, \tag{8.61}$$

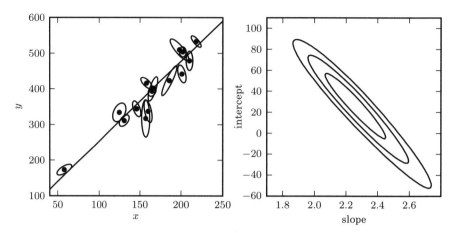

Figure 8.6. A linear fit to data with correlated errors in x and y. In the literature, this is often referred to as *total least squares* or *errors-in-variables* fitting. The left panel shows the lines of best fit; the right panel shows the likelihood contours in slope/intercept space. The points are the same set used for the examples in [8].

where $\theta_1 = \arctan(\alpha)$ and α is the angle between the line and the x-axis. The covariance matrix projects onto this space as

$$S_i^2 = \mathbf{n}^T \Sigma_i \mathbf{n} \tag{8.62}$$

and the distance between a point and the line is given by (see [8])

$$\Delta_i = \mathbf{n}^T z_i - \theta_0 \cos\alpha, \tag{8.63}$$

where z_i represents the data point (x_i, y_i). The log-likelihood is then

$$\ln L = -\sum_i \frac{\Delta_i^2}{2S_i^2}. \tag{8.64}$$

Maximizing this likelihood for the regression parameters, θ_0 and θ_1 is shown in figure 8.6, where we use the data from [8] with correlated uncertainties on the x and y components, and recover the underlying linear relation. For a single parameter search (θ_1) the regression can be undertaken in a brute-force manner. As we increase the complexity of the model or the dimensionality of the data, the computational cost will grow and techniques such as MCMC must be employed (see [4]).

8.9. Regression That Is Robust to Outliers

A fact of experimental life is that if you can measure an attribute you can also measure it incorrectly. Despite the increase in fidelity of survey data sets, any regression or model fitting must be able to account for outliers from the fit. For the standard least-squares regression the use of an L_2 norm results in outliers that have substantial leverage in any fit (contributing as the square of the systematic deviation). If we knew

$e(y_i|y)$ for all of the points in our sample (e.g., they are described by an exponential distribution where we would use the L_1 norm to define the error) then we would simply include the error distribution when defining the likelihood. When we do not have a priori knowledge of $e(y_i|y)$, things become more difficult. We can either model $e(y_i|y)$ as a mixture model (see § 5.6.7) or assume a form for $e(y_i|y)$ that is less sensitive to outliers. An example of the latter would be the adoption of the L_1 norm, $\sum_i ||y_i - w_i x_i||$, which we introduced in § 8.3, which is less sensitive to outliers than the L_2 norm (and was, in fact, proposed by Rudjer Bošković prior to the development of least-squares regression by Legendre, Gauss, and others [2]). Minimizing the L_1 norm is essentially finding the median. The drawback of this least absolute value regression is that there is no closed-form solution and we must minimize the likelihood space using an iterative approach.

Other approaches to robust regression adopt an approach that seeks to reject outliers. In the astronomical community this is usually referred to as "sigma clipping" and is undertaken in an iterative manner by progressively pruning data points that are not well represented by the model. Least-trimmed squares formalizes this, somewhat ad hoc approach, by searching for the subset of K points which minimize $\sum_i^K (y_i - \theta_i x_i)^2$. For large N the number of combinations makes this search expensive.

Complementary to outlier rejection are the Theil–Sen [20] or the Kendall robust line-fit method and associated techniques. In these cases the regression is determined from the median of the slope, θ_1, calculated from all pairs of points within the data set. Given the slope, the offset or zero point, θ_0, can be defined from the median of $y_i - \theta_1 x_i$. Each of these techniques is simple to estimate and scales to large numbers.

M estimators (M stands for "maximum-likelihood-type") approach the problem of outliers by modifying the underlying likelihood estimator to be less sensitive than the classic L_2 norm. M estimators are a class of estimators that include many maximum-likelihood approaches (including least squares). They replace the standard least squares, which minimizes the sum of the squares of the residuals between a data value and the model, with a different function. Ideally the M estimator has the property that it increases less than the square of the residual and has a unique minimum at zero.

Huber loss function

An example of an M estimator that is common in robust regression is that of the Huber loss (or cost) function [9]. The Huber estimator minimizes

$$\sum_{i=1}^{N} e(y_i|y),\tag{8.65}$$

where $e(y_i|y)$ is modeled as

$$\phi(t) = \begin{cases} \frac{1}{2}t^2 & \text{if } |t| \leq c, \\ c|t| - \frac{1}{2}c^2 & \text{if } |t| \geq c, \end{cases}\tag{8.66}$$

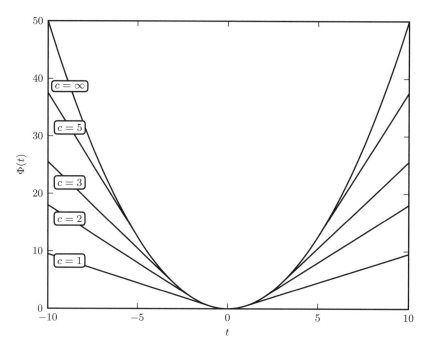

Figure 8.7. The Huber loss function for various values of c.

and $t = y_i - y$ with a constant c that must be chosen. Therefore, $e(t)$ is a function which acts like t^2 for $|t| \leq c$ and like $|t|$ for $|t| > c$ and is continuous and differentiable (see figure 8.7). The transition in the Huber function is equivalent to assuming a Gaussian error distribution for small excursions from the true value of the function and an exponential distribution for large excursions (its behavior is a compromise between the mean and the median). Figure 8.8 shows an application of the Huber loss function to data with outliers. Outliers do have a small effect, and the slope of the Huber loss fit is closer to that of standard linear regression.

8.9.1. Bayesian Outlier Methods

From a Bayesian perspective, one can use the techniques developed in chapter 5 within the context of a regression model in order to account for, and even to individually identify outliers (recall § 5.6.7). Figure 8.9 again shows the data set used in figure 8.8, which contains three clear outliers. In a standard straight-line fit to the data, the result is strongly affected by these points. Though this standard linear regression problem is solvable in closed form (as it is in figure 8.8), here we compute the best-fit slope and intercept using MCMC sampling (and show the resulting contours in the upper-right panel).

The remaining two panels show two different Bayesian strategies for accounting for outliers. The main idea is to enhance the model such that it can naturally explain the presence of outliers. In the first model, we account for the outliers through the use of a mixture model, adding a background Gaussian component to our data. This is the regression analog of the model explored in § 5.6.5, with the difference that here we are modeling the background as a wide Gaussian rather than a uniform distribution.

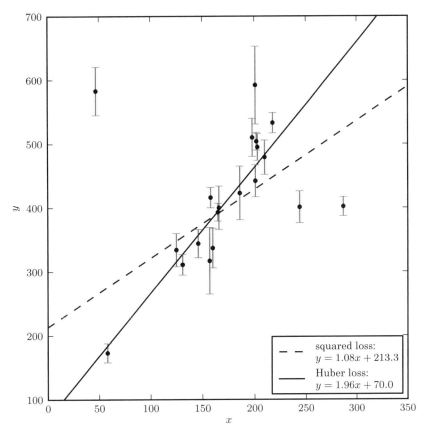

Figure 8.8. An example of fitting a simple linear model to data which includes outliers (data is from table 1 of [8]). A comparison of linear regression using the squared-loss function (equivalent to ordinary least-squares regression) and the Huber loss function, with $c = 1$ (i.e., beyond 1 standard deviation, the loss becomes linear).

The mixture model includes three additional parameters: μ_b and V_b, the mean and variance of the background, and p_b, the probability that any point is an outlier. With this model, the likelihood becomes (cf. eq. 5.83; see also [8])

$$p(\{y_i\}|\{x_i\}, \{\sigma_i\}, \theta_0, \theta_1, \mu_b, V_b, p_b) \propto \prod_{i=1}^{N} \left[\frac{1 - p_b}{\sqrt{2\pi \sigma_i^2}} \exp\left(-\frac{(y_i - \theta_1 x_i - \theta_0)^2}{2\sigma_i^2} \right) \right.$$
$$\left. + \frac{p_b}{\sqrt{2\pi(V_b + \sigma_i^2)}} \exp\left(-\frac{(y_i - \mu_b)^2}{2(V_b + \sigma_i^2)} \right) \right].$$
(8.67)

Using MCMC sampling and marginalizing over the background parameters yields the dashed-line fit in figure 8.9. The marginalized posterior for this model is shown in the lower-left panel. This fit is much less affected by the outliers than is the simple regression model used above.

Finally, we can go further and perform an analysis analogous to that of § 5.6.7, in which we attempt to identify bad points individually. In analogy with eq. 5.94 we

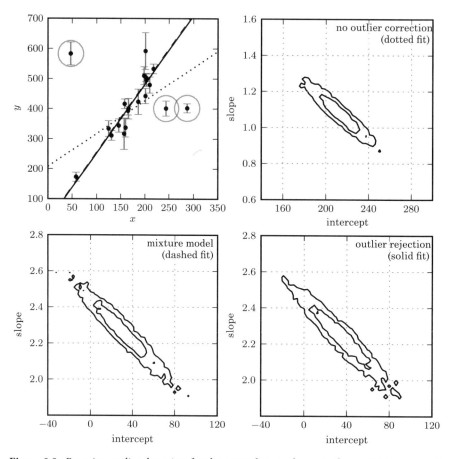

Figure 8.9. Bayesian outlier detection for the same data as shown in figure 8.8. The top-left panel shows the data, with the fits from each model. The top-right panel shows the 1σ and 2σ contours for the slope and intercept with no outlier correction: the resulting fit (shown by the dotted line) is clearly highly affected by the presence of outliers. The bottom-left panel shows the marginalized 1σ and 2σ contours for a mixture model (eq. 8.67). The bottom-right panel shows the marginalized 1σ and 2σ contours for a model in which points are identified individually as "good" or "bad" (eq. 8.68). The points which are identified by this method as bad with a probability greater than 68% are circled in the first panel.

can fit for nuisance parameters g_i, such that if $g_i = 1$, the point is a "good" point, and if $g_i = 0$ the point is a "bad" point. With this addition our model becomes

$$p(\{y_i\}|\{x_i\}, \{\sigma_i\}, \{g_i\}, \theta_0, \theta_1, \mu_b, V_b) \propto \prod_{i=1}^{N} \left[\frac{g_i}{\sqrt{2\pi\sigma_i^2}} \exp\left(-\frac{(y_i - \theta_1 x_i - \theta_0)^2}{2\sigma_i^2}\right) \right.$$
$$\left. + \frac{1 - g_i}{\sqrt{2\pi(V_b + \sigma_i^2)}} \exp\left(-\frac{(y_i - \mu_b)^2}{2(V_b + \sigma_i^2)}\right) \right].$$

$$(8.68)$$

This model is very powerful: by marginalizing over all parameters but a particular g_i, we obtain a posterior estimate of whether point i is an outlier. Using this procedure, the "bad" points have been marked with a circle in the upper-left panel of figure 8.9. If instead we marginalize over the nuisance parameters, we can compute a two-dimensional posterior for the slope and intercept of the straight-line fit. The lower-right panel shows this posterior, after marginalizing over $\{g_i\}$, μ_b, and V_b. In this example, the result is largely consistent with the simpler Bayesian mixture model used above.

8.10. Gaussian Process Regression

Another powerful class of regression algorithms is Gaussian process regression. Despite its name, Gaussian process regression is widely applicable to data that are not generated by a Gaussian process, and can lead to very flexible regression models that are more data driven than other parametric approaches. There is a rich literature on the subject, and we will only give a cursory treatment here. An excellent book-length treatment of Gaussian processes can be found in [19].

A Gaussian process is a collection of random variables in parameter space, any subset of which is defined by a joint Gaussian distribution. It can be shown that a Gaussian process can be completely specified by its mean and covariance function. Gaussian processes can be defined in any number of dimensions for any positive covariance function. For simplicity, in this section we will consider one dimension and use a familiar squared-exponential covariance function,

$$\text{Cov}(x_1, x_2; h) = \exp\left(\frac{-(x_1 - x_2)^2}{2h^2}\right), \tag{8.69}$$

where h is the bandwidth. Given a chosen value of h, this covariance function specifies the statistics of an infinite set of possible functions $f(x)$. The upper-left panel of figure 8.10 shows three of the possible functions drawn from a zero-mean Gaussian process[3] with $h = 1.0$. This becomes more interesting when we specify constraints on the Gaussian process: that is, we select only those functions $f(x)$ which pass through particular points in the space. The remaining panels of figure 8.10 show this Gaussian process constrained by points without error (upper-right panel), points with error (lower-left panel), and a set of 20 noisy observations drawn from the function $f_{\text{true}}(x) = \cos x$. In each, the shaded region shows the 2σ contour in which 95% of all possible functions $f(x)$ lie.

These constrained Gaussian process examples hint at how these ideas could be used for standard regression tasks. The Gaussian process regression problem can be formulated similarly to the other regression problems discussed above. We assume our data is drawn from an underlying model $f(x)$: that is, our observed data is $\{x_i, y_i = f(x_i) + \sigma_i\}$. Given the observed data, we desire an estimate of the mean value $\bar{f}_j^*$ and variance Σ_{jk}^* for a new set of measurements x_j^*. In Bayesian terms, we want to compute the posterior pdf

$$p(f_j | \{x_i, y_i, \sigma_i\}, x_j^*). \tag{8.70}$$

---

[3] These functions are drawn by explicitly creating the $N \times N$ covariance matrix C from the functional form in eq. 8.69, and drawing N correlated Gaussian random variables with mean 0 and covariance C.

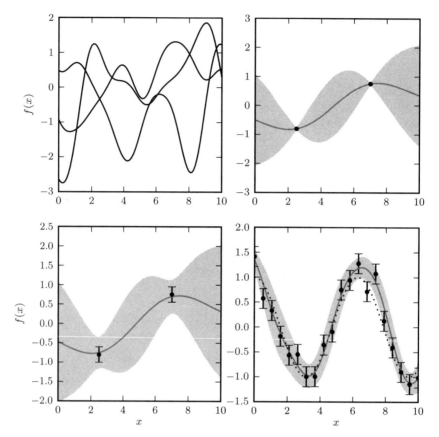

Figure 8.10. An example of Gaussian process regression. The upper-left panel shows three functions drawn from an unconstrained Gaussian process with squared-exponential covariance of bandwidth $h = 1.0$. The upper-right panel adds two constraints, and shows the 2σ contours of the constrained function space. The lower-left panel shows the function space constrained by the points with error bars. The lower-right panel shows the function space constrained by 20 noisy points drawn from $f(x) = \cos x$.

This amounts to averaging over the *entire set* of possible functions $f(x)$ which pass through our constraints. This seemingly infinite calculation can be made tractable using a "kernel trick," as outlined in [19], transforming from the infinite function space to a finite covariance space. The result of this mathematical exercise is that the Gaussian process regression problem can be formulated as follows.

Using the assumed form of the covariance function (eq. 8.69), we compute the covariance matrix

$$K = \begin{pmatrix} K_{11} & K_{12} \\ K_{12}^T & K_{22} \end{pmatrix}, \tag{8.71}$$

where K_{11} is the covariance between the input points x_i with observational errors σ_i^2 added in quadrature to the diagonal, K_{12} is the cross-covariance between the input points x_i and the unknown points x_j^*, and K_{22} is the covariance between the unknown points x_j^*. Then for observed vectors $\mathbf{x}$ and $\mathbf{y}$, and a vector of unknown points $\mathbf{x}^*$, it can be shown that the posterior in eq. 8.70 is given by

$$p(f_j|\{x_i, y_i, \sigma_i\}, x_j^*) = \mathcal{N}(\mu, \Sigma), \tag{8.72}$$

where

$$\mu = K_{12}K_{11}^{-1}\mathbf{y}, \tag{8.73}$$

$$\Sigma = K_{22} - K_{12}^T K_{11}^{-1} K_{12}. \tag{8.74}$$

Then μ_j gives the expected value $\bar{f}_j^*$ of the result, and Σ_{jk} gives the error covariance between any two unknown points. Note that the physics of the underlying process enters through the assumed form of the covariance function (e.g., eq. 8.69; for an analysis of several plausible covariance functions in the astronomical context of quasar variability; see [24]).

In figure 8.11, we show a Gaussian process regression analysis of the supernova data set used above. The model is very well constrained near $z = 0.6$, where there is a lot of data, but not well constrained at higher redshifts. This is an important feature of Gaussian process regression: the analysis produces not only a best-fit model, but an uncertainty at each point, as well as a full covariance estimate of the result at unknown points.

A powerful and general framework for Gaussian process regression is available in Scikit-learn. It can be used as follows:

```
import numpy as np
from sklearn.gaussian_process import GaussianProcess

X = np.random.random((100, 2)) # 100 pts in 2 dims
y = np.sin(10 * X[:, 0] + X[:, 1])
gp = GaussianProcess(corr='squared_exponential')
gp.fit(X, y)
y_pred, dy_pred = gp.predict(X, eval_MSE=True)
```

The predict method can optionally return the mean square error, which is the diagonal of the covariance matrix of the fit. For more details, see the source code of the figures in this chapter, as well as the Scikit-learn documentation.

The results shown in figure 8.11 depend on correctly tuning the correlation function bandwidth h. If h is too large or too small, the particular Gaussian process model will be unsuitable for the data and the results will be biased. Figure 8.11 uses a simple cross-validation approach to decide on the best value of h: we will discuss cross-validation in the following section. Sometimes, the value of h is scientifically an interesting outcome of the analysis; such an example is discussed in the context of the damped random walk model in § 10.5.4.

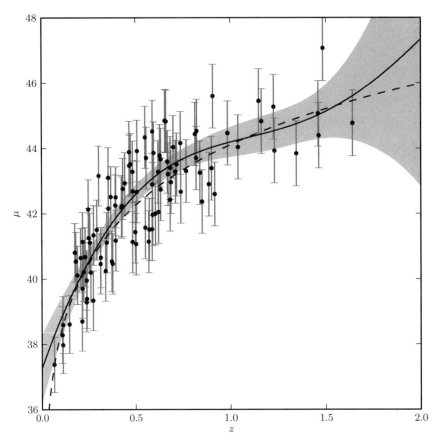

Figure 8.11. A Gaussian process regression analysis of the simulated supernova sample used in figure 8.2. This uses a squared-exponential covariance model, with bandwidth learned through cross-validation.

8.11. Overfitting, Underfitting, and Cross-Validation

When using regression, whether from a Bayesian or maximum likelihood perspective, it is important to recognize some of the potential pitfalls associated with these methods. As noted above, the optimality of the regression is contingent on correct model selection. In this section we explore cross-validation methods which can help determine whether a potential model is a good fit to the data. These techniques are complementary to the model selection techniques such as AIC and BIC discussed in § 4.3. This section will introduce the important topics of overfitting and underfitting, bias and variance, and introduces the frequentist tool of cross-validation to understand these.

Here, for simplicity, we will consider the example of a simple one-dimensional model with homoscedastic errors, though the results of this section naturally generalize to more sophisticated models. As above, our observed data is x_i, and we're trying to predict the dependent variable y_i. We have a training sample in which we have observed both x_i and y_i, and an unknown sample for which only x_i is measured.

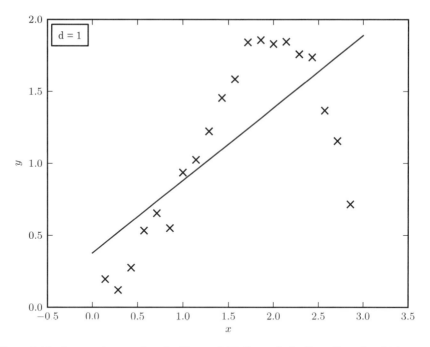

Figure 8.12. Our toy data set described by eq. 8.75. Shown is the line of best fit, which quite clearly underfits the data. In other words, a linear model in this case has high bias.

For example, you may be looking at the fundamental plane for elliptical galaxies, and trying to predict a galaxy's central black hole mass given the velocity dispersion and surface brightness of the stars. Here y_i is the mass of the black hole, and x_i is a vector of a length two consisting of velocity dispersion and surface brightness measurements.

Throughout the rest of this section, we will use a simple model where x and y satisfy the following:

$$0 \leq x_i \leq 3,$$
$$y_i = x_i \sin(x_i) + \epsilon_i, \tag{8.75}$$

where the noise is drawn from a normal distribution $\epsilon_i \sim \mathcal{N}(0, 0.1)$. The values for 20 regularly spaced points are shown in figure 8.12.

We will start with a simple straight-line fit to our data. The model is described by two parameters, the slope of the line, θ_1, and the y-axis intercept, θ_0, and is found by minimizing the mean square error,

$$\epsilon = \frac{1}{N} \sum_{i=1}^{N} (y_i - \theta_0 - \theta_1 x_i)^2. \tag{8.76}$$

The resulting best-fit line is shown in figure 8.12. It is clear that a straight line is not a good fit: it does not have enough flexibility to accurately model the data. We say in this case that the model is biased, and that it underfits the data.

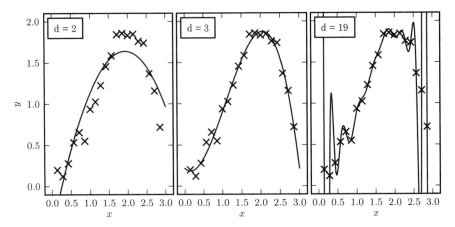

Figure 8.13. Three models of increasing complexity applied to our toy data set (eq. 8.75). The $d = 2$ model, like the linear model in figure 8.12, suffers from high bias, and underfits the data. The $d = 19$ model suffers from high variance, and overfits the data. The $d = 3$ model is a good compromise between these extremes.

What can be done to improve on this? One possibility is to make the model more sophisticated by increasing the degree of the polynomial (see § 8.2.2). For example, we could fit a quadratic function, or a cubic function, or in general a d-degree polynomial. A more complicated model with more free parameters should be able to fit the data much more closely. The panels of figure 8.13 show the best-fit polynomial model for three different choices of the polynomial degree d.

As the degree of the polynomial increases, the best-fit curve matches the data points more and more closely. In the extreme of $d = 19$, we have 20 degrees of freedom with 20 data points, and the training error given by eq. 8.76 can be reduced to zero (though numerical issues can prevent this from being realized in practice). Unfortunately, it is clear that the $d = 19$ polynomial is not a better fit to our data as a whole: the wild swings of the curve in the spaces between the training points are not a good description of the underlying data model. The model suffers from high variance; it overfits the data. The term variance is used here because a small perturbation of one of the training points in the $d = 19$ model can change the best-fit model by a large magnitude. In a high-variance model, the fit varies strongly depending on the exact set or subset of data used to fit it.

The center panel of figure 8.13 shows a $d = 3$ model which balances the trade-off between bias and variance: it does not display the high bias of the $d = 2$ model, and does not display high variance like the $d = 19$ model. For simple two-dimensional data like that seen here, the bias/variance trade-off is easy to visualize by plotting the model along with the input data. But this strategy is not as fruitful as the number of data dimensions grows. What we need is a general measure of the "goodness of fit" of different models to the training data. As displayed above, the mean square error does not paint the whole picture: increasing the degree of the polynomial in this case can lead to smaller and smaller training errors, but this reflects overfitting of the data rather than an improved approximation of the underlying model.

An important practical aspect of regression analysis lies in addressing this deficiency of the training error as an evaluation of goodness of fit, and finding a model which best compromises between high bias and high variance. To this end, the process of *cross-validation* can be used to quantitatively evaluate the bias and variance of a regression model.

8.11.1. Cross-Validation

There are several possible approaches to cross-validation. We will discuss one approach in detail here, and list some alternative approaches at the end of the section. The simplest approach to cross-validation is to split the training data into three parts: the training set, the cross-validation set, and the test set. As a rule of thumb, the training set should comprise 50–70% of the original training data, while the remainder is divided equally into the cross-validation set and test set.

The training set is used to determine the parameters of a given model (i.e., the optimal values of θ_j for a given choice of d). Using the training set, we evaluate the training error ϵ_{tr} using eq. 8.76. The cross-validation set is used to evaluate the cross-validation error ϵ_{cv} of the model, also via eq. 8.76. Because this cross-validation set was not used to construct the fit, the cross-validation error will be large for a high-bias (overfit) model, and better represents the true goodness of fit of the model. With this in mind, the model which minimizes this cross-validation error is likely to be the best model in practice. Once this model is determined, the test error is evaluated using the test set, again via eq. 8.76. This test error gives an estimate of the reliability of the model for an unlabeled data set.

Why do we need a test set as well as a cross-validation set? In one sense, just as the parameters (in this case, θ_j) are learned from the training set, the so-called hyperparameters—those parameters which describe the complexity of the model (in this case, d)—are learned from the cross-validation set. In the same way that the parameters can be overfit to the training data, the hyperparameters can be overfit to the cross-validation data, and the cross-validation error gives an overly optimistic estimate of the performance of the model on an unlabeled data set. The test error is a better representation of the error expected for a new set of data. This is why it is recommended to use both a cross-validation set and a test set in your analysis.

A useful way to use the training error and cross-validation error to evaluate a model is to look at the results graphically. Figure 8.14 shows the training error and cross-validation error for the data in figure 8.13 as a function of the polynomial degree d. For reference, the dotted line indicates the level of intrinsic scatter added to our data.

The broad features of this plot reflect what is generally seen as the complexity of a regression model is increased: for small d, we see that both the training error and cross-validation error are very high. This is the tell-tale indication of a high-bias model, in which the model underfits the data. Because the model does not have enough complexity to describe the intrinsic features of the data, it performs poorly for both the training and cross-validation sets.

For large d, we see that the training error becomes very small (smaller than the intrinsic scatter we added to our data) while the cross-validation error becomes very large. This is the telltale indication of a high-variance model, in which the model overfits the data. Because the model is overly complex, it can match subtle variations in the training set which do not reflect the underlying distribution. Plotting this sort of information is a very straightforward way to settle on a suitable model. Of course,

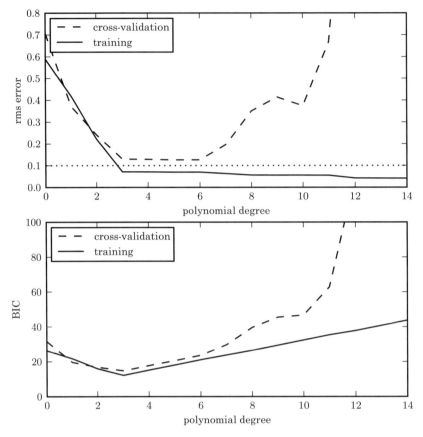

Figure 8.14. The top panel shows the root-mean-square (rms) training error and cross-validation error for our toy model (eq. 8.75) as a function of the polynomial degree d. The horizontal dotted line indicates the level of intrinsic scatter in the data. Models with polynomial degree from 3 to 5 minimize the cross-validation rms error. The bottom panel shows the Bayesian information criterion (BIC) for the training and cross-validation subsamples. According to the BIC, a degree-3 polynomial gives the best fit to this data set.

AIC and BIC provide another way to choose optimal d. Here both methods would choose the model with the best possible cross-validation error: $d = 3$.

8.11.2. Learning Curves

One question that cross-validation does not directly address is that of how to improve a model that is not giving satisfactory results (e.g., the cross-validation error is much larger than the known errors). There are several possibilities:

1. **Get more training data**. Often, using more data to train a model can lead to better results. Surprisingly, though, this is not always the case.
2. **Use a more/less complicated model**. As we saw above, the complexity of a model should be chosen as a balance between bias and variance.
3. **Use more/less regularization**. Including regularization, as we saw in the discussion of ridge regression (see § 8.3.1) and other methods above, can help

with the bias/variance trade-off. In general, increasing regularization has a similar effect to decreasing the model complexity.

4. **Increase the number of features**. Adding more observations of each object in your set can lead to a better fit. But this may not always yield the best results.

The choice of which route to take is far beyond a simple philosophical matter: for example, if you desire to improve your photometric redshifts for a large astronomical survey, it is important to evaluate whether you stand to benefit more from increasing the size of the training set (i.e., gathering spectroscopic redshifts for more galaxies) or from increasing the number of observations of each galaxy (i.e., reobserving the galaxies through other passbands). The answer to this question will inform the allocation of limited, and expensive, telescope time.

Note that this is a fundamentally different question than that explored above. There, we had a fixed data set, and were trying to determine the best model. Here, we assume a fixed model, and are asking how to improve the data set. One way to address this question is by plotting learning curves. A learning curve is the plot of the training and cross-validation error as a function of the number of training points. The details are important, so we will write this out explicitly.

Let our model be represented by the set of parameters $\boldsymbol{\theta}$. In the case of our simple example, $\boldsymbol{\theta} = \{\theta_0, \theta_1, \ldots, \theta_d\}$. We will denote by $\boldsymbol{\theta}^{(n)} = \{\theta_0^{(n)}, \theta_1^{(n)}, \ldots, \theta_d^{(n)}\}$ the model parameters which best fit the first n points of the training data: here $n \leq N_{\text{train}}$, where N_{train} is the total number of training points. The truncated training error for $\theta^{(n)}$ is given by

$$\epsilon_{\text{tr}}^{(n)} = \sqrt{\frac{1}{n} \sum_{i=1}^{n} \left[y_i - \sum_{m=0}^{d} \theta_0^{(n)} x_i^m \right]}. \tag{8.77}$$

Note that the training error $\epsilon_{\text{tr}}^{(n)}$ is evaluated using only the n points on which the model parameters $\theta^{(n)}$ were trained, not the full set of N_{train} points. Similarly, the truncated cross-validation error is given by

$$\epsilon_{\text{cv}}^{(n)} = \sqrt{\frac{1}{n} \sum_{i=1}^{N_{\text{cv}}} \left[y_i - \sum_{m=0}^{d} \theta_0^{(n)} x_i^m \right]}, \tag{8.78}$$

where we sum over *all* of the cross-validation points. A learning curve is the plot of the truncated training error and truncated cross-validation error as a function of the size n of the training set used. For our toy example, this plot is shown in figure 8.15 for models with $d = 2$ and $d = 3$. The dotted line in each panel again shows for reference the intrinsic error added to the data.

The two panels show some common features, which are reflective of the features of learning curves for any regression model:

1. As we increase the size of the training set, the training error increases. The reason for this is simple: a model of a given complexity can better fit a small set of data than a large set of data. Moreover, aside from small random fluctuations, we expect this training error to always increase with the size of the training set.

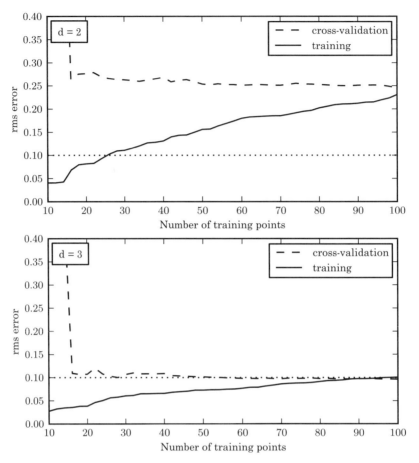

Figure 8.15. The learning curves for the data given by eq. 8.75, with $d = 2$ and $d = 3$. Both models have high variance for a few data points, visible in the spread between training and cross-validation error. As the number of points increases, it is clear that $d = 2$ is a high-bias model which cannot be improved simply by adding training points.

2. As we increase the size of the training set, the cross-validation error decreases. The reason for this is easy to see: a smaller training set leads to overfitting the model, meaning that the model is less representative of the cross-validation data. As the training set grows, overfitting is reduced and the cross-validation error decreases. Again, aside from random fluctuations, and as long as the training set and cross-validation set are statistically similar, we expect the cross-validation error to always decrease as the training set size grows.

3. The training error is everywhere less than or equal to the cross-validation error, up to small statistical fluctuations. We expect the model on average to better describe the data used to train it.

4. The logical outcome of the above three observations is that as the size N of the training set becomes large, the training and cross-validation curves will converge to the same value.

By plotting these learning curves, we can quickly see the effect of adding more training data. When the curves are separated by a large amount, the model error is dominated by variance, and additional training data will help. For example, in the lower panel of figure 8.15, at $N = 15$, the cross-validation error is very high and the training error is very low. Even if we only had 15 points and could not plot the remainder of the curve, we could infer that adding more data would improve the model.

On the other hand, when the two curves have converged to the same value, the model error is dominated by bias, and adding additional training data cannot improve the results *for that model*. For example, in the top panel of figure 8.15, at $N = 100$ the errors have nearly converged. Even without additional data, we can infer that adding data will not decrease the error below about 0.23. Improving the error in this case requires a more sophisticated model, or perhaps more features measured for each point.

To summarize, plotting learning curves can be very useful for evaluating the efficiency of a model and potential paths to improving your data. There are two possible situations:

1. **The training error and cross-validation error have converged**. In this case, increasing the number of training points under the same model is futile: the error cannot improve further. This indicates a model error dominated by bias (i.e., it is underfitting the data). For a high-bias model, the following approaches may help:

   • Add additional features to the data.
   • Increase the model complexity.
   • Decrease the regularization.

2. **The training error is much smaller than the cross-validation error**. In this case, increasing the number of training points is likely to improve the model. This condition indicates that the model error is dominated by variance (i.e., it is overfitting the data). For a high-variance model, the following approaches may help:

   • Increase the training set size.
   • Decrease the model complexity.
   • Increase the amplitude of the regularization.

Finally, we note a few caveats: first, the learning curves seen in figure 8.15 and the model complexity evaluation seen in figure 8.14 are actually aspects of a three-dimensional space. Changing the data and changing the model go hand in hand, and one should always combine the two diagnostics to seek the best match between model and data.

Second, this entire discussion assumes that the training data, cross-validation data, and test data are statistically similar. This analysis will fail if the samples are drawn from different distributions, or have different measurement errors or different observational limits.

8.11.3. Other Cross-Validation Techniques

There are numerous cross-validation techniques available which are suitable for different situations. It is easy to generalize from the above discussion to these various cross-validation strategies, so we will just briefly mention them here.

Twofold cross-validation

Above, we split the data into a training set d_1, a cross-validation set d_2, and a test set d_0. Our simple tests involved training the model on d_0 and cross-validating the model on d_1. In twofold cross-validation, this process is repeated, training the model on d_1 and cross-validating the model on d_0. The training error and cross-validation error are computed from the mean of the errors in each fold. This leads to more robust determination of the cross-validation error for smaller data sets.

K-fold cross-validation

A generalization of twofold cross-validation is K-fold cross-validation. Here we split the data into $K + 1$ sets: the test set d_0, and the cross-validation sets $d_1, d_2, \ldots, d_K$. We train K different models, each time leaving out a single subset to measure the cross-validation error. The final training error and cross-validation error can be computed using the mean or median of the set of results. The median can be a better statistic than the mean in cases where the subsets d_i contain few points.

Leave-one-out cross-validation

At the extreme of K-fold cross-validation is leave-one-out cross-validation. This is essentially the same as K-fold cross-validation, but this time our sets $d_1, d_2, \ldots, d_K$ have only one data point each. That is, we repeatedly train the model, leaving out only a single point to estimate the cross-validation error. Again, the final training error and cross-validation error are estimated using the mean or median of the individual trials. This can be useful when the size of the data set is very small, so that significantly reducing the number of data points leads to much different model characteristics.

Random subset cross-validation

In this approach, the cross-validation set and training set are selected by randomly partitioning the data, and repeating any number of times until the error statistics are well sampled. The disadvantage here is that not every point is guaranteed to be used both for training and for cross-validation. Thus, there is a finite chance that an outlier can lead to spurious results. For N points and P random samplings of the data, this situation becomes very unlikely for $N/2^P \ll 1$.

8.11.4. Summary of Cross-Validation and Learning Curves

In this section we have shown how to evaluate how well a model fits a data set through cross-validation. This is one practical route to the model selection ideas presented in chapters 4–5. We have covered how to determine the best model given a data set (§ 8.11.1) and how to address both the model and the data together to improve results (§ 8.11.2).

Cross-validation is one place where machine learning and data mining may be considered more of an art than a science. The optimal route to improving a model is not always straightforward. We hope that by following the suggestions in this chapter, you can apply this art successfully to your own data.

8.12. Which Regression Method Should I Use?

As we did in the last two chapters, we will use the axes of "accuracy," "interpretability," "simplicity," and "speed" (see § 6.6 for a description of these terms) to provide a rough guide to the trade-offs in choosing between the different regression methods described in this chapter.

What are the most* accurate *regression methods ? The starting point is basic linear regression. Adding ridge or LASSO regularization in principle increases accuracy, since as long as $\lambda = 0$ is among the options tried for λ, it provides a superset of possible complexity trade-offs to be tried. This goes along with the general principle that models with more parameters have a greater chance of fitting the data well. Adding the capability to incorporate measurement errors should in principle increase accuracy, assuming of course that the error estimates are accurate enough. Principal component regression should generally increase accuracy by treating collinearity and effectively denoising the data. All of these methods are of course linear—the largest leap in accuracy is likely to come when going from linear to nonlinear models. A starting point for increasing accuracy through nonlinearity is a linear model on nonlinear transformations of the original data. The extent to which this approach increases accuracy depends on the sensibility of the transformations done, since the nonlinear functions introduced are chosen manually rather than automatically. The sensibility can be diagnosed in various ways, such as checking the Gaussianity of the resulting errors. Truly nonlinear models, in the sense that the variable interactions can be nonlinear as well, should be more powerful in general. The final significant leap of accuracy will generally come from going to nonparametric methods such as kernel regression, starting with Nadaraya–Watson regression and more generally, local polynomial regression. Gaussian process regression is typically even more powerful in principle, as it effectively includes an aspect similar to that of kernel regression through the covariance matrix, but also learns coefficients on each data point.

What are the most* interpretable *regression methods ? Linear methods are easy to interpret in terms of understanding the relative importance of each variable through the coefficients, as well as reasoning about how the model will react to different inputs. Ridge or LASSO regression in some sense increase interpretability by identifying the most important features. Because feature selection happens as the result of an overall optimization, reasoning deeply about why particular features were kept or eliminated is not necessarily fruitful. Bayesian formulations, as in the case of models that incorporate errors, add to interpretability by making assumptions clear. PCR begins to decrease interpretability in the sense that the columns are no longer identifiable in their original terms. Moving to nonlinear methods, generalized linear models maintain the identity of the original columns, while general nonlinear

TABLE 8.1.
A summary of the practical properties of different regression methods.

Method	Accuracy	Interpretability	Simplicity	Speed
Linear regression	L	H	H	H
Linear basis function regression	M	M	M	M
Ridge regression	L	H	M	H
LASSO regression	L	H	M	L
PCA regression	M	M	M	M
Nadaraya–Watson regression	M/H	M	H	L/M
Local linear/polynomial regression	H	M	M	L/M
Nonlinear regression	M	H	L	L/M

models tend to become much less interpretable. Kernel regression methods can arguably be relatively interpretable, as their behavior can be reasoned about in terms of distances between points. Gaussian process regression is fairly opaque, as it is relatively difficult to reason about the behavior of the inverse of the data covariance matrix.

What are the **simplest** *regression methods* ? Basic linear regression has no tunable parameters, so it qualifies as the simplest out-of-the-box method. Generalized linear regression is similar, aside from the manual preparation of the nonlinear features. Ridge and LASSO regression require that only one parameter is tuned, and the optimization is convex, so that there is no need for random restarts, and cross-validation can make learning fairly automatic. PCR is similar in that there is only one critical parameter and the optimization is convex. Bayesian formulations of regression models which result in MCMC are subject to our comments in § 5.9 regarding fiddliness (as well as computational cost). Nadaraya–Watson regression typically has only one critical parameter, the bandwidth. General local polynomial regression can be regarded as having another parameter, the polynomial order. Basic Gaussian process regression has only one critical parameter, the bandwidth of the covariance kernel, and is convex, though extensions with several additional parameters are often used.

What are the **most** *scalable regression methods* ? Methods based on linear regression are fairly tractable using state-of-the-art linear algebra methods, including basic linear regression, ridge regression, and generalized linear regression, assuming the dimensionality is not excessively high. LASSO requires the solution of a linear program, which becomes expensive as the dimension rises. For PCR, see our comments in § 8.4 regarding PCA and SVD. Kernel regression methods are naively $\mathcal{O}(N^2)$ but can be sped up by fast tree-based algorithms in ways similar to those discussed in § 2.5.2 and chapter 6. Certain approximate algorithms exist for Gaussian process regression, which is naively $\mathcal{O}(N^3)$, but GP is by far the most expensive among the methods we have discussed, and difficult to speed up algorithmically while maintaining high predictive accuracy.

Other considerations, and taste Linear methods can be straightforwardly augmented to handle missing values. The Bayesian approach to linear regression incorporates measurement errors, while standard versions of machine learning methods do not incorporate errors. Gaussian process regression typically comes with posterior uncertainty bands, though confidence bands can be obtained for any method, as we have discussed in previous chapters. Regarding taste, LASSO is embedded in much recent work on sparsity and is related to work on compressed sensing, and Gaussian process regression has interesting interpretations in terms of priors in function space and in terms of kernelized linear regression.

Simple summary. We summarize our discussion in table 8.1, in terms of "accuracy," "interpretability," "simplicity," and "speed" (see § 6.6 for a description of these terms), given in simple terms of high (H), medium (M), and low (L).

References

[1] Astier, P., J. Guy, N. Regnault, and others (2005). The supernova legacy survey: Measurement of ω_m, ω_λ and w from the first year data set. *Astronomy & Astrophysics 447*(1), 24.

[2] Boscovich, R. J. (1757). De litteraria expeditione per pontificiam ditionem, et synopsis amplioris operis, ac habentur plura ejus ex exemplaria etiam sensorum impressa. *Bononiensi Scientarum et Artum Instituto Atque Academia Commentarii IV*, 353–396.

[3] Cleveland, W. S. (1979). Robust locally weighted regression and smoothing scatterplots. *Journal of the American Statistical Association 74*(368), 829–836.

[4] Dellaportas, P. and D. A. Stephens (1995). Bayesian analysis of errors-in-variables regression models. *Biometrics 51*(3), 1085–1095.

[5] Efron, B., T. Hastie, I. Johnstone, and R. Tibshirani (2004). Least angle regression. *Annals of Statistics 32*(2), 407–451.

[6] Gauss, K. F. (1809). *Theoria Motus Corporum Coelestium in Sectionibus Conicis Solem Ambientium*. Hamburg: Sumtibus F. Perthes et I. H. Besser.

[7] Hogg, D. W., I. K. Baldry, M. R. Blanton, and D. J. Eisenstein (2002). The K correction. *ArXiv:astro-ph/0210394*.

[8] Hogg, D. W., J. Bovy, and D. Lang (2010). Data analysis recipes: Fitting a model to data. *ArXiv:astro-ph/1008.4686*.

[9] Huber, P. J. (1964). Robust estimation of a local parameter. *Annals of Mathematical Statistics 35*, 73–101.

[10] Jolliffe, I. T. (1986). *Principal Component Analysis*. Springer.

[11] Kelly, B. C. (2011). Measurement error models in astronomy. *ArXiv:astro-ph/1112.1745*.

[12] Kim, J., Y. Kim, and Y. Kim (2008). A gradient-based optimization algorithm for LASSO. *Journal of Computational and Graphical Statistics 17*(4), 994–1009.

[13] Krajnović, D. (2011). A Jesuit anglophile: Rogerius Boscovich in England. *Astronomy and Geophysics 52*(6), 060000–6.

[14] Legendre, A. M. (1805). *Nouvelles méthodes pour la détermination des orbites des comètes*. Paris: Courcier.

[15] Levenberg, K. (1944). A method for the solution of certain non-linear problems in least squares. *Quarterly Applied Mathematics II(2)*, 164–168.

[16] Marquardt, D. W. (1963). An algorithm for least-squares estimation of non-linear parameters. *Journal of the Society of Industrial and Applied Mathematics 11*(2), 431–441.

[17] Mertens, B., T. Fearn, and M. Thompson (1995). The efficient cross-validation of principal components applied to principal component regression. *Statistics and Computing 5*, 227–235. 10.1007/BF00142664.

[18] Nadaraya, E. A. (1964). On estimating regression. *Theory of Probability and its Applications 9*, 141–142.

[19] Rasmussen, C. and C. Williams (2005). *Gaussian Processes for Machine Learning*. Adaptive Computation And Machine Learning. MIT Press.

[20] Theil, H. (1950). A rank invariant method of linear and polynomial regression analysis, I, II, III. *Proceedings of the Koninklijke Nederlandse Akademie Wetenschappen, Series A – Mathematical Sciences 53*, 386–392, 521–525, 1397–1412.

[21] Tibshirani, R. J. (1996). Regression shrinkage and selection via the Lasso. *Journal of the Royal Statistical Society, Series B 58*(1), 267–288.

[22] Tikhonov, A. N. (1995). *Numerical Methods for the Solution of Ill-Posed Problems*, Volume 328 of *Mathematics and Its Applications*. Kluwer.

[23] Watson, G. S. (1964). Smooth regression analysis. *Sankhyā Ser. 26*, 359–372.

[24] Zu, Y., C. S. Kochanek, S. Kozłowski, and A. Udalski (2012). Is quasar variability a damped random walk? *ArXiv:astro-ph/1202.3783*.

9 Classification

"One must always put oneself in a position to choose between two alternatives."
(Talleyrand)

In chapter 6 we described techniques for estimating joint probability distributions from multivariate data sets and for identifying the inherent clustering within the properties of sources. We can think of this approach as the *unsupervised classification* of data. If, however, we have labels for some of these data points (e.g., an object is tall, short, red, or blue) we can utilize this information to develop a relationship between the label and the properties of a source. We refer to this as *supervised classification* .

The motivation for supervised classification comes from the long history of classification in astronomy. Possibly the most well known of these classification schemes is that defined by Edwin Hubble for the morphological classification of galaxies based on their visual appearance; see [7]. This simple classification scheme, subdividing the types of galaxies into seven categorical subclasses, was broadly adopted throughout extragalactic astronomy. Why such a simple classification became so predominant when subsequent works on the taxonomy of galaxy morphology (often with a better physical or mathematical grounding) did not, argues for the need to keep the models for classification simple. This agrees with the findings of George Miller who, in 1956, proposed that the number of items that people are capable of retaining within their short term memory was 7 ± 2 ("The magical number 7 ± 2" [10]). Subsequent work by Herbert Simon suggested that we can increase seven if we implement a partitioned classification system (much like telephone numbers) with a chunk size of three. Simple schemes have more impact—a philosophy we will adopt as we develop this chapter.

9.1. Data Sets Used in This Chapter

In order to demonstrate the strengths and weaknesses of these classification techniques, we will use two astronomical data sets throughout this chapter.

RR Lyrae

First is the set of photometric observations of RR Lyrae stars in the SDSS [8]. The data set comes from SDSS Stripe 82, and combines the Stripe 82 standard stars (§1.5.8), which represent observations of nonvariable stars; and the RR Lyrae variables, pulled from the same observations as the standard stars, and selected based on their variability using supplemental data; see [16]. The sample is further constrained to a smaller region of the overall color–color space following [8] ($0.7 < u - g < 1.35$, $-0.15 < g - r < 0.4$, $-0.15 < r - i < 0.22$, and $-0.21 < i - z < 0.25$). These selection criteria lead to a sample of 92,658 nonvariable stars, and 483 RR Lyraes. Two features of this combined data set make it a good candidate for testing classification algorithms:

1. The RR Lyrae stars and main sequence stars occupy a very similar region in u, g, r, i, z color space. The distributions overlap slightly, which makes the choice of decision boundaries subject to the completeness and contamination trade-off discussed in §4.6.1 and §9.2.1.
2. The extreme imbalance between the number of sources and the number of background objects is typical of real-world astronomical studies, where it is often desirable to select rare events out of a large background. Such unbalanced data aptly illustrates the strengths and weaknesses of various classification methods.

We will use these data in the context of the classification techniques discussed below.

Quasars and stars

As a second data set for photometric classification, we make use of two catalogs of quasars and stars from the SDSS Spectroscopic Catalog. The quasars are derived from the DR7 Quasar Catalog (§1.5.6), while the stars are derived from the SEGUE Stellar Parameters Catalog (§1.5.7). The combined data has approximately 100,000 quasars and 300,000 stars. In this chapter, we use the $u - g$, $g - r$, $r - i$, and $i - z$ colors to demonstrate photometric classification of these objects. We stress that because of the different selection functions involved in creating the two catalogs, the combined sample does not reflect a real-world sample of the objects: we use it for purposes of illustration only.

Photometric redshifts

While photometric redshifts are technically a regression problem which belongs to chapter 8, they offer an excellent test case for decision trees and random forests, introduced in §9.7. The data for the photometric redshifts come from the SDSS spectroscopic database (§1.5.5). The magnitudes used are the model magnitudes mentioned above, while the true redshift measurements come from the spectroscopic pipeline.

9.2. Assigning Categories: Classification

Supervised classification takes a set of features and relates them to predefined sets of classes. Choosing the optimal set of features was touched on in the discussion of dimensionality reduction in §7.3. We will not address how we define the labels or taxonomy for the classification other than noting that the time-honored system of having a graduate student label data does not scale to the size of today's data.[1] We start by assuming that we have a set of predetermined labels that have been assigned to a subset of the data we are considering. Our goal is to characterize the relation between the features in the data and their classes and apply these classifications to a larger set of unlabeled data.

As we go we will illuminate the connections between classification, regression, and density estimation. Classification can be posed in terms of density estimation— this is called *generative classification* (so-called since we will have a full model of the density for each class, which is the same as saying we have a model which describes how data could be generated from each class). This will be our starting point, where we will visit a number of methods. Among the advantages of this approach is a high degree of interpretability.

Starting from the same principles we will go to classification methods that focus on finding the decision boundary that separates classes directly, avoiding the step of modeling each class's density, called *discriminative classification*, which can often be better in high-dimensional problems.

9.2.1. Classification Loss

Perhaps the most common loss (cost) function in classification is *zero-one loss*, where we assign a value of one for a misclassification and zero for a correct classification. With $\widehat{y}$ representing the best guess value of y, we can write this classification loss, $L(y, \widehat{y})$, as

$$L(y, \widehat{y}) = \delta(y \neq \widehat{y}), \tag{9.1}$$

which means

$$L(y, \widehat{y}) = \begin{cases} 1 & \text{if } y \neq \widehat{y}, \\ 0 & \text{otherwise.} \end{cases} \tag{9.2}$$

The classification *risk* of a model (defined to be the expectation value of the loss—see §4.2.8) is given by

$$\mathbb{E}\left[L(y, \widehat{y})\right] = p(y \neq \widehat{y}), \tag{9.3}$$

or the probability of misclassification. This can be compared to the case of regression, where the most common loss function is $L(y, \widehat{y}) = (y - \widehat{y})^2$, leading to the risk $\mathbb{E}[(y - \widehat{y})^2]$. For the zero-one loss of classification, the risk is equal to the *misclassification rate* or *error rate*.

One particularly common case of classification in astronomy is that of "detection," where we wish to assign objects (i.e., regions of the sky or groups of pixels on a CCD) into one of two classes: a detection (usually with label 1) and a nondetection (usually with label 0). When thinking about this sort of problem, we may wish to

---

[1]If, however, you can enlist thousands of citizen scientists in this proposition then you can fundamentally change this aspect; see [9].

distinguish between the two possible kinds of error: assigning a label 1 to an object whose true class is 0 (a "false positive"), and assigning the label 0 to an object whose true class is 1 (a "false negative").

As in §4.6.1, we will define the *completeness*,

$$\text{completeness} = \frac{\text{true positives}}{\text{true positives} + \text{false negatives}}, \quad (9.4)$$

and *contamination*,

$$\text{contamination} = \frac{\text{false positives}}{\text{true positives} + \text{false positives}}. \quad (9.5)$$

The completeness measures the fraction of total detections identified by our classifier, while the contamination measures the fraction of detected objects which are misclassified. Depending on the nature of the problem and the goal of the classification, we may wish to optimize one or the other.

Alternative names for these measures abound: in some fields the completeness and contamination are respectively referred to as the "sensitivity" and the "Type I error." In astronomy, one minus the contamination is often referred to as the "efficiency." In machine learning communities, the efficiency and completeness are respectively referred to as the "precision" and "recall."

9.3. Generative Classification

Given a set of data $\{\mathbf{x}\}$ consisting of N points in D dimensions, such that x_i^j is the jth feature of the ith point, and a set of discrete labels $\{y\}$ drawn from K classes, with values y_k, Bayes' theorem describes the relation between the labels and features:

$$p(y_k|\mathbf{x}_i) = \frac{p(\mathbf{x}_i|y_k)p(y_k)}{\sum_i p(\mathbf{x}_i|y_k)p(y_k)}. \quad (9.6)$$

If we knew the full probability densities $p(\mathbf{x}, y)$ it would be straightforward to estimate the classification likelihoods directly from the data. If we chose not to fully sample $p(\mathbf{x}, y)$ with our training set we can still define the classifications by drawing from $p(y|\mathbf{x})$ and comparing the likelihood ratios between classes (in this way we can focus our labeling on the specific, and rare, classes of source rather than taking a brute-force random sample).

In generative classifiers we are modeling the class-conditional densities explicitly, which we can write as $p_k(\mathbf{x})$ for $p(\mathbf{x}|y = y_k)$, where the class variable is, say, $y_k = 0$ or $y_k = 1$. The quantity $p(y = y_k)$, or π_k for short, is the probability of any point having class k, regardless of which point it is. This can be interpreted as the *prior* probability of the class k. If these are taken to include subjective information, the whole approach is Bayesian (chapter 5). If they are estimated from data, for example by taking the proportion in the training set that belong to class k, this can be considered as either a frequentist or as an empirical Bayes (see §5.2.4).

The task of learning the best classifier then becomes the task of estimating the p_k's. This approach means we will be doing multiple separate *density estimates* using many of the techniques introduced in chapter 6. The most powerful (accurate) classifier of this type then, corresponds to the most powerful density estimator used for the p_k models. Thus the rest of this section will explore various models and approximations for the $p_k(\mathbf{x})$ in eq. 9.6. We will start with the simplest kinds of models, and gradually build the model complexity from there. First, though, we will discuss several illuminating aspects of the generative classification model.

9.3.1. General Concepts of Generative Classification

Discriminant function

With slightly more effort, we can formally relate the classification task to two of the major machine learning tasks we have seen already: density estimation (chapter 6) and regression (chapter 8). Recall, from chapter 8, the regression function $\widehat{y} = f(y|\mathbf{x})$: it represents the best guess value of y given a specific value of $\mathbf{x}$. Classification is simply the analog of regression where y is categorical, for example $y = \{0, 1\}$. We now call $f(y|\mathbf{x})$ the *discriminant function*:

$$g(\mathbf{x}) = f(y|\mathbf{x}) = \int y\, p(y|\mathbf{x})\, dy \tag{9.7}$$

$$= 1 \cdot p(y = 1|\mathbf{x}) + 0 \cdot p(y = 0|\mathbf{x}) = p(y = 1|\mathbf{x}). \tag{9.8}$$

If we now apply Bayes' rule (eq. 3.10), we find (cf. eq. 9.6)

$$g(\mathbf{x}) = \frac{p(\mathbf{x}|y = 1)\, p(y = 1)}{p(\mathbf{x}|y = 1)\, p(y = 1) + p(\mathbf{x}|y = 0)\, p(y = 0)} \tag{9.9}$$

$$= \frac{\pi_1 p_1(\mathbf{x})}{\pi_1 p_1(\mathbf{x}) + \pi_0 p_0(\mathbf{x})}. \tag{9.10}$$

Bayes classifier

Making the discriminant function yield a binary prediction gives the abstract template called a *Bayes classifier*. It can be formulated as

$$\widehat{y} = \begin{cases} 1 & \text{if } g(\mathbf{x}) > 1/2, \\ 0 & \text{otherwise}, \end{cases} \tag{9.11}$$

$$= \begin{cases} 1 & \text{if } p(y = 1|\mathbf{x}) > p(y = 0|\mathbf{x}), \\ 0 & \text{otherwise}, \end{cases} \tag{9.12}$$

$$= \begin{cases} 1 & \text{if } \pi_1 p_1(\mathbf{x}) > \pi_0 p_0(\mathbf{x}), \\ 0 & \text{otherwise}. \end{cases} \tag{9.13}$$

This is easily generalized to any number of classes K, since we can think of a $g_k(\mathbf{x})$ for each class (in a two-class problem it is sufficient to consider $g(\mathbf{x}) = g_1(\mathbf{x})$). The Bayes classifier is a template in the sense that one can plug in different types of model for the p_k's and the π's. Furthermore, the Bayes classifier can be shown to be optimal if the p_k's and π's are chosen to be the true distributions: that is, lower error cannot

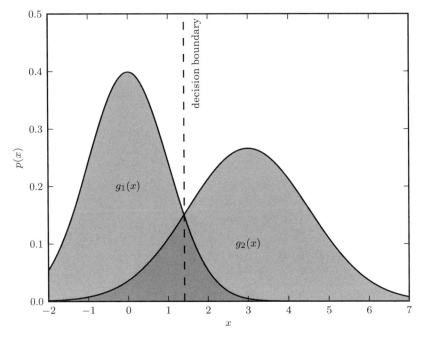

Figure 9.1. An illustration of a decision boundary between two Gaussian distributions.

be achieved. The Bayes classification template as described is an instance of empirical Bayes (§5.2.4).

Again, keep in mind that so far this is "Bayesian" only in the sense of utilizing Bayes' rule, an identity based on the definition of conditional distributions (§3.1.1), not in the sense of Bayesian inference. The interpretation/usage of the π_k quantities is what will make the approach either Bayesian or frequentist.

Decision boundary

The *decision boundary* between two classes is the set of x values at which each class is equally likely; that is,

$$\pi_1 p_1(\mathbf{x}) = \pi_2 p_2(\mathbf{x}); \tag{9.14}$$

that is, $g_1(\mathbf{x}) = g_2(\mathbf{x})$; that is, $g_1(\mathbf{x}) - g_2(\mathbf{x}) = 0$; that is, $g(\mathbf{x}) = 1/2$; in a two-class problem. Figure 9.1 shows an example of the decision boundary for a simple model in one dimension, where the density for each class is modeled as a Gaussian. This is very similar to the concept of hypothesis testing described in §4.6.

9.3.2. Naive Bayes

The Bayes classifier formalism presented above is conceptually simple, but can be very difficult to compute: in practice, the data {$\mathbf{x}$} above may be in many dimensions, and have complicated probability distributions. We can dramatically reduce the complexity of the problem by making the assumption that all of the attributes we measure are conditionally independent. This means that

$$p(x^i, x^j | y_k) = p(x^i | y) p(x^j | y_k), \tag{9.15}$$

where, recall, the superscript indexes the feature of the vector **x**. For data in many dimensions, this assumption can be expressed as

$$p(x^0, x^1, x^2, \ldots, x^N | y_k) = \prod_i p(x^i | y_k). \tag{9.16}$$

Again applying Bayes' rule, we rewrite eq. 9.6 as

$$p(y_k | x^0, x^1, \ldots, x^N) = \frac{p(x^0, x^1, \ldots, x^N | y_k) p(y_k)}{\sum_j p(x^0, x^1, \ldots, x^N | y_j) p(y_j)}. \tag{9.17}$$

With conditional independence this becomes

$$p(y_k | x^0, x^1, \ldots, x^N) = \frac{\prod_i p(x^i | y_k) p(y_k)}{\sum_j \prod_i p(x^i | y_j) p(y_j)}. \tag{9.18}$$

Using this expression, we can calculate the most likely value of y by maximizing over y_k,

$$\widehat{y} = \arg\max_{y_k} \frac{\prod_i p(x^i | y_k) p(y_k)}{\sum_j \prod_i p(x^i | y_j) p(y_j)}, \tag{9.19}$$

or, using our shorthand notation,

$$\widehat{y} = \arg\max_{y_k} \frac{\prod_i p_k(x^i) \pi_k}{\sum_j \prod_i p_j(x^i) \pi_j}. \tag{9.20}$$

This gives a general prescription for the naive Bayes classification. Once sufficient models for $p_k(x^i)$ and π_k are known, the estimator $\widehat{y}$ can be computed very simply. The challenge, then, is to determine $p_k(x^i)$ and π_k, most often from a set of training data. This can be accomplished in a variety of ways, from fitting parametrized models using the techniques of chapters 4 and 5, to more general parametric and nonparametric density estimation techniques discussed in chapter 6.

The determination of $p_k(x^i)$ and μ_k can be particularly simple when the features x^i are categorical rather than continuous. In this case, assuming that the training set is a fair sample of the full data set (which may not be true), for each label y_k in the training set, the maximum likelihood estimate of the probability for feature x^i is simply equal to the number of objects with a particular value of x^i, divided by the total number of objects with $y = y_k$. The prior probabilities π_k are given by the fraction of training data with $y = y_k$.

Almost immediately, a complication arises. If the training set does not cover the full parameter space, then this estimate of the probability may lead to $p_k(x^i) = 0$ for some value of y_k and x^i. If this is the case, then the posterior probability in eq. 9.20 is

$p(y_k|\{x^i\}) = 0/0$ which is undefined! A particularly simple solution in this case is to use *Laplace smoothing*: an offset α is added to the probability of each bin $p_k(x^i)$ for all i, k, leading to well-defined probabilities over the entire parameter space. Though this may seem to be merely a heuristic trick, it can be shown to be equivalent to the addition of a Bayesian prior to the naive Bayes classifier.

9.3.3. Gaussian Naive Bayes and Gaussian Bayes Classifiers

It is rare in astronomy that we have discrete measurements for **x** even if we have categorical labels for y. The estimator for $\widehat{y}$ given in eq. 9.20 can also be applied to continuous data, given a sufficient estimate of $p_k(x^i)$. In Gaussian naive Bayes, each of these probabilities $p_k(x^i)$ is modeled as a one-dimensional normal distribution, with means μ_k^i and widths σ_k^i determined, for example, using the frequentist techniques in §4.2.3. In this case the estimator in eq. 9.20 can be expressed as

$$\widehat{y} = \arg\max_{y_k} \left[\ln \pi_k - \frac{1}{2} \sum_{i=1}^{N} \left(2\pi(\sigma_k^i)^2 + \frac{(x^i - \mu_k^i)^2}{(\sigma_k^i)^2} \right) \right], \tag{9.21}$$

where for simplicity we have taken the log of the Bayes criterion, and omitted the normalization constant, neither of which changes the result of the maximization.

The Gaussian naive Bayes estimator of eq. 9.21 essentially assumes that the multivariate distribution $p(\mathbf{x}|y_k)$ can be modeled using an axis-aligned multivariate Gaussian distribution. In figure 9.2, we perform a Gaussian naive Bayes classification on a simple, well-separated data set. Though examples like this one make classification straightforward, data in the real world is rarely so clean. Instead, the distributions often overlap, and categories have hugely imbalanced numbers. These features are seen in the RR Lyrae data set.

Scikit-learn has an estimator which performs fast Gaussian Naive Bayes classification:

```
import numpy as np
from sklearn.naive_bayes import GaussianNB

X = np.random.random((100, 2)) # 100 pts in 2 dims
y = (X[:, 0] + X[:, 1] > 1).astype(int)
# simple division
gnb = GaussianNB()
gnb.fit(X, y)
y_pred = gnb.predict(X)
```

For more details see the Scikit-learn documentation.

In figure 9.3, we show the naive Bayes classification for RR Lyrae stars from SDSS Stripe 82. The completeness and contamination for the classification are shown in the right panel, for various combinations of features. Using all four colors, the Gaussian

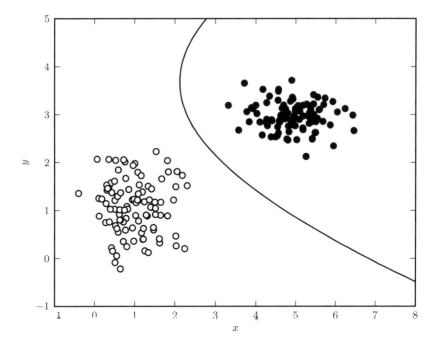

Figure 9.2. A decision boundary computed for a simple data set using Gaussian naive Bayes classification. The line shows the decision boundary, which corresponds to the curve where a new point has equal posterior probability of being part of each class. In such a simple case, it is possible to find a classification with perfect completeness and contamination. This is rarely the case in the real world.

naive Bayes classifier in this case attains a completeness of 87.6%, at the cost of a relatively high contamination rate of 79.0%.

A logical next step is to relax the assumption of conditional independence in eq. 9.16, and allow the Gaussian probability model for each class to have arbitrary correlations between variables. Allowing for covariances in the model distributions leads to the *Gaussian Bayes classifier* (i.e., it is no longer naive). As we saw in §3.5.4, a multivariate Gaussian can be expressed as

$$p_k(\mathbf{x}) = \frac{1}{|\Sigma_k|^{1/2}(2\pi)^{D/2}} \exp\left\{-\frac{1}{2}(\mathbf{x} - \mu_{\mathbf{k}})^T \Sigma_k^{-1}(\mathbf{x} - \mu_{\mathbf{k}})\right\}, \tag{9.22}$$

where Σ_k is a $D \times D$ symmetric covariance matrix with determinant $\det(\Sigma_k) \equiv |\Sigma_k|$, and $\mathbf{x}$ and $\mu_{\mathbf{k}}$ are D-dimensional vectors. For this generalized Gaussian Bayes classifier, the estimator $\widehat{y}$ is (cf. eq. 9.21)

$$\widehat{y} = \arg\max_k \left\{-\frac{1}{2}\log|\Sigma_k| - \frac{1}{2}(\mathbf{x} - \mu_{\mathbf{k}})^T \Sigma_k^{-1}(\mathbf{x} - \mu_{\mathbf{k}}) + \log \pi_k\right\} \tag{9.23}$$

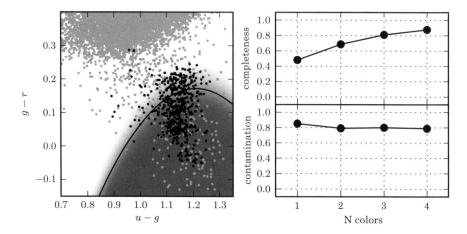

Figure 9.3. Gaussian naive Bayes classification method used to separate variable RR Lyrae stars from nonvariable main sequence stars. In the left panel, the light gray points show nonvariable sources, while the dark points show variable sources. The classification boundary is shown by the black line, and the classification probability is shown by the shaded background. In the right panel, we show the completeness and contamination as a function of the number of features used in the fit. For the single feature, $u - g$ is used. For two features, $u - g$ and $g - r$ are used. For three features, $u - g$, $g - r$, and $r - i$ are used. It is evident that the $g - r$ color is the best discriminator. With all four colors, naive Bayes attains a completeness of 0.876 and a contamination of 0.790.

or equivalently,

$$\widehat{y} = \begin{cases} 1 & \text{if } m_1^2 < m_0^2 + 2\log\left(\frac{\pi_1}{\pi_0}\right) + \left(\frac{|\Sigma_1|}{|\Sigma_0|}\right), \\ 0 & \text{otherwise,} \end{cases} \tag{9.24}$$

where $m_k^2 = (x - \mu_k)^T \Sigma_k^{-1}(x - \mu_k)$ is known as the *Mahalanobis distance*.

This step from Gaussian naive Bayes to a more general Gaussian Bayes formalism can include a large jump in computational cost: to fit a D-dimensional multivariate normal distribution to observed data involves estimation of $D(D + 3)/2$ parameters, making a closed-form solution (like that for $D = 2$ in §3.5.2) increasingly tedious as the number of features D grows large. One efficient approach to determining the model parameters μ_k and Σ_k is the expectation maximization algorithm discussed in §4.4.3, and again in the context of Gaussian mixtures in §6.3. In fact, we can use the machinery of Gaussian mixture models to extend Gaussian naive Bayes to a more general Gaussian Bayes formalism, simply by fitting to each class a "mixture" consisting of a single component. We will explore this approach, and the obvious extension to multiple component mixture models, in §9.3.5 below.

9.3.4. Linear Discriminant Analysis and Relatives

Linear discriminant analysis (LDA), like Gaussian naive Bayes, relies on some simplifying assumptions about the class distributions $p_k(\mathbf{x})$ in eq. 9.6. In particular, it assumes that these distributions have identical covariances for all K classes. This makes all classes a set of shifted Gaussians. The optimal classifier can then be derived from the log of the class posteriors to be

$$g_k(\mathbf{x}) = \mathbf{x}^T \Sigma^{-1} \mu_\mathbf{k} - \frac{1}{2} \mu_\mathbf{k}^T \Sigma^{-1} + \log \pi_k, \tag{9.25}$$

with $\mu_\mathbf{k}$ the mean of class k and Σ the covariance of the Gaussians (which, in general, does not need to be diagonal). The class dependent covariances that would normally give rise to a quadratic dependence on $\mathbf{x}$ cancel out if they are assumed to be constant. The Bayes classifier is, therefore, linear with respect to $\mathbf{x}$.

The discriminant boundary between classes is the line that minimizes the overlap between Gaussians:

$$g_k(\mathbf{x}) - g_\ell(\mathbf{x}) = \mathbf{x}^T \Sigma^{-1} (\mu_k - \mu_\ell) - \frac{1}{2} (\mu_k - \mu_\ell)^T \Sigma^{-1} (\mu_k - \mu_\ell) + \log \left(\frac{\pi_k}{\pi_\ell} \right) = 0. \tag{9.26}$$

If we were to relax the requirement that the covariances of the Gaussians are constant, the discriminant function for the classes becomes quadratic in x:

$$g(\mathbf{x}) = -\frac{1}{2} \log |\Sigma_k| - \frac{1}{2} (\mathbf{x} - \mu_k)^T C^{-1} (\mathbf{x} - \mu_k) + \log \pi_k. \tag{9.27}$$

This is sometimes known as *quadratic discriminant analysis* (QDA), and the boundary between classes is described by a quadratic function of the features $\mathbf{x}$.

A related technique is called *Fisher's linear discriminant* (FLD). It is a special case of the above formalism where the priors are set equal but without the requirement that the covariances be equal. Geometrically, it attempts to project all data onto a single line, such that a decision boundary can be found on that line. By minimizing the loss over all possible lines, it arrives at a classification boundary. Because FLD is so closely related to LDA and QDA, we will not explore it further.

Scikit-learn has estimators which perform both LDA and QDA. They have a very similar interface:

```python
import numpy as np
from sklearn.lda import LDA
from sklearn.qda import QDA

X = np.random.random((100, 2)) # 100 pts in 2 dims
y = (X[:, 0] + X[:, 1] > 1).astype(int)
    # simple division
lda = LDA()
lda.fit(X, y)
y_pred = lda.predict(X)

qda = QDA()
qda.fit(X, y)
y_pred = qda.predict(X)
```

For more details see the Scikit-learn documentation.

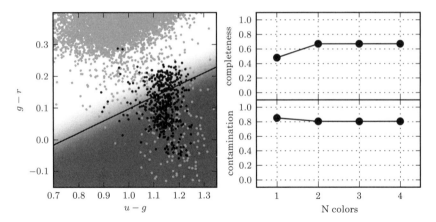

Figure 9.4. The linear discriminant boundary for RR Lyrae stars (see caption of figure 9.3 for details). With all four colors, LDA achieves a completeness of 0.672 and a contamination of 0.806.

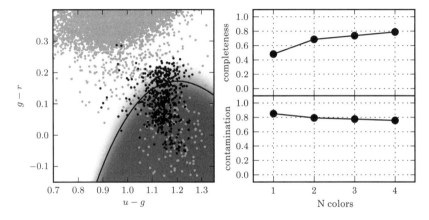

Figure 9.5. The quadratic discriminant boundary for RR Lyrae stars (see caption of figure 9.3 for details). With all four colors, QDA achieves a completeness of 0.788 and a contamination of 0.757.

The results of linear discriminant analysis and quadratic discriminant analysis on the RR Lyrae data from figure 9.3 are shown in figures 9.4 and 9.5, respectively. Notice that, true to their names, linear discriminant analysis results in a linear boundary between the two classes, while quadratic discriminant analysis results in a quadratic boundary. As may be expected with a more sophisticated model, QDA yields improved completeness and contamination in comparison to LDA.

9.3.5. More Flexible Density Models: Mixtures and Kernel Density Estimates

The above methods take the very general result expressed in eq. 9.6 and introduce simplifying assumptions which make the classification more computationally feasible. However, assumptions regarding conditional independence (as in naive Bayes) or Gaussianity of the distributions (as in Gaussian Bayes, LDA, and QDA) are not necessary parts of the model. With a more flexible model for the probability distribution, we could more closely model the true distributions and improve on our ability to classify the sources. To this end, many of the techniques from chapter 6 can be applicable.

The next common step up in representation power for each $p_k(x)$, beyond a single Gaussian with arbitrary covariance matrix, is to use a Gaussian mixture model (GMM) (described in §6.3). Let us call this the *GMM Bayes classifier* for lack of a standard term. Each of the components may be constrained to a simple case (such as diagonal-covariance-only Gaussians etc.) to ease the computational cost of model fitting. Note that the number of Gaussian components K must be chosen, ideally, for each class independently, in addition to the cost of model fitting for each value of K tried. Adding the ability to account for measurement errors in Gaussian mixtures was described in §6.3.3.

AstroML contains an implementation of GMM Bayes classification based on the Scikit-learn Gaussian mixture model code:

```
import numpy as np
from astroML.classification import GMMBayes

X = np.random.random((100, 2)) # 100 pts in 2 dims
y = (X[:, 0] + X[:, 1] > 1).astype(int)
    # simple division

gmmb = GMMBayes(3) # 3 clusters per class
gmmb.fit(X, y)
y_pred = gmmb.predict(X)
```

For more details see the AstroML documentation, or the source code of figure 9.6.

Figure 9.6 shows the GMM Bayes classification of the RR Lyrae data. The results with one component are similar to those of naive Bayes in figure 9.3. The difference is that here the Gaussian fits to the densities are allowed to have arbitrary covariances between dimensions. When we move to a density model consisting of three components, we significantly decrease the contamination with only a small effect on completeness. This shows the value of using a more descriptive density model.

For the ultimate in flexibility, and thus accuracy, we can model each class with a kernel density estimate. This *nonparametric* Bayes classifier is sometimes called *kernel discriminant analysis*. This method can be thought of as taking Gaussian mixtures to its natural limit, with one mixture component centered at each training point. It can also be generalized from the Gaussian to any desired kernel function. It turns out that even though the model is more complex (able to represent more complex functions), by going to this limit things become computationally simpler: unlike the typical GMM case, there is no need to optimize over the locations of the mixture components; the locations are simply the training points themselves. The optimization is over only one variable, the bandwidth of the kernel. One advantage of this approach is that when such flexible density models are used in the setting of

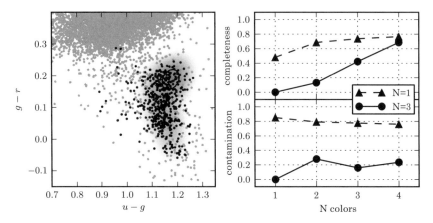

Figure 9.6. Gaussian mixture Bayes classifier for RR Lyrae stars (see caption of figure 9.3 for details). Here the left panel shows the decision boundary for the three-component model, and the right panel shows the completeness and contamination for both a one- and three-component mixture model. With all four colors and a three-component model, GMM Bayes achieves a completeness of 0.686 and a contamination of 0.236.

classification, their parameters can be chosen to maximize *classification* performance directly rather than density estimation performance.

The cost of the extra accuracy of kernel discriminant analysis is the high computational cost of evaluating kernel density estimates. We briefly discussed fast algorithms for kernel density estimates in §6.1.1, but an additional idea can be used to accelerate them in this setting. A key observation is that our computational problem can be solved more efficiently than by actually computing the full kernel summation needed for each class—to determine the class label for each query point, we need only determine the greater of the two kernel summations. The idea is to use the distance bounds obtained from tree nodes to bound $p_1(x)$ and $p_0(x)$: if at any point it can be proven that $\pi_1 p_1(x) > \pi_0 p_0(x)$ for all x in a query node, for example, then the actual class probabilities at those query points need not be evaluated.

Realizing this pruning idea most effectively motivates an alternate way of traversing the tree—a hybrid breadth and depth expansion pattern rather than the common depth-first traversal, where query nodes are expanded in a depth-first fashion and reference nodes are expanded in a breadth-first fashion. Pruning occurs when the upper and lower bounds are tight enough to achieve the correct classification, otherwise either the query node or all of the reference nodes are expanded. See [14] for further details.

9.4. K-Nearest-Neighbor Classifier

It is now easy to see the intuition behind one of the most widely used and powerful classifiers, the *nearest-neighbor* classifier: that is, just use the class label of the nearest point. The intuitive justification is that $p(y|x) \approx p(y|x')$ if x' is very close to x. It can also be understood as an approximation to kernel discriminant analysis where a variable-bandwidth (where the bandwidth is based on the distance to the nearest neighbor) kernel density estimate is used. The simplicity of K-nearest-neighbor is that it assumes nothing about the form of the conditional density distribution, that is, it is completely nonparametric. The resulting decision boundary between the nearest-neighbor points is a Voronoi tessellation of the attribute space.

A smoothing parameter, the number of neighbors K, is typically used to regulate the complexity of the classification by acting as a smoothing of the data. In its simplest form a majority rule classification is adopted, where each of the K points votes on the classification. Increasing K decreases the variance in the classification but at the expense of an increase in the bias. Choosing K such that it minimizes the classification error rate can be achieved using cross-validation (see §8.11).

Weights can be assigned to the individual votes by weighting the vote by the distance to the nearest point, similar in spirit to kernel regression. In fact, the K-nearest-neighbor classifier is directly related to kernel regression discussed in §8.5, where the regressed value was weighted by a distance-dependent kernel.

In general a Euclidean distance is used for the distance metric. This can, however, be problematic when comparing attributes with no defined distance metric (e.g., comparing morphology with color). An arbitrary rescaling of any one of the axes can lead to a mixing of the data and alter the nearest-neighbor classification. Normalization of the features (i.e., scaling from [0–1]), weighting the importance of features based on cross-validation (including a 0/1 weighting which is effectively a feature selection), and use of the Mahalanobis distance $D(x, x_0) = (x - x_0)^T C^{-1}(x - x_0)$ which weights by the covariance of the data, are all approaches that have been adopted to account for this effect.

Like all nonparametric methods, nearest-neighbor classification works best when the number of samples is large; when the number of data is very small, parametric methods which "fill in the blanks" with model-based assumptions are often best. While it is simple to parallelize, the computational time for searching for the neighbors (even using kd-trees) can be expensive and particularly so for high-dimensional data sets. Sampling of the training samples as a function of source density can reduce the computational requirements.

Scikit-learn contains a fast K-neighbors classifier built on a ball-tree for fast neighbor searches:

```
import numpy as np
from sklearn.neighbors import KNeighborsClassifier

X = np.random.random((100, 2)) # 100 pts in 2 dims
y = (X[:, 0] + X[:, 1] > 1).astype(int)
    # simple division

knc = KNeighborsClassifier(5) # use 5 nearest
    neighbors
knc.fit(X, y)
y_pred = knc.predict(X)
```

For more details see the AstroML documentation, or the source code of figure 9.7.

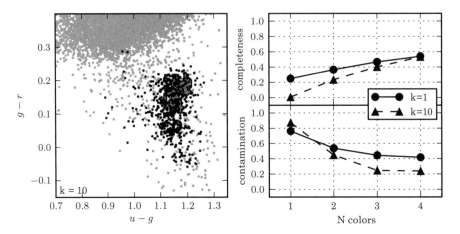

Figure 9.7. K-nearest-neighbor classification for RR Lyrae stars (see caption of figure 9.3 for details). Here the left panel shows the decision boundary for the model based on $K = 10$ neighbors, and the right panel shows the completeness and contamination for both $K = 1$ and $K = 10$. With all four colors and $K = 10$, K-neighbors classification achieves a completeness of 0.533 and a contamination of 0.240.

Figure 9.7 shows K-nearest-neighbor classification applied to the RR Lyrae data set with both $K = 1$ and $K = 10$. Note the complexity of the decision boundary, especially in regions where the density distributions overlap. This shows that even for our large data sets, K-nearest-neighbor may be overfitting the training data, leading to a suboptimal estimator. The discussion of overfitting in the context of regression (§8.11) is applicable here as well: KNN is unbiased, but is prone to very high variance when the parameter space is undersampled. This can be remedied by increasing the number of training points to better fill the space. When this is not possible, simple K-nearest-neighbor classification is probably not the best estimator, and other classifiers discussed in this chapter may provide a better solution.

9.5. Discriminative Classification

With nearest-neighbor classifiers we started to see a subtle transition—while clearly related to Bayes classifiers using variable-bandwidth kernel estimators, the class density estimates were skipped in favor of a simple classification decision. This is an example of *discriminative classification*, where we directly model the decision boundary between two or more classes of source. Recall, for $y \in \{0, 1\}$, the discriminant function is given by $g(x) = p(y = 1|x)$. Once we have it, no matter how we obtain it, we can use the rule

$$\widehat{y} = \begin{cases} 1 & \text{if } g(x) > 1/2, \\ 0 & \text{otherwise,} \end{cases} \tag{9.28}$$

to perform classification.

9.5.1. Logistic Regression

Logistic regression can be in the form of two (binomial) or more (multinomial) classes. For the initial discussion we will consider binomial logistic regression and consider the linear model

$$p(y = 1|x) = \frac{\exp\left[\sum_j \theta_j x^j\right]}{1 + \exp\left[\sum_j \theta_j x^j\right]}$$

$$= p(\boldsymbol{\theta}), \tag{9.29}$$

where we define the logit function as

$$\text{logit}(p_i) = \log\left(\frac{p_i}{1 - p_i}\right) = \sum_j \theta_j x_i^j. \tag{9.30}$$

The name logistic regression comes from the fact that the function $e^x/(1 + e^x)$ is called the logistic function. Its name is due to its roots in regression, even though it is a method for classification.

Because y is binary, it can be modeled as a Bernoulli distribution (see §3.3.3), with (conditional) likelihood function

$$L(\beta) = \prod_{i=1}^{N} p_i(\beta)^{y_i} (1 - p_i(\beta))^{1-y_i}. \tag{9.31}$$

Linear models we saw earlier under the generative (Bayes classifier) paradigm are related to logistic regression: linear discriminant analysis (LDA) uses the same model. In LDA,

$$\log\left(\frac{p(y = 1|x)}{p(y = 0|x)}\right) = -\frac{1}{2}(\mu_0 + \mu_1)^T \Sigma^{-1}(\mu_1 - \mu_0)$$

$$+ \log\left(\frac{\pi_0}{\pi_1}\right) + x^T \Sigma^{-1}(\mu_1 - \mu_0)$$

$$= \alpha_0 + \alpha^T x. \tag{9.32}$$

In logistic regression the model is by assumption

$$\log\left(\frac{p(y = 1|x)}{p(y = 0|x)}\right) = \beta_0 + \beta^T x. \tag{9.33}$$

The difference is in how they estimate parameters—in logistic regression they are chosen to effectively minimize classification error rather than density estimation error.

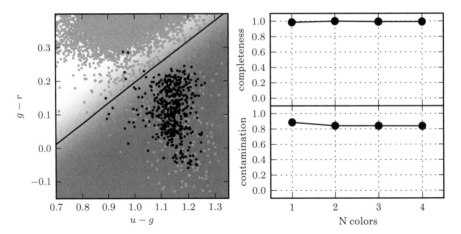

Figure 9.8. Logistic regression for RR Lyrae stars (see caption of figure 9.3 for details). With all four colors, logistic regression achieves a completeness of 0.993 and a contamination of 0.838.

Scikit-learn contains an implementation of logistic regression, which can be used as follows:

```
import numpy as np
from sklearn.linear_model import LogisticRegression

X = np.random.random((100, 2)) # 100 pts in 2 dims
y = (X[:, 0] + X[:, 1] > 1).astype(int)
    # simple division

logr = LogisticRegression(penalty='l2')
logr.fit(X, y)
y_pred = logr.predict(X)
```

For more details see the Scikit-learn documentation, or the source code of figure 9.8.

Figure 9.8 shows an example of logistic regression on the RR Lyrae data sets.

9.6. Support Vector Machines

Now let us look at yet another way of choosing a linear decision boundary, which leads off in an entirely different direction, that of *support vector machines*.

Consider finding the hyperplane that maximizes the distance of the closest point from either class (see figure 9.9). We call this distance the *margin*. Points on the margin are called *support vectors*. Let us begin by assuming the classes are linearly separable. Here we will use $y \in \{-1, 1\}$, as it will make things notationally cleaner.

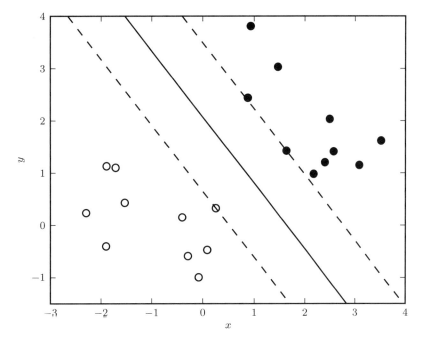

Figure 9.9. Illustration of SVM. The region between the dashed lines is the *margin*, and the points which the dashed lines touch are called the *support vectors*.

The hyperplane which maximizes the margin is given by finding

$$\max_{\beta_0, \beta}(m) \quad \text{subject to} \quad \frac{1}{\|\beta\|} y_i(\beta_0 + \beta^T x_i) \geq m \ \forall i. \tag{9.34}$$

Equivalently, the constraints can be written as $y_i(\beta_0 + \beta^T x_i) \geq m \|\beta\|$. Since for any β_0 and β satisfying these inequalities, any positively scaled multiple satisfies them too, we can arbitrarily set $\|\beta\| = 1/m$.

Thus the optimization problem is equivalent to minimizing

$$\frac{1}{2} \|\beta\| \quad \text{subject to} \quad y_i(\beta_0 + \beta^T x_i) \geq 1 \ \forall i. \tag{9.35}$$

It turns out this optimization problem is a *quadratic programming* problem (quadratic objective function with linear constraints), a standard type of optimization problem for which methods exist for finding the global optimum. The theory of convex optimization tells us there is an equivalent way to write this optimization problem (its *dual formulation*).

Let $g^*(x)$ denote the optimal (maximum margin) hyperplane. Let $\langle x_i, x_{i'} \rangle$ denote the inner product of x_i and $x_{i'}$. Then

$$\beta_j^* = \sum_{i=1}^{N} \alpha_i \, y_i \, x_{ij}, \tag{9.36}$$

where α is the vector of weights that maximizes

$$\sum_{i=1}^{N} \alpha_i - \frac{1}{2} \sum_{i=1}^{N} \sum_{i'=1}^{N} \alpha_i \, \alpha_{i'} \, y_i \, y_{i'} \langle x_i, x_{i'} \rangle \qquad (9.37)$$

$$\text{subject to} \quad \alpha_i \geq 0 \quad \text{and} \quad \sum_i \alpha_i y_i = 0. \qquad (9.38)$$

For realistic problems, however, we must relax the assumption that the classes are linearly separable. In the primal formulation, instead of minimizing

$$\frac{1}{2} \, \| \beta \| \quad \text{subject to} \quad y_i (\beta_0 + \beta^T x_i) \geq 1 \quad \forall \, i, \qquad (9.39)$$

we will now minimize

$$\frac{1}{2} \, \| \beta \| \quad \text{subject to} \quad y_i (\beta_0 + \beta^T x_i) \geq 1 - \xi_i \quad \forall \, i, \qquad (9.40)$$

where the ξ_i are called *slack variables* and we limit the amount of slack by adding the constraints

$$\xi_i \geq 0 \quad \text{and} \quad \sum_i \xi_i \leq C. \qquad (9.41)$$

This effectively bounds the total number of misclassifications at C, which becomes a tuning parameter of the support vector machine. The points x_i for which $\alpha_i \neq 0$ are the support vectors.

The discriminant function can be rewritten as

$$g(x) = \beta_0 + \sum_{i=1}^{N} \alpha_i \, y_i \, \langle x, x_i \rangle \qquad (9.42)$$

and the final classification rule is $\widehat{c}(x) = \text{sgn}[g(x)]$.

It turns out the SVM optimization is equivalent to minimizing

$$\sum_{i=1}^{N} (1 - y_i \, g(x_i))_+ + \lambda \, \| \beta \|^2, \qquad (9.43)$$

where λ is related to the tuning parameter C and the index $+$ stands for $x_+ = \max(0, x)$. Notice the similarity here to the ridge regularization discussed in §8.3.1: the tuning parameter λ controls the strength of an L_2 regularization over the parameters β.

Figure 9.10 shows the SVM decision boundary computed for the RR Lyrae data sets. Note that because SVM uses a metric which maximizes the margin rather than a measure over all points in the data sets, it is in some sense similar in spirit to the rank-based estimators discussed in chapter 3. The median of a distribution is unaffected

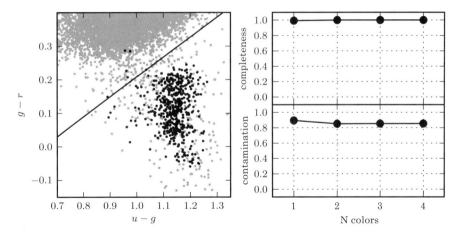

Figure 9.10. SVM applied to the RR Lyrae data (see caption of figure 9.3 for details). With all four colors, SVM achieves a completeness of 1.0 and a contamination of 0.854.

by even large perturbations of outlying points, as long as those perturbations do not cross the median. In the same way, once the support vectors are determined, changes to the positions or numbers of points beyond the margin will not change the decision boundary. For this reason, SVM can be a very powerful tool for discriminative classification. This is why figure 9.10 shows such a high completeness compared to the other methods discussed above: it is not swayed by the fact that the background sources outnumber the RR Lyrae stars by a factor of ∼200 to 1: it simply determines the best boundary between the small RR Lyrae clump and the large background clump. This completeness, however, comes at the cost of a relatively large contamination level.

Scikit-learn includes a fast SVM implementation for both classification and regression tasks. The SVM classifier can be used as follows:

```
import numpy as np
from sklearn.svm import LinearSVC

X = np.random.random((100, 2)) # 100 pts in 2 dims
y = (X[:, 0] + X[:, 1] > 1).astype(int)
    # simple division

model = LinearSVC(loss='12')
model.fit(X, y)
y_pred = model.predict(X)
```

For more details see the Scikit-learn documentation, or the source code of figure 9.11.

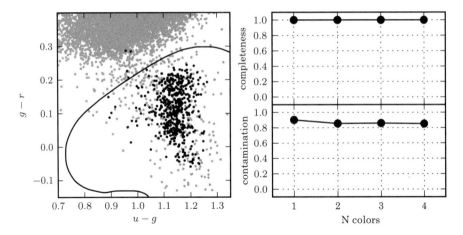

Figure 9.11. Kernel SVM applied to the RR Lyrae data (see caption of figure 9.3 for details). This example uses a Gaussian kernel with $\gamma = 20$. With all four colors, kernel SVM achieves a completeness of 1.0 and a contamination of 0.852.

One major limitation of SVM is that it is limited to linear decision boundaries. The idea of *kernelization* is a simple but powerful way to take a support vector machine and make it nonlinear—in the dual formulation, one simply replaces each occurrence of $\langle x_i, x_{i'} \rangle$ with a kernel function $K(x_i, x_{i'})$ with certain properties which allow one to think of the SVM as operating in a higher-dimensional space. One such kernel is the Gaussian kernel

$$K(x_i, x_{i'}) = e^{-\gamma \|x_i - x_{i'}\|^2}, \tag{9.44}$$

where γ is a parameter to be learned via cross-validation. An example of applying kernel SVM to the RR Lyrae data is shown in figure 9.11. This nonlinear classification improves over the linear version only slightly. For this particular data set, the contamination is not driven by nonlinear effects.

9.7. Decision Trees

The decision boundaries that we discussed in §9.5 can be applied hierarchically to a data set. This observation leads to a powerful methodology for classification that is known as the *decision tree*. An example decision tree used for the classification of our RR Lyrae stars is shown in figure 9.12. As with the tree structures described in §2.5.2, the top node of the decision tree contains the entire data set. At each branch of the tree these data are subdivided into two child nodes (or subsets), based on a predefined decision boundary, with one node containing data below the decision boundary and the other node containing data above the decision boundary. The boundaries themselves are usually axis aligned (i.e., the data are split along one feature at each level of the tree). This splitting process repeats, recursively, until we achieve a predefined stopping criteria (see §9.7.1).

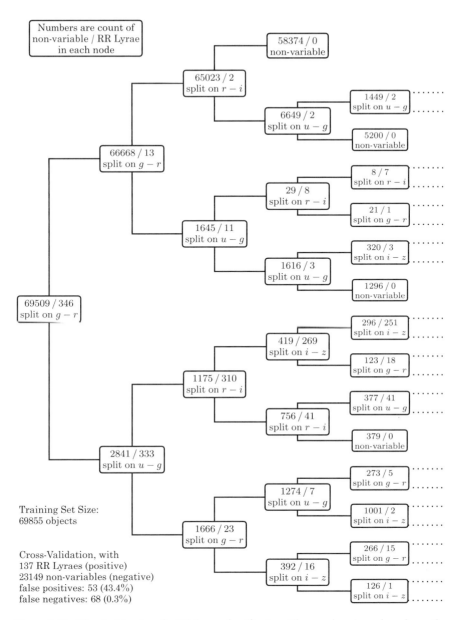

Figure 9.12. The decision tree for RR Lyrae classification. The numbers in each node are the statistics of the *training* sample of ∼70,000 objects. The cross-validation statistics are shown in the bottom-left corner of the figure. See also figure 9.13.

For the two-class decision tree shown in figure 9.12, the tree has been learned from a training set of standard stars (§1.5.8), and RR Lyrae variables with known classifications. The terminal nodes of the tree (often referred to as "leaf nodes") record the fraction of points contained within that node that have one classification or the other, that is, the fraction of standard stars or RR Lyrae.

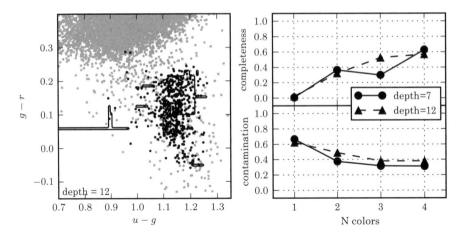

Figure 9.13. Decision tree applied to the RR Lyrae data (see caption of figure 9.3 for details). This example uses tree depths of 7 and 12. With all four colors, this decision tree achieves a completeness of 0.569 and a contamination of 0.386.

Scikit-learn includes decision-tree implementations for both classification and regression. The decision-tree classifier can be used as follows:

```
import numpy as np
from sklearn.tree import DecisionTreeClassifier

X = np.random.random((100, 2)) # 100 pts in 2 dims
y = (X[:, 0] + X[:, 1] > 1).astype(int)
    # simple division

model = DecisionTreeClassifier(max_depth=6)
model.fit(X, y)
y_pred = model.predict(X)
```

For more details see the Scikit-learn documentation, or the source code of figure 9.11.

The result of the full decision tree as a function of the number of features used is shown in figure 9.13. This classification method leads to a completeness of 0.569 and a contamination of 0.386. The depth of the tree also has an effect on the precision and accuracy. Here, going to a depth of 12 (with a maximum of $2^{12} = 4096$ nodes) slightly overfits the data: it divides the parameter space into regions which are too small. Using fewer nodes prevents this, and leads to a better classifier.

Application of the tree to classifying data is simply a case of following the branches of the tree through a series of binary decisions (one at each level of the tree) until we reach a leaf node. The relative fraction of points from the training set classified as one class or the other defines the class associated with that leaf node.

Decision trees are, therefore, classifiers that are simple, and easy to visualize and interpret. They map very naturally to how we might interrogate a data set by hand (i.e., a hierarchy of progressively more refined questions).

9.7.1. Defining the Split Criteria

In order to build a decision tree we must choose the feature and value on which we wish to split the data. Let us start by considering a simple split criteria based on the information content or entropy of the data; see [11]. In §5.2.2, we define the entropy, $E(x)$, of a data set, x, as

$$E(x) = -\sum_i p_i(x) \ln(p_i(x)), \tag{9.45}$$

where i is the class and $p_i(x)$ is the probability of that class given the training data. We can define information gain as the reduction in entropy due to the partitioning of the data (i.e., the difference between the entropy of the parent node and the sum of entropies of the child nodes). For a binary split with $i = 0$ representing those points below the split threshold and $i = 1$ for those points above the split threshold, the information gain, $IG(x)$, is

$$IG(x|x_i) = E(x) - \sum_{i=0}^{1} \frac{N_i}{N} E(x_i), \tag{9.46}$$

where N_i is the number of points, x_i, in the ith class, and $E(x_i)$ is the entropy associated with that class (also known as Kullback–Leibler divergence in the machine learning community).

Finding the optimal decision boundary on which to split the data is generally considered to be a computationally intractable problem. The search for the split is, therefore, undertaken in a greedy fashion where each feature is considered one at a time and the feature that provides the largest information gain is split. The value of the feature at which to split the data is defined in an analogous manner, whereby we sort the data on feature i and maximize the information gain for a given split point, s,

$$IG(x|s) = E(x) - \arg\max_s \left(\frac{N(x|x < s)}{N} E(x|x < s) - \frac{N(x|x \geq s)}{N} E(x|x \geq s) \right). \tag{9.47}$$

Other loss functions common in decision trees include the Gini coefficient (see §4.7.2) and the misclassification error. The Gini coefficient estimates the probability that a source would be incorrectly classified if it was chosen at random from a data set and the label was selected randomly based on the distribution of classifications within the data set. The Gini coefficient, G, for a k-class sample is given by

$$G = \sum_i^k p_i(1 - p_i), \tag{9.48}$$

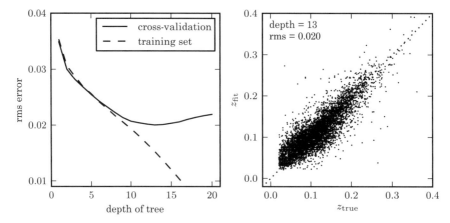

Figure 9.14. Photometric redshift estimation using decision-tree regression. The data is described in §1.5.5. The training set consists of u, g, r, i, z magnitudes of 60,000 galaxies from the SDSS spectroscopic sample. Cross-validation is performed on an additional 6000 galaxies. The left panel shows training error and cross-validation error as a function of the maximum depth of the tree. For a number of nodes $N > 13$, overfitting is evident.

where p_i is the probability of finding a point with class i within a data set. The misclassification error, MC, is the fractional probability that a point selected at random will be misclassified and is defined as

$$MC = 1 - \max_i(p_i). \tag{9.49}$$

The Gini coefficient and classification error are commonly used in classification trees where the classification is categorical.

9.7.2. Building the Tree

In principle, the recursive splitting of the tree could continue until there is a single point per node. This is, however, inefficient as it results in $\mathcal{O}(N)$ computational cost for both the construction and traversal of the tree. A common criterion for stopping the recursion is, therefore, to cease splitting the nodes when either a node contains only one class of object, when a split does not improve the information gain or reduce the misclassifications, or when the number of points per node reaches a predefined value.

As with all model fitting, as we increase the complexity of the model we run into the issue of overfitting the data. For decision trees the complexity is defined by the number of levels or depth of the tree. As the depth of the tree increases, the error on the training set will decrease. At some point, however, the tree will cease to represent the correlations within the data and will reflect the noise within the training set. We can, therefore, use the cross-validation techniques introduced in §8.11 and either the entropy, Gini coefficient, or misclassification error to optimize the depth of the tree. Figure 9.14 illustrates this cross-validation using a decision tree that predicts photometric redshifts. For a training sample of approximately 60,000 galaxies, with the rms error in estimated redshift used as the misclassification criterion, the optimal depth is 13. For this depth there are roughly $2^{13} \approx 8200$ leaf nodes. Splitting beyond this level leads to overfitting, as evidenced by an increased cross-validation error.

A second approach for controlling the complexity of the tree is to grow the tree until there are a predefined number of points in a leaf node (e.g., five) and then use the cross-validation or test data set to prune the tree. In this method we take a greedy approach and, for each node of the tree, consider whether terminating the tree at that node (i.e., making it a leaf node and removing all subsequent branches of the tree) improves the accuracy of the tree. Pruning of the decision tree using an independent test data set is typically the most successful of these approaches. Other approaches for limiting the complexity of a decision tree include random forests (see §9.7.3), which effectively limits the number of attributes on which the tree is constructed.

9.7.3. Bagging and Random Forests

Two of the most successful applications of *ensemble learning* (the idea of combining the outputs of multiple models through some kind of voting or averaging) are those of *bagging* and *random forests* [1]. Bagging (from bootstrap aggregation) averages the predictive results of a series of bootstrap samples (see §4.5) from a training set of data. Often applied to decision trees, bagging is applicable to regression and many nonlinear model fitting or classification techniques. For a sample of N points in a training set, bagging generates K equally sized bootstrap samples from which to estimate the function $f_i(x)$. The final estimator, defined by bagging, is then

$$f(x) = \frac{1}{K} \sum_i^K f_i(x). \tag{9.50}$$

Random forests expand upon the bootstrap aspects of bagging by generating a set of decision trees from these bootstrap samples. The features on which to generate the tree are selected at random from the full set of features in the data. The final classification from the random forest is based on the averaging of the classifications of each of the individual decision trees. In so doing, random forests address two limitations of decision trees: the overfitting of the data if the trees are inherently deep, and the fact that axis-aligned partitioning of the data does not accurately reflect the potentially correlated and/or nonlinear decision boundaries that exist within data sets.

In generating a random forest we define n, the number of trees that we will generate, and m, the number of attributes that we will consider splitting on at each level of the tree. For each decision tree a subsample (bootstrap sample) of data is selected from the full data set. At each node of the tree, a set of m variables are randomly selected and the split criteria is evaluated for each of these attributes; a different set of m attributes are used for each node. The classification is derived from the mean or mode of the results from all of the trees. Keeping m small compared to the number of features controls the complexity of the model and reduces the concerns of overfitting.

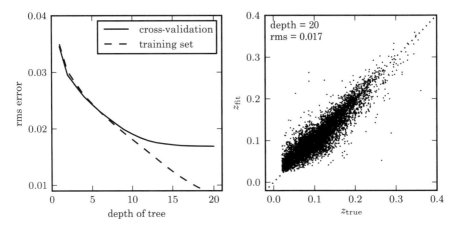

Figure 9.15. Photometric redshift estimation using random forest regression, with ten random trees. Comparison to figure 9.14 shows that random forests correct for the overfitting evident in very deep decision trees. Here the optimal depth is 20 or above, and a much better cross-validation error is achieved.

Scikit-learn contains a random forest implementation which can be used for classification or regression. For example, classification tasks can be approached as follows:

```
import numpy as np
from sklearn.ensemble import RandomForestClassifier

X = np.random.random((100, 2)) # 100 pts in 2 dims
y = (X[:, 0] + X[:, 1] > 1).astype(int)
    # simple division

model = RandomForestClassifier(10)
# forest of 10 trees
model.fit(X, y)
y_pred = model.predict(X)
```

For more details see the Scikit-learn documentation, or the source code of figure 9.15.

Figure 9.15 demonstrates the application of a random forest of regression trees to photometric redshift data (using a forest of ten random trees—see [2] for a more detailed discussion). The left panel shows the cross-validation results as a function of the depth of each tree. In comparison to the results for a single tree (figure 9.14), the use of randomized forests reduces the effect of overfitting and leads to a smaller rms error.

Similar to the cross-validation technique used to arrive at the optimal depth of the tree, cross-validation can also be used to determine the number of trees, n, and the number of random features m, simply by optimizing over all free parameters. With random forests, n is typically increased until the cross-validation error plateaus, and m is often chosen to be $\sim \sqrt{K}$, where K is the number of attributes in the sample.

9.7.4. Boosting Classification

Boosting is an ensemble approach that was motivated by the idea that combining many weak classifiers can result in an improved classification. This idea differs fundamentally from that illustrated by random forests: rather than create the models separately on different data sets, which can be done all in parallel, boosting creates each new model to attempt to correct the errors of the ensemble so far. At the heart of boosting is the idea that we reweight the data based on how incorrectly the data were classified in the previous iteration.

In the context of classification (boosting is also applicable in regression) we can run the classification multiple times and each time reweight the data based on the previous performance of the classifier. At the end of this procedure we allow the classifiers to vote on the final classification. The most popular form of boosting is that of adaptive boosting [4]. For this case, imagine that we had a weak classifier, $h(x)$, that we wish to apply to a data set and we want to create a strong classifier, $f(x)$, such that

$$f(x) = \sum_m^K \theta_m h_m(x), \tag{9.51}$$

where m indicates the number of the iteration of the weak classifier and θ_m is the weight of the mth iteration of the classifier.

If we start with a set of data, x, with known classifications, y, we can assign a weight, $w_m(x)$, to each point (where the initial weight is uniform, $1/N$, for the N points in the sample). After the application of the weak classifier, $h_m(x)$, we can estimate the classification error, e_m, as

$$e_m = \sum_{i=1}^N w_m(x_i) I(h_m(x_i) \neq y_i), \tag{9.52}$$

where $I(h_m(x_i) \neq y_i)$ is the indicator function (with $I(h_m(x_i) \neq y_i)$ equal to 1 if $h_m(x_i) \neq y_i$ and equal to 0 otherwise). From this error we define the weight of that iteration of the classifier as

$$\theta_m = \frac{1}{2} \log \left(\frac{1 - e_m}{e_m} \right) \tag{9.53}$$

and update the weights on the points,

$$w_{m+1}(x_i) = w_m(x_i) \times \begin{cases} e^{-\theta_m} & \text{if } h_m(x_i) = y_i, \\ e^{\theta_m} & \text{if } h_m(x_i) \neq y_i, \end{cases} \tag{9.54}$$

$$= \frac{w_m(x_i) e^{-\theta_m y_i h_m(x_i)}}{\sum_{i=1}^N w_m(x_i) e^{-\theta_m y_i h_m(x_i)}}. \tag{9.55}$$

The effect of updating $w(x_i)$ is to increase the weight of the misclassified data. After K iterations the final classification is given by the weighted votes of each classifier given by eq. 9.51. As the total error, e_m, decreases, the weight of that iteration in the final classification increases.

A fundamental limitation of the boosted decision tree is the computation time for large data sets. Unlike random forests, which can be trivially parallelized, boosted decision trees rely on a chain of classifiers which are each dependent on the last. This may limit their usefulness on very large data sets. Other methods for boosting have been developed such as gradient boosting; see [5]. Gradient boosting involves approximating a steepest descent criterion after each simple evaluation, such that an additional weak classification can improve the classification score and may scale better to larger data sets.

Scikit-learn contains several flavors of boosted decision trees, which can be used for classification or regression. For example, boosted classification tasks can be approached as follows:

```python
import numpy as np
from sklearn.ensemble import
GradientBoostingClassifier

X = np.random.random((100, 2)) # 2 pts in 100 dims
y = (X[:, 0] + X[:, 1] > 1).astype(int)
    # simple division

model = GradientBoostingClassifier()
model.fit(X, y)
y_pred = model.predict(X)
```

For more details see the Scikit-learn documentation, or the source code of figure 9.16.

Figure 9.16 shows the results for a gradient-boosted decision tree for the SDSS photometric redshift data. For the weak estimator, we use a decision tree with a maximum depth of 3. The cross-validation results are shown as a function of boosting iteration. By 500 steps, the cross-validation error is beginning to level out, but there are still no signs of overfitting. The fact that the training error and cross-validation error remain very close indicates that a more complicated model (i.e., deeper trees or more boostings) would likely allow improved errors. Even so, the rms error recovered with these suboptimal parameters is comparable to that of the random forest classifier.

9.8. Evaluating Classifiers: ROC Curves

Comparing the performance of classifiers is an important part of choosing the best classifier for a given task. "Best" in this case can be highly subjective: for some

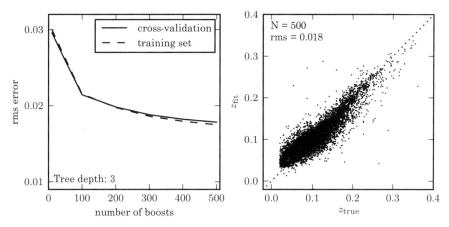

Figure 9.16. Photometric redshift estimation using gradient-boosted decision trees, with 100 boosting steps. As with random forests (figure 9.15), boosting allows for improved results over the single tree case (figure 9.14). Note, however, that the computational cost of boosted decision trees is such that it is computationally prohibitive to use very deep trees. By stringing together a large number of very naive estimators, boosted trees improve on the underfitting of each individual estimator.

problems, one might wish for high completeness at the expense of contamination; at other times, one might wish to minimize contamination at the expense of completeness. One way to visualize this is to plot receiver operating characteristic (ROC) curves (see §4.6.1). An ROC curve usually shows the true-positive rate as a function of the false-positive rate as the discriminant function is varied. How the function is varied depends on the model: in the example of Gaussian naive Bayes, the curve is drawn by classifying data using relative probabilities between 0 and 1.

A set of ROC curves for a selection of classifiers explored in this chapter is shown in the left panel of figure 9.17. The curves closest to the upper left of the plot are the best classifiers: for the RR Lyrae data set, the ROC curve indicates that GMM Bayes and K-nearest-neighbor classification outperform the rest. For such an unbalanced data set, however, ROC curves can be misleading. Because there are fewer than five sources for every 1000 background objects, a false-positive rate of even 0.05 means that false positives outnumber true positives ten to one! When sources are rare, it is often more informative to plot the efficiency (equal to one minus the contamination, eq. 9.5) vs. the completeness (eq. 9.5). This can give a better idea of how well a classifier is recovering rare data from the background.

The right panel of figure 9.17 shows the completeness vs. efficiency for the same set of classifiers. A striking feature is that the simpler classifiers reach a maximum efficiency of about 0.25: this means that at their best, only 25% of objects identified as RR Lyrae are actual RR Lyrae. By the completeness–efficiency measure, the GMM Bayes model outperforms all others, allowing for higher completeness at virtually any efficiency level. We stress that this is not a general result, and that the best classifier for any task depends on the precise nature of the data.

As an example where the ROC curve is a more useful diagnostic, figure 9.18 shows ROC curves for the classification of stars and quasars from four-color photometry (see the description of the data set in §9.1). The stars and quasars in

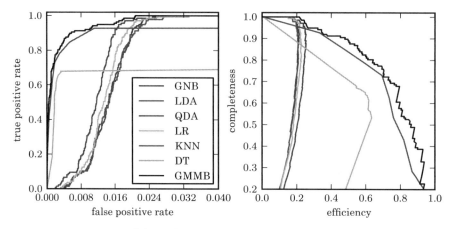

Figure 9.17. ROC curves (left panel) and completeness–efficiency curves (right panel) for the four-color RR Lyrae data using several of the classifiers explored in this chapter: Gaussian naive Bayes (GNB), linear discriminant analysis (LDA), quadratic discriminant analysis (QDA), logistic regression (LR), K-nearest-neighbor classification (KNN), decision tree classification (DT), and GMM Bayes classification (GMMB). See color plate 7.

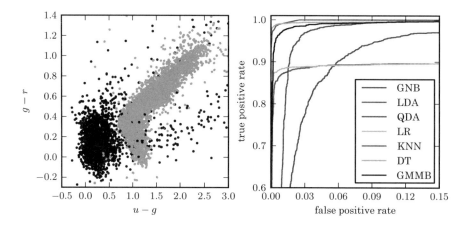

Figure 9.18. The left panel shows data used in color-based photometric classification of stars and quasars. Stars are indicated by gray points, while quasars are indicated by black points. The right panel shows ROC curves for quasar identification based on $u - g$, $g - r$, $r - i$, and $i - z$ colors. Labels are the same as those in figure 9.17. See color plate 8.

this sample are selected with differing selection functions: for this reason, the data set does not reflect a realistic sample. We use it for purposes of illustration only. The stars outnumber the quasars by only a factor of 3, meaning that a false-positive rate of 0.3 corresponds to a contamination of ∼50%. Here we see that the best-performing classifiers are the neighbors-based and tree-based classifiers, both of which approach 100% true positives with a very small number of false positives. An interesting feature is that classifiers with linear discriminant functions (LDA and logistic regression) plateau at a true-positive rate of 0.9. These simple classifiers, while useful in some situations, do not adequately explain these photometric data.

Scikit-learn has some built-in tools for computing ROC curves and completeness–efficiency curves (known as precision-recall curves in the machine learning community). They can be used as follows:

```
import numpy as np
from sklearn.naive_bayes import GaussianNB
from sklearn import metrics

X = np.random.random((100, 2))# 100 points in 2 dims
y = (X[:, 0] + X[:, 1] > 1).astype(int)
  # simple boundary

gnb = GaussianNB().fit(X, y)
prob = gnb.predict_proba(X)

# Compute precision / recall curve
pr, re, thresh = metrics.precision_recall_curve
        (y, prob[:, 0])

# Compute ROC curve: true positives / false positives
tpr, fpr, thresh = metrics.roc_curve(y, prob[:, 0])
```

The thresholds are automatically determined based on the probability levels within the data. For more information, see the Scikit-learn documentation.

9.9. Which Classifier Should I Use?

Continuing as we have in previous chapters, we will answer this question by decomposing the notion of "best" along the axes of "accuracy," "interpretability," "simplicity," and "speed."

What are the most *accurate* classifiers? A bit of thought will lead to the obvious conclusion that no single type of model can be known in advance to be the best classifier for all possible data sets, as each data set can have an arbitrarily different underlying distribution. This is famously formalized in the no free lunch theorem [17]. The conclusion is even more obvious for regression, where the same theorem applies, though it was popularized in the context of classification. However, we can still draw upon a few useful rules of thumb.

As we have said in the summaries of earlier chapters, in general, the more parameters a model has, the more complex a function it can fit (whether that be the probability density functions, in the case of a generative classifier, or the decision boundary, in the case of a discriminative classifier), and thus the more likely it is to yield high predictive accuracy. Parametric methods, which have a fixed number of parameters with respect to the number of data points N, include (roughly in increasing order of typical accuracy) naive Bayes, linear discriminant analysis, logistic regression, linear support vector machines, quadratic discriminant analysis, and linear ensembles of linear models. Nonparametric methods, which have a number of parameters that grows as the number of data points N grows, include (roughly in order of typical accuracy) decision trees, K-nearest-neighbor, neural networks, kernel discriminant analysis, kernelized support vector machines, random forests, and boosting with nonlinear methods (a model like K-nearest-neighbor for this purpose is considered to have $\mathcal{O}(N)$ parameters, as the model consists of the training points themselves—however, there is only one parameter, K, that in practice is *optimized* through cross-validation). While there is no way to know for sure which method is best for a given data set until it is tried empirically, generally speaking the most accurate method is most likely to be nonparametric.

Some models including neural networks and GMM Bayes classifiers are parametric for a fixed number of hidden units or components, but if that number is chosen in a way that can grow with N, such a model becomes nonparametric. The practical complication of trying every possible number of parameters typically prevents a disciplined approach to model selection, making such approaches difficult to call either nonparametric or strictly parametric. Among the nonparametric methods, the ensemble methods (e.g., bagging, boosting) are effectively models with many more parameters, as they combine hundreds or thousands of base models. Thus, as we might expect based on our general rule of thumb, an ensemble of models will generally yield the highest possible accuracy.

The generally simpler parametric methods can shine in some circumstances, however. When the dimensionality is very high compared to the number of data points (as is typical in text data where for each document the features are the counts of each of thousands of words, or in bioinformatics data where for each patient the features are the expression levels of each of thousands of genes), the points in such a space are effectively so dispersed that a linear hyperplane can (at least mostly) separate them. Linear classifiers have thus found great popularity in such applications, though this situation is fairly atypical in astronomy problems. Another setting that favors parametric methods is that of low sample size. When there are few data points, a simple function is all that is needed to achieve a good fit, and the principle of Occam's razor (§5.4.2) dictates favoring the simplest model that works. In such a regime, the paucity of data points makes the ability of domain knowledge-based assumptions to fill in the gaps compelling, leading one toward Bayesian approaches and carefully hand-constructed parametric models rather than relatively "blind," or assumption-free nonparametric methods. Hierarchical modeling, as possibly assisted by the formalism of Bayesian networks, becomes the mode of thinking in this regime.

Complexity control in the form of model selection is more critical for nonparametric methods—it is important to remember that the accuracy benefits of such methods are only realized assuming that model selection is done properly, via cross-validation or other measures (in order to avoid being misled by overly optimistic-looking accuracies). Direct comparisons between different ML methods, which are most common in the context of classification, can also be misleading when the amount of human and/or computational effort spent on obtaining the best parameter settings was not equal between methods. This is typically the case, for example, in a paper proposing one method over others.

What are the most *interpretable* classifiers? In many cases we would like to know *why* the classifier made the decision it did, or in general, what sort of discriminatory pattern the classifier has found in general, for example, which dimensions are the primary determinants of the final prediction. In general, this is where parametric methods tend to be the most useful. Though it is certainly possible to make a complex and unintelligible parametric model, for example by using the powerful general machinery of graphical models, the most popular parametric methods tend to be simple and easy to reason about. In particular, in a linear model, the coefficients on the dimensions can be interpreted to indicate the importance of each variable in the model.

Nonparametric methods are often too large to be interpretable, as they typically scale in size (number of parameters) with the number of data points. However, among nonparametric methods, certain ones can be intepretable, in different senses. A decision tree can be explained in plain English terms, by reading paths from the root to each leaf as a rule containing if-then tests on the variables (see, e.g., figure 9.12). The interpretability of decision trees was used to advantage in understanding star–galaxy classification [3]. A nearest-neighbor classifier's decisions can be understood by simply examining the k neighbors returned, and their class labels. A probabilistic classifier which explicitly implements Bayes' rule, such as kernel discriminant analysis, can be explained in terms of the class under which the test point was more likely—an explanation that is typically natural for physicists in particular. Neural networks, kernelized support vector machines, and ensembles such as random forests and boosting are among the least interpretable methods.

What are the most *scalable* classifiers? Naive Bayes and its variants are by far the easiest to compute, requiring in principle only one pass through the data. Learning a logistic regression model via standard unconstrained optimization methods such as conjugate gradient requires only a modest number of relatively cheap iterations through the data. Though the model is still linear, linear support vector machines are more expensive, though several fast algorithms exist. We have discussed K-nearest-neighbor computations in §2.5.2, but K-nearest-neighbor *classification* is in fact a slightly easier problem, and this can be exploited algorithmically. Kernel discriminant analysis can also be sped up by fast tree-based algorithms, reducing its cost from $\mathcal{O}(N^2)$ to $\mathcal{O}(N)$. Decision trees are relatively efficient, requiring $\mathcal{O}(N \log N)$ time to build and $\mathcal{O}(\log N)$ time to classify, as are neural networks. Random forests and boosting with trees simply multiplies the cost of decision trees by the number of trees, which is typically very large, but this can be easily parallelized. Kernelized support vector machines are $\mathcal{O}(N^3)$ in the worst case, and have resisted attempts at appreciably fast algorithms. Note that the need to use fast algorithms ultimately counts against a method in terms of its simplicity, at least in implementation.

What are the *simplest* classifiers? Naive Bayes classifiers are possibly the simplest in terms of both implementation and learning, requiring no tuning parameters to be tried. Logistic regression and decision tree are fairly simple to implement, with no tuning parameters. K-nearest-neighbor classification and KDA are simple methods that have only one critical parameter, and cross-validation over it is generally straightforward. Kernelized support vector machines also have only one

TABLE 9.1.
Summary of the practical properties of different classifiers.

Method	Accuracy	Interpretability	Simplicity	Speed
Naive Bayes classifier	L	H	H	H
Mixture Bayes classifier	M	H	H	M
Kernel discriminant analysis	H	H	H	M
Neural networks	H	L	L	M
Logistic regression	L	M	H	M
Support vector machines: linear	L	M	M	M
Support vector machines: kernelized	H	L	L	L
K-nearest-neighbor	H	H	H	M
Decision trees	M	H	H	M
Random forests	H	M	M	M
Boosting	H	L	L	L

or two critical parameters which are easy to cross-validate over, though the method is not as simple to understand as KNN or KDA. Mixture Bayes classifiers inherit the properties of Gaussian mixtures discussed in §6.6 (i.e., they are fiddly). Neural networks are perhaps the most fiddly methods. The onus of having to store and manage the typically hundreds or thousands of trees in random forests counts against it, though it is conceptually simple.

Other considerations, and taste. If obtaining good class probability estimates rather than just the correct class labels is of importance, KDA is a good choice as it is based on the most powerful approach for density estimation, kernel density estimation. For this reason, as well as its state-of-the-art accuracy and ability to be computed efficiently, once a fast KDA algorithm had been developed (see [14]), it produced a dramatic leap in the size and quality of quasar catalogs [12, 13], and associated scientific results [6, 15]. Optimization-based approaches, such as logistic regression, can typically be augmented to handle missing values, whereas distance-based methods such as K-nearest-neighbor, kernelized SVMs, and KDA cannot. When both continuous and discrete attributes are present, decision trees are one of the few methods that seamlessly handles both without modification.

Simple summary. Using our axes of "accuracy," "interpretability," "simplicity," and "speed," a summary of each of the methods considered in this chapter, in terms of high (H), medium (M), and low (L) is given in table 9.1.

Note that it is hard to say in advance how a method will behave in terms of accuracy, for example, on a particular data set—and no method, in general, always beats other methods. Thus our table should be considered an initial guideline based on extensive experience.

References

[1]   Breiman, L. (2001). Random forests. *Mach. Learn.* 45(1), 5–32.

[2]   Carliles, S., T. Budavári, S. Heinis, C. Priebe, and A. S. Szalay (2010). Random forests for photometric redshifts. *ApJ 712*, 511–515.

[3]   Fayyad, U. M., N. Weir, and S. G. Djorgovski (1993). SKICAT: A machine learning system for automated cataloging of large scale sky surveys. In *International Conference on Machine Learning (ICML)*, pp. 112–119.

[4]   Freund, Y. and R. E. Schapire (1997). A decision-theoretic generalization of on-line learning and an application to boosting. *J. Comput. Syst. Sci. 55*(1), 119–139. Special issue for EuroCOLT '95.

[5]   Friedman, J. H. (2000). Greedy function approximation: A gradient boosting machine. *Annals of Statistics 29*, 1189–1232.

[6]   Giannantonio, T., R. Crittenden, R. Nichol, and others (2006). A high redshift detection of the integrated Sachs-Wolfe effect. *Physical Review D 74*.

[7]   Hubble, E. P. (1926). Extragalactic nebulae. *ApJ 64*, 321–369.

[8]   Ivezić, Ž., A. K. Vivas, R. H. Lupton, and R. Zinn (2005). The selection of RR Lyrae stars using single-epoch data. *AJ 129*, 1096–1108.

[9]   Lintott, C. J., K. Schawinski, A. Slosar, and others (2008). Galaxy zoo: morphologies derived from visual inspection of galaxies from the Sloan Digital Sky Survey. *MNRAS 389*, 1179–1189.

[10]   Miller, G. A. (1956). The magical number seven, plus or minus two: Some limits on our capacity for processing information. *The Psychological Review 63*(2), 81–97.

[11]   Quinlan, J. R. (1992). *C4.5: Programs for Machine Learning* (first ed.). Morgan Kaufmann Series in Machine Learning. Morgan Kaufmann.

[12]   Richards, G., R. Nichol, A. G. Gray, and others (2004). Efficient photometric selection of quasars from the Sloan Digital Sky Survey: 100,000 $z < 3$ quasars from Data Release One. *ApJS 155*, 257–269.

[13]   Richards, G. T., A. D. Myers, A. G. Gray, and others (2009). Efficient photometric selection of quasars from the Sloan Digital Sky Survey II. ~1,000,000 quasars from Data Release Six. *ApJS 180*, 67–83.

[14]   Riegel, R., A. G. Gray, and G. Richards (2008). Massive-scale kernel discriminant analysis: Mining for quasars. In *SDM*, pp. 208–218. SIAM.

[15]   Scranton, R., B. Menard, G. Richards, and others (2005). Detection of cosmic magnification with the Sloan Digital Sky Survey. *ApJ 633*, 589–602.

[16]   Sesar, B., Ž. Ivezić, S. H. Grammer, and others (2010). Light curve templates and galactic distribution of RR Lyrae stars from Sloan Digital Sky Survey Stripe 82. *ApJ 708*, 717–741.

[17]   Wolpert, D. H. and W. G. Macready (1997). No free lunch theorems for optimization. *IEEE Transactions on Evolutionary Computation 1*, 67–82.

10 Time Series Analysis

"It is my feeling that Time ripens all things; with Time all things are revealed; Time is the father of truth." (Francois Rabelais)

This chapter summarizes the fundamental concepts and tools for analyzing time series data. Time series analysis is a branch of applied mathematics developed mostly in the fields of signal processing and statistics. Contributions to this field, from an astronomical perspective, have predominantly focused on unevenly sampled data, low signal-to-noise data, and heteroscedastic errors. There are more books written about time series analysis than pages in this book and, by necessity, we can only address a few common use cases from contemporary astronomy. Even when limited to astronomical data sets, the diversity of potential applications is enormous. The most common applications range from the detection of variability and periodicity to the treatment of nonperiodic variability and searches for localized events.

Within time-domain data, measurement errors can range from as small as one part in 100,000 (e.g., the photometry from the Kepler mission [32]), to potential events buried in noise with a signal-to-noise ratio per data point of, at best, a few (e.g., searches for gravitational waves using the Laser Interferometric Gravitational Observatory (LIGO) data[1] [1]). Data sets can include many billions of data points, and sample sizes can be in the millions (e.g., the LINEAR data set with 20 million light curves, each with a few hundred measurements [50]). Upcoming surveys, such as Gaia and LSST, will increase existing data sets by large factors; the Gaia satellite will measure about a billion sources about 70 times during its five-year mission, and the ground-based LSST will obtain about 800 measurements each for about 20 billion sources over its ten years of operation. Scientific utilization of such data sets will include searches for extrasolar planets; tests of stellar astrophysics through studies of variable stars and supernova explosions; distance determination (e.g., using standard candles such as Cepheids, RR Lyrae, and supernovas); and fundamental physics such

---

[1]LIGO aims to detect gravitational waves, the ripples in the fabric of space and time produced by astrophysical events such as black hole and neutron star collisions, predicted by the general theory of relativity. The two 4 km large detectors in Hanford, Washington and Livingston, Louisiana utilize laser interferometers to measure a fractional change of distance between two mirrors with a precision of about 1 part in 10^{18}.

as tests of general relativity with radio pulsars, cosmological studies with supernovas and searches for gravitational wave events.

We start with a brief introduction to the main concepts in time series analysis, and then discuss the main tools from the modeling toolkit for time series analysis. Despite being set in the context of time series, many tools and results are readily applicable in other domains, and for this reason our examples will not be strictly limited to time-domain data. Armed with the modeling toolkit, we will then discuss the analysis of periodic time series, search for temporally localized signals, and conclude with a brief discussion of stochastic processes. The main data sets used in this chapter include light curves obtained by the LINEAR survey (§1.5.9) and a variety of simulated examples.

10.1. Main Concepts for Time Series Analysis

The time series discussed here will be limited to two-dimensional scalar data sets: pairs of random variables, $(t_1, y_1), \ldots, (t_N, y_N)$, with no assumptions about the sampling of the time coordinate t (with the exception of the so-called arrival time data that consists of detection times for individual photons, discussed in §10.3.5). In many ways, the analysis methods discussed here are closely related to the parameter estimation and model selection problems discussed in the context of regression in chapter 8; when the temporal variable t is replaced by x, this connection becomes more obvious. Nevertheless, there are some important differences encountered in time series analysis, such as models that have a sense of the directionality of time in them (discussed in §10.5.3). Unlike regression problems where different y measurements are typically treated as independent random variables, in such models the value of y_{i+1} directly depends on the preceding value y_i.

The main tasks of time series analysis are (1) to characterize the presumed temporal correlation between different values of y, including its significance, and (2) to forecast (predict) future values of y. In many astronomical cases, the characterization of the underlying physical processes that produced the data, which is typically addressed by learning parameters for a model, is the key goal. For example, analysis of a light curve can readily differentiate between pulsating and eclipsing variable stars. Good examples of the second task are the solar activity forecasting, or prediction of the time and place of a potential asteroid impact on Earth.

10.1.1. Is My time Series Just Noise?

Given a time series, we often first want to determine whether we have detected variability, irrespective of the details of the underlying process. This is equivalent to asking whether the data are consistent with the null hypothesis described by a model consisting of a constant signal plus measurement noise.

From the viewpoint of classical (frequentist) statistics, this question can be treated as a case of hypothesis testing: What is the probability that we would obtain our data by chance if the null hypothesis of no variability were correct? If the errors are known and Gaussian, we can simply compute χ^2 and the corresponding p values.

If the errors are unknown, or non-Gaussian, the modeling and model selection tools, such as those introduced in chapter 5 for treating exponential noise or outliers, can be used instead.

Consider a simple example of $y(t) = A \sin(\omega t)$ sampled by $N \sim 100$ data points with homoscedastic Gaussian errors with standard deviation σ. The variance of a well-sampled time series given by this model is $V = \sigma^2 + A^2/2$. For a model with $A = 0$, $\chi^2_{\rm dof} = N^{-1} \sum_j (y_j/\sigma)^2 \sim V/\sigma^2$. When $A = 0$ is true, the $\chi^2_{\rm dof}$ has an expectation value of 1 and a standard deviation of $\sqrt{2/N}$. Therefore, if variability is present (i.e., $|A| > 0$), the computed $\chi^2_{\rm dof}$ will be larger than its expected value of 1. The probability that $\chi^2_{\rm dof} > 1 + 3\sqrt{2/N}$ is about 1 in 1000. If this false-positive rate is acceptable (recall §4.6; for example, if the expected fraction of variable stars in a sample is 1%, this false-positive rate will result in a sample contamination rate of $\sim$10%), then the minimum detectable amplitude is $A > 2.9\sigma/N^{1/4}$ (derived from $V/\sigma^2 = 1 + 3\sqrt{2/N}$). For example, for $N = 100$ data points, the minimum detectable amplitude is $A = 0.92\sigma$, and $A = 0.52\sigma$ for $N = 1000$. However, we will see that in all cases of specific models, our ability to discover variability is *greatly improved* compared to this simple $\chi^2_{\rm dof}$ selection. For illustration, for the single harmonic model, the minimum detectable variability levels for the false-positive rate of 1 in 1000 are $A = 0.42\sigma$ for $N = 100$ and $A = 0.13\sigma$ for $N = 1000$ (derived using $\sigma_A = \sigma\sqrt{2/N}$; see eq. 10.39). We will also see, in the case of periodic models, that such a simple harmonic fit performs even better than what we might expect a priori (i.e., even in cases of much more complex underlying variations).

This improvement in ability to detect a signal using a model is not limited to periodic variability—this is a general feature of model fitting (sometimes called "matched filter" extraction). Within the Bayesian framework, we cannot even begin our analysis without specifying an alternative model to the constant signal model. If underlying variability is not periodic, it can be roughly divided into two other families: stochastic variability, where variability is always there but the changes are not predictable for an indefinite period (e.g., quasar variability), and temporally localized events such as bursts (e.g., flares from stars, supernova explosions, gamma-ray bursts, or gravitational wave events). The various tools and methods to perform such time series analysis are discussed in the next section.

10.2. Modeling Toolkit for Time Series Analysis

The main tools for time series analysis belong to either the time domain or the frequency domain. Many of the tools and methods discussed in earlier chapters play a prominent role in the analysis of time series data. In this section, we first revisit methods introduced earlier (mostly applicable to the time-domain analysis) and discuss parameter estimation, model selection, and classification in the context of time series analysis. We then extend this toolkit by introducing tools for analysis in the frequency domain, such as Fourier analysis, discrete Fourier transform, wavelet analysis, and digital filtering. Nondeterministic (stochastic) time series are briefly discussed in §10.5.

10.2.1. Parameter Estimation, Model Selection, and Classification for Time Series Data

Detection of a signal, whatever it may be, is essentially a hypothesis testing or model selection problem. The quantitative description of a signal belongs to parameter estimation and regression problems. Once such a description is available for a set of time series data (e.g., astronomical sources from families with distinctive light curves), their classification utilizes essentially the same methods as discussed in the preceding chapter.

In general, we will fit a model to a set of N data points (t_j, y_j), $j = 1, \ldots, N$ with known errors for y,

$$y_j(t_j) = \sum_{m=1}^{M} \beta_m \, T_m(t_j | \boldsymbol{\theta}_m) + \epsilon_j, \tag{10.1}$$

where the functions $T_m(t | \boldsymbol{\theta}_m)$ need not be periodic, nor do the times t_j need to be evenly sampled. As before, the vector $\boldsymbol{\theta}_m$ contains model parameters that describe each $T_m(t)$ (here we use the symbol $|$ to mean "given parameters $\boldsymbol{\theta}_m$," and not in the sense of a conditional pdf). Common deterministic models for the underlying process that generates data include $T(t) = \sin(\omega t)$ and $T(t) = \exp(-\alpha t)$, where the frequency ω and decay rate α are model parameters to be estimated from data. Another important model is the so-called "chirp signal," $T(t) = \sin(\phi + \omega t + \alpha t^2)$. In eq. 10.1, ϵ stands for noise, which is typically described by heteroscedastic Gaussian errors with zero mean and parametrized by known σ_j. Note that in this chapter, we have changed the index for data values from i to j because we will frequently encounter the imaginary unit $i = \sqrt{-1}$.

Finding whether data favor such a model over the simplest possibility of no variability ($y(t)$=constant plus noise) is no different from model selection problems discussed earlier, and can be addressed via the Bayesian model odds ratio, or approximately using AIC and BIC criteria (see §5.4). Given a quantitative description of time series $y(t)$, the best-fit estimates of model parameters $\boldsymbol{\theta}_m$ can then be used as attributes for various supervised and unsupervised classification methods (possibly with additional attributes that are not extracted from the analyzed time series).

Depending on the amount of data, the noise behavior (and our understanding of it), sampling, and the complexity of a specific model, such analyses can range from nearly trivial to quite complex and computationally intensive. Despite this diversity, there are only a few new concepts needed for the analysis that were not introduced in earlier chapters.

10.2.2. Fourier Analysis

Fourier analysis plays a major role in the analysis of time series data. In Fourier analysis, general functions are represented or approximated by integrals or sums of simpler trigonometric functions. As first shown in 1822 by Fourier himself in his analysis of heat transfer, this representation often greatly simplifies analysis. Figure 10.1 illustrates how an RR Lyrae light curve can be approximated by a sum of sinusoids (details are discussed in §10.2.3). The more terms that are included in

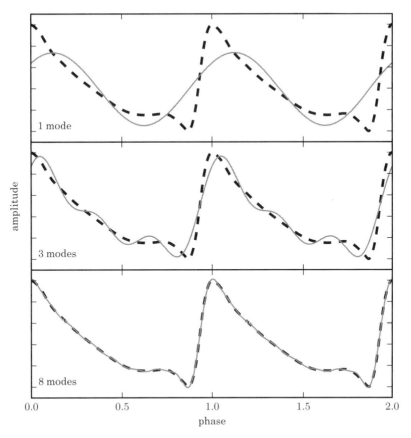

Figure 10.1. An example of a truncated Fourier representation of an RR Lyrae light curve. The thick dashed line shows the true curve; the gray lines show the approximation based on 1, 3, and 8 Fourier modes (sinusoids).

the sum, the better is the resulting approximation. For periodic functions, such as periodic light curves in astronomy, it is often true that a relatively small number of terms (less than 10) suffices to reach an approximation precision level similar to the measurement precision.

The most useful applications of Fourier analysis include convolution and deconvolution, filtering, correlation and autocorrelation, and power spectrum estimation (practical examples are interspersed throughout this chapter). The use of these methods is by no means limited to time series data; for example, they are often used to analyze spectral data or in characterizing the distributions of points. When the data are evenly sampled and the signal-to-noise ratio is high, Fourier analysis can be a powerful tool. When the noise is high compared to the signal, or the signal has a complex shape (i.e., it is not a simple harmonic function), a probabilistic treatment (e.g., Bayesian analysis) offers substantial improvements, and for irregularly (unevenly) sampled data probabilistic treatment becomes essential. For these reasons, in the analysis of astronomical time series, which are often irregularly sampled with heteroscedastic errors, Fourier analysis is often replaced by other methods (such as the periodogram analysis discussed in §10.3.1). Nevertheless, most of the main concepts introduced in Fourier analysis carry over to those other methods and thus Fourier analysis is an indispensable tool when analyzing time series.

A periodic signal such as the one in figure 10.1 can be decomposed into Fourier modes using the fast Fourier transform algorithm available in `scipy.fftpack`:

```
from scipy import fftpack\index{fftpack}
from astroML.datasets import fetch_rrlyrae_templates

templates = fetch_rrlyrae_templates()
x, y = templates['115r'].T

k = 3  # reconstruct using 3 frequencies
y_fft = fftpack.fft(y)  # compute the Fourier
        # transform
y_fft[k + 1:-k] = 0 # zero-out frequencies higher
                # than k
y_fit = fftpack.ifft(y_fft).real
# reconstruct using k modes
```

The resulting array is the reconstruction with k modes: this procedure was used to generate figure 10.1. For more information on the fast Fourier transform, see §10.2.3 and appendix E.

Numerous books about Fourier analysis are readily available. An excellent concise summary of the elementary properties of the Fourier transform is available in NumRec (see also the appendix of Greg05 for a very illustrative summary). Here, we will briefly summarize the main features of Fourier analysis and limit our discussion to the concepts used in the rest of this chapter.

The Fourier transform of function $h(t)$ is defined as

$$H(f) = \int_{-\infty}^{\infty} h(t) \exp(-i2\pi f t)\, dt, \tag{10.2}$$

with inverse transformation

$$h(t) = \int_{-\infty}^{\infty} H(f) \exp(i2\pi f t)\, df, \tag{10.3}$$

where t is time and f is frequency (for time in seconds, the unit for frequency is hertz, or Hz; the units for $H(f)$ are the product of the units for $h(t)$ and inverse hertz; note that in this chapter f is *not* a symbol for the empirical pdf as in the preceding chapters). We note that NumRec and most physics textbooks define the argument of the exponential function in the inverse transform with the minus sign; the above definitions are consistent with SciPy convention and most engineering literature. Another notational detail is that angular frequency, $\omega = 2\pi f$, is often used instead of frequency (the unit for ω is radians per second) and the extra factor of 2π due to the change of variables is absorbed into either $h(t)$ or $H(f)$, depending on convention.

For a real function $h(t)$, $H(f)$ is in general a complex function.[2] In the special case when $h(t)$ is an even function such that $h(-t) = h(t)$, $H(f)$ is real and even as well. For example, the Fourier transform of a pdf of a zero-mean Gaussian $\mathcal{N}(0, \sigma)$ in the time domain is a Gaussian $H(f) = \exp(-2\pi^2\sigma^2 f^2)$ in the frequency domain. When the time axis of an arbitrary function $h(t)$ is shifted by Δt, then the Fourier transform of $h(t + \Delta t)$ is

$$\int_{-\infty}^{\infty} h(t + \Delta t) \exp(-i2\pi f t)\, dt = H(f) \exp(i2\pi f \Delta t), \qquad (10.4)$$

where $H(f)$ is given by eq. 10.2. Therefore, the Fourier transform of a Gaussian $\mathcal{N}(\mu, \sigma)$ is

$$H_{\mathrm{Gauss}}(f) = \exp(-2\pi^2\sigma^2 f^2)\, [\cos(2\pi f \mu) + i\, \sin(2\pi f \mu)]. \qquad (10.5)$$

This result should not be confused with a Fourier transform of Gaussian noise with time-independent variance σ^2, which is simply a constant. This is known as "white noise" since there is no frequency dependence (also known as "thermal noise" or "Johnson's noise"). The cases known as "pink noise" and "red noise" are discussed in §10.5.

An important quantity in time series analysis is the one-sided power spectral density (PSD) function (or power spectrum) defined for $0 \le f < \infty$ as

$$\mathrm{PSD}(f) \equiv |H(f)|^2 + |H(-f)|^2. \qquad (10.6)$$

The PSD gives the amount of power contained in the frequency interval between f and $f + df$ (i.e., the PSD is a quantitative statement about the "importance" of each frequency mode). For example, when $h(t) = \sin(2\pi t/T)$, $P(f)$ is a δ function centered on $f = 1/T$.

The total power is the same whether computed in the frequency or the time domain:

$$P_{\mathrm{tot}} \equiv \int_0^{\infty} \mathrm{PSD}(f)\, df = \int_{-\infty}^{\infty} |h(t)|^2\, dt. \qquad (10.7)$$

This result is known as Parseval's theorem.

Convolution theorem

Another important result is the convolution theorem: A convolution of two functions $a(t)$ and $b(t)$ is given by (we already introduced it as eq. 3.44)

$$(a \star b)(t) \equiv \int_{-\infty}^{\infty} a(t')b(t - t')\, dt'. \qquad (10.8)$$

---

[2]Recall Euler's formula, $\exp(ix) = \cos x + i \sin x$.

Convolution is an unavoidable result of the measurement process because the measurement resolution, whether in time, spectral, spatial, or any other domain, is never infinite. For example, in astronomical imaging the true intensity distribution on the sky is convolved with the atmospheric seeing for ground-based imaging, or the telescope diffraction pattern for space-based imaging (radio astronomers use the term "beam convolution"). In the above equation, the function a can be thought of as the "convolving pattern" of the measuring apparatus, and the function b is the signal. In practice, we measure the convolved (or smoothed) version of our signal, $[a * b](t)$, and seek to uncover the original signal b using the presumably known a.

The convolution theorem states that if $h = a \star b$, then the Fourier transforms of h, a, and b are related by their pointwise products:

$$H(f) = A(f)B(f). \tag{10.9}$$

Thus a convolution of two functions is transformed into a simple multiplication of the associated Fourier representations. Therefore, to obtain b, we can simply take the inverse Fourier transform of the ratio $H(f)/A(f)$. In the absence of noise, this operation is exact. The convolution theorem is a very practical result; we shall consider further examples of its usefulness below.

A schematic representation of the convolution theorem is shown in figure 10.2. Note that we could have started from the convolved function shown in the bottom-left panel and uncovered the underlying signal shown in the top-left panel. When noise is present we can, however, never fully recover all the detail in the signal shape. The methods for the deconvolution of noisy data are many and we shall review a few of them in §10.2.5.

10.2.3. Discrete Fourier Transform

In practice, data are always discretely sampled. When the spacing of the time interval is constant, the discrete Fourier transform is a powerful tool. In astronomy, temporal data are rarely sampled with uniform spacing, though we note that LIGO data are a good counterexample (an example of LIGO data is shown and discussed in figure 10.6). Nevertheless, uniformly sampled data is a good place to start, because of the very fast algorithms available for this situation, and because the primary concepts also extend to unevenly sampled data.

When computing the Fourier transform for discretely and uniformly sampled data, the Fourier integrals from eqs. 10.2 and 10.3 are translated to sums. Let us assume that we have a continuous real function $h(t)$ which is sampled at N equal intervals $h_j = h(t_j)$ with $t_j \equiv t_0 + j\Delta t$, $j = 0, \ldots, (N - 1)$, where the sampling interval Δt and the duration of data taking T are related via $T = N\Delta t$ (the binning could have been done by the measuring apparatus, e.g., CCD imaging, or during the data analysis).

The discrete Fourier transform of the vector of values h_j is a complex vector of length N defined by

$$H_k = \sum_{j=0}^{N-1} h_j \exp[-i2\pi jk/N], \tag{10.10}$$

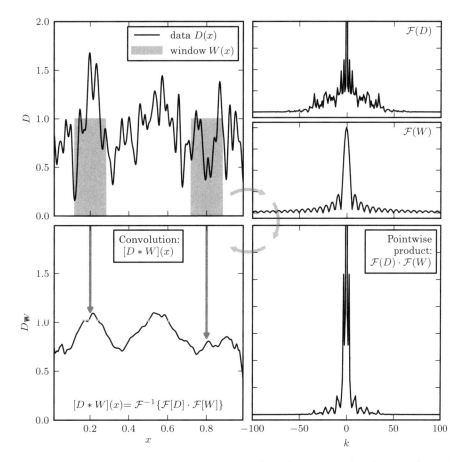

Figure 10.2. A schematic of how the convolution of two functions works. The top-left panel shows simulated data (black line); this time series is convolved with a top-hat function (gray boxes); see eq. 10.8. The top-right panels show the Fourier transform of the data and the window function. These can be multiplied together (bottom-right panel) and inverse transformed to find the convolution (bottom-left panel), which amounts to integrating the data over copies of the window at all locations. The result in the bottom-left panel can be viewed as the signal shown in the top-left panel smoothed with the window (top-hat) function.

where $k = 0, \ldots, (N - 1)$. The corresponding inverse discrete Fourier transform is defined by

$$h_j = \frac{1}{N} \sum_{k=0}^{N-1} H_k \exp[i 2\pi j k / N], \tag{10.11}$$

where $j = 0, \ldots, (N - 1)$. Unlike the continuous transforms, here the units for H_k are the same as the units for h_j. Given H_k, we can represent the function described by h_j as a sum of sinusoids, as was done in figure 10.1.

The Nyquist sampling theorem

What is the relationship between the transforms defined by eqs. 10.2 and 10.3, where integration limits extend to infinity, and the discrete transforms given by eqs. 10.10 and 10.11, where sums extend over sampled data? For example, can we estimate the PSD given by eq. 10.6 using a discrete Fourier transform? The answer to these questions is provided by the Nyquist sampling theorem (also known as the Nyquist–Shannon theorem, and as the cardinal theorem of interpolation theory), an important result developed within the context of signal processing.

Let us define $h(t)$ to be *band limited* if $H(f) = 0$ for $|f| > f_c$, where f_c is the band limit, or the Nyquist critical frequency. If $h(t)$ is band limited, then there is some "resolution" limit in t space, $t_c = 1/(2f_c)$ below which $h(t)$ appears "smooth." When $h(t)$ is band limited, then according to the Nyquist sampling theorem we can *exactly reconstruct $h(t)$ from evenly sampled data when $\Delta t \leq t_c$*, as

$$h(t) = \frac{\Delta t}{t_c} \sum_{k=-\infty}^{k=\infty} h_k \frac{\sin[2\pi f_c(t - k\Delta t)]}{2\pi f_c(t - k\Delta t)}. \tag{10.12}$$

This result is known as the Whittaker–Shannon, or often just Shannon, interpolation formula (or "sinc-shifting" formula). Note that the summation goes to infinity, but also that the term multiplying h_k vanishes for large values of $|t - k\Delta t|$. For example, $h(t) = \sin(2\pi t/P)$ has a period P and is band limited with $f_c = 1/P$. If it is sampled with Δt not larger than $P/2$, it can be fully reconstructed at any t (it is important to note that this entire discussion assumes that there is no noise associated with sampled values h_j). On the other hand, when the sampled function $h(t)$ is not band limited, or when the sampling rate is not sufficient (i.e., $\Delta t > t_c$), an effect called "aliasing" prevents us from exactly reconstructing $h(t)$ (see figure 10.3). In such a case, all of the power spectral density from frequencies $|f| > f_c$ is aliased (falsely transferred) into the $-f_c < f < f_c$ range. The aliasing can be thought of as inability to resolve details in a time series at a finer detail than that set by f_c. The aliasing effect can be recognized if the Fourier transform is nonzero at $|f| = 1/(2\Delta t)$, as is shown in the lower panels of figure 10.3.

Therefore, the discrete Fourier transform is a good estimate of the true Fourier transform for properly sampled band limited functions. Eqs. 10.10 and 10.11 can be related to eqs. 10.2 and 10.3 by approximating $h(t)$ as constant outside the sampled range of t, and assuming $H(f) = 0$ for $|f| > 1/(2\Delta t)$. In particular,

$$|H(f_k)| \approx \Delta t\, |H_k|, \tag{10.13}$$

where $f_k = k/(N\Delta t)$ for $k \leq N/2$ and $f_k = (k - N)/(N\Delta t)$ for $k \geq N/2$ (see appendix E for a more detailed discussion of this result). The discrete analog of eq. 10.6 can now be written as

$$\mathrm{PSD}(f_k) = (\Delta t)^2 \left(|H_k|^2 + |H_{N-k}|^2\right), \tag{10.14}$$

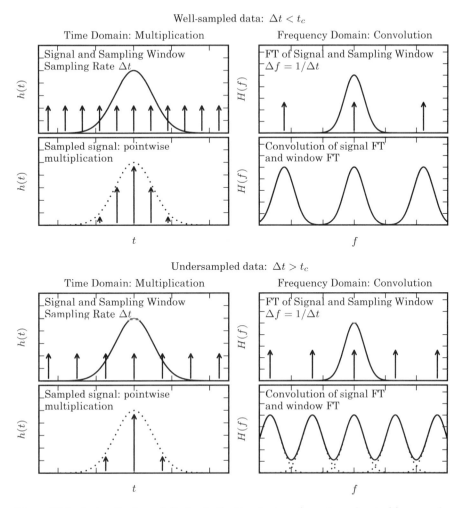

Figure 10.3. A visualization of aliasing in the Fourier transform. In each set of four panels, the top-left panel shows a signal and a regular sampling function, the top-right panel shows the Fourier transform of the signal and sampling function, the bottom-left panel shows the sampled data, and the bottom-right panel shows the convolution of the Fourier-space representations (cf. figure 10.2). In the top four panels, the data is well sampled, and there is little to no aliasing. In the bottom panels, the data is not well sampled (the spacing between two data points is larger) which leads to aliasing, as seen in the overlap of the convolved Fourier transforms (figure adapted from Greg05).

and explicitly

$$\mathrm{PSD}(f_k) = 2 \left(\frac{T}{N}\right)^2 \left[\left(\sum_{j=0}^{N-1} h_j \, \cos(2\pi f_k t_j) \right)^2 + \left(\sum_{j=0}^{N-1} h_j \, \sin(2\pi f_k t_j) \right)^2 \right].$$

(10.15)

Using these results, we can estimate the Fourier transform and PSD of any discretely and evenly sampled function. As discussed in §10.3.1, these results are

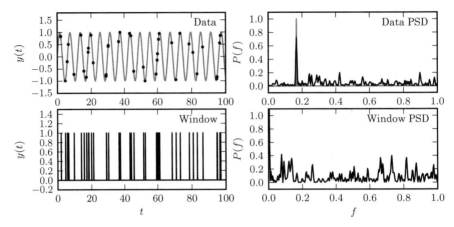

Figure 10.4. An illustration of the impact of a sampling window function of resulting PSD. The top-left panel shows a simulated data set with 40 points drawn from the function $y(t|P) = \sin t$ (i.e., $f = 1/(2\pi) \sim 0.16$). The sampling is random, and illustrated by the vertical lines in the bottom-left panel. The PSD of sampling times, or spectral window, is shown in the bottom-right panel. The PSD computed for the data set from the top-left panel is shown in the top-right panel; it is equal to a convolution of the single peak (shaded in gray) with the window PSD shown in the bottom-right panel (e.g., the peak at $f \sim 0.42$ in the top-right panel can be traced to a peak at $f \sim 0.26$ in the bottom-right panel).

strictly true only for noiseless data (although in practice they are often applied, sometimes incorrectly, to noisy data).

The window function

Figure 10.3 shows the relationship between sampling and the window function: the sampling window function in the time domain can be expressed as the sum of delta functions placed at sampled observation times. In this case the observations are regularly spaced. The Fourier transform of a set of delta functions with spacing Δt is another set of delta functions with spacing $1/\Delta t$; this result is at the core of the Nyquist sampling theorem. By the convolution theorem, pointwise multiplication of this sampling window with the data is equivalent to the convolution of their Fourier representations, as seen in the right-hand panels.

When data are nonuniformly sampled, the impact of sampling can be understood using the same framework. The sampling window is the sum of delta functions, but because the delta functions are not regularly spaced, the Fourier transform is a more complicated, and in general complex, function of f. The PSD can be computed using the discrete Fourier transform by constructing a fine grid of times and setting the window function to one at the sampled times and zero otherwise. The resulting PSD is called the spectral window function, and models how the Fourier-space signal is affected by the sampling. As discussed in detail in [19], the observed PSD is a convolution of the true underlying PSD and this spectral window function.

An example of an irregular sampling window is shown in figure 10.4: here the true Fourier transform of the sinusoidal data is a localized spike. The Fourier transform of the function viewed through the sampling window is a convolution of the true FT and the FT of the window function. This type of analysis of the spectral window function can be a convenient way to summarize the sampling properties of a given data set, and can be used to understand aliasing properties as well; see [23].

The fast Fourier transform

The Fast Fourier transform (FFT) is an algorithm for computing discrete Fourier transforms in $\mathcal{O}(N \log N)$ time, rather than $\mathcal{O}(N^2)$ using a naive implementation. The algorithmic details for the FFT can be found in NumRec. The speed of FFT makes it a widespread tool in the analysis of evenly sampled, high signal-to-noise ratio, time series data.

The FFT and various related tools are available in Python through the submodules numpy.fft and scipy.fftpack:

```
import numpy as np
from scipy import fftpack

x = np.random.normal(size=1000)  # white noise
x_fft = fftpack.fft(x)  # Fourier transform
x2 = fftpack.ifft(x_fft)  # inverse: x2=x to
    # numerical precision
```

For more detailed examples of using the FFT in practice, see appendix E or the source code of many of the figures in this chapter.

An example of such analysis is shown in figure 10.5 for a function with a single dominant frequency: a sine wave whose amplitude is modulated by a Gaussian. The figure shows the results in the presence of noise, for two different noise levels. For the high noise level, the periodic signal is hard to recognize in the time domain. Nevertheless, the dominant frequency is easily discernible in the bottom panel for both noise realizations. One curious property is that the expected value of the peak heights are the same for both noise realizations. Another curious feature of the discrete PSD given by eq. 10.14 is that its precision as an estimator of the PSD given by eq. 10.6 does not depend on the number of data values, N (i.e., the discrete PSD is an inconsistent estimator of the true PSD). For example, if N is doubled by doubling the data-taking interval T, then the resulting discrete PSD is defined at twice as many frequencies, but the value of PSD at a given frequency does not change. Alternatively, if N is doubled by doubling the sampling rate such that $\Delta t \rightarrow \Delta t/2$, then the Nyquist frequency increases by a factor of 2 to accommodate twice as many points, again without a change in PSD at a given frequency. We shall discuss PSD peaks in more detail in §10.3.1, when we generalize the concept of the PSD to unevenly sampled data.

The discrete Fourier transform can be a powerful tool even when data are not periodic. A good example is estimating power spectrum for noise that is not white. In figure 10.6 we compute the noise power spectrum for a stream of time series data from LIGO. The measurement noise is far from white: it has a minimum at frequencies of a few hundred hertz (the minimum level is related to the number

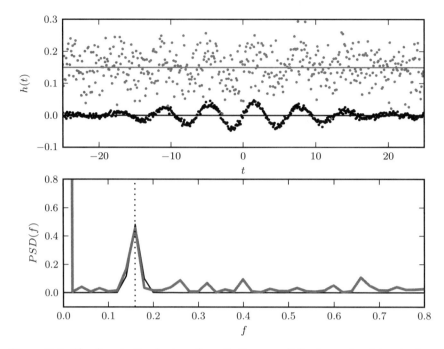

Figure 10.5. The discrete Fourier transform (bottom panel) for two noisy data sets shown in the top panel. For 512 evenly sampled times t ($\Delta t = 0.0977$), points are drawn from $h(t) = a + \sin(t) G(t)$, where $G(t)$ is a Gaussian $\mathcal{N}(\mu = 0, \sigma = 10)$. Gaussian noise with $\sigma = 0.05$ (top data set) and 0.005 (bottom data set) is added to signal $h(t)$. The value of the offset a is 0.15 and 0, respectively. The discrete Fourier transform is computed as described in §10.2.3. For both noise realizations, the correct frequency $f = (2\pi)^{-1} \approx 0.159$ is easily discernible in the bottom panel. Note that the height of peaks is the same for both noise realizations. The large value of $|H(f = 0)|$ for data with larger noise is due to the vertical offset.

of photons traveling through the interferometers), and increases rapidly at smaller frequencies due to seismic effects, and at higher frequencies due to a number of instrumental effects. The predicted signal strengths are at best a few times stronger than the noise level and thus precise noise characterization is a prerequisite for robust detection of gravitational waves.

For noisy data with many samples, more sophisticated FFT-based methods can be used to improve the signal-to-noise ratio of the resulting PSD, at the expense of frequency resolution. One well-known method is Welch's method [62], which computes multiple Fourier transforms over overlapping windows of the data to smooth noise effects in the resulting spectrum; we used this method and two window functions (top-hat and the Hanning, or cosine window) to compute PSDs shown in figure 10.6. The Hanning window suppresses noise and better picks up features at high frequencies, at the expense of affecting the shape of the continuum (note that computations are done in linear frequency space, while the figure shows a logarithmic frequency axis).

For detailed discussion of these effects and other methods to analyze gravitational wave data, see the literature provided at the LIGO website.

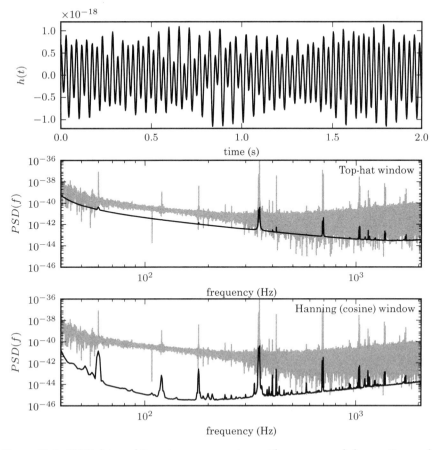

Figure 10.6. LIGO data and its noise power spectrum. The upper panel shows a 2-second-long stretch of data (∼8000 points; essentially noise without signal) from LIGO Hanford. The middle and bottom panels show the power spectral density computed for 2048 seconds of data, sampled at 4096 Hz (∼8 million data values). The gray line shows the PSD computed using a naive FFT approach; the dark line uses Welch's method of overlapping windows to smooth noise [62]; the middle panel uses a 1-second-wide top-hat window and the bottom panel the so-called Hanning (cosine) window with the same width.

10.2.4. Wavelets

The trigonometric basis functions used in the Fourier transform have an infinite extent and for this reason the Fourier transform may not be the best method to analyze nonperiodic time series data, such as the case of a localized event (e.g., a burst that decays over some timescale so that the PSD is also varying with time). Although we can evaluate the PSD for finite stretches of time series and thus hope to detect its eventual changes, this approach (called spectrogram, or dynamical power spectra analysis) suffers from degraded spectral resolution and is sensitive to the specific choice of time series segmentation length. With basis functions that are localized themselves, this downside of the Fourier transform can be avoided and the ability to identify signal, filter, or compress data, significantly improved.

An increasingly popular family of basis functions is called wavelets. A good introduction is available in NumRec.[3] By construction, wavelets are localized in *both* frequency and time domains. Individual wavelets are specified by a set of *wavelet filter coefficients*. Given a wavelet, a complete orthonormal set of basis functions can be constructed by scalings and translations. Different wavelet families trade the localization of a wavelet with its smoothness. For example, in the frequently used Daubechies wavelets [13], members of a family range from highly localized to highly smooth. Other popular wavelets include "Mexican hat" and Haar wavelets. A famous application of wavelet-based compression is the FBI's 200 TB database, containing 30 million fingerprints.

The discrete wavelet transform (DWT) can be used to analyze the power spectrum of a time series as a function of time. While similar analysis could be performed using the Fourier transform evaluated in short sliding windows, the DWT is superior. If a time series contains a localized event in time and frequency, DWT may be used to discover the event and characterize its power spectrum. A toolkit with wavelet analysis implemented in Python, PyWavelets, is publicly available.[4] A well-written guide to the use of wavelet transforms in practice can be found in [57].

Figures 10.7 and 10.8 show examples of using a particular wavelet to compute a wavelet PSD as a function of time t_0 and frequency f_0. The wavelet used is of the form

$$w(t|t_0, f_0, Q) = A \exp[i2\pi f_0(t - t_0)] \exp[-f_0^2(t - t_0)^2/Q^2], \qquad (10.16)$$

where t_0 is the central time, f_0 is the central frequency, and the dimensionless parameter Q is a model parameter which controls the width of the frequency window. Several examples of this wavelet are shown in figure 10.9. The Fourier transform of eq. 10.16 is given by

$$W(f|t_0, f_0, Q) = \left(\frac{\pi}{f_0^2/Q^2}\right)^{1/2} \exp(-i2\pi f t_0) \exp\left[\frac{-\pi^2 Q^2(f - f_0)^2}{Q f_0^2}\right]. \quad (10.17)$$

We should be clear here: the form given by eqs. 10.16–10.17 is not technically a wavelet because it does not meet the admissibility criterion (the equivalent of orthogonality in Fourier transforms). This form is closely related to a true wavelet, the *Morlet wavelet*, through a simple scaling and offset. Because of this, eqs. 10.16–10.17 should probably be referred to as "matched filters" rather than "wavelets." Orthonormality considerations aside, however, these functions display quite nicely one main property of wavelets: the localization of power in both time and frequency. For this reason, we will refer to these functions as "wavelets," and explore their ability to localize frequency signals, all the while keeping in mind the caveat about their true nature.

---

[3] A compendium of contemporary materials on wavelet analysis can be found at http://www.wavelet.org/
[4] http://www.pybytes.com/pywavelets/

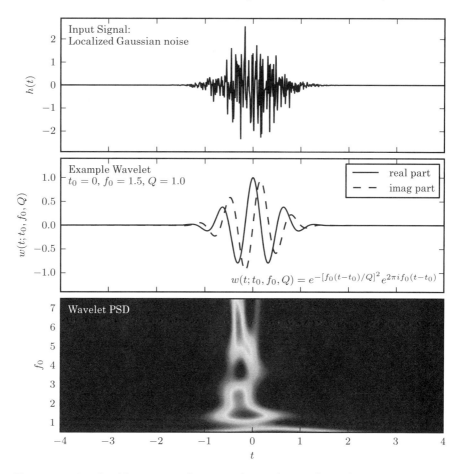

Figure 10.7. Localized frequency analysis using the wavelet transform. The upper panel shows the input signal, which consists of localized Gaussian noise. The middle panel shows an example wavelet. The lower panel shows the power spectral density as a function of the frequency f_0 and the time t_0, for $Q = 1.0$. See color plate 9.

The wavelet transform applied to data $h(t)$ is given by

$$H_w(t_0; f_0, Q) = \int_{-\infty}^{\infty} h(t)\, w(t|t_0, f_0, Q). \tag{10.18}$$

This is a convolution; by the convolution theorem (eq. 10.9), we can write the Fourier transform of H_w as the pointwise product of the Fourier transforms of $h(t)$ and $w^*(t; t_0, f_0, Q)$. The first can be approximated using the discrete Fourier transform as shown in appendix E; the second can be found using the analytic formula for $W(f)$ (eq. 10.17). This allows us to quickly evaluate H_w as a function of t_0 and f_0, using two $\mathcal{O}(N \log N)$ fast Fourier transforms.

Figures 10.7 and 10.8 show the wavelet PSD, defined by $\mathrm{PSD}_w(f_0, t_0; Q) = |H_w(t_0; f_0, Q)|^2$. Unlike the typical Fourier-transform PSD, the wavelet PSD allows

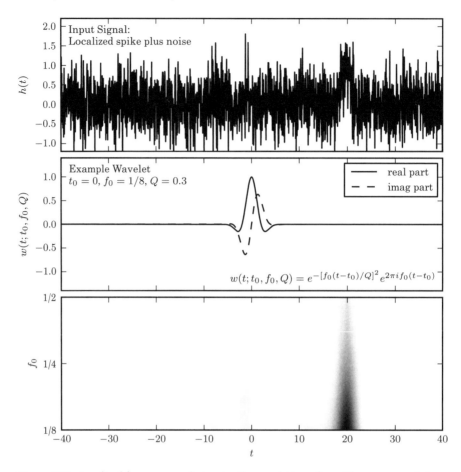

Figure 10.8. Localized frequency analysis using the wavelet transform. The upper panel shows the input signal, which consists of a Gaussian spike in the presence of white (Gaussian) noise (see figure 10.10). The middle panel shows an example wavelet. The lower panel shows the power spectral density as a function of the frequency f_0 and the time t_0, for $Q = 0.3$.

detection of frequency information which is localized in time. This is one approach used in the LIGO project to detect gravitational wave events. Because of the noise level in the LIGO measurements (see figure 10.6), rather than a standard wavelet like that seen in eq. 10.16, LIGO instead uses functions which are tuned to the expected form of the signal (i.e., matched filters). Another example of wavelet application is discussed in figure 10.28.

A related method for time-frequency analysis when PSD is not constant, called matching pursuit, utilizes a large redundant set of nonorthogonal functions; see [38]. Unlike the wavelet analysis, which assumes a fixed set of basis functions, in this method the data themselves are used to derive an appropriate large set of basis functions (called a dictionary). The matching pursuit algorithm has been successful in sound analysis, and recently in astronomy; see [34].

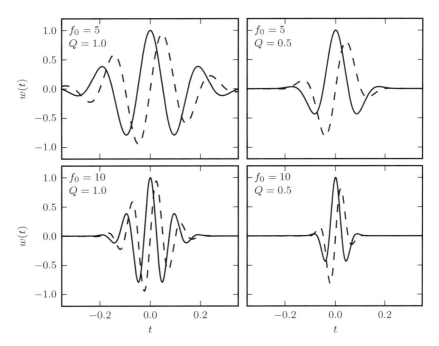

Figure 10.9. Wavelets for several values of wavelet parameters Q and f_0. Solid lines show the real part and dashed lines show the imaginary part (see eq. 10.16).

For exploring signals suspected to have time-dependent and frequency-dependent power, there are several tools available. Matplotlib implements a basic sliding-window spectrogram, using the function `matplotlib.mlab.specgram`. Alternatively, AstroML implements the wavelet PSD described above, which can be used as follows:

```
import numpy as np
from astroML.fourier import wavelet_PSD

t = np.linspace(0, 1, 1000)  # times of signal
x = np.random.normal(size=1000)   # white noise
f0 = np.linspace(0.01, 1, 100)   # candidate
      # frequencies

WPSD = wavelet_PSD(t, x, f0, Q=1)   # 100 x 1000 PSD
```

For more detailed examples, see the source code used to generate the figures in this chapter.

10.2.5. Digital Filtering

Digital filtering aims to reduce noise in time series data, or to compress data. Common examples include low-pass filtering, where high frequencies are suppressed, high-pass filtering, where low frequencies are suppressed, passband filtering, where only a finite range of frequencies is admitted, and a notch filter, where a finite range of frequencies is blocked. Fourier analysis is one of the most useful tools for performing filtering. We will use a few examples to illustrate the most common applications of filtering. Numerous other techniques can be found in signal processing literature, including approaches based on the wavelets discussed above.

We emphasize that filtering always decreases the information content of data (despite making it appear less noisy). As we have already learned throughout previous chapters, when model parameters are estimated from data, raw (unfiltered) data should be used. In some sense, this is an analogous situation to binning data to produce a histogram—while very useful for visualization, estimates of model parameters can become biased if one is not careful. This connection will be made explicit below for the Wiener filter, where we show its equivalence to kernel density estimation (§6.1.1), the generalization of histogram binning.

Low-pass filters

The power spectrum for common Gaussian noise is flat and will extend to frequencies as high as the Nyquist limit, $f_N = 1/(2\Delta t)$. If the data are band limited to a lower frequency, $f_c < f_N$, then they can be smoothed without much impact by suppressing frequencies $|f| > f_c$. Given a filter in frequency space, $\Phi(f)$, we can obtain a smoothed version of data by taking the inverse Fourier transform of

$$\hat{Y}(f) = Y(f)\,\Phi(f), \tag{10.19}$$

where $Y(f)$ is the discrete Fourier transform of data. At least in principle, we could simply set $\Phi(f)$ to zero for $|f| > f_c$, but this approach would result in ringing (i.e., unwanted oscillations) in the signal. Instead, the optimal filter for this purpose is constructed by minimizing the MISE between $\hat{Y}(f)$ and $Y(f)$ (for detailed derivation see NumRec) and is called the Wiener filter:

$$\Phi(f) = \frac{P_S(f)}{P_S(f) + P_N(f)}. \tag{10.20}$$

Here $P_S(f)$ and $P_N(f)$ represent components of a two-component (signal and noise) fit to the PSD of input data, $\mathrm{PSD}_Y(f) = P_S(f) + P_N(f)$, which holds as long as the signal and noise are uncorrelated. Given some assumed form of signal and noise, these terms can be determined from a fit to the observed PSD, as illustrated by the example shown in figure 10.10. Even when the fidelity of the PSD fit is not high, the resulting filter performs well in practice (the key features are that $\Phi(f) \sim 1$ at small frequencies and that it drops to zero at high frequencies for a band-limited signal).

There is a basic Wiener filter implementation in scipy.signal.wiener, based on assumptions of the local data mean and variance. AstroML implements a Wiener

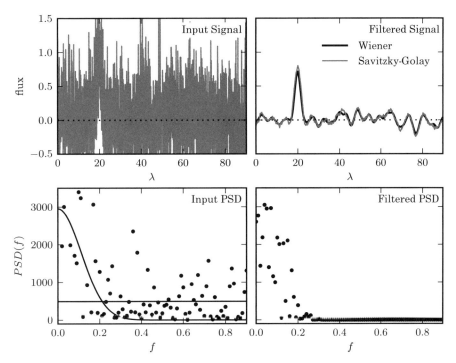

Figure 10.10. An example of data filtering using a Wiener filter. The upper-left panel shows noisy input data (200 evenly spaced points) with a narrow Gaussian peak centered at $x = 20$. The bottom panels show the input (left) and Wiener-filtered (right) power spectral density (PSD) distributions. The two curves in the bottom-left panel represent two-component fit to PSD given by eq. 10.20. The upper-right panel shows the result of the Wiener filtering on the input: the Gaussian peak is clearly seen. For comparison, we also plot the result of a fourth-order Savitzky–Golay filter with a window size of $\Delta\lambda = 10$.

filter based on the more sophisticated procedure outlined above, using user-defined priors regarding the signal and noise:

```
import numpy as np
from astroML.filters import wiener_filter

t = np.linspace(0, 1, 1000)
y = np.random.normal(size=1000)  # white noise
y_smooth = wiener_filter(t, y, signal='gaussian',
                         noise='flat')
```

For a more detailed example, see the source code of figure 10.10.

There is an interesting connection between the kernel density estimation method discussed in §6.1.1 and Wiener filtering. By the convolution theorem, the Wiener-filtered result is equivalent to the convolution of the unfiltered signal with the inverse Fourier transform of $\Phi(f)$: this is the kernel shown in figure 10.11. This convolution is equivalent to kernel density estimation. When Wiener filtering is viewed in this way, it effectively says that we believe the signal is as wide as the

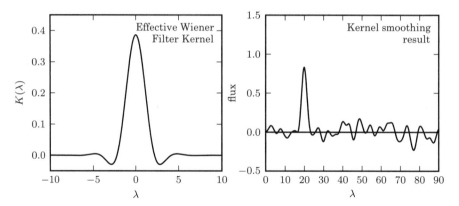

Figure 10.11. The left panel shows the inverse Fourier transform of the Wiener filter $\Phi(f)$ applied in figure 10.10. By the convolution theorem, the Wiener-filtered result is equivalent to the convolution of the unfiltered signal with the kernel shown above, and thus Wiener filtering and kernel density estimation (KDE) are directly related. The right panel shows the data smoothed by this kernel, which is equivalent to the Wiener filter smoothing in figure 10.10.

central peak shown in figure 10.11, and the statistics of the noise are such that the minor peaks in the wings work to cancel out noise in the major peak. Hence, the modeling of the PSD in the frequency domain via eq. 10.20 corresponds to choosing the optimal kernel width. Just as detailed modeling of the Wiener filter is not of paramount importance, the choice of kernel is not either.

When data are not evenly sampled, the above Fourier techniques cannot be used. There are numerous alternatives discussed in NumRec and digital signal processing literature. As a low-pass filter, a very simple but powerful method is the Savitzky–Golay filter. It fits low-order polynomials to data (in the time domain) using sliding windows (it is also known as the least-squares filter). For a detailed discussion, see NumRec. The results of a fourth-order Savitzky–Golay filter with a window function of size $\Delta\lambda = 10$ is shown beside the Wiener filter result in figure 10.10.

High-pass filters

The most common example of high-pass filtering in astronomy is baseline estimation in spectral data. Unlike the case of low-pass filtering, here there is no universal filter recipe. Baseline estimation is usually the first step toward the estimation of model parameters (e.g., location, width, and strength of spectral lines). In such cases, the best approach might be full modeling and marginalization of baseline parameters as nuisance parameters at the end of analysis.

A simple iterative technique for high-pass filtering, called *minimum component filtering*, is discussed in detail in WJ03. These are the main steps:

1. Determine baseline: exclude or mask regions where signal is clearly evident and fit a baseline model (e.g., a low-order polynomial) to the unmasked regions.
2. Get FT for the signal: after subtracting the baseline fit in the unmasked regions (i.e., a linear regression fit), apply the discrete Fourier transform.
3. Filter the signal: remove high frequencies using a low-pass filter (e.g., Wiener filter), and inverse Fourier transform the result.

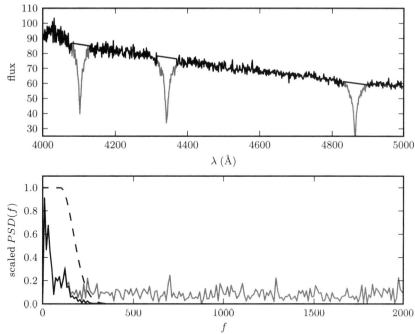

Figure 10.12. The intermediate steps of the minimum component filter procedure applied to the spectrum of a white dwarf from the SDSS data set (mjd= 52199, plate=659, fiber=381). The top panel shows the input spectrum; the masked sections of the input spectrum are shown by thin lines (i.e., step 1 of the process in §10.2.5). The bottom panel shows the PSD of the masked spectrum, after the linear fit has been subtracted (gray line). A simple low-pass filter (dashed line) is applied, and the resulting filtered spectrum (dark line) is used to construct the result shown in figure 10.13.

4. Recombine the baseline and the filtered signal: add the baseline fit subtracted in step 2 to the result from step 3. This is the minimum component filtering estimate of baseline.

A minimum component filter applied to the spectrum of a white dwarf from the Sloan Digital Sky Survey is shown in figures 10.12 and 10.13.

AstroML contains an implementation of the minimum component filter described above:

```
import numpy as np
from astroML.filters import min_component_filter

t = np.linspace(0, 1, 1000)
x = np.exp(-500 * (t - 0.5) ** 2)   # a spike at 0.5
x += np.random.random(size=1000)   # white noise
mask = (t > 0.4) & (t < 0.6)   # mask the signal

x_smooth = min_component_filter(t, x, mask)
```

For a more detailed example, see the source code for figures 10.12 and 10.13.

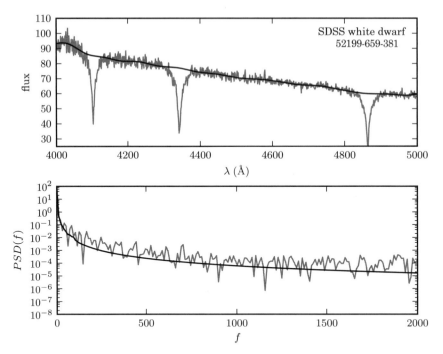

Figure 10.13. A minimum component filter applied to the spectrum of a white dwarf from SDSS data set (mjd= 52199, plate=659, fiber=381). The upper panel shows a portion of the input spectrum, along with the continuum computed via the minimum component filtering procedure described in §10.2.5 (see figure 10.12). The lower panel shows the PSD for both the input spectrum and the filtered result.

10.3. Analysis of Periodic Time Series

We shall now focus on characterization of periodic time series. Many types of variable stars show periodic flux variability; analysis of such stars is important both for understanding stellar evolution and for using such stars as distance indicators (e.g., Cepheids and RR Lyrae stars); for a good summary of the variable star zoo, see [24]. The main goal of the analysis is to detect variability and to estimate the period and its uncertainty.

A periodic time series satisfies $y(t + P) = y(t)$, where P is the period (assuming no noise). In the context of periodic variability, a convenient concept is the so-called phased light curve, where the data (and models) are plotted as function of phase,

$$\phi = \frac{t}{P} - \text{int}\left(\frac{t}{P}\right),$$ (10.21)

where the function int(x) returns the integer part of x.

We begin discussion with analysis of a simple single harmonic model, including its relationship to the discrete Fourier transform and the Lomb–Scargle periodogram. We then extend discussion to analysis of truncated Fourier series and provide an example of classification of periodic light curves. We conclude with methods for analysis of arrival time data.

10.3.1. A Single Sinusoid Model

Given time series data $(t_1, y_1), \ldots, (t_N, y_N)$, we want to test whether it is consistent with periodic variability and, if so, to estimate the period. In order to compute the posterior pdf for the frequency (or period) of a periodic variability sought in data, we need to adopt a specific model. We will first consider a simple model based on a single harmonic with angular frequency $\omega \, (= 2\pi f = 2\pi P)$,

$$y(t) = A \, \sin(\omega t + \phi) + \epsilon, \tag{10.22}$$

where the first term models the underlying process that generated the data and ϵ is measurement noise. Instead of using the phase ϕ, it is possible to shift the time axis and write the argument as $\omega \, (t - t_o)$. In the context of subsequent analysis, it is practical to use trigonometric identities to rewrite this model as

$$y(t) = a \, \sin(\omega t) + b \, \cos(\omega t), \tag{10.23}$$

where $A = (a^2 + b^2)^{1/2}$ and $\phi = \tan^{-1}(b/a)$. The model is now linear with respect to coefficients a and b, and nonlinear only with respect to frequency ω. Determination of these three parameters from the data is the main goal of the following derivation.

We fit this model to a set of data points (t_j, y_j), $j = 1, \ldots, N$ with noise ϵ described by homoscedastic Gaussian errors parametrized by σ. We will consider cases of both known and unknown σ. Note that there is no assumption that the times t_j are evenly sampled. Below, we will generalize this model to a case with heteroscedastic errors and an additional constant term in the assumed model (here, we will assume that the mean value was subtracted from "raw" data values to obtain y_j, that is, $\overline{y} = 0$; this may not work well in practice, as discussed below). We begin with this simplified case for pedagogical reasons, to better elucidate choices to be made in Bayesian analysis and its connections to classical power spectrum analysis. For the same reasons, we provide a detailed derivation.

Following the methodology from chapters 4 and 5, we can write the data likelihood as

$$L \equiv p(\{t, y\}|\omega, a, b, \sigma) = \prod_{j=1}^{N} \frac{1}{\sqrt{2\pi}\sigma} \, \exp\left(\frac{-[y_j - a \sin(\omega t_j) - b \cos(\omega t_j)]^2}{2\sigma^2} \right). \tag{10.24}$$

Although we *assumed* a Gaussian error distribution, if the only information about noise was a known value for the variance of its probability distribution, we would still end up with a Gaussian distribution via the principle of maximum entropy (see §5.2.2).

We shall retrace the essential steps of a detailed analysis developed by Bretthorst [4, 6, 7]. We shall assume uniform priors for a, b, ω, and σ. Note that this choice of priors leads to nonuniform priors on A and ϕ if we choose to parametrize the model via eq. 10.22. Nevertheless, the resulting pdfs are practically equal when data overwhelms the prior information; for a more detailed discussion see [3]. We will also assume that ω and σ must be positive. The posterior pdf is

$$p(\omega, a, b, \sigma | \{t, y\}) \propto \sigma^{-N} \exp\left(\frac{-NQ}{2\sigma^2}\right), \tag{10.25}$$

where

$$Q = V - \frac{2}{N}\left[a\, I(\omega) + b\, R(\omega) - a\, b\, M(\omega) - \frac{1}{2}a^2\, S(\omega) - \frac{1}{2}b^2\, C(\omega)\right]. \tag{10.26}$$

The following terms depend only on data and frequency ω:

$$V = \frac{1}{N}\sum_{j=1}^{N} y_j^2, \tag{10.27}$$

$$I(\omega) = \sum_{j=1}^{N} y_j \sin(\omega t_j), \quad R(\omega) = \sum_{j=1}^{N} y_j \cos(\omega t_j), \tag{10.28}$$

$$M(\omega) = \sum_{j=1}^{N} \sin(\omega t_j)\, \cos(\omega t_j), \tag{10.29}$$

and

$$S(\omega) = \sum_{j=1}^{N} \sin^2(\omega t_j), \quad C(\omega) = \sum_{j=1}^{N} \cos^2(\omega t_j). \tag{10.30}$$

The expression for Q can be considerably simplified. When $N \gg 1$ (and unless $\omega t_N \ll 1$, which is a low-frequency case corresponding to a period of oscillation longer than the data-taking interval and will be considered below) we have that $S(\omega) \approx C(\omega) \approx N/2$ and $M(\omega) \ll N/2$ (using identities $\sin^2(\omega t_j) = [1 - \cos(2\omega t_j)]/2$, $\cos^2(\omega t_j) = [1 + \cos(2\omega t_j)]/2$ and $\sin(\omega t_j)\cos(\omega t_j) = \sin(2\omega t_j)/2$), and thus

$$Q \approx V - \frac{2}{N}[a\, I(\omega) + b\, R(\omega)] + \frac{1}{2}(a^2 + b^2). \tag{10.31}$$

When quantifying the evidence for periodicity, we are not interested in specific values of a and b. To obtain the two-dimensional posterior pdf for ω and σ, we marginalize over the four-dimensional pdf given by eq. 10.25,

$$p(\omega, \sigma | \{t, y\}) \propto \int p(\omega, a, b, \sigma | \{t, y\})\, da\, db, \tag{10.32}$$

where the integration limits for a and b are sufficiently large for the integration to be effectively limited by the exponential (and not by the adopted limits for a and b, whose absolute values should be at least several times larger than σ/N). It is easy to derive (by completing the square of the arguments in the exponential)

$$p(\omega, \sigma | \{t, y\}) \propto \sigma^{-(N-2)} \exp\left(\frac{-NV}{2\sigma^2} + \frac{P(\omega)}{\sigma^2}\right), \tag{10.33}$$

where the periodogram $P(\omega)$ is given by

$$P(\omega) = \frac{1}{N}[I^2(\omega) + R^2(\omega)]. \tag{10.34}$$

In the case when the noise level σ is known, this result further simplifies to

$$p(\omega | \{t, y\}, \sigma) \propto \exp\left(\frac{P(\omega)}{\sigma^2}\right). \tag{10.35}$$

Alternatively, when σ is unknown, $p(\omega, \sigma | \{t, y\})$ can be marginalized over σ to obtain (see [3])

$$p(\omega | \{t, y\}) \propto \left[1 - \frac{2P(\omega)}{NV}\right]^{1-N/2}. \tag{10.36}$$

The best-fit amplitudes

Marginalizing over amplitudes a and b is distinctively Bayesian. We now determine MAP estimates for a and b (which are identical to maximum likelihood estimates because we assumed uniform priors) using

$$\left.\frac{d\,p(\omega, a, b, \sigma | \{t, y\})}{da}\right|_{a=a_0} = 0, \tag{10.37}$$

and analogously for b, yielding

$$a_0 = \frac{2I(\omega)}{N}, \quad b_0 = \frac{2R(\omega)}{N}. \tag{10.38}$$

By taking second derivatives of $p(\omega, a, b, \sigma | \{t, y\})$ with respect to a and b, it is easy to show that uncertainties for MAP estimates of amplitudes, a_0 and b_0, in the case of known σ are

$$\sigma_a = \sigma_b = \sigma\sqrt{2/N}. \tag{10.39}$$

Therefore, for a given value of ω, the best-fit amplitudes (a and b) from eq. 10.23 are given by eqs. 10.38 and 10.39 (in case of known σ).

The meaning of periodogram

We have not yet answered what is the best value of ω supported by the data, and whether the implied periodic variability is statistically significant. We can compute $\chi^2(\omega)$ for a fit with $a = a_0$ and $b = b_0$ as

$$\chi^2(\omega) \equiv \frac{1}{\sigma^2} \sum_{j=1}^{N} [y_j - y(t_j)]^2 = \frac{1}{\sigma^2} \sum_{j=1}^{N} [y_j - a_0 \sin(\omega t_j) - b_0 \cos(\omega t_j)]^2. \quad (10.40)$$

It can be easily shown that

$$\chi^2(\omega) = \chi_0^2 \left[1 - \frac{2}{NV} P(\omega) \right], \quad (10.41)$$

where $P(\omega)$ is the periodogram given by eq. 10.34, and χ_0^2 corresponds to a model $y(t) = $ constant (recall that here we assumed $\bar{y} = 0$),

$$\chi_0^2 = \frac{1}{\sigma^2} \sum_{j=1}^{N} y_j^2 = \frac{NV}{\sigma^2}. \quad (10.42)$$

This result motivates a renormalized definition of the periodogram as

$$P_{LS}(\omega) = \frac{2}{NV} P(\omega), \quad (10.43)$$

where index LS stands for Lomb–Scargle periodogram, introduced and discussed below. With this renormalization, $0 \leq P_{LS}(\omega) \leq 1$, and thus the reduction in χ^2 for the harmonic model, relative to χ^2 for the pure noise model, χ_0^2, is

$$\frac{\chi^2(\omega)}{\chi_0^2} = 1 - P_{LS}(\omega). \quad (10.44)$$

The relationship between $\chi^2(\omega)$ and $P(\omega)$ can be used to assess how well $P(\omega)$ estimates the true power spectrum. If the model is correct, then we expect that χ^2 corresponding to the peak with maximum height, at $\omega = \omega_0$, is N, with a standard deviation of $\sqrt{2N}$ (assuming that N is sufficiently large so that this Gaussian approximation is valid). It is easy to show that the expected height of the peak is

$$P(\omega_0) = \frac{N}{4}(a_0^2 + b_0^2), \quad (10.45)$$

with a standard deviation

$$\sigma_P(\omega_0) = \frac{\sqrt{2}}{2} \sigma^2, \quad (10.46)$$

where a_0 and b_0 are evaluated using eq. 10.38 and $\omega = \omega_0$.

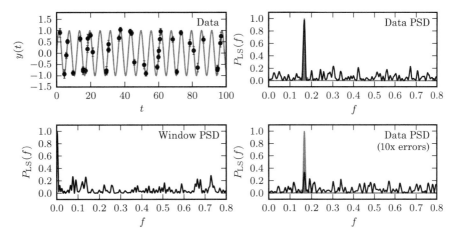

Figure 10.14. An illustration of the impact of measurement errors on P_{LS} (cf. figure 10.4). The top-left panel shows a simulated data set with 40 points drawn from the function $y(t|P) = \sin t$ (i.e., $f = 1/(2\pi) \sim 0.16$) with random sampling. Heteroscedastic Gaussian noise is added to the observations, with a width drawn from a uniform distribution with $0.1 \leq \sigma \leq 0.2$ (this error level is negligible compared to the amplitude of variation). The spectral window function (PSD of sampling times) is shown in the bottom-left panel. The PSD (P_{LS}) computed for the data set from the top-left panel is shown in the top-right panel; it is equal to a convolution of the single peak (shaded in gray) with the window PSD shown in the bottom-left panel (e.g., the peak at $f \sim 0.42$ in the top-right panel can be traced to a peak at $f \sim 0.26$ in the bottom-left panel). The bottom-right panel shows the PSD for a data set with errors increased by a factor of 10. Note that the peak $f \sim 0.16$ is now much shorter, in agreement with eq. 10.47. In addition, errors now exceed the amplitude of variation and the data PSD is no longer a simple convolution of a single peak and the spectral window.

As is evident from eq. 10.45, the expected height of the peaks in a periodogram does not depend on σ, as we already observed in figure 10.5. On the other hand, its variation from the expected height depends only on noise σ, and not on the sample size N. Alternatively, the expected height of P_{LS}, which is bound to the 0–1 range, is

$$P_{LS}(\omega_0) = 1 - \frac{\sigma^2}{V}. \qquad (10.47)$$

As noise becomes negligible, $P_{LS}(\omega_0)$ approaches its maximum value of 1. As noise increases, $P_{LS}(\omega_0)$ decreases and eventually the peak becomes too small and "buried" in the background periodogram noise. Of course, these results are only correct if the model is correct; if it is not, the PSD peaks are shorter (because χ^2 is larger; see eq. 10.44).

An illustration of the impact of measurement errors σ on P_{LS} is shown in figure 10.14. The measured PSD is a convolution of the true underlying PSD and the spectral window (the PSD of the sampling window function; recall §10.2.3). As the measurement noise increases, the peak corresponding to the underlying frequency in the data can become as small as the peaks in the spectral window; in this case, the underlying periodic variability becomes hard to detect.

Finally, we can use the results of this section to quantify the detailed behavior of frequency peaks around their maximum, and to estimate the uncertainty in ω of the highest peak. When the single harmonic model is appropriate and well constrained by data, the posterior pdf for ω given by eq. 10.35 can be approximated as a Gaussian $\mathcal{N}(\omega_0, \sigma_\omega)$. The uncertainty σ_ω can be obtained by taking the second derivative of $P(\omega)$,

$$\sigma_\omega = \left| \frac{d^2 P(\omega)}{d\omega^2} \right|_{\omega=\omega_0}^{-1/2}. \tag{10.48}$$

The Gaussian approximation implies that $P_{\rm LS}$ can be approximated by a parabola around its maximum,

$$P_{\rm LS}(\omega) \approx 1 - \frac{\sigma^2}{V} - \frac{(\omega - \omega_0)^2}{N V \sigma_\omega^2}. \tag{10.49}$$

Note that the height of the peak, $P_{\rm LS}(\omega_0)$, does *not* signify the precision with which ω_0 is estimated; instead, σ_ω is related to the peak width. It can be easily shown that the full width at half maximum of the peak, $\omega_{1/2}$, is related to σ_ω as

$$\sigma_\omega = \omega_{1/2} \left[2N(V - \sigma^2) \right]^{-1/2}. \tag{10.50}$$

For a fixed length of time series, T, $\omega_{1/2} \propto T^{-1}$, and $\omega_{1/2}$ does not depend on the number of data points N when there are on average at least a few points per cycle. Therefore, for a fixed T, $\sigma_\omega \propto N^{-1/2}$ (note that fractional errors in ω_0 and the period are equal).

We can compute σ_ω, the uncertainty of ω_0, from data using eq. 10.48 and

$$\left| \frac{d^2 P(\omega)}{d\omega^2} \right|_{\omega=\omega_0} = \frac{2}{N} \left[\left(R'(\omega_0) \right)^2 + R(\omega_0) R''(\omega_0) + \left(I'(\omega_0) \right)^2 + I(\omega_0) I''(\omega_0) \right], \tag{10.51}$$

where

$$R'(\omega) = -\sum_{j=1}^{N} y_j \, t_j \, \sin(\omega t_j), \quad R''(\omega) = -\sum_{j=1}^{N} y_j \, t_j^2 \, \cos(\omega t_j), \tag{10.52}$$

and

$$I'(\omega) = \sum_{j=1}^{N} y_j \, t_j \, \cos(\omega t_j), \quad I''(\omega) = -\sum_{j=1}^{N} y_j \, t_j^2 \, \sin(\omega t_j). \tag{10.53}$$

The significance of periodogram peaks

For a given ω, the peak height, as shown by eq. 10.44, is a measure of the reduction in χ^2 achieved by the model, compared to χ^2 for a pure noise model. We can use BIC and AIC information criteria to compare these two models (see eqs. 4.17 and 5.35).

The difference in BIC is

$$\Delta\mathrm{BIC} = \chi_0^2 - \chi^2(\omega_0) - (k_0 - k_\omega)\ln N, \qquad (10.54)$$

where the number of free parameters is $k_0 = 1$ for the no-variability model (the mean value was subtracted) and $k_\omega = 4$ for a single harmonic model (it is assumed that the uncertainty for all free parameters decreases proportionally to $N^{-1/2}$). For homoscedastic errors,

$$\Delta\mathrm{BIC} = \frac{NV}{\sigma^2}\, P_{\mathrm{LS}}(\omega_0) - 3\ln N, \qquad (10.55)$$

and similarly

$$\Delta\mathrm{AIC} = \frac{NV}{\sigma^2}\, P_{\mathrm{LS}}(\omega_0) - 6. \qquad (10.56)$$

There is an important caveat here: it was assumed that ω_0 was given (i.e., known). When we need to find ω_0 using data, we evaluate $P_{\mathrm{LS}}(\omega)$ for many ω and thus we have the case of multiple hypothesis testing (recall §4.6). We return to this point below (§10.3.2).

When errors are heteroscedastic, the term NV/σ^2 is replaced by $\chi_0^2 = \sum_j (y_j/\sigma_j)^2$. Using the approximation given by eq. 10.47, and assuming a single harmonic with amplitude A ($V = \sigma^2 + A^2/2$), the first term becomes $(A/\sigma)^2/2$. If we adopt a difference of 10 as a threshold for evidence in favor of harmonic behavior for both information criteria, the minimum A/σ ratio to detect periodicity is approximately

$$\frac{A}{\sigma} > \left(\frac{20 + 6\ln N}{N}\right)^{1/2} \qquad (10.57)$$

using BIC, and with $\ln N$ replaced by 2 for AIC. For example, with $N = 100$, periodicity can be found for $A \sim 0.7\sigma$, and when $N = 1000$ even for $A \sim 0.2\sigma$. At the same time, the fractional accuracy of estimated A is about 20–25% (i.e., the signal-to-noise ratio for measuring A is $A/\sigma_A \sim 4$–5).

Therefore, to answer the question "Did my data come from a periodic process?", we need to compute $P_{\mathrm{LS}}(\omega)$ first, and then the model odds ratio for a single sinusoid model vs. no-variability model via eq. 10.55. These results represent the foundations for analysis of unevenly periodic time series. Practical examples of this analysis are discussed in the next section.

Bayesian view of Fourier analysis

Now we can understand the results of Fourier analysis from a Bayesian viewpoint. The discrete Fourier PSD given by eq. 10.15 corresponds to the periodogram $P(\omega)$ from eq. 10.34, and *the highest peak in the discrete Fourier PSD is an optimal frequency estimator for the case of a single harmonic model and homoscedastic Gaussian noise.* As discussed in more detail in [3], the discrete PSD gives optimal results if the following conditions are met:

1. The underlying variation is a single harmonic with constant amplitude and phase.

2. The data are evenly sampled and N is large.
3. Noise is Gaussian and homoscedastic.

The performance of the discrete PSD when these conditions are not met varies from suboptimal to simply impossible to use, as in cases of unevenly sampled data. In the rest of this chapter, we will consider examples that violate all three of these conditions.

10.3.2. The Lomb–Scargle Periodogram

As we already discussed, one of the most popular tools for analysis of regularly (evenly) sampled time series is the discrete Fourier transform (§10.2.3). However, it cannot be used when data are unevenly (irregularly) sampled (as is often the case in astronomy). The Lomb–Scargle periodogram [35, 45] is a standard method to search for periodicity in unevenly sampled time series data. A normalized Lomb–Scargle periodogram,[5] with heteroscedastic errors, is defined as

$$P_{\rm LS}(\omega) = \frac{1}{V} \left[\frac{R^2(\omega)}{C(\omega)} + \frac{I^2(\omega)}{S(\omega)} \right], \tag{10.58}$$

where data-based quantities independent of ω are

$$\overline{y} = \sum_{j=1}^{N} w_j\, y_j, \tag{10.59}$$

and

$$V = \sum_{j=1}^{N} w_j\, (y_j - \overline{y})^2, \tag{10.60}$$

with weights (for homoscedastic errors $w_j = 1/N$)

$$w_j = \frac{1}{W}\frac{1}{\sigma_j^2}, \quad W = \sum_{j=1}^{N} \frac{1}{\sigma_j^2}. \tag{10.61}$$

Quantities which depend on ω are defined as

$$R(\omega) = \sum_{j=1}^{N} w_j\, (y_j - \overline{y})\, \cos[\omega(t_j - \tau)], \quad I(\omega) = \sum_{j=1}^{N} w_j\, (y_j - \overline{y})\, \sin[\omega(t_j - \tau)], \tag{10.62}$$

[5] An analogous periodogram in the case of uniformly sampled data was introduced in 1898 by Arthur Schuster with largely intuitive justification. Parts of the method attributed to Lomb and Scargle were also used previously by Gottlieb et al. [27].

$$C(\omega) = \sum_{j=1}^{N} w_j \, \cos^2[\omega(t_j - \tau)], \quad S(\omega) = \sum_{j=1}^{N} w_j \, \sin^2[\omega(t_j - \tau)]. \quad (10.63)$$

The offset τ makes $P_{LS}(\omega)$ invariant to translations of the t-axis, and is defined by

$$\tan(2\omega\tau) = \frac{\sum_{j=1}^{N} w_j \sin(2\omega t_j)}{\sum_{j=1}^{N} w_j \cos(2\omega t_j)}. \quad (10.64)$$

For the purposes of notational simplicity in the derivations below, we will also define the quantity

$$M(\omega) = \sum_{j=1}^{N} w_j \sin[\omega(t_j - \tau)] \, \cos[\omega(t_j - \tau)]. \quad (10.65)$$

If τ is instead set to zero, then eq. 10.58 becomes slightly more involved, though still based only on the sums defined above; see [63]. We note that the definition of the Lomb–Scargle periodogram in NumRec contains an additional factor of 2 before V, and does not account for heteroscedastic errors. The above normalization follows Lomb [35], and produces $0 \le P_{LS}(\omega) < 1$.

The meaning of the Lomb–Scargle periodogram

The close similarity of the Lomb–Scargle periodogram and the results obtained for a single harmonic model in the previous section is evident. The main differences are inclusion of heteroscedastic (but still Gaussian!) errors in the Lomb–Scargle periodogram and slightly different expressions for the periodograms. When terms $C(\omega)$ and $S(\omega)$ in eq. 10.58 are approximated as 1/2, eq. 10.43 follows from eq. 10.58. Without these approximations, the exact solutions for MAP estimates of a and b are (cf. approximations from eq. 10.38)

$$a(\omega) = \frac{I(\omega) \, C(\omega) - R(\omega) \, M(\omega)}{C(\omega) \, S(\omega) - M^2(\omega)} \quad (10.66)$$

and

$$b(\omega) = \frac{R(\omega) \, S(\omega) - I(\omega) \, M(\omega)}{C(\omega) \, S(\omega) - M^2(\omega)}. \quad (10.67)$$

Therefore, the Lomb–Scargle periodogram corresponds to a single sinusoid model, and it is directly related to the χ^2 of this model (via eq. 10.44) evaluated with MAP estimates for a and b; see [12, 63]. It can be thought of as an "inverted" plot of the $\chi^2(\omega)$ normalized by the "no-variation" χ_0^2.

It is often misunderstood that the Lomb–Scargle periodogram somehow saves computational effort because it purportedly avoids explicit model fitting. However, the coefficients a and b can be computed using eqs. 10.66 and 10.67 with little extra effort. Instead, the key point of using the periodogram is that the significance of each peak can be assessed, as discussed in the previous section.

Practical application of the Lomb–Scargle periodogram

The underlying model of the Lomb–Scargle periodogram is nonlinear in frequency and basis functions at different frequencies are not orthogonal. As a result, the periodogram has many local maxima and thus in practice the global maximum of the periodogram is found by grid search. The searched frequency range can be bounded by $\omega_{min} = 2\pi/T_{data}$, where $T_{data} = t_{max} - t_{min}$ is the interval sampled by the data, and by ω_{max}. As a good choice for the maximum search frequency, a pseudo-Nyquist frequency $\omega_{max} = \overline{\pi/\Delta t}$, where $\overline{1/\Delta t}$ is the median of the inverse time interval between data points, was proposed by [18] (in the case of even sampling, ω_{max} is equal to the Nyquist frequency). In practice, this choice may be a gross underestimate because unevenly sampled data can detect periodicity with frequencies even higher than $2\pi/(\Delta t)_{min}$ (see [23]). An appropriate choice of ω_{max} thus depends on sampling (the phase coverage at a given frequency is the relevant quantity) and needs to be carefully chosen: a hard limit on maximum detectable frequency is of course given by the time interval over which individual measurements are performed, such as imaging exposure time.

The frequency step can be taken as proportional to ω_{min}, $\Delta\omega = \eta\omega_{min}$, with $\eta \sim 0.1$ (see [18]). A linear regular grid for ω is a good choice because the width of peaks in $P_{LS}(\omega)$ does not depend on ω_0. Note that in practice the ratio $\omega_{max}/\omega_{min}$ can be very large (often exceeding 10^5) and thus lead to many trial frequencies (the grid step must be sufficiently small to resolve the peak; that is, $\Delta\omega$ should not be larger than σ_ω). The use of trigonometric identities can speed up computations, as implemented in the astroML code used in the following example. Another approach to speeding up the evaluation for a large number of frequencies is based on Fourier transforms, and is described in NumRec.

SciPy contains a fast Lomb–Scargle implementation, which works only for homoscedastic errors: `scipy.signal.spectral.lombscargle`. AstroML implements both the standard and generalized Lomb–Scargle periodograms, correctly accounting for heteroscedastic errors:

```
import numpy as np
from astroML.time_series import lomb_scargle

t = 100 * np.random.random(1000)
# irregular observations
dy = 1 + np.random.random(1000)
# heteroscedastic errors
y = np.sin(t) + np.random.normal(0, dy)
omega = np.linspace(0.01, 2, 1000)

P_LS = lomb_scargle(t, y, 1, omega, generalized=True)
```

For more details, see the online source code of the figures in this chapter.

Figure 10.15 shows the Lomb–Scargle periodogram for a relatively small sample with $N = 30$ and $\sigma \sim 0.8A$, where σ is the typical noise level and A is the amplitude

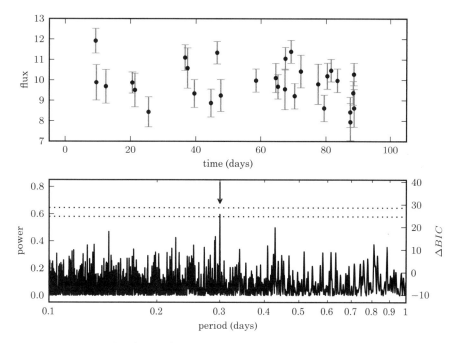

Figure 10.15. Example of a Lomb–Scargle periodogram. The data include 30 points drawn from the function $y(t|P) = 10 + \sin(2\pi t/P)$ with $P = 0.3$. Heteroscedastic Gaussian noise is added to the observations, with a width drawn from a uniform distribution with $0.5 \leq \sigma \leq 1.0$. Data are shown in the top panel and the resulting Lomb–Scargle periodogram is shown in the bottom panel. The arrow marks the location of the true period. The dotted lines show the 1% and 5% significance levels for the highest peak, determined by 1000 bootstrap resamplings (see §10.3.2). The change in BIC compared to a nonvarying source (eq. 10.55) is shown on the right y-axis. The maximum power corresponds to a $\Delta BIC = 26.1$, indicating the presence of a periodic signal. Bootstrapping indicates the period is detected at $\sim 5\%$ significance.

of a single sinusoid model. The data are sampled over $\sim$300 cycles. Due to large noise and poor sampling, the data do not reveal any obvious pattern of periodic variation. Nevertheless, the correct period is easily discernible in the periodogram, and corresponds to $\Delta BIC = 26.1$.

False alarm probability

The derivation of eq. 10.54 assumed that ω_0 was given (i.e., known). However, to find ω_0 using data, $P_{LS}(\omega)$ is evaluated for many different values of ω and thus the false alarm probability (FAP, the probability that $P_{LS}(\omega_0)$ is due to chance) will reflect the multiple hypothesis testing discussed in §4.6. Even when the noise in the data is homoscedastic and Gaussian, an analytic estimator for the FAP for general uneven sampling does not exist (a detailed discussion and references can be found in FB2012; see also [25] and [49]).

A straightforward method for computing the FAP that relies on nonparametric bootstrap resampling was recently discussed in [54]. The times of observations are kept fixed and the values of y are drawn B times from observed values with replacement. The periodogram is computed for each resample and the maximum value found. The distribution of B maxima is then used to quantify the FAP. This method was used to estimate the 1% and 5% significance levels for the highest peak shown in figure 10.15.

Generalized Lomb–Scargle periodogram

There is an important practical deficiency in the original Lomb–Scargle method described above: it is implicitly assumed that the mean of data values, $\overline{y}$, is a good estimator of the mean of $y(t)$. In practice, the data often do not sample all the phases equally, the data set may be small, or it may not extend over the whole duration of a cycle: the resulting error in mean can cause problems such as aliasing; see [12]. A simple remedy proposed in [12] is to add a constant offset term to the model from eq. 10.22. Zechmeister and Kürster [63] have derived an analytic treatment of this approach, dubbed the "generalized" Lomb–Scargle periodogram (it may be confusing that the same terminology was used by Bretthorst for a very different model [5]). The resulting expressions have a similar structure to the equations corresponding to the standard Lomb–Scargle approach listed above and are not reproduced here. Zechmeister and Kürster also discuss other methods, such as the floating-mean method and the date-compensated discrete Fourier transform, and show that they are by and large equivalent to the generalized Lomb–Scargle method.

Both the standard and generalized Lomb–Scargle methods are implemented in AstroML. Figure 10.16 compares the two in a worst-case scenario where the data sampling is such that the standard method grossly overestimates the mean. While the standard approach fails to detect the periodicity due to the unlucky data sampling, the generalized Lomb–Scargle approach still recovers the expected signal. Though this example is quite contrived, it is not entirely artificial: in practice one could easily end up in such a situation if the period of the object in question were on the order of one day, such that minima occur only during daylight hours during the period of observation.

10.3.3. Truncated Fourier Series Model

What happens if data have an underlying variability that is more complex than a single sinusoid? Is the Lomb–Scargle periodogram still an appropriate model to search for periodicity? We address these questions by considering a multiple harmonic model.

Figure 10.17 shows phased (recall eq. 10.21) light curves for six stars from the LINEAR data set, with periods estimated using the Lomb–Scargle periodogram. In most cases the phased light curves are smooth and indicate that a correct period has been found, despite significant deviation from a single sinusoid shape. A puzzling case can be seen in the top-left panel where something is clearly wrong: at $\phi \sim 0.6$ the phased light curve has two branches! We will first introduce a tool to treat such cases, and then discuss it in more detail.

The single sinusoid model can be extended to include M Fourier terms,

$$y(t) = b_0 + \sum_{m=1}^{M} a_m \sin(m\omega t) + b_m \cos(m\omega t). \tag{10.68}$$

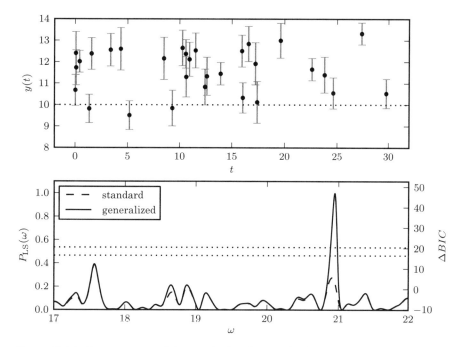

Figure 10.16. A comparison of the standard and generalized Lomb–Scargle periodograms for a signal $y(t) = 10 + \sin(2\pi t/P)$ with $P = 0.3$, corresponding to $\omega_0 \approx 21$. This example is, in some sense, a worst-case scenario for the standard Lomb–Scargle algorithm because there are no sampled points during the times when $y_{\rm true} < 10$, which leads to a gross overestimation of the mean. The bottom panel shows the Lomb–Scargle and generalized Lomb–Scargle periodograms for these data; the generalized method recovers the expected peak as the highest peak, while the standard method incorrectly chooses the peak at $\omega \approx 17.6$ (because it is higher than the true peak at $\omega_0 \approx 21$). The dotted lines show the 1% and 5% significance levels for the highest peak in the generalized periodogram, determined by 1000 bootstrap resamplings (see §10.3.2).

Following the steps from the single harmonic case, it can be easily shown that in this case the periodogram is (normalized to the 0–1 range)

$$P_M(\omega) = \frac{2}{V} \sum_{m=1}^{M} \left[R_m^2(\omega) + I_m^2(\omega) \right] , \qquad (10.69)$$

where

$$I_m(\omega) = \sum_{j=1}^{N} w_j\, y_j\, \sin(m\omega t_j) \qquad (10.70)$$

and

$$R_m(\omega) = \sum_{j=1}^{N} w_j\, y_j\, \cos(m\omega t_j), \qquad (10.71)$$

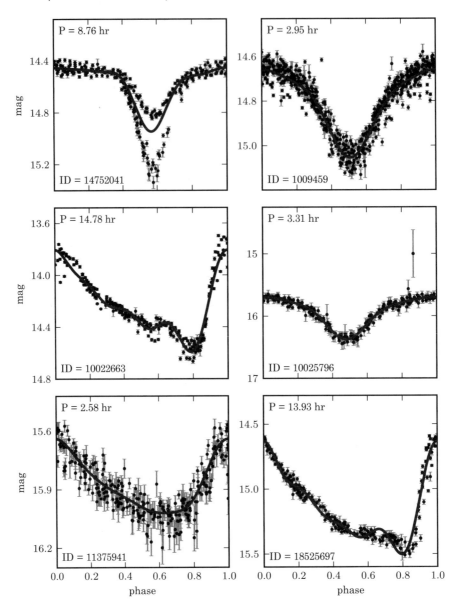

Figure 10.17. Phased light curves for six of the periodic objects from the LINEAR data set. The lines show the best fit to the phased light curve using the first four terms of the Fourier expansion (eq. 10.68), with the ω_0 selected using the Lomb–Scargle periodogram.

where weights w_j are given by eq. 10.61 and V by eq. 10.60. Trigonometric functions with argument $m\omega t_j$ can be expressed in terms of functions with argument ωt_j so fitting M harmonics is not M times the computational cost of fitting a single harmonic (for detailed discussion see [40]). If M harmonics are indeed a better fit to data than a single harmonic, the peak of $P(\omega)$ around the true frequency will be enhanced relative to the peak for $M = 1$.

In the limit of large N, the MAP values of amplitudes can be estimated from

$$a_m = \frac{2I_m(\omega)}{N}, \quad b_m = \frac{2R_m(\omega)}{N}. \tag{10.72}$$

These expressions are only approximately correct (see discussion after eq. 10.30). The errors for coefficients a_m and b_m for $m > 0$ remain $\sigma\sqrt{2/N}$ as in the case of a single harmonic. The MAP value for b_0 is simply $\bar{y}$.

It is clear from eq. 10.69 that the periodogram $P_M(\omega)$ increases with M at all frequencies ω. The reason for this increase is that more terms allow for more fidelity and thus produce a smaller χ^2. Indeed, the input data could be exactly reproduced with $M = N/2 - 1$.

AstroML includes a routine for computing the multiterm periodogram, for any choice of M terms. It has an interface similar to the `lomb_scargle` function discussed above:

```
import numpy as np
from astroML.time_series import multiterm_periodogram

t = 100 * np.random.random(1000)
# irregular observations
dy = 1 + np.random.random(1000)
# heteroscedastic errors
y = np.sin(t) + np.random.normal(0, dy)
omega = np.linspace(0.01, 2, 1000)

P_M = multiterm_periodogram(t, y, dy, omega)
```

For more details, see the online source code of the figures in this chapter.

Figure 10.18 compares the periodograms and phased light curves for the problematic case from the top-left panel in figure 10.17 using $M = 1$ and $M = 6$. The single sinusoid model ($M = 1$) is so different from the true signal shape that it results in an incorrect period equal to $1/2$ of the true period. The reason is that the underlying light curve has two minima (this star is an Algol-type eclipsing binary star) and a single sinusoid model produces a smaller χ^2 than for the pure noise model when the two minima are aligned, despite the fact that they have different depths. The $M = 6$ model is capable of modeling the two different minima, as well as flat parts of the light curve, and achieves a lower χ^2 for the correct period than for its alias favored by the $M = 1$ model. Indeed, the correct period is essentially unrecognizable in the power spectrum of the $M = 1$ model. Therefore, *when the signal shape significantly differs from a single sinusoid, the Lomb–Scargle periodogram may easily fail* (this is true both for the original and generalized implementations).

As this example shows, a good method for recognizing that there might be a problem with the best period is to require the phased light curve to be smooth. This requirement forms the basis for the so-called minimum string length (MSL) method

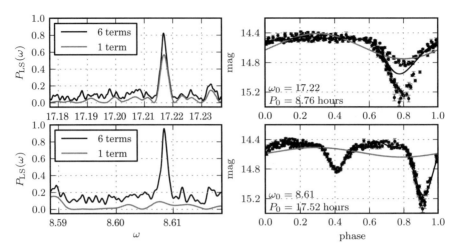

Figure 10.18. Analysis of a light curve where the standard Lomb–Scargle periodogram fails to find the correct period (the same star as in the top-left panel in figure 10.17). The two top panels show the periodograms (left) and phased light curves (right) for the truncated Fourier series model with $M = 1$ and $M = 6$ terms. Phased light curves are computed using the incorrect aliased period favored by the $M = 1$ model. The correct period is favored by the $M = 6$ model but unrecognized by the $M = 1$ model (bottom-left panel). The phased light curve constructed with the correct period is shown in the bottom-right panel. This case demonstrates that the Lomb–Scargle periodogram may easily fail when the signal shape significantly differs from a single sinusoid.

(see [21]) and the phase dispersion minimization (PDM) method (see [53]). Both methods are based on analysis of the phased light curve: the MSL measures the length of the line connecting the points, and the PDM compares the interbin variance to the sample variance. Both metrics are minimized for smooth phased light curves.

The key to avoiding such pitfalls is to use a more complex model, such as a truncated Fourier series (or a template, if known in advance, or a nonparametric model, such as discussed in the following section). How do we choose an appropriate M for a truncated Fourier series? We can extend the analysis from the previous section and compare the BIC and AIC value for the model M to those for the no-variability model $y(t) = b_0$. The difference in BIC is

$$\Delta\text{BIC}_M = \chi_0^2 \, P_M(\omega_0^M) - (2M + 1) \ln N, \qquad (10.73)$$

where $\chi_0^2 = \sum_j (y_j/\sigma_j)^2$ ($\chi_0^2 = NV/\sigma^2$ in homoscedastic case) and similarly for AIC (with $\ln N$ replaced by 2). Figure 10.19 shows the value of ΔBIC as a function of the number of frequency components, using the same two peaks as shown in figure 10.18. With many Fourier terms in the fit, the BIC strongly favors the lower-frequency peak, which agrees with our intuition based on figure 10.18.

Finally, we note that while the Lomb–Scargle periodogram is perhaps the most popular method of finding periodicity in unevenly sampled time series data, it is not the only option. For example, nonparametric Bayesian estimation based on Gaussian processes (see §8.10) has recently been proposed in [61]. The MSL and PDM methods introduced above, as well as the Bayesian blocks algorithm (§5.7.2), are good choices

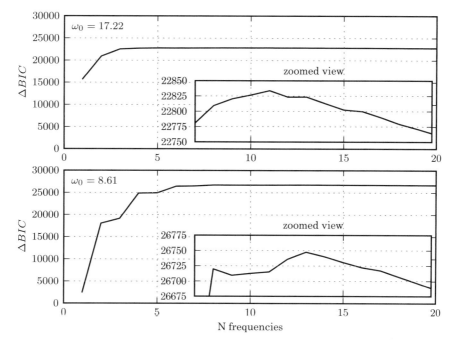

Figure 10.19. BIC as a function of the number of frequency components for the light curve shown in figure 10.18. BIC for the two prominent frequency peaks is shown. The inset panel details the area near the maximum. For both frequencies, the BIC peaks at between 10 and 15 terms; note that a high value of BIC is achieved already with 6 components. Comparing the two, the longer period model (bottom panel) is much more significant.

when the shape of the underlying light curve cannot be approximated with a small number of Fourier terms.

10.3.4. Classification of Periodic Light Curves

As illustrated in Fig 10.17, stellar light curves often have distinctive shapes (e.g., such as skewed light curves of RR Lyrae type ab stars, or eclipsing binary stars). In addition to shapes, the period and amplitude of the light curve also represent distinguishing characteristics. With large data sets, it is desirable and often unavoidable to use machine learning methods for classification (as opposed to manual/visual classification). In addition to light curves, other data such as colors are also used in classification.

As discussed in chapters 6 and 9, classification methods can be divided into supervised and unsupervised. With supervised methods we provide a training sample, with labels such as "RR Lyrae", "Algol type", "Cepheid" for each light curve, and then seek to assign these labels to another data set (essentially, we ask, "Find me more light curves such as this one in the new sample."). With unsupervised methods, we provide a set of attributes and ask if the data set displays clustering in the multidimensional space spanned by these attributes. As practical examples, below we discuss unsupervised clustering and classification of variable stars with light curves found in the LINEAR data set, augmented with photometric (color) data from the SDSS and 2MASS surveys.

The Lomb–Scargle periodogram fits a single harmonic (eq. 10.23). If the underlying time series includes higher harmonics, a more general model than a single sinusoid should be used to better describe the data and obtain a more robust period, as discussed in the preceding section. As an added benefit of the improved modeling, the amplitudes of Fourier terms can be used to efficiently classify light curves; for example, see [29, 41]. In some sense, fitting a low-M Fourier series to data represents an example of the dimensionality reduction techniques discussed in chapter 7. Of course, it is not necessary to use Fourier series and other methods have been proposed, such as direct analysis of folded light curves using PCA; see [17]. For an application of PCA to analyze light curves measured in several passbands simultaneously, see [55].

Given the best period, $P = 2\pi/\omega_0$, determined from the M-term periodogram $P_M(\omega)$ given by eq. 10.69 (with M either fixed a priori, or determined in each case using BIC/AIC criteria), a model based on the first M Fourier harmonics can be fit to the data,

$$y(t) = b_0 + \sum_{m=1}^{M} a_m \sin(m\omega_0 t) + b_m \cos(m\omega_0 t). \tag{10.74}$$

Since ω_0 is assumed known, this model is linear in terms of $(2M + 1)$ unknown coefficients a_j and b_j and thus the fitting can be performed rapidly (approximate solutions given by eq. 10.72 are typically accurate enough for classification purposes).

Given a_m and b_m, useful attributes for the classification of light curves are the amplitudes of each harmonic

$$A_m = (a_m^2 + b_m^2)^{1/2}, \tag{10.75}$$

and phases

$$\phi_m = \text{atan}(b_m, a_m), \tag{10.76}$$

with $-\pi < \phi_m \leq \pi$. It is customary to define the zero phase to correspond to the maximum, or the minimum, of a periodic light curve. This convention can be accomplished by setting ϕ_1 to the desired value (0 or $\pi/2$), and redefining phases of other harmonics as

$$\phi_m^0 = \phi_m - m\phi_1. \tag{10.77}$$

It is possible to extend this model to more than one fundamental period, for example, as done by Debosscher et al. in analysis of variable stars [18]. They subtract the best-fit model given by eq. 10.74 from data and recompute the periodogram to obtain next best period, find the best-fit model again, and then once again repeat all the steps to obtain three best periods. Their final model for a light curve is thus

$$y(t) = b_0 + \sum_{k=1}^{3} \sum_{m=1}^{M} a_{km} \left[\sin(m\omega_k t) + b_{km} \cos(m\omega_k t) \right], \tag{10.78}$$

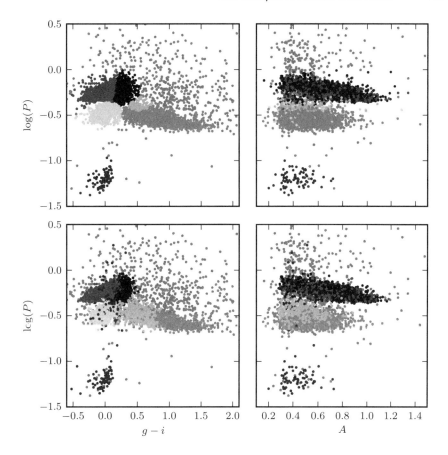

Figure 10.20. Unsupervised clustering analysis of periodic variable stars from the LINEAR data set. The top row shows clusters derived using two attributes ($g - i$ and $\log P$) and a mixture of 12 Gaussians. The colorized symbols mark the five most significant clusters. The bottom row shows analogous diagrams for clustering based on seven attributes (colors $u - g$, $g - i$, $i - K$, and $J - K$; $\log P$, light-curve amplitude, and light-curve skewness), and a mixture of 15 Gaussians. See figure 10.21 for data projections in the space of other attributes for the latter case. See color plate 10.

where $\omega_k = 2\pi / P_k$. With three fixed periods, there are $6M + 1$ free parameters to be fit. Again, finding the best-fit parameters is a relatively easy linear regression problem when period(s) are assumed known. This and similar approaches to the classification of variable stars are becoming a standard in the field [2, 44]. A multistaged treelike classification scheme, with explicit treatment of outliers, seems to be an exceptionally powerful and efficient approach, even in the case of sparse data [2, 43].

We now return to the specific example of the LINEAR data (see §1.5.9). Figures 10.20 and 10.21 show the results of a Gaussian mixture clustering analysis which attempts to find self-similar (or compact) classes of about 6000 objects without using any training sample. The main idea is that different physical classes of objects (different types of variable stars) might be clustered in the multidimensional attribute space. If we indeed identify such clusters, then we can attempt to assign them a physical meaning.

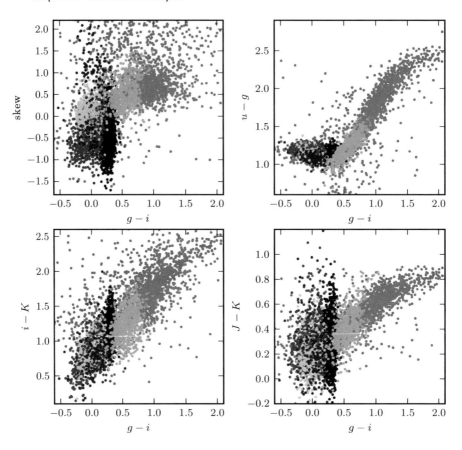

Figure 10.21. Unsupervised clustering analysis of periodic variable stars from the LINEAR data set. Clusters are derived using seven attributes (colors $u - g$, $g - i$, $i - K$, and $J - K$; $\log P$, light-curve amplitude, and light-curve skewness), and a mixture of 15 Gaussians. The $\log P$ vs. $g - i$ diagram and $\log P$ vs. light-curve amplitude diagram for the same clusters are shown in the lower panels of figure 10.20. See color plate 11.

The top panels of figure 10.20 show a 12-component Gaussian mixture fit to only two data attributes: the $g - i$ color and $\log P$, the base 10 logarithm of the best Lomb–Scargle period measured in days. The number of mixture components is determined by minimizing the BIC. The six most compact clusters are color coded; other clusters attempt to describe the background. Three clusters can be identified with ab- and c-type RR Lyrae stars. Interestingly, ab-type RR Lyrae stars are separated into two clusters. The reason is that the $g - i$ color is a single-epoch color from SDSS that corresponds to a random phase. Since ab-type RR Lyrae stars spend more time close to minimum than to maximum light, when their colors are red compared to colors at maximum light, their color distribution deviates strongly from a Gaussian. The elongated sequence populated by various types of eclipsing binary stars is also split into two clusters because its shape cannot be described by a single Gaussian either. The cluster with $\log P < -1$ is dominated by the so-called δ Scu and SX Phe variable stars; see [24]. The upper-right panel of figure 10.20 shows the clusters in a different

TABLE 10.1.
The means and standard deviations for each cluster and each attribute.

	$u - g$	$g - i$	$i - K$	$J - K$	$\log P$	Amplitude	Skew
1	1.19 ± 0.05	0.00 ± 0.12	0.88 ± 0.18	0.23 ± 0.14	-0.47 ± 0.06	0.43 ± 0.07	-0.00 ± 0.26
2	1.16 ± 0.04	0.30 ± 0.05	1.15 ± 0.17	0.29 ± 0.15	-0.24 ± 0.05	0.69 ± 0.17	-0.45 ± 0.37
3	1.20 ± 0.05	0.01 ± 0.14	0.84 ± 0.22	0.30 ± 0.17	-0.24 ± 0.05	0.71 ± 0.18	-0.44 ± 0.37
4	1.50 ± 0.33	0.78 ± 0.26	1.49 ± 0.33	0.50 ± 0.14	-0.53 ± 0.07	0.51 ± 0.13	0.63 ± 0.28
5	1.09 ± 0.09	0.09 ± 0.25	1.01 ± 0.32	0.23 ± 0.11	-0.35 ± 0.14	0.40 ± 0.06	0.86 ± 0.63
6	1.13 ± 0.08	-0.01 ± 0.13	0.88 ± 0.26	0.31 ± 0.29	-0.84 ± 0.44	0.48 ± 0.12	0.12 ± 0.84

projection: $\log P$ vs. light curve amplitude. The top four clusters are still fairly well localized in this projection due to $\log P$ carrying significant discriminative power, but there is some mixing between the background and the clusters.

The bottom panels of figure 10.20 show a 15-component Gaussian mixture fit to seven data attributes. The number of mixture components is again determined by minimizing BIC. The clustering attributes include four photometric colors based on SDSS and 2MASS measurements ($u - g, g - i, i - K, J - K$) and three parameters determined from the LINEAR light curve data ($\log P$, amplitude, and light-curve skewness). The clusters derived from all seven attributes are remarkably similar to the clusters derived from just two attributes; this shows that the additional data adds very little new information. Figure 10.21 shows the locations of these clusters in the space of other attributes. The means and standard deviations of the distribution of points assigned to each cluster for the seven-attribute clustering are shown in table 10.1.

As is evident from visual inspection of figures 10.20 and 10.21, the most discriminative attribute is the period. Clusters 2 and 3, which have very similar period distributions, are separated by the $g - i$ and $i - K$ colors, which are a measure of the star's effective temperature; see [11].

To contrast this unsupervised clustering with classification based on a training sample, light-curve types for the LINEAR data set based on visual (manual) classification by domain experts are utilized with two machine learning methods: a Gaussian mixture model Bayes classifier (GMMB; see §9.3.5) and support vector machines (SVM; see §9.6). As above, both two-attribute and seven-attribute cases are considered. Figure 10.22 shows GMMB results and figure 10.23 shows SVM results. The training sample of 2000 objects includes five input classes and the methods assign the most probable classification, among these five classes, to each object not in the training sample (about 4000). Since the input expert classification is known for all 6000 objects, it can be used to compute completeness and contamination for automated methods, as summarized in tables 10.2 and 10.3.

The performance of the GMMB and SVM methods is very similar and only a quantitative analysis of completeness and contamination listed in these tables reveals some differences. Both methods assign a large fraction of the EA class (Algol-type eclipsing binaries) to the EB/EW class (contact binaries). This is not necessarily a problem with automated methods because these two classes are hard to distinguish even by experts. While GMMB achieves a higher completeness for c-type RR Lyrae and EB/EW classes, SVM achieves a much higher completeness for the SX Phe class. Therefore, neither method is clearly better, but both achieve a reasonably satisfactory performance level relative to domain expert classification.

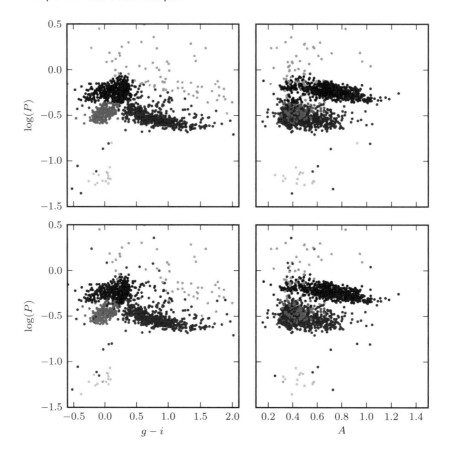

Figure 10.22. Supervised classification of periodic variable stars from the LINEAR data set using a Gaussian mixture model Bayes classifier. The training sample includes five input classes. The top row shows clusters derived using two attributes ($g - i$ and $\log P$) and the bottom row shows analogous diagrams for classification based on seven attributes (colors $u-g$, $g-i$, $i-K$, and $J-K$; $\log P$, light-curve amplitude, and light-curve skewness). See table 10.2 for the classification performance.

10.3.5. Analysis of Arrival Time Data

Discussion of periodic signals in the preceding sections assumed that the data included a set of N data points (t_j, y_j), $j = 1, \ldots, N$ with known errors for y. An example of such a data set is the optical light curve of an astronomical source where many photons are detected and the measurement error is typically dominated by photon counting statistics and background noise. Very different data sets are collected at X-ray and shorter wavelengths where individual photons are detected and background contamination is often negligible. In such cases, the data set consists of the arrival times of individual photons t_j, $j = 1, \ldots, N$, where it can be typically assumed that errors are negligible. Given such a data set, how do we search for a periodic signal, and more generally, how do we test for any type of variability?

The best known classical test for variability in arrival time data is the Rayleigh test, and it bears some similarity to the analysis of periodograms (its applicability

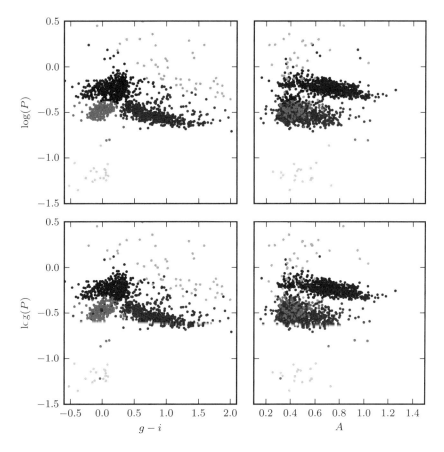

Figure 10.23. Supervised classification of periodic variable stars from the LINEAR data set using a support vector machines method. The training sample includes five input classes. The top row shows clusters derived using two attributes ($g - i$ and $\log P$) and the bottom row shows analogous diagrams for classification based on seven attributes (colors $u - g$, $g - i$, $i - K$, and $J - K$; $\log P$, light-curve amplitude, and light-curve skewness). See table 10.3 for the classification performance.

goes far beyond this context). Given a trial period, the phase ϕ_j corresponding to each datum is evaluated using eq. 10.21 and the following statistic is formed:

$$R^2 = \left(\sum_{j=1}^{N} \cos(2\pi\phi_j) \right)^2 + \left(\sum_{j=1}^{N} \sin(2\pi\phi_j) \right)^2. \qquad (10.79)$$

This expression can be understood in terms of a random walk, where each angle ϕ_j defines a unit vector, and R is the length of the resulting vector. For random data R^2 is small, and for periodic data R^2 is large when the correct period is chosen. Similarly to the analysis of the Lomb–Scargle periodogram, R^2 is evaluated for a grid of P, and the best period is chosen as the value that maximizes R^2. For $N > 10$, $2R^2/N$ is distributed as χ^2 with two degrees of freedom (this easily follows from the random walk interpretation), and this fact can be used to assess the significance of the best-fit

TABLE 10.2.

The performance of supervised classification using a Gaussian mixture model Bayes classifier. Each row corresponds to an input class listed in the first column (ab RRL: ab-type RR Lyrae; c RRL: c-type RR Lyrae; EA: Algol-type eclipsing binaries; EB/EW: contact eclipsing binaries; SX Phe: SX Phe and δ Scu candidates). The second column lists the number of objects in each input class, and the remaining columns list the percentage of sources classified into classes listed in the top row. The bottom row lists classification contamination in percent for each class listed in the top row.

class	N	ab RRL	c RRL	EA	EB/EW	SX Phe
ab RRL	1772	96.4	0.5	1.2	1.9	0.0
c RRL	583	0.2	95.3	0.2	4.3	0.0
EA	228	1.3	0.0	63.2	35.5	0.0
EB/EW	1507	0.7	0.8	2.9	95.6	0.0
SX Phe	56	0.0	0.0	0.0	26.8	73.2
contam.	—	0.9	3.6	31.4	9.7	0.0

TABLE 10.3.

The performance of supervised classification using a support vector machines method. Each row corresponds to an input class listed in the first column (ab RRL: ab-type RR Lyrae; c RRL: c-type RR Lyrae; EA: Algol-type eclipsing binaries; EB/EW: contact eclipsing binaries; SX Phe: SX Phe and δ Scu candidates). The second column lists the number of objects in each input class, and the remaining columns list the percentage of sources classified into classes listed in the top row. The bottom row lists classification contamination in percent for each class listed in the top row.

class	N	ab RRL	c RRL	EA	EB/EW	SX Phe
ab RRL	1772	95.9	0.3	1.4	2.4	0.0
c RRL	583	1.5	91.3	0.2	7.0	0.0
EA	228	5.3	1.3	67.5	25.9	0.0
EB/EW	1507	2.1	4.0	3.1	90.7	0.1
SX Phe	56	0.0	1.8	0.0	1.8	96.4
contam.	—	3.0	11.6	31.6	9.5	1.8

period (i.e., the probability that a value that large would happen by chance when the signal is stationary). A more detailed discussion of classical tests can be found in [14].

An alternative solution to this problem was derived by Gregory and Loredo [28] and here we will retrace their analysis. First, we divide the time interval $T = t_N - t_1$ into many arbitrarily small steps, Δt, so that each interval contains either 1 or 0 detections. If the event rate (e.g., the number of photons per unit time) is $r(t)$, then the expectation value for the number of events during Δt is

$$\mu(t) = r(t)\,\Delta t. \tag{10.80}$$

Following the Poisson statistics, the probability of detecting no events during Δt is

$$p(0) = e^{-r(t)\Delta t}, \tag{10.81}$$

and the probability of detecting a single event is

$$p(1) = r(t) \, \Delta t \, e^{-r(t)\Delta t}. \tag{10.82}$$

The data likelihood becomes

$$p(D|r, I) = (\Delta t)^N \, e^{-\int_{(T)} r(t)\, dt} \prod_{j=1}^{N} r(t_j). \tag{10.83}$$

The integral of $r(t)$ over time should be performed only over the intervals when the data were collected. For simplicity, hereafter we assume that the data were collected in a single stretch of time with no gaps.

With an appropriate choice of model, and priors for the model parameters, analysis of arrival time data is no different than any other model selection and parameter estimation problem. For example, figure 10.24 shows the posterior pdf for a model based on periodic $r(t)$ and arrival times for 104 photons. Input model parameters and their uncertainties are easily evaluated using MCMC and the data likelihood from eq. 10.83 (though MCMC may not be necessary in such a low-dimensional problem).

Instead of fitting a parametrized model, such as a Fourier series, Gregory and Loredo used a nonparametric description of the rate function $r(t)$. They described the shape of the phased light curve using a piecewise constant function, f_j, with M steps of the same width, and $\sum_j f_j = 1$. The rate is described as

$$r(t_j) \equiv r_j = M \, A \, f_j, \tag{10.84}$$

where A is the average rate, and bin j corresponding to t_j is determined from the phase corresponding to t_j and the trial period. In addition to the frequency ω (or period), phase offset, and the average rate A, their model includes $M - 1$ parameters f_j (not M, because of the normalization constraint). They marginalize the resulting pdf to produce an analog of the periodogram, expressions for computing the model odds ratio for signal detection, and for estimating the light curve shape. In the case when little is known about the signal shape, this method is superior to the more popular Fourier series expansion.

Bayesian blocks

Scargle has developed a nonparametric Bayesian method similar in spirit to the Gregory–Loredo method for treating arrival time data from periodic time series [47, 48]. Scargle's method works with both arrival time and binned data, and can be used to detect bursts and characterize their shapes. The algorithm produces the most probable segmentation of the observation into time intervals during which the signal has no statistically significant variations, dubbed Bayesian blocks (we previously discussed Bayesian blocks in the context of histograms; see §5.7.2). In this case, the underlying model is a piecewise constant variation, and the position and number of blocks is determined by data (blocks are not of uniform duration). In some sense, Bayesian blocks improves on the simple idea of the phase dispersion minimization method by using nonuniform adaptive bins.

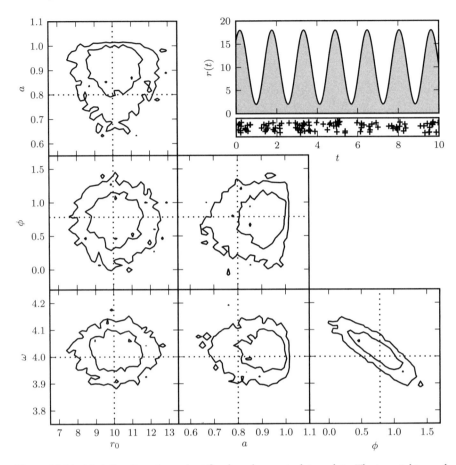

Figure 10.24. Modeling time-dependent flux based on arrival time data. The top-right panel shows the rate $r(t) = r_0[1 + a \sin(\omega t + \phi)]$, along with the locations of the 104 detected photons. The remaining panels show the model contours calculated via MCMC; dotted lines indicate the input parameters. The likelihood used is from eq. 10.83. Note the strong covariance between ϕ and ω in the bottom-right panel.

10.4. Temporally Localized Signals

A case frequently encountered in practice is a stationary signal with an event localized in time. Astronomical examples include the magnification due to gravitational microlensing, and bursts of emission (where the source brightness increases and then decreases to the original level over a finite time interval), and the signature of a gravitational wave in data from LIGO (and other gravitational wave detectors).

When the noise properties are understood, and the expected shape of the signal is known, a tool of choice is full forward modeling. That is, here too the analysis includes model selection and parameter estimation steps, and is often called a *matched filter* search. Even when the shape of the matched filter is not known, it can be treated in a nonparametrized form as was discussed in the context of arrival time data. Similarly, even when a full understanding of the noise is missing, it is possible to marginalize over unknown noise when some weak assumptions are made about its properties (recall the example from §5.8.5).

We will discuss two simple parametric models here: a burst signal and a chirp signal. In both examples we assume Gaussian known errors. The generalization to nonparametric models and more complex models can be relatively easily implemented by modifying the code developed for these two examples.

10.4.1. Searching for a Burst Signal

Consider a model where the signal is stationary, $y(t) = b_0 + \epsilon$, and at some unknown time, T, it suddenly increases, followed by a decay to the original level b_0 over some unknown time period. Let us describe such a burst by

$$y_B(t|T, A, \boldsymbol{\theta}) = A\, g_B(t - T|\boldsymbol{\theta}), \qquad (10.85)$$

where the function g_B describes the shape of the burst signal ($g_B(t < 0) = 0$). This function is specified by a vector of parameters $\boldsymbol{\theta}$ and can be analytic, tabulated in the form of a template, or treated in a nonparametric form. Typically, MCMC methods are used to estimate model parameters.

For illustration, we consider here a case with $g_B(t|\alpha) = \exp(-\alpha t)$. Figure 10.25 shows the simulated data and projections of posterior pdf for the four model parameters (b_0, T, A, and α). Other models for the burst shape can be readily analyzed using the same code with minor modifications.

Alternatively, the burst signal could be treated in the case of arrival time data, using the approach outlined in §10.3.5. Here, the rate function is not periodic, and can be obtained as $r(t) = (\Delta t)^{-1} y(t)$, where $y(t)$ is the sum of the stationary signal and the burst model (eq. 10.85).

10.4.2. Searching for a Chirp Signal

Here we consider a chirp signal, added to a stationary signal b_0,

$$y(t) = b_0 + A \sin[\omega t + \beta t^2], \qquad (10.86)$$

and analyze it using essentially the same code as for the burst signal. Figure 10.26 shows the simulated data and projections of posterior pdf for the four model parameters (b_0, A, ω, and β). Note that here the second term in the argument of the sine function above (βt^2) produces the effect of increasing frequency in the signal seen in the top-right panel. The resulting fit shows a strong inverse correlation between β and ω. This is expected because they both act to increase the frequency: starting from a given model, slightly increasing one while slightly decreasing the other leads to a very similar prediction.

Figure 10.27 illustrates a more complex ten-parameter case of chirp modeling. The chirp signal is temporally localized and it decays exponentially for $t > T$:

$$y_C(t|T, A, \phi, \omega, \beta) = A\, \sin[\phi + \omega(t - T) + \beta(t - T)^2]\, \exp[-\alpha(t - T)]. \quad (10.87)$$

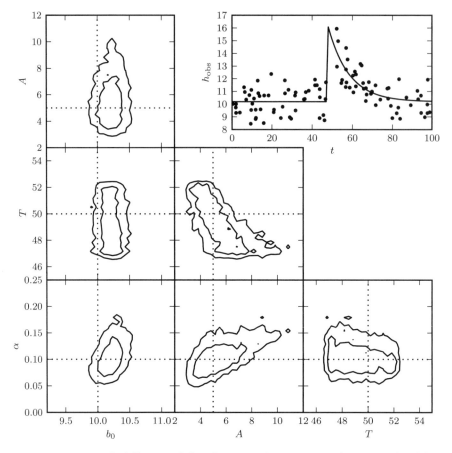

Figure 10.25. A matched filter search for a burst signal in time series data. A simulated data set generated from a model of the form $y(t) = b_0$ for $t < T$ and $y = b_0 + A\exp[-\alpha(t - T)]$ for $t > T$, with homoscedastic Gaussian errors with $\sigma = 2$, is shown in the top-right panel. The posterior pdf for the four model parameters is determined using MCMC and shown in the other panels.

The signal in the absence of chirp is taken as

$$y(t) = b_0 + b_1 \sin(\Omega_1 t)\sin(\Omega_2 t). \tag{10.88}$$

Here, we can consider parameters A, ω, β, and α as "interesting," and other parameters can be treated as "nuisance." Despite the model complexity, the MCMC-based analysis is not much harder than in the first simpler case, as illustrated in figure 10.28.

In both examples of a matched filter search for a signal, we assumed white Gaussian noise. When noise power spectrum is not flat (e.g., in the case of LIGO data; see figure 10.6), the analysis becomes more involved. For signals that are localized not only in time, but in frequency as well, the wavelet-based analysis discussed in §10.2.4 is a good choice. A simple example of such an analysis is shown in figure 10.28. The two-dimensional wavelet-based PSD easily recovers the increase of characteristic

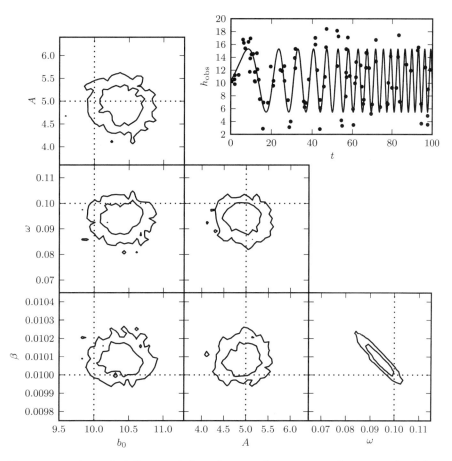

Figure 10.26. A matched filter search for a chirp signal in time series data. A simulated data set generated from a model of the form $y = b_0 + A \sin[\omega t + \beta t^2]$, with homoscedastic Gaussian errors with $\sigma = 2$, is shown in the top-right panel. The posterior pdf for the four model parameters is determined using MCMC and shown in the other panels.

chirp frequency with time. To learn more about such types of analysis, we refer the reader to the rapidly growing body of tools and publications developed in the context of gravitational wave analysis.[6]

10.5. Analysis of Stochastic Processes

Stochastic variability includes behavior that is not predictable forever as in the periodic case, but unlike temporally localized events, variability is always there. Typically, the underlying physics is so complex that we cannot deterministically predict future values (i.e., the stochasticity is inherent in the process, rather than due to measurement noise). Despite their seemingly irregular behavior, stochastic

---

[6]See, for example, http://www.ligo.caltech.edu/

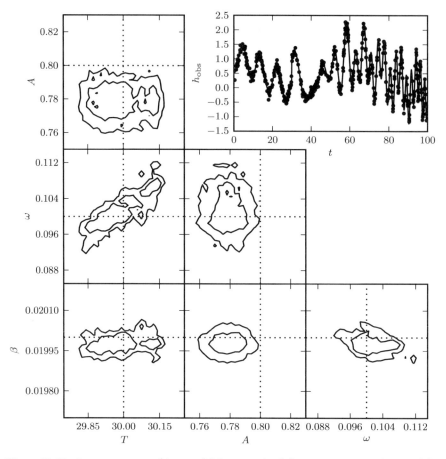

Figure 10.27. A ten-parameter chirp model (see eq. 10.87) fit to a time series. Seven of the parameters can be considered nuisance parameters, and we marginalize over them in the likelihood contours shown here.

processes can be quantified too, as briefly discussed in this section. References to more in-depth literature on stochastic processes are listed in the final section.

10.5.1. The Autocorrelation and Structure Functions

One of the main statistical tools for the analysis of stochastic variability is the autocorrelation function. It represents a specialized case of the correlation function of two functions, $f(t)$ and $g(t)$, scaled by their standard deviations, and defined at time lag Δt as

$$\mathrm{CF}(\Delta t) = \frac{\lim_{T \to \infty} \frac{1}{T} \int_{(T)} f(t)\, g(t + \Delta t)\, dt}{\sigma_f\, \sigma_g}, \tag{10.89}$$

where σ_f and σ_g are standard deviations of $f(t)$ and $g(t)$, respectively. With this normalization, the correlation function is unity for $\Delta t = 0$ (without normalization by standard deviation, the above expression is equal to the covariance function). It is

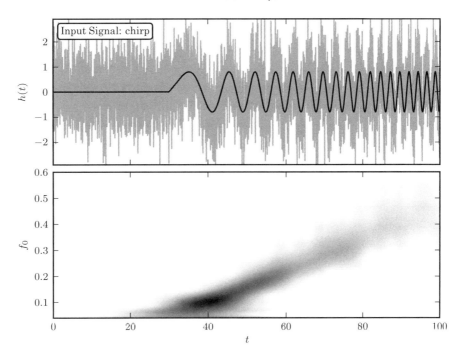

Figure 10.28. A wavelet PSD of the ten-parameter chirp signal similar to that analyzed in figure 10.27. Here, the signal with an amplitude of $A = 0.8$ is sampled in 4096 evenly spaced bins, and with Gaussian noise with $\sigma = 1$. The two-dimensional wavelet PSD easily recovers the increase of characteristic chirp frequency with time.

assumed that both f and g are statistically weakly stationary functions, which means that their mean and autocorrelation function (see below) do not depend on time (i.e., they are statistically the same irrespective of the time interval over which they are evaluated). The correlation function yields information about the time delay between two processes. If one time series is produced from another one by simply shifting the time axis by t_{lag}, their correlation function has a peak at $\Delta t = t_{\text{lag}}$.

With $f(t) = g(t) = y(t)$, the autocorrelation of $y(t)$ defined at time lag Δt is

$$\text{ACF}(\Delta t) = \frac{\lim_{T \to \infty} \frac{1}{T} \int_{(T)} y(t)\, y(t + \Delta t)\, dt}{\sigma_y^2}. \qquad (10.90)$$

The autocorrelation function yields information about the variable timescales present in a process. When y values are uncorrelated (e.g., due to white noise without any signal), $\text{ACF}(\Delta t) = 0$, except for $\text{ACF}(0) = 1$. For processes that "retain memory" of previous states only for some characteristic time τ, the autocorrelation function vanishes for $\Delta t \gg \tau$. In other words, the predictability of future behavior for such a process is limited to times up to $\sim \tau$. One such process is damped random walk, discussed in more detail in §10.5.4.

The autocorrelation function and the PSD of function $y(t)$ (see eq. 10.6) are Fourier pairs; this fact is known as the Wiener–Khinchin theorem and applies to stationary random processes. The former represents an analysis method in the time domain, and the latter in the frequency domain. For example, for a periodic process with a period P, the autocorrelation function oscillates with the same period, while for processes that retain memory of previous states for some characteristic time τ, ACF drops to zero for $t \sim \tau$.

The structure function is another quantity closely related to the autocorrelation function,

$$\mathrm{SF}(\Delta t) = \mathrm{SF}_{\infty} \left[1 - \mathrm{ACF}(\Delta t) \right]^{1/2}, \qquad (10.91)$$

where SF_{∞} is the standard deviation of the time series evaluated over an infinitely large time interval (or at least much longer than any characteristic timescale τ). The structure function, as defined by eq. 10.91, is equal to the standard deviation of the distribution of the difference of $y(t_2) - y(t_1)$ evaluated at many different t_1 and t_2 such that time lag $\Delta t = t_2 - t_1$, and divided by $\sqrt{2}$ (because of differencing). When the structure function $\mathrm{SF} \propto t^{\alpha}$, then PSD $\propto 1/f^{(1+2\alpha)}$. In the statistics literature, the structure function given by eq. 10.91 is called the second-order structure function (or variogram) and is defined without the square root (e.g., see FB2012). Although the early use in astronomy followed the statistics literature, for example, [52], we follow here the convention used in recent studies of quasar variability, for example, [58] and [16] (the appeal of taking the square root is that SF then has the same unit as the measured quantity). Note, however, that definitions in the astronomical literature are not consistent regarding the $\sqrt{2}$ factor discussed above.

Therefore, a stochastic time series can be analyzed using the autocorrelation function, the PSD, or the structure function. They can reveal the statistical properties of the underlying process, and distinguish processes such as white noise, random walk (see below) and damped random walk (discussed in §10.5.4). They are mathematically equivalent and all are used in practice; however, due to issues of noise and sampling, they may not always result in equivalent inferences about the data.

Examples of stochastic processes: $1/f$ and $1/f^2$ processes

For a given autocorrelation function or PSD, the corresponding time series can be generated using the algorithm described in [56]. Essentially, the amplitude of the Fourier transform is given by the PSD, and phases are assigned randomly; the inverse Fourier transform then generates time series.

The connection between the PSD and the appearance of time series is illustrated in figure 10.29 for two power-law PSDs: $1/f$ and $1/f^2$. The PSD normalization is such that both cases have similar power at low frequencies. For this reason, the overall amplitudes (more precisely, the variance) of the two time series are similar. The power at high frequencies is much larger for the $1/f$ case, and this is why the corresponding time series has the appearance of noisy data (the top-left panel in figure 10.29). The structure function for the $1/f$ process is constant, and proportional to $t^{1/2}$ for the $1/f^2$ process (remember that we defined structure function with a square root).

The $1/f^2$ process is also known as Brownian motion and as random walk (or "drunkard's walk"). For an exellent introduction from a physicist's perspective, see [26]. Processes whose PSD is proportional to $1/f$ are sometimes called long-term memory processes (mostly in the statistical literature), "flicker noise" and "red noise." The latter is not unique as sometimes the $1/f^2$ process is called "red noise," while the $1/f$ process is then called "pink noise." The $1/f$ processes have infinite variance; the variance of an observed time series of a finite length increases logarithmically

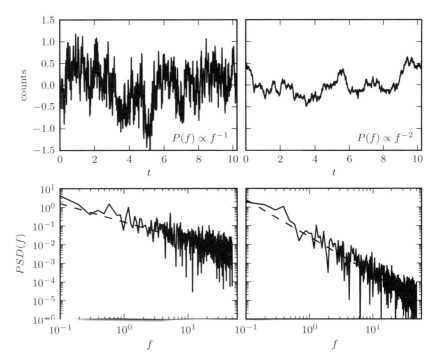

Figure 10.29. Examples of stochastic time series generated from power-law PSDs (left: $1/f$; right: $1/f^2$) using the method from [56]. The top panels show the generated data, while the bottom panels show the corresponding PSD (dashed lines: input PSD; solid lines: determined from time series shown in the top panels).

with the length (for more details, see [42]). Similarly to the behavior of the mean for Cauchy distribution (see §5.6.3), the variance of the mean for the $1/f$ process does not decrease with the sample size. Another practical problem with the $1/f$ process is that the Fourier transform of its autocovariance function does not produce a reliable estimate of the power spectrum in the distribution's tail. Difficulties with estimating properties of power-law distributions (known as Pareto distribution in the statistics literature) in general cases (i.e., not only in the context of time series analysis) are well summarized in [10].

AstroML includes a routine which generates power-law light curves based on the method of [56]. It can be used as follows:

```
import numpy as np
from astroML.time_series import generate_power_law

# beta gives the power-law index: P ~ f^-beta
y = generate_power_law(N=1024, dt=0.01, beta=2)
```

This routine is used to generate the data shown in figure 10.29.

10.5.2. Autocorrelation and Structure Function for Evenly and Unevenly Sampled Data

In the case of evenly sampled data, with $t_i = (i - 1)\Delta t$, the autocorrelation function of a discretely sampled $y(t)$ is defined as

$$\mathrm{ACF}(j) = \frac{\sum_{i=1}^{N-j} \left[(y_i - \overline{y})(y_{i+j} - \overline{y}) \right]}{\sum_{i=1}^{N} (y_i - \overline{y})^2}. \tag{10.92}$$

With this normalization the autocorrelation function is dimensionless and ACF(0)=1. The normalization by variance is sometimes skipped (see [46]), in which case a more appropriate name is the covariance function.

When a time series has a nonvanishing ACF, the uncertainty of its mean is larger than for an uncorrelated data set (cf. eq. 3.34),

$$\sigma_{\overline{x}} = \frac{\sigma}{\sqrt{N}} \left[1 + 2 \sum_{j=1}^{N} \left(1 - \frac{j}{N} \right) \mathrm{ACF}(j) \right]^{1/2}, \tag{10.93}$$

where σ is the homoscedastic measurement error. This fact is often unjustifiably neglected in analysis of astronomical data.

When data are unevenly sampled, the ACF cannot be computed using eq. 10.92. For the case of unevenly sampled data, Edelson and Krolik [22] proposed the "discrete correlation function" (DCF) in an astronomical context (called the "slot autocorrelation function" in physics). For discrete unevenly sampled data with homoscedastic errors, they defined a quantity

$$\mathrm{UDCF}_{ij} = \frac{(y_i - \overline{y})(g_j - \overline{g})}{\left[(\sigma_y^2 - e_y^2)(\sigma_g^2 - e_g^2) \right]^{1/2}}, \tag{10.94}$$

where e_y and e_g are homoscedastic measurement errors for time series y and g. The associated time lag is $\Delta t_{ij} = t_i - t_j$. The discrete correlation function at time lag Δt is then computed by binning and averaging UDCF_{ij} over M pairs of points for which $\Delta t - \delta t/2 \leq \Delta t_{ij} \leq \Delta t + \delta t/2$, where δt is the bin size. The bin size is a trade-off between accuracy of $\mathrm{DCF}(\Delta t)$ and its resolution. Edelson and Krolik showed that even uncorrelated time series will produce values of the cross-correlation $\mathrm{DCF}(\Delta t) \sim \pm 1/\sqrt{M}$.

With its binning, this method is similar to procedures for computing the structure function used in studies of quasar variability [15, 52]. The main downside of the DCF method is the assumption of homoscedastic error. Nevertheless, heteroscedastic errors can be easily incorporated by first computing the structure function, and then obtaining the ACF using eq. 10.91. The structure function is equal to the intrinsic distribution width divided by $\sqrt{2}$ for a bin of Δt_{ij} (just as when computing the DCF above). This width can be estimated for heteroscedastic data using eq. 5.69, or the corresponding exact solution given by eq. 5.64.

Scargle has developed different techniques to evaluate the discrete Fourier transform, correlation function and autocorrelation function of unevenly sampled time series (see [46]). In particular, the discrete Fourier transform for unevenly sampled data and the Wiener–Khinchin theorem are used to estimate the autocorrelation function. His method also includes a prescription for correcting the effects of uneven sampling, which results in leakage of power to nearby frequencies (the so-called sidelobe effect). Given an unevenly sampled time series, $y(t)$, the essential steps of Scargle's procedure are as follows:

1. Compute the generalized Lomb–Scargle periodogram for $y(t_i), i = 1, \ldots, N,$ namely $P_{LS}(\omega)$.
2. Compute the sampling window function using the generalized Lomb–Scargle periodogram using $z(t_i) = 1, i = 1, \ldots, N,$ namely $P_{LS}^W(\omega)$.
3. Compute inverse Fourier transforms for $P_{LS}(\omega)$ and $P_{LS}^W(\omega)$, namely $\rho(t)$ and $\rho^W(t)$, respectively.
4. The autocorrelation function at lag t is $\mathrm{ACF}(t) = \rho(t)/\rho^W(t)$.

AstroML includes tools for computing the ACF using both Scargle's method and the Edelson and Krolik method:

```
import numpy as np
from astroML.time_series import generate_damped_RW
from astroML.time_series import ACF_scargle, ACF_EK

t = np.arange(0, 1000)
y = generate_damped_RW(t, tau=300)
dy = 0.1
y = np.random.normal(y, dy)

# Scargle's method
ACF, bins = ACF_scargle(t, y, dy)

# Edelson-Krolik method
ACF, ACF_err, bins = ACF_EK(t, y, dy)
```

For more detail, see the source code of figure 10.30

Figure 10.30 illustrates the use of Edelson and Krolik's DCF method and the Scargle method. They produce similar results; errors are easier to compute for the DCF method and this advantage is crucial when fitting models to the autocorrelation function.

Another approach to estimating the autocorrelation function is direct modeling of the correlation matrix, as discussed in the next section.

10.5.3. Autoregressive Models

Autocorrelated time series can be analyzed and characterized using stochastic "autoregressive models." Autoregressive models provide a good general description of processes that "retain memory" of previous states (but are not periodic). An example

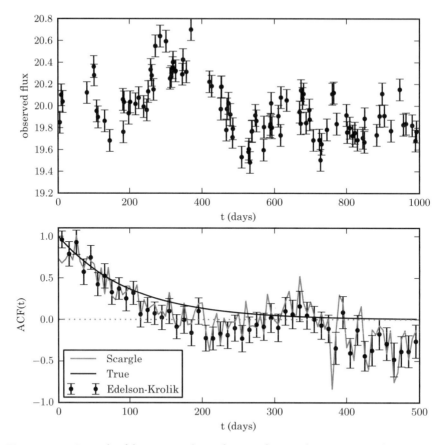

Figure 10.30. Example of the autocorrelation function for a stochastic process. The top panel shows a simulated light curve generated using a damped random walk model (§10.5.4). The bottom panel shows the corresponding autocorrelation function computed using Edelson and Krolik's DCF method and the Scargle method. The solid line shows the input autocorrelation function used to generate the light curve.

of such a model is the random walk, where each new value is obtained by adding noise to the preceding value:

$$y_i = y_{i-1} + e_i. \tag{10.95}$$

When y_{i-1} is multiplied by a constant factor greater than 1, the model is known as a geometric random walk model (used extensively to model stock market data). The noise need not be Gaussian; white noise consists of uncorrelated random variables with zero mean and constant variance, and Gaussian white noise represents the most common special case of white noise.

The random walk can be generalized to the linear autoregressive (AR) model with dependencies on k past values (i.e., not just one as in the case of random walk). An autoregressive process of order k, AR(k), for a discrete data set is defined by

$$y_i = \sum_{j=1}^{k} a_j\, y_{i-j} + e_i. \tag{10.96}$$

That is, the latest value of y is expressed as a linear combination of the k previous values of y, with the addition of noise (for random walk, $k = 1$ and $a_1 = 1$). If the data are drawn from a stationary process, coefficients a_j satisfy certain conditions. The ACF for an AR(k) process is nonzero for all lags, but it decays quickly.

The literature on autoregressive models is abundant because applications vary from signal processing and general engineering to stock-market modeling. Related modeling frameworks include the moving average (MA, where y_i depends only on past values of noise), autoregressive moving average (ARMA, a combination of AR and MA processes), autoregressive integrated moving average (ARIMA, a combination of ARMA and random walk), and state-space or dynamic linear modeling (so-called Kalman filtering). More details and references about these stochastic autoregressive models can be found in FB2012. Alternatively, modeling can be done in the frequency domain (per the Wiener–Khinchin theorem).

For example, a simple but astronomically very relevant problem is distinguishing a random walk from pure noise. That is, given a time series, the question is whether it better supports the hypothesis that $a_1 = 0$ (noise) or that $a_1 = 1$ (random walk). For comparison, in stock market analysis this pertains to predicting the next data value based on the current data value and the historic mean. If a time series is a random walk, values higher and lower than the current value have equal probabilities. However, if a time series is pure noise, there is a useful asymmetry in probabilities due to regression toward the mean (see §4.7.1). A standard method for answering this question is to compute the Dickey–Fuller statistic; see [20].

An autoregressive process defined by eq. 10.96 applies only to evenly sampled time series. A generalization is called the continuous autoregressive process, CAR(k); see [31]. The CAR(1) process has recently received a lot of attention in the context of quasar variability and is discussed in more detail in the next section.

In addition to autoregressive models, data can be modeled using the covariance matrix (e.g., using Gaussian process; see §8.10). For example, for the CAR(1) process,

$$S_{ij} = \sigma^2 \exp(-|t_{ij}|/\tau), \tag{10.97}$$

where σ and τ are model parameters; σ^2 controls the short timescale covariance ($t_{ij} \ll \tau$), which decays exponentially on a timescale given by τ. A number of other convenient models and parametrizations for the covariance matrix are discussed in the context of quasar variability in [64].

10.5.4. Damped Random Walk Model

The CAR(1) process is described by a stochastic differential equation which includes a damping term that pushes $y(t)$ back to its mean (see [31]); hence, it is also known as damped random walk (another often-used name is the Ornstein–Uhlenbeck process, especially in the context of Brownian motion; see [26]). In analogy with calling random walk "drunkard's walk," damped random walk could be called "married drunkard's walk" (who always comes home instead of drifting away).

Following eq. 10.97, the autocorrelation function for a damped random walk is

$$\text{ACF}(t) = \exp(-t/\tau), \tag{10.98}$$

where τ is the characteristic timescale (relaxation time, or damping timescale). Given the ACF, it is easy to show that the structure function is

$$\text{SF}(t) = \text{SF}_\infty \left[1 - \exp(-t/\tau)\right]^{1/2}, \tag{10.99}$$

where SF_∞ is the asymptotic value of the structure function (equal to $\sqrt{2}\sigma$, where σ is defined in eq. 10.97, when the structure function applies to differences of the analyzed process; for details see [31, 37]) and

$$\text{PSD}(f) = \frac{\tau^2 \, \text{SF}_\infty^2}{1 + (2\pi f \tau)^2}. \tag{10.100}$$

Therefore, the damped random walk is a $1/f^2$ process at high frequencies, just as ordinary random walk. The "damped nature" is seen as the flat PSD at low frequencies ($f \ll 2\pi/\tau$). An example of a light curve generated using a damped random walk is shown in figure 10.30.

For evenly sampled data, the CAR(1) process is equivalent to the AR(1) process with $a_1 = \exp(-1/\tau)$, that is, the next value of y is the damping factor times the previous value plus noise. The noise for the AR(1) process, σ_{AR} is related to SF_∞ via

$$\sigma_{\text{AR}} = \frac{\text{SF}_\infty}{\sqrt{2}} \left[1 - \exp(-2/\tau)\right]^{1/2}. \tag{10.101}$$

A damped random walk provides a good description of the optical continuum variability of quasars; see [31, 33, 37]. Indeed, this model is so successful that it has been used to distinguish quasars from stars (both are point sources in optical images, and can have similar colors) based solely on variability behavior; see [9, 36]. Nevertheless, at short timescales of the order a month or less (at high frequencies from 10^{-6} Hz up to 10^{-5} Hz), the PSD is closer to $1/f^3$ behavior than to $1/f^2$ predicted by the damped random walk model; see [39, 64].

Scikit-learn contains a utility which generates damped random walk light curves given a random seed:

```
import numpy as np
from astroML.time_series import generate_damped_RW

t = np.arange(0, 1000)
y = generate_damped_RW(t, tau=300, random_state=0)
```

For a more detailed example, see the source code associated with figure 10.30.

10.6. Which Method Should I Use for Time Series Analysis?

Despite extensive literature developed in the fields of signal processing, statistics, and econometrics, there are no universal methods that always work. This is even more so in astronomy where uneven sampling, low signal-to-noise, and heteroscedastic errors often prevent the use of standard methods drawn from other fields.

The main tools for time series analysis belong to either the time domain or the frequency domain. When searching for periodic variability, tools from the frequency domain are usually better because the signal becomes more "concentrated." This is a general feature of model fitting, where a "matched filter" approach can greatly improve the ability to detect a signal. For typical astronomical periodic time series, the generalized Lomb–Scargle method is a powerful method; when implemented to model several terms in a truncated Fourier series, instead of a single sinusoid, it works well when analyzing variable stars. It is well suited to unevenly sampled data with low signal-to-noise and heteroscedastic errors. Nevertheless, when the shape of the underlying light curve cannot be approximated with a small number of Fourier terms, nonparametric methods such as the minimum string length method, the phase dispersion minimization method, or the Bayesian blocks algorithm may perform better. Analysis of arrival time data represents different challenges; the Gregory and Loredo algorithm is a good general method in this case.

We have only briefly reviewed methods for the analysis of stochastic time series. Tools such as autocorrelation and structure functions are becoming increasingly used in the context of nonperiodic and stochastic variability. We have not discussed several topics of growing importance, such as state-space models, Kalman filters, Markov chains, and stochastic differential equations (for astronomical discussion of the latter, see [59] and [60]). For a superb book on stochastic dynamical systems written in a style accessible to astronomers, see [30]. Excellent monographs by statisticians that discuss forecasting, ARMA and ARIMA models, state-space models, Kalman filtering, and related time series topics are references [8] and [51].

References

[1] Abbott, B. P., R. Abbott, R. Adhikari, and others (2009). LIGO: the Laser Interferometer Gravitational-Wave Observatory. *Reports on Progress in Physics 72*(7), 076901.

[2] Blomme, J., L. M. Sarro, F. T. O'Donovan, and others (2011). Improved methodology for the automated classification of periodic variable stars. *MNRAS 418*, 96–106.

[3] Bretthorst, G. (1988). *Bayesian Spectrum Analysis and Parameter Estimation*. Lecture Notes in Statistics. Springer.

[4] Bretthorst, G. L. (2001a). Generalizing the Lomb-Scargle periodogram. In A. Mohammad-Djafari (Ed.), *Bayesian Inference and Maximum Entropy Methods in Science and Engineering*, Volume 568 of *American Institute of Physics Conference Series*, pp. 241–245.

[5] Bretthorst, G. L. (2001b). Generalizing the Lomb-Scargle periodogram–the nonsinusoidal case. In A. Mohammad-Djafari (Ed.), *Bayesian Inference and Maximum Entropy Methods in Science and Engineering*, Volume 568 of *American Institute of Physics Conference Series*, pp. 246–251.

[6] Bretthorst, G. L. (2001c). Nonuniform sampling: Bandwidth and aliasing. Volume 567 of *American Institute of Physics Conference Series*, pp. 1–28.

[7] Bretthorst, G. L. (2003). Frequency estimation, multiple stationary nonsinusoidal resonances with trend. In C. J. Williams (Ed.), *Bayesian Inference and Maximum Entropy Methods in Science and Engineering*, Volume 659 of *American Institute of Physics Conference Series*, pp. 3–22.

[8] Brockwell, P. and R. Davis (2006). *Time Series: Theory and Methods*. Springer.

[9] Butler, N. R. and J. S. Bloom (2011). Optimal time-series selection of quasars. *AJ 141*, 93.

[10] Clauset, A., C. R. Shalizi, and M. Newman (2009). Power-law distributions in empirical data. *SIAM Review 51*, 661–703.

[11] Covey, K. R., Ž. Ivezić, D. Schlegel, and others (2007). Stellar SEDs from 0.3 to 2.5 μm: Tracing the stellar locus and searching for color outliers in the SDSS and 2MASS. *AJ 134*, 2398–2417.

[12] Cumming, A., G. W. Marcy, and R. P. Butler (1999). The Lick Planet Search: Detectability and mass thresholds. *ApJ 526*, 890–915.

[13] Daubechies, I. (1992). *Ten Lectures on Wavelets*. SIAM: Society for Industrial and Applied Mathematics.

[14] de Jager, O. C., B. C. Raubenheimer, and J. Swanepoel (1989). A powerful test for weak periodic signals with unknown light curve shape in sparse data. *A&A 221*, 180–190.

[15] de Vries, W. H., R. H. Becker, and R. L. White (2003). Long-term variability of Sloan Digital Sky Survey quasars. *AJ 126*, 1217–1226.

[16] de Vries, W. H., R. H. Becker, R. L. White, and C. Loomis (2005). Structure function analysis of long-term quasar variability. *AJ 129*, 615–629.

[17] Deb, S. and H. P. Singh (2009). Light curve analysis of variable stars using Fourier decomposition and principal component analysis. *A&A 507*, 1729–1737.

[18] Debosscher, J., L. M. Sarro, C. Aerts, and others (2007). Automated supervised classification of variable stars. I. Methodology. *A&A 475*, 1159–1183.

[19] Deeming, T. J. (1975). Fourier analysis with unequally-spaced data. *Ap&SS 36*, 137–158.

[20] Dickey, D. A. and W. A. Fuller (1979). Distribution of the estimators for autoregressive time series with a unit root. *Journal of the American Statistical Association 74*, 427–431.

[21] Dworetsky, M. M. (1983). A period-finding method for sparse randomly spaced observations of "How long is a piece of string?". *MNRAS 203*, 917–924.

[22] Edelson, R. A. and J. H. Krolik (1988). The discrete correlation function – A new method for analyzing unevenly sampled variability data. *ApJ 333*, 646–659.

[23] Eyer, L. and P. Bartholdi (1999). Variable stars: Which Nyquist frequency? *A&AS 135*, 1–3.

[24] Eyer, L. and N. Mowlavi (2008). Variable stars across the observational HR diagram. *Journal of Physics Conference Series 118*(1), 012010.

[25] Frescura, F., C. A. Engelbrecht, and B. S. Frank (2008). Significance of periodogram peaks and a pulsation mode analysis of the Beta Cephei star V403Car. *MNRAS 388*, 1693–1707.

[26] Gillespie, D. T. (1996). The mathematics of Brownian motion and Johnson noise. *American Journal of Physics 64*, 225–240.

[27] Gottlieb, E. W., E. L. Wright, and W. Liller (1975). Optical studies of UHURU sources. XI. A probable period for Scorpius X-1 = V818 Scorpii. *ApJL 195*, L33–L35.

[28] Gregory, P. C. and T. J. Loredo (1992). A new method for the detection of a periodic signal of unknown shape and period. *ApJ 398*, 146–168.

[29] Hoffman, D. I., T. E. Harrison, and B. J. McNamara (2009). Automated variable star classification using the Northern Sky Variability Survey. *AJ 138*, 466–477.

[30] Honerkamp, J. (1994). *Stochastic Dynamical Systems*. John Wiley.

[31] Kelly, B. C., J. Bechtold, and A. Siemiginowska (2009). Are the variations in quasar optical flux driven by thermal fluctuations? *ApJ 698*, 895–910.

[32] Koch, D. G., W. J. Borucki, G. Basri, and others (2010). Kepler mission design, realized photometric performance, and early science. *ApJL 713*, L79–L86.

[33] Kozłowski, S., C. S. Kochanek, A. Udalski, and others (2010). Quantifying quasar variability as part of a general approach to classifying continuously varying sources. *ApJ 708*, 927–945.

[34] Lachowicz, P. and C. Done (2010). Quasi-periodic oscillations under wavelet microscope: the application of matching pursuit algorithm. *A&A 515*, A65.

[35] Lomb, N. R. (1976). Least-squares frequency analysis of unequally spaced data. *Ap&SS 39*, 447–462.

[36] MacLeod, C. L., K. Brooks, Ž. Ivezić, and others (2011). Quasar selection based on photometric variability. *ApJ 728*, 26.

[37] MacLeod, C. L., Ž. Ivezić, C. S. Kochanek, and others (2010). Modeling the time variability of SDSS Stripe 82 quasars as a damped random walk. *ApJ 721*, 1014–1033.

[38] Mallat, S. G. and Z. Zhang (1992). Matching pursuits with time-frequency dictionaries. *IEEE Transactions on Signal Processing 41*(12), 3397–3415.

[39] Mushotzky, R. F., R. Edelson, W. Baumgartner, and P. Gandhi (2011). Kepler observations of rapid optical variability in Active Galactic Nuclei. *ApJL 743*, L12.

[40] Palmer, D. M. (2009). A fast chi-squared technique for period search of irregularly sampled data. *ApJ 695*, 496–502.

[41] Pojmanski, G. (2002). The All Sky Automated Survey. Catalog of variable stars. I. 0h - 6h Quarter of the southern hemisphere. *Acta Astronomica 52*, 397–427.

[42] Press, W. H. (1978). Flicker noises in astronomy and elsewhere. *Comments on Astrophysics 7*, 103–119.

[43] Richards, J. W., D. L. Starr, N. R. Butler, and others (2011). On machine-learned classification of variable stars with sparse and noisy time-series data. *ApJ 733*, 10.

[44] Sarro, L. M., J. Debosscher, M. López, and C. Aerts (2009). Automated supervised classification of variable stars. II. Application to the OGLE database. *A&A 494*, 739–768.

[45] Scargle, J. D. (1982). Studies in astronomical time series analysis. II – Statistical aspects of spectral analysis of unevenly spaced data. *ApJ 263*, 835–853.

[46] Scargle, J. D. (1989). Studies in astronomical time series analysis. III – Fourier transforms, autocorrelation functions, and cross-correlation functions of unevenly spaced data. *ApJ 343*, 874–887.

[47] Scargle, J. D. (1998). Studies in astronomical time series analysis. V. Bayesian blocks, a new method to analyze structure in photon counting data. *ApJ 504*, 405.

[48] Scargle, J. D., J. P. Norris, B. Jackson, and J. Chiang (2012). Studies in astronomical time series analysis. VI. Bayesian block representations. *ArXiv:astro-ph/1207.5578*.

[49] Schwarzenberg-Czerny, A. (1998). The distribution of empirical periodograms: Lomb-Scargle and PDM spectra. *MNRAS 301*, 831–840.

[50] Sesar, B., J. S. Stuart, Ž. Ivezić, and others (2011). Exploring the variable sky with LINEAR. I. Photometric recalibration with the Sloan Digital Sky Survey. *AJ 142*, 190.

[51] Shumway, R. and D. Stoffer (2000). *Time Series Analysis and Its Applications*. Springer.

[52] Simonetti, J. H., J. M. Cordes, and D. S. Heeschen (1985). Flicker of extragalactic radio sources at two frequencies. *ApJ 296*, 46–59.

[53] Stellingwerf, R. F. (1978). Period determination using phase dispersion minimization. *ApJ 224*, 953–960.

[54] Süveges, M. (2012). False alarm probability based on bootstrap and extreme-value methods for periodogram peaks. In J.-L. Starck and C. Surace (Eds.), *ADA7 – Seventh Conference on Astronomical Data Analysis*.

[55] Süveges, M., B. Sesar, M. Váradi, and others (2012). Search for high-amplitude Delta Scuti and RR Lyrae stars in Sloan Digital Sky Survey Stripe 82 using principal component analysis. *ArXiv:astro-ph/1203.6196*.

[56] Timmer, J. and M. Koenig (1995). On generating power law noise. *A&A 300*, 707.

[57] Torrence, C. and G. P. Compo (1998). A practical guide to wavelet analysis. *Bull. Amer. Meteor. Soc. 79*(1), 61–78.

[58] Vanden Berk, D. E., B. C. Wilhite, R. G. Kron, and others (2004). The ensemble photometric variability of ∼25,000 quasars in the Sloan Digital Sky Survey. *ApJ 601*, 692–714.

[59] Vio, R., S. Cristiani, O. Lessi, and A. Provenzale (1992). Time series analysis in astronomy - An application to quasar variability studies. *ApJ 391*, 518–530.

[60] Vio, R., N. R. Kristensen, H. Madsen, and W. Wamsteker (2005). Time series analysis in astronomy: Limits and potentialities. *A&A 435*, 773–780.

[61] Wang, Y., R. Khardon, and P. Protopapas (2012). Nonparametric Bayesian estimation of periodic light curves. *ApJ 756*, 67.

[62] Welch, P. (1967). The use of fast Fourier transform for the estimation of power spectra: A method based on time averaging over short, modified periodograms. *IEEE Transactions on Audio and Electroacoustics 15*(2), 70–73.

[63] Zechmeister, M. and M. Kürster (2009). The generalised Lomb-Scargle periodogram. A new formalism for the floating-mean and Keplerian periodograms. *A&A 496*, 577–584.

[64] Zu, Y., C. S. Kochanek, S. Kozłowski, and A. Udalski (2012). Is quasar variability a damped random walk? *ArXiv:astro-ph/1202.3783*.

PART IV
Appendices

A. An Introduction to Scientific Computing with Python

"The world's a book, writ by the eternal art – Of the great author printed in man's heart, 'Tis falsely printed, though divinely penned, And all the errata will appear at the end." (Francis Quarles)

In this appendix we aim to give a brief introduction to the Python language[1] and its use in scientific computing. It is intended for users who are familiar with programming in another language such as IDL, MATLAB, C, C++, or Java. It is beyond the scope of this book to give a complete treatment of the Python language or the many tools available in the modules listed below. The number of tools available is large, and growing every day. For this reason, no single resource can be complete, and even experienced scientific Python programmers regularly reference documentation when using familiar tools or seeking new ones. For that reason, this appendix will emphasize the best ways to access documentation both on the web and within the code itself.

We will begin with a brief history of Python and efforts to enable scientific computing in the language, before summarizing its main features. Next we will discuss some of the packages which enable efficient scientific computation: NumPy, SciPy, Matplotlib, and IPython. Finally, we will provide some tips on writing efficient Python code. Some recommended resources are listed in §A.10.

A.1. A Brief History of Python

Python is an open-source, interactive, object-oriented programming language, created by Guido Van Rossum in the late 1980s. Python is loosely based on the earlier ABC, a language intended to be taught to scientists and other computer users with no formal background in computer science or software development. Python was created to emulate the strengths of ABC—its ease of use, uncluttered syntax, dynamic typing, and interactive execution —while also improving on some of ABC's weaknesses. It was designed primarily as a second language for experienced programmers—a full-featured scripting framework to "bridge the gap between the shell and C."[2] But because of its heritage in ABC, Python quickly became a popular language choice for introductory programming courses, as well as for nonprogrammers who require a computing tool. Increasingly, the same beauty and ease of use that

---

[1] http://python.org

[2] Guido Van Rossum, "The Making of Python," interview available at http://www.artima.com/intv/pythonP.html; accessed October 2012.

make Python appealing to beginning programmers have drawn more experienced converts to the language as well.

One important strength of Python is its extensible design: third-party users can easily extend Python's type system to suit their own applications. This feature is a key reason that Python has developed into a powerful tool for a large range of applications, from network control, to web design, to high-performance scientific computing. Unlike proprietary systems like MATLAB or IDL, development in the Python universe is driven by the users of the language, most of whom volunteer their time. Partly for this reason, the user base of Python has grown immensely since its creation, and the language has evolved as well. Guido Van Rossum is still actively involved in Python's development, and is affectionately known in the community as the "BDFL"—the Benevolent Dictator For Life. He regularly gives keynote talks at the Python Software Foundation's annual *PyCon* conferences, which now annually attract several thousand attendees from a wide variety of fields and backgrounds.

A.2. The SciPy Universe

Though Python provides a sound linguistic foundation, the language alone would be of little use to scientists. Scientific computing with Python today relies primarily on the *SciPy* ecosystem, an evolving set of open-source packages built on Python which implement common scientific tasks, and are maintained by a large and active community.

A.2.1. NumPy

The central package of the SciPy ecosystem is *NumPy* (pronounced "Num-Pie"), short for "Numerical Python." NumPy's core object is an efficient N-dimensional array implementation, and it includes tools for common operations such as sorting, searching, elementwise operations, subarray access, random number generation, fast Fourier transforms, and basic linear algebra. NumPy was created by Travis Oliphant in 2005, when he unified the features of two earlier Python array libraries, *Numeric* and *NumArray*. NumPy is now at the core of most scientific tools written in Python. Find more information at http://www.numpy.org.

A.2.2. SciPy

One step more specialized than NumPy is *SciPy* (pronounced "Sigh-Pie"), short for "Scientific Python." SciPy is a collection of common computing tools built upon the NumPy array framework. It contains wrappers of much of the well-tested and highly optimized code in the NetLib archive, much of which is written in Fortran (e.g., BLAS, LAPACK, FFTPACK, FITPACK, etc.). SciPy contains routines for operations as diverse as numerical integration, spline interpolation, linear algebra, statistical analysis, tree-based searching, and much more. SciPy traces its roots to Travis Oliphant's *Multipack* package, a collection of Python interfaces to scientific modules written mainly in Fortran. In 2001, Multipack was combined with scientific toolkits created by Eric Jones and Pearu Peterson, and the expanded package was renamed SciPy. NumPy was created by the SciPy developers, and was originally envisioned to be a part of the SciPy package. For ease of maintenance and development, the two projects have different release cycles, but their development communities remain closely tied. More information can be found at http://scipy.org.

A.2.3. IPython

One popular aspect of well-known computing tools such as IDL, MATLAB, and Mathematica is the ability to develop code and analyze data in an interactive fashion. This is possible to an extent with the standard Python command-line interpreter, but *IPython* (short for Interactive Python) extends these capabilities in many convenient ways. It allows tab completion of Python commands, allows quick access to command history, features convenient tools for documentation, provides time-saving "magic" commands, and much more. Conceived by Fernando Perez in 2001, and building on functionality in the earlier *IPP* and *LazyPython* packages, IPython has developed from a simple enhanced command line into a truly indispensable tool for interactive scientific computing. The recently introduced parallel processing functionality and the notebook feature are already changing the way that many scientists share and execute their Python code. As of the writing of this book in 2013, the IPython team had just received a $1.15 million grant from the Sloan Foundation to continue their development of this extremely useful research tool. Find more information at http://ipython.org

A.2.4. Matplotlib

Scientific computing would be lost without a simple and powerful system for data visualization. In Python this is provided by *Matplotlib*, a multiplatform system for plotting and data visualization, which is built on the NumPy framework. Matplotlib allows quick generation of line plots, scatter plots, histograms, flow charts, three-dimensional visualizations, and much more. It was conceived by John Hunter in 2002, and originally intended to be a patch to enable basic MATLAB-style plotting in IPython. Fernando Perez, the main developer of IPython, was then in the final stretch of his PhD, and unable to spend the time to incorporate Hunter's code and ideas. Hunter decided to set out on his own, and version 0.1 of Matplotlib was released in 2003. It received a boost in 2004 when the Space Telescope Science Institute lent institutional support to the project, and the additional resources led to a greatly expanded package. Matplotlib is now the de facto standard for scientific visualization in Python, and is cleanly integrated with IPython. Find more information at http://matplotlib.org.

A.2.5. Other Specialized Packages

There are a host of available Python packages which are built upon a subset of these four core packages and provide more specialized scientific tools. These include *Scikit-learn* for machine learning, *scikits-image* for image analysis and manipulation, *statsmodels* for statistical tests and data exploration, *Pandas* for storage and analysis of heterogeneous labeled data, *Chaco* for enhanced interactive plotting, *MayaVi* for enhanced three-dimensional visualization, *SymPy* for symbolic mathematics, *NetworkX* for graph and network analysis, and many others which are too numerous to list here. A repository of Python modules can be found in the Python Package

Index at http://pypi.python.org. For more information on these and other packages, see the reference list in §A.10.

A.3. Getting Started with Python

In this section we will give a brief overview of the main features of the Python syntax, geared toward a newcomer to the language. A complete introduction to the Python language is well beyond the scope of this small section; for this purpose many resources are available both online and in print. For some suggestions, see the references in §A.10.

A.3.1. Installation

Python is open source and available to download for free at http://python.org. One important note when installing Python is that there are currently two branches of Python available: Python 2 and Python 3. Python 3.x includes some useful enhancements, but is not backward compatible with Python 2.x. At the time of this book's publication, many of the numerical and computational packages available for Python have not yet moved to full compatibility with Python 3.x. For that reason, the code in this book is written using the syntax of Python 2.4+, the dependency for the NumPy package. Note, however, that the Matplotlib package currently requires Python 2.5+, and Scikit-learn currently requires Python 2.6+. We recommend installing Python 2.7 if it is available.

Third-party packages such as NumPy and SciPy can be installed by downloading the source code or binaries, but there are also tools which can help to automate the process. One option is `pip`, the Python Package Index installer, which streamlines the process of downloading and installing new modules: for example, NumPy can be installed on a computer with a working C compiler by simply typing `pip install numpy`. Find out more at http://pypi.python.org/. Refer to individual package documentation for installation troubleshooting.

Another good option is to use Linux's Advanced Packaging Tool (a core utility in many Linux distributions), which gives users access to a repository of software, including Python packages that are precompiled and optimized for the particular Linux system, and also keeps track of any dependencies needed for new packages. See your system's documentation for more information.

There are also several third-party Python builds which contain many or all of these tools within a single installation. Using one of these can significantly streamline setting up the above tools on a new system. Some examples of these are PythonXY[3] and Extension Packages[4] for Windows, the SciPy Superpack[5] for Mac, and the Enthought Python Distribution (EPD),[6] and Anaconda[7] for multiple platforms. The latter two are proprietary package distributions, but have free versions available.

---

[3] http://www.pythonxy.com
[4] http://www.lfd.uci.edu/ gohlke/pythonlibs/
[5] http://fonnesbeck.github.com/ScipySuperpack/
[6] http://enthought.com/
[7] http://continuum.io/

A.3.2. Running a Python Script

There are several ways to run a piece of Python code. One way is to save it in a text file and pass this filename as an argument to the Python executable. For example, the following command can be saved to the file hello_world.py:

```
# simple script
print 'hello world'
```

and then executed using the command-line argument `python hello_world.py`. (Note that comments in Python are indicated by the # character: anything after this character on a line will be ignored by the interpreter.) Alternatively, one can simply run python with no arguments and enter the interactive shell, where commands can be typed at the >>> prompt:

```
>>> print 'hello world'
hello world
```

In §A.4, we will also introduce the IPython interpreter, which is similar to Python's command-line interface but contains added features.

A.3.3. Variables and Operators

Python, like many programming languages, centers around the use of **variables**. Variables can be of several built-in types, which are inferred dynamically (i.e., Python variables do not require any type information in declaration):

```
>>> x = 2                   # integer value
>>> pi = 3.1415             # floating-point variable
>>> label = "Addition:"     # string variable
>>> print label, x, '+', pi, "=", (x + pi)
Addition: 2 + 3.1415 = 5.1415
>>>
>>> label = 'Division:'
>>> print label, x, '/', pi, "=", (x / pi)
Division: 2 / 3.1415 = 0.636638548464
```

Variable names are case sensitive, so that `pi` and `PI` may refer to two different objects. Note the seamless use of integers, floating-point (decimal) values, strings (indicated by single quotes '...' or double-quotes "..."), and that simple arithmetic expressions (+, -, *, and /) behave as expected.[8]

---

[8]One potential point of confusion is the division operator with integers: in Python 2.x, if two integers are divided, an integer is always returned, ignoring any remainder. This can be remedied by explicitly converting one value to floating point, using `float(x)`. Alternatively, to use the Python 3 semantics where integer division returns a float, you may use the statement `from future import __division__`

Also available are the ** operator for exponentiation, and the modulus operator % which finds the remainder between two numbers:

```
>>> 2 ** 3   # exponentiation
8
>>> 7 % 3   # modulus (remainder after division)
1
```

Python also provides the operators +=, -=, /=, *=, **=, and %=, which combine arithmetic operations with assignment:

```
>>> x = 4
>>> x += 2   # equivalent to x = x + 2
>>> x
6
```

A.3.4. Container Objects

Besides numeric and string variables, Python also provides several built-in container types, which include **lists**, **tuples**, **sets**, and **dictionaries**.

Lists

A list holds an ordered sequence of objects, which can be accessed via the zero-based item index using square brackets [ ]:

```
>>> L = [1, 2, 3]
>>> L[0]
1
```

Lists items can be any mix of types, which makes them very flexible:

```
>>> pi = 3.14
>>> L = [5, 'dog', pi]
>>> L[-1]
3.14
```

Here we have used a negative index to access items at the end of the list. Another useful list operation is **slicing**, which allows access to sublists. Slices can start and end anywhere in the list, and can be contiguous or noncontiguous:

```
>>> L = ['Larry', 'Goran', 'Curly', 'Peter', 'Paul',
         'Mary']
>>> L[0:3]   # slice containing the first three items
['Larry', 'Goran', 'Curly']
```

```
>>> L[:3]   # same as above: the zero is implied
['Larry', 'Goran', 'Curly']
>>> L[-2:]   # last two items
['Paul', 'Mary']
>>> L[1:4]   # items 1 (inclusive) through 4
        # (non-inclusive)
['Goran', 'Curly', 'Peter']
>>> L[::2]   # every second item
['Larry', 'Curly', 'Paul']
>>> L[::-1]   # all items, in reverse order
['Mary', 'Paul', 'Peter', 'Curly', 'Goran', 'Larry']
```

A general slice is of the form [start:stop:stride]. start defaults to 0, stop defaults to none, and stride, which indicates the number of steps to take between each new element, defaults to 1. Slicing will become even more important when working with N-dimensional NumPy arrays below.

Tuples

Tuples are similar to lists, but are indicated by parentheses (1, 2, 3) rather than square brackets [1, 2, 3]. They support item access and slicing using the same syntax as lists. The primary difference is that tuples are immutable: once they're created the items in them cannot be changed. Tuples are most commonly used in functions which return multiple values.

Sets

Sets, available since Python 2.6, act as unordered lists in which items are not repeated. They can be very convenient to use in circumstances where no repetition of elements is desired:

```
>>> S = set([1, 1, 3, 2, 1, 3])
>>> S
set([1, 2, 3])
```

Dictionaries

A dictionary is another container type, which stores an unordered sequence of key-value pairs. It can be defined using curly brackets {}, and like lists allows mixes of types:

```
>>> D = {'a':1, 'b':2.5, 'L':[1, 2, 3]}
>>> D['a']
1
>>> D['L']
[1, 2, 3]
```

Internally, Python uses dictionaries for many built-in aspects of the language: for example, the function globals() will return a dictionary object containing all currently defined global variables.

A.3.5. Functions

For operations which will be repeatedly executed, it is often convenient to define a **function** which implements the desired operation. A function can optionally accept one or several **parameters**, and can optionally return a computed value:

```
>>> def convert_to_radians(deg):
...     pi = 3.141592653
...     return deg * pi / 180.0
...
>>> convert_to_radians(90)
1.5707963265
```

Notice the key elements of a function definition: the def keyword, the function name, the arguments in parentheses (), the colon : marking the beginning of a code block, the **local variable** pi defined in the function, and the optional return statement which returns a computed value to the point of the function call. Here we see our first use of **indentation** in Python: unlike many languages, *white space in Python has meaning*. This can be difficult to become accustomed to for a programmer with background in other languages, but Python proponents point out that this feature can make code very easy to read and write. Indentation in Python can consist of tabs or any number of spaces; standard practice is to use four spaces, and to never use tabs.

Python has many built-in functions which implement common tasks. For example, it is often convenient to be able to quickly define a sequential list of integers: Python allows this through the built-in range function:

```
>>> x = range(10)
>>> x
[0, 1, 2, 3, 4, 5, 6, 7, 8, 9]
```

Notice that range(N) starts at zero and has N elements, and therefore does not include N itself. There are a large number of other useful built-in functions: for more information, refer to the references listed in §A.10.

A.3.6. Logical Operators

Another of Python's built-in object types are the boolean values, True and False (case sensitive). A number of boolean operators are available, as specified in table A.1.

Nonboolean variables can be coerced into boolean types: for example, a nonzero integer evaluates to True, while a zero integer evaluates to False; an empty string

TABLE A.1.
Boolean operators.

Operation	Description
x or y	True if either x or y, or both evaluate to True; otherwise False
x and y	True only if both x and y evaluate to True; otherwise False
not x	True only if x evaluates to False; otherwise False.

TABLE A.2.
Comparison expressions.

Operator	Description	Operator	Description
<	Strictly less than	<=	Less than or equal
>	Strictly greater than	>=	Greater than or equal
==	Equal	!=	Not Equal

evaluates to False in a boolean expression, while a nonempty string evaluates to True:

```
>>> bool(''), bool('hello')
(False, True)
```

Hand in hand with the boolean operators are the comparison expressions, which are summarized in table A.2.

As a simple example, one can use comparisons and boolean expressions in combination:

```
>>> x = 2
>>> y = 4
>>> (x == 2) and (y >= 3)
True
```

These boolean expressions become very important when used with control flow statements, which we will explore next.

A.3.7. Control Flow

Control flow statements include **conditionals** and **loops** which allow the programmer to control the order of code execution.

For conditional statements, Python implements the if, elif, and else commands:

```
>>> def check_value(x):
...     if x < 0:
...         return 'negative'
...     elif x == 0:
...         return 'zero'
```

```
...         else:
...             return 'positive'
...
>>> check_value(0)
zero
>>> check_value(123.4)
positive
```

The keyword elif is a contraction of else if, and allows multiple conditionals in series without excessive indentation. There can be any number of elif statements strung together, and the series may or may not end with else.

Python contains two types of loop statements: for loops and while loops. The syntax of the for loop is as follows:

```
>>> for drink in ['coffee', 'slivovitz', 'water']:
...     print drink
...
coffee
slivovitz
water
```

In practice, for loops are often used with the built-in range function, which was introduced above:

```
>>> words = ['twinkle', 'twinkle', 'little', 'star']
>>> for i in range(4):
...     print i, words[i]
...
0 twinkle
1 twinkle
2 little
3 star
```

The second type of loop, while loops, operate similarly to while loops in other languages:

```
>>> i = 0
>>> while i < 10:
...     i += 3
...
>>> i
12
```

In loops, it can often be useful to skip the remainder of a loop iteration or to break out of the loop. These tasks can be accomplished using the continue and break statements:

```
>>> i = 0
>>> while True:
...     i += 3
...     if i > 10:
...         break   # break out of the loop
...
>>> print i
12
>>> for i in range(6):
...     if i % 2 == 0:
...         continue   # continue from beginning of
...                    # loop block
...     print i, 'only odd numbers get here'
...
1 only odd numbers get here
3 only odd numbers get here
5 only odd numbers get here
```

Another useful statement is the pass statement, which is a null operation that can be useful as a placeholder:

```
>>> word = 'python'
>>> for char in word:
...     if char == 'o':
...         pass   # this does nothing
...     else:
...         print char,
...
p y t h n
```

A.3.8. Exceptions and Exception Handling

When something goes wrong in a Python script, an **exception** is raised. This can happen, for example, when a list index is out of range, or when a mathematical expression is undefined:

```
>>> L = [1, 2, 3]
>>> L[5]
Traceback (most recent call last):
  File "<stdin>", line 1, in <module>
IndexError: list index out of range
>>> 0 / 0
Traceback (most recent call last):
```

```
    File "<stdin>", line 1, in <module>
ZeroDivisionError: integer division or modulo by zero
```

In these cases, it is often useful to **catch** the exceptions in order to decide what to do. This is accomplished with the try, except statement:

```
>>> def safe_getitem(L, i):
...     try:
...         return L[i]
...     except IndexError:
...         return 'index out of range'
...     except TypeError:
...         return 'index of wrong type'
...
>>> L = [1, 2, 3]
>>> safe_getitem(L, 1)
2
>>> safe_getitem(L, 100)
index out of range
>>> safe_getitem(L, 'cat')
index of wrong type
```

In addition to try and except, the else and finally keywords can be used to fine-tune the behavior of your exception handling blocks. For more information on the meaning of these keywords, see the language references in §A.10.

In your own code, you may wish to raise an exception under certain circumstances. This can be done with the raise keyword:

```
>>> def laugh(n):
...     if n <= 0:
...         raise ValueError('n must be positive')
...     else:
...         return n * 'ha! '
>>> laugh(6)
ha! ha! ha! ha! ha! ha!
>>> laugh(-2)
Traceback (most recent call last):
    File "<stdin>", line 1, in <module>
ValueError: n must be positive
```

Python has numerous built-in exception types, and users may also define custom exception types for their own projects.

A.3.9. Modules and Packages

A key component of the extensibility of Python is the existence of **modules** and **packages**, which may be built in or defined by the user. A module is a collection of functions and variables, and a package is a collection of modules. These can be accessed using the `import` statement. There are several useful built-in modules in Python, for example the `math` module can be accessed this way:

```
>>> import math
>>> math.sqrt(2)
1.41421356237
```

Alternatively, one can import specific variables, functions, or classes from modules using the `from`, `import` idiom:

```
>>> from math import pi, sin
>>> sin(pi / 2)
1.0
```

At times, you'll see the use of, for example, `from math import *`, which imports everything in the module into the global namespace. This should generally be avoided, because it can easily lead to name conflicts if multiple modules or packages are used. More information on any module can be found by calling the built in `help` function with the module as the argument:

```
>>> help(math)
```

Another useful function is the `dir()` function, which lists all the attributes of a module or object. For example, typing the following will list all the operations available in the `math` module:

```
>>> dir(math)
```

A.3.10. Objects in Python

Python can be used as an object-oriented language, and the basic building blocks of the language are objects. Every variable in Python is an object. A variable containing an integer is simply an object of type `int`; a list is simply an object of type `list`. Any object may have **attributes**, **methods**, and/or **properties** associated with it. Attributes and methods are analogous to variables and functions, respectively, except that attributes and methods are associated with particular objects. Properties act as a hybrid of an attribute and a method, and we won't discuss them further here.

As an example of using attributes and methods, we'll look at the `complex` data type:

```
>>> c = complex(1.0, 2.0)
>>> c
(1+2j)
>>> c.real   # attribute
```

```
1.0
>>> c.imag   # attribute
2.0
>>> c.conjugate()   # method
(1-2j)
```

Here we have created a complex number c, and accessed two of its attributes, c.real (the real part) and c.imag (the imaginary part). These attributes are variables which are associated with the object c. Similarly, we call the method c.conjugate(), which computes the complex conjugate of the number. This method is a function which is associated with the object c.

Most built-in object types have methods which can be used to view or modify the objects. For example, the append method of list objects can be used to extend the array:

```
>>> L = [4, 5, 6]
>>> L.append(8)
>>> L
[4, 5, 6, 8]
```

and the sort method can be used to sort a list in place:

```
>>> numbers = [5, 2, 6, 3]
>>> numbers.sort()
>>> numbers
[2, 3, 5, 6]
>>> words = ['we', 'will', 'alphabetize', 'these',
    'words']
>>> words.sort()
>>> words
['alphabetize', 'these', 'we', 'will', 'words']
```

To learn more about other useful attributes and methods of built-in Python types, use the help() or dir() commands mentioned above, or see the references in §A.10.

A.3.11. Classes: User-Defined Objects

Python provides a syntax for users to create their own object types. These class objects can be defined in the following way:

```
>>> class MyObject(object):   # derive from the
    # base-class `object`
...     def __init__(self, x):
...         print 'initializing with x =', x
...         self.x = x
...
...     def x_squared(self):
```

```
...             return self.x ** 2
...
>>> obj = MyObject(4)
initializing with x = 4
>>> obj.x  # access attribute
4
>>> obj.x_squared()  # invoke method
16
```

The special method `__init__()` is what is called when the object is initialized. Its first argument is the object itself, which by convention is named `self`. The remaining arguments are up to the user. Special methods like `__init__` are marked by the double underscore, and have specific purposes. In addition, the user can define any number of custom attributes and methods.

This class interface in Python is very powerful: it allows Python scripts and applications to make use of *inheritance* and other object-oriented design principles, which if used well can make code both very flexible and very easy to write, read, and understand. An introduction to object-oriented design principles is beyond the scope of this appendix, but an excellent introduction can be found in several of the references in §A.10.

A.3.12. Documentation Strings

One key feature of Python objects is that they can have built-in documentation strings, usually referred to as *docstrings*. Docstrings can be accessed by calling the `help()` function on the object, as we saw above. When writing your own code, it is good practice to write good docstrings. This is accomplished by creating a string (most often using the triple quote `"""` to indicate a multiline string) in the first line of the function:

```
>>> def radians(deg):
...     """
...     Convert an angle to radians
...     """
...     pi = 3.141592653
...     return deg * pi / 180.0
...
>>> help(radians)
```

This help command will open a text reader showing the docstring defined at the top of the function.

A.3.13. Summary

In this section, we have given a brief introduction to the basic syntax of Python. In the next section, we will introduce some of the powerful open-source scientific-computing tools which are built on this framework. First, though, we will highlight some features of IPython which will make our job easier.

A.4.   IPython: The Basics of Interactive Computing

In the example above, we used Python's standard interactive computing environment for line-by-line interpreted coding. For the remainder of this introduction, we'll switch to using IPython's enhanced interactive interface. There are many useful features of IPython, but we'll just mention a few of the most important ones here. The IPython environment can be started by running `ipython` with no arguments: the default prompt is different than that of the normal Python interpreter, but acts similarly:

```
In [1]: print 'hello world'
hello world
In [2]: 4 + 6
Out[2]: 10
```

A.4.1. Command History

IPython stores the command history in the special global variables In and Out. So, for example, to see the first command which was typed above, we can simply use In:

```
In [3]: print In[1]
print 'hello world'
```

Any output is also stored. So, the result of our input in line 2, can be recovered:

```
In [4]: print Out[2]
10
```

A.4.2. Command Completion

IPython allows tab completion of commands, both for commands in the user history, and for arbitrary commands in the namespace. For example, if you type `for i in ran` and then type the tab key, IPython will fill in the rest of the command, giving you `for i in range`. If there are multiple completion possibilities, it will display a list of those possibilities:

```
In [5]: x = ra<TAB>
raise          range          raw_input
```

Similarly, using the up arrow allows you to cycle through your command history. Typing a partial command followed by the up arrow will cycle through any commands which start with those characters. For example, in the current session, typing the partial command `prin` followed by the up arrow will automatically fill in the command `print Out[2]`, the most recent statement which begins with those characters. Typing the up arrow again will change this to `print In[1]` and then `print 'hello world'`. In interactive coding, when complicated commands are often repeated, this command completion feature is very helpful.

A.4.3. Help and Documentation

Above we saw how the `help()` function can be used to access any objects docstring. In IPython, this can be accomplished using the special character ?. For example, the documentation of the `range` function is simply displayed:

```
In [6]: range?
Type:          builtin_function_or_method
String Form:<built-in function range>
Namespace:  ipython builtin
Docstring:
range([start,] stop[, step]) -> list of integers

Return a list containing an arithmetic progression of
    integers.
range(i, j) returns [i, i+1, i+2, ..., j-1]; start (!)
    defaults to 0.
When step is given, it specifies the increment (or
    decrement).
For example, range(4) returns [0, 1, 2, 3].
The end point is omitted!
These are exactly the valid indices for a list of 4
    elements.
```

For pure Python functions or classes (i.e., ones which are not derived from compiled libraries), the double question mark ?? allows one to examine the source code itself:

```
In [7]: def myfunc(x):
   ...:        return 2 * x
   ...:
In [8]: myfunc??   # note: no space between object
         # name and ??
Type:          function
String Form:<function myfunc at 0x2a6c140>
File:          /home/<ipython-input-15-21fdc7f0ea27>
Definition: myfunc(x)
Source:
def myfunc(x):
    return 2 * x
```

This can be very useful when exploring functions and classes from third-party modules.

A.4.4. Magic Functions

The IPython interpreter also includes **magic functions** which give shortcuts to useful operations. They are marked by a % sign, and are too numerous to list here. For example, the %run command allows the code within a file to be run from within IPython. We can run the hello_world.py file created above as follows:

```
In [9]: %run hello_world.py
hello world
```

There are many more such magic functions, and we will encounter a few of them below. You can see a list of what is available by typing % followed by the tab key. The IPython documentation features also work for magic functions, so that documentation can be viewed by typing, for example, %run?.

In addition to the interactive terminal, IPython provides some powerful tools for parallel processing; for creating interactive html notebooks combining code, formatted text, and graphics; and much more. For more information about IPython, see the references in §A.10, especially IPython's online documentation and tutorials. For a well-written practical introduction to IPython, we especially recommend the book *Python for Data Analysis*, listed in §A.10.

A.5. Introduction to NumPy

In this section, we will give an introduction to the core concepts of scientific computing in Python, which is primarily built around the NumPy array. Becoming fluent in the methods and operations of NumPy arrays is vital to writing effective and efficient Python code. This section is too short to contain a complete introduction, and the interested reader should make use of the references at the end of this appendix for more information. NumPy is contained in the numpy module, which by convention is often imported under the shortened name np.

A.5.1. Array Creation

NumPy arrays are objects of the type np.ndarray, though this type specifier is rarely used directly. Instead there are several array creation routines that are often used:

```
In [1]: import numpy as np

In [2]: np.array([5, 2, 6, 7])   # create an array
           # from a python list
Out[2]: array([5, 2, 6, 7])

In [3]: np.arange(4)   # similar to the built-in
           # range() function
Out[3]: array([0, 1, 2, 3])
```

```
In [4]: np.linspace(1, 2, 5)   # 5 evenly-spaced steps
           # from 1 to 2
Out[4]: array([ 1.  ,  1.25,  1.5 ,  1.75,  2.  ])

In [5]: np.zeros(5)   # array of zeros
Out[5]: array([ 0.,  0.,  0.,  0.,  0.])

In [6]: np.ones(6)   # array of ones
Out[6]: array([ 1.,  1.,  1.,  1.,  1.,  1.])
```

Arrays can also be multidimensional:

```
In [7]: np.zeros((2, 4))   # 2 X 4 array of zeros
Out[7]:
array([[ 0.,  0.,  0.,  0.],
       [ 0.,  0.,  0.,  0.]])
```

NumPy also has tools for random array creation:

```
In [9]: np.random.random(size=4)
          # uniform between 0 and 1
Out[9]: array([ 0.28565   ,  0.3614929 ,
          0.95431006,  0.24266193])

In [10]: np.random.normal(size=4) # standard-norm
           # distributed
Out[10]: array([-0.62332252,  0.09098354,
          0.40975753,  0.53407146])
```

Other array creation routines are available, but the examples above summarize the most useful options. With any of them, the ? functionality of IPython can be used to determine the usage and arguments.

A.5.2. Element Access

Elements of arrays can be accessed using the same indexing and slicing syntax as in lists:

```
In [11]: x = np.arange(10)

In [12]: x[0]
Out[12]: 0

In [13]: x[::2]
Out[13]: array([0, 2, 4, 6, 8])
```

Note that array slices are implemented as **views** rather than copies of the original array, so that operations such as the following are possible:

```
In [14]: x = np.zeros(8)  # array of eight zeros

In [15]: x[::2] = np.arange(1, 5) # [::2] selects
            # every 2nd entry

In [16]: x
Out[16]: array([ 1.,   0.,   2.,   0.,   3.,   0.,   4.,
            0.])
```

For multidimensional arrays, element access uses multiple indices:

```
In [17]: X = np.zeros((2, 2))  # 2 X 2 array of zeros

In [18]: X[0, 0] = 1

In [19]: X[0, 1] = 2

In [20]: X
Out[20]:
array([[ 1.,   2.],
       [ 0.,   0.]])

In [21]: X[0, :]  # first row of X. X[0] is
            # equivalent
Out[21]: array([ 1.,   2.])

In [22]: X[:, 1]  # second column of X.
Out[22]: array([ 2.,   0.])
```

Combinations of the above operations allow very complicated arrays to be created easily and efficiently.

A.5.3. Array Data Type

NumPy arrays are typed objects, and often the type of the array is very important. This can be controlled using the dtype parameter. The following example is on a 64-bit machine:

```
In [23]: x = np.zeros(4)

In [24]: x.dtype
Out[24]: dtype('float64')

In [25]: y = np.zeros(4, dtype=int)
```

```
In [26]: y.dtype
Out[26]: dtype('int64')
```

Allocating arrays of the correct type can be very important; for example, a floating-point number assigned to an integer array will have its decimal part truncated:

```
In [27]: x = np.zeros(1, dtype=int)

In [28]: x[0] = 3.14  # converted to an integer!

In [29]: x[0]
Out[29]: 3
```

Other possible values for dtype abound: for example, you might use bool for boolean values or complex for complex numbers. When we discuss *structured arrays* below, we'll see that even more sophisticated data types are possible.

A.5.4. Universal Functions and Broadcasting

A powerful feature of NumPy is the concept of a ufunc, short for "universal function." These functions operate on every value of an array. For example, trigonometric functions are implemented in NumPy as **unary ufuncs**—ufuncs with a single parameter:

```
In [30]: x = np.arange(3)  # [0, 1, 2]

In [31]: np.sin(x)  # take the sine of each element
Out[31]: array([ 0.    ,  0.84147098,  0.90929743])
```

Arithmetic operations like +, -, *, / are implemented as **binary ufuncs**—ufuncs with two parameters:

```
In [32]: x * x  # multiply each element of x by
              # itself
Out[32]: array([0, 1, 4])
```

Ufuncs can also operate between arrays and scalars, applying the operation and the scalar to each array value:

```
In [33]: x + 5  # add 5 to each element of x
Out[33]: array([5, 6, 7])
```

This is a simple example of **broadcasting**, where one argument of a ufunc has a shape upgraded to the shape of the other argument. A more complicated example of broadcasting is adding a vector to a matrix:

```
In [34]: x = np.ones((3, 3))   # 3 X 3 array of ones

In [35]: y = np.arange(3)   # [0, 1, 2]

In [36]: x + y   # add y to each row of x
Out[36]:
array([[ 1.,   2.,   3.],
       [ 1.,   2.,   3.],
       [ 1.,   2.,   3.]])
```

Sometimes, both arrays are broadcast at the same time:

```
In [37]: x = np.arange(3)

In [38]: y = x.reshape((3, 1))   # create a 3 X 1
            # column array

In [39]: x + y
Out[39]:
array([[0, 1, 2],
       [1, 2, 3],
       [2, 3, 4]])
```

Here a row array and a column array are added together, and both are broadcast to complete the operation. A visualization of these broadcasting operations is shown in figure A.1.

Broadcasting can seem complicated, but it follows simple rules:

1. If the two arrays differ in their number of dimensions, the shape of the array with fewer dimensions is padded with ones on its leading (left) side.
2. If the shape of the two arrays does not match in any dimension, the array with shape equal to 1 in that dimension is broadcast up to match the other shape.
3. If in any dimension the sizes disagree and neither is equal to 1, an error is raised.

All of this takes place efficiently without actually allocating any extra memory for these temporary arrays. Broadcasting in conjunction with universal functions allows some very fast and flexible operations on multidimensional arrays.

A.5.5. Structured Arrays

Often data sets contain a mix of variables of different types. For this purpose, NumPy contains **structured arrays**, which store compound data types. For example, imagine

$np.arange(3) + 5$

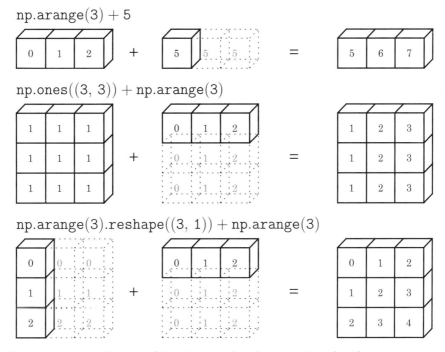

$np.ones((3, 3)) + np.arange(3)$

$np.arange(3).reshape((3, 1)) + np.arange(3)$

Figure A.1. A visualization of NumPy array broadcasting. Note that the extra memory indicated by the dotted boxes is never allocated, but it can be convenient to think about the operations as if it is.

we have a set of data where each object has an integer ID, a five-character string name,[9] and a floating-point measurement. We can store this all in one array:

```
In [40]: dtype = [('ID', int), ('name', (str, 5)),
            ('value', float)]

In [41]: data = np.zeros(3, dtype=dtype)

In [42]: print data
[(0, '', 0.0) (0, '', 0.0) (0, '', 0.0)]
```

The fields of the array (it may be convenient to think of them as "columns") can be accessed via the field name:

```
In [43]: data['ID' ] = [154, 207, 669]

In [44]: data['name'] = ['obj_1', 'obj_2', 'obj_3']

In [45]: data['value'] = [0.1, 0.2, 0.3]
```

[9] Because NumPy arrays use fixed memory blocks, we will also need to specify the maximum length of the string argument.

```
In [46]: print data[0]
(154, 'obj_1', 0.1)

In [47]: print data['value']
[ 0.1  0.2  0.3]
```

The data sets used in this book are loaded into structured arrays for convenient data manipulation.

A.5.6. Summary

This has been just a brief overview of the fundamental object for most scientific computing in Python: the NumPy array. The references at the end of the appendix include much more detailed tutorials on all the powerful features of NumPy array manipulation.

A.6. Visualization with Matplotlib

Matplotlib offers a well-supported framework for producing publication-quality plots using Python and NumPy. Every figure in this book has been created using Matplotlib, and the source code for each is available on the website http://www.astroML.org. Here we will show some basic plotting examples to get the reader started.

IPython is built to interact cleanly with Matplotlib. The magic command %pylab allows figures to be created and manipulated interactively, and we will use the pylab environment in the examples below. Here is how we'll prepare the session:

```
In [1]: %pylab
Welcome to pylab, a matplotlib-based Python
environment [backend: TkAgg].
For more information, type 'help(pylab)'.

In [2]: import numpy as np

In [3]: import matplotlib.pyplot as plt
```

Lines 2 and 3 above are not strictly necessary: the pylab environment includes these, but we explicitly show them here for clarity.

Matplotlib figures center around the figure and axes objects. A figure is essentially a plot window, and can contain any number of axes. An axes object is a subwindow which can contain lines, images, shapes, text, and other graphical elements. Figures and axes can be created by the plt.figure and plt.axes objects, but the pylab interface takes care of these details in the background.

A useful command to be aware of is the plt.show() command. It should be called once at the end of a script to tell the program to keep the plot open before terminating the program. It is not needed when using interactive plotting in IPython.

Let's start by creating and labeling a simple plot:[10]

```
In [4]: x = np.linspace(0, 2 * np.pi, 1000)
        # 1000 values from 0 to 2pi

In [5]: y = np.sin(x)

In [6]: ax = plt.axes()

In [7]: ax.plot(x, y)

In [8]: ax.set_xlim(0, 2 * np.pi)  # set x limits

In [9]: ax.set_ylim(-1.3, 1.3)  # set y limits

In [10]: ax.set_xlabel('x')

In [11]: ax.set_ylabel('y')

In [12]: ax.set_title('Simple Sinusoid Plot')
```

The resulting plot is shown in figure A.2. The style of the line (dashed, dotted, etc.) or the presence of markers at each point (circles, squares, triangles, etc.) can be controlled using parameters in the plt.plot() command. Use plt.plot? in IPython to see the documentation, and experiment with the possibilities.

Another useful plot type is the error bar plot. Matplotlib implements this using plt.errorbar:

```
In [13]: x_obs = 2 * np.pi * np.random.random(50)

In [14]: y_obs = np.sin(x_obs)

In [15]: y_obs += np.random.normal(0, 0.1, 50)
         # add some error

In [16]: ax.errorbar(x_obs, y_obs, 0.1, fmt='.',
         color='black')
```

The result is shown in figure A.3. We have used the same axes object to overplot these points on the previous line. Notice that we have set the format to '.', indicating a single point for each item, and set the color to 'black'.

---

[10]Here we will use Matplotlib's object/method interface to plotting. There is also a MATLAB-style interface available, where for example plt.plot(x, y) could be called directly. This is sometimes convenient, but is less powerful. For more information, see the references in §A.10.

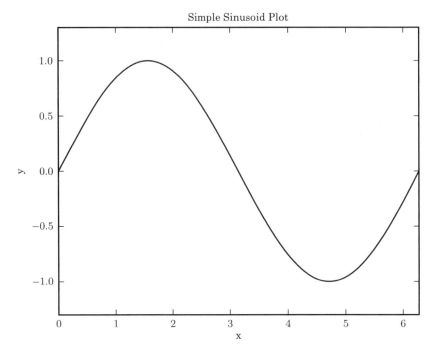

Figure A.2. Output of the simple plotting example.

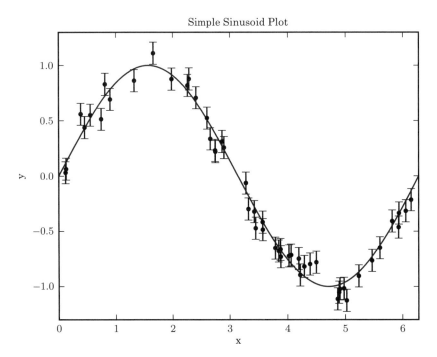

Figure A.3. Output of the error bar plotting example.

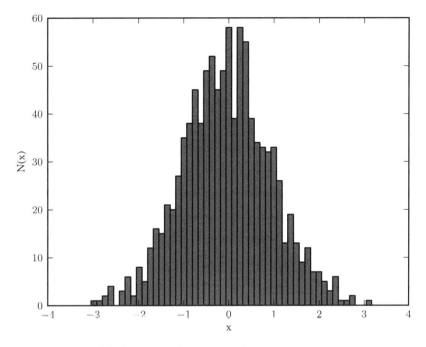

Figure A.4. Output of the histogram plotting example.

Another basic type of plot that is very useful is the histogram. This is implemented in Matplotlib through the ax.hist() command:

```
In [17]: fig = plt.figure()  # create a new figure
         # window

In [18]: ax = plt.axes()  # create a new axes

In [19]: x = np.random.normal(size=1000)

In [20]: ax.hist(x, bins=50)

In [21]: ax.set_xlabel('x')

In [22]: ax.set_ylabel('N(x)')
```

The result is shown in figure A.4. Again, there are many options to control the appearance of the histogram. Use plt.hist? in IPython to access the histogram documentation for details.

Matplotlib has many more features, including tools to create scatter plots (plt.scatter), display images (plt.imshow), display contour plots (plt.contour and plt.contourf), create hexagonal tessellations (plt.hexbin), draw filled regions (plt.fill and plt.fill_between), create multiple subplots (plt.subplot and plt.axes), and much more. For information about these routines, use the help feature in IPython. For additional documentation and examples, the main Matplotlib website (especially the *gallery* section) is a very useful resource: http://matplotlib.org. Additionally, the reader may wish to browse the source code of the figures in this book, available at http://www.astroML.org.

A.7. Overview of Useful NumPy/SciPy Modules

NumPy and SciPy contain a large collection of useful routines and tools, and we will briefly summarize some of them here. More information can be found in the references at the end of this appendix.

A.7.1. Reading and Writing Data

NumPy has tools for reading and writing both text-based data and binary format data. To save an array to an ASCII file, use np.savetxt; text files can be loaded using np.loadtxt:

```
In [1]: import numpy as np

In [2]: x = np.random.random((10, 3))

In [3]: np.savetxt('x.txt', x)

In [4]: y = np.loadtxt('x.txt')

In [5]: np.all(x == y)
Out[5]: True
```

A more customizable text loader can be found in np.genfromtxt. Arrays can similarly be written to binary files using np.save and np.load for single arrays, or np.savez to store multiple arrays in a single zipped file. More information is available in the documentation of each of these functions.

A.7.2. Pseudorandom Number Generation

NumPy and SciPy provide powerful tools for pseudorandom number generation. They are based on the Mersenne twister algorithm [5], one of the most sophisticated and well-tested pseudorandom number algorithms available. The np.random submodule contains routines for random number generation from a variety of distributions. For reproducible results, the random seed can explicitly be set:

```
In [6]: np.random.seed(0)

In [7]: np.random.random(4) # uniform between 0 and 1
Out[7]: array([ 0.5488135 ,  0.71518937,
         0.60276338,  0.54488318])
```

```
In [8]: np.random.normal(loc=0, scale=1, size=4)
        # standard norm
Out[8]: array([ 1.86755799, -0.97727788,
        0.95008842, -0.15135721])

In [9]: np.random.randint(low=0, high=10, size=4)
        # random integers
Out[9]: array([8, 1, 6, 7])
```

Many other distributions are available as well, and are listed in the documentation of np.random. Another convenient way to generate random variables is using the distributions submodule in scipy.stats:

```
In [10]: from scipy.stats import distributions

In [11]: distributions.poisson.rvs(10, size=4)
Out[11]: array([10, 16,  5, 13])
```

Here we have used the poisson object to generate random variables from a Poisson distribution with $\mu = 10$. The distribution objects in scipy.stats offer much more than just random number generation: they contain tools to compute the probability density, the cumulative probability density, the mean, median, standard deviation, nth moments, and much more. For more information, type scipy.distributions? in IPython to view the module documentation. See also the many distribution examples in chapter 3.

A.7.3. Linear Algebra

NumPy and SciPy contain some powerful tools for efficient linear algebraic computations. The basic dot (i.e., matrix-vector) product is implemented in NumPy:

```
In [12]: M = np.arange(9).reshape((3, 3))

In [13]: M
Out[13]:
array([[0, 1, 2],
       [3, 4, 5],
       [6, 7, 8]])

In [14]: x = np.arange(1, 4)   # [1, 2, 3]

In [15]: np.dot(M, x)
Out[15]: array([ 8, 26, 44])
```

Many common linear algebraic operations are available in the submodules numpy.linalg and scipy.linalg. It is possible to compute matrix inverses (linalg.inv), singular value decompositions (linalg.svd), eigenvalue decompositions (`linalg.eig`), least-squares solutions (`linalg.lstsq`) and much more. The linear algebra submodules in NumPy and SciPy have many duplicate routines, and have a common interface. The SciPy versions of the routines often have more options to control output, and `scipy.linalg` contains a number of less common algorithms that are not a part of `numpy.linalg`. For more information, enter `numpy.linalg?` or `scipy.linalg?` in IPython to see the documentation.

A.7.4. Fast Fourier Transforms

The fast Fourier transform is an important algorithm in many aspects of data analysis, and these routines are available in both NumPy (`numpy.fft` submodule) and SciPy (`scipy.fftpack` submodule). Like the linear algebra routines discussed above, the two implementations differ slightly, and the SciPy version has more options to control the results.

A simple example is as follows:

```
In [16]: x = np.ones(4)

In [17]: np.fft.fft(x)   # forward transform
Out[17]: array([ 4.+0.j,   0.+0.j,   0.+0.j,   0.+0.j])

In [18]: np.fft.ifft(Out[17])   # inverse recovers
            # input
Out[18]: array([ 1.+0.j,   1.+0.j,   1.+0.j,   1.+0.j])
```

The forward transform is implemented with `fft()`, while the inverse is implemented with `ifft`. To learn more, use `scipy.fftpack?` or `numpy.fft?` to view the documentation. See appendix E for some more discussion and examples of Fourier transforms.

A.7.5. Numerical Integration

Often it is useful to perform numerical integration of numerical functions. The literature on numerical integration techniques is immense, and SciPy implements many of the more popular routines in the module `scipy.integrate`. Here we'll use a function derived from QUADPACK, an optimized numerical integration package written in Fortran, to integrate the function $\sin x$ from 0 to π:

```
In [19]: from scipy import integrate

In [20]: result, error = integrate.quad(np.sin, 0,
            np.pi)

In [21]: result, error
Out[21]: (2.0, 2.220446049250313e-14)
```

As expected, the algorithm finds that $\int_0^\pi \sin x \, dx = 2$. The integrators accept any function of one variable as the first argument, even one defined by the user. Other integration options (including double and triple integration) are also available. Use `integrate?` in IPython to see the documentation for more details.

A.7.6. Optimization

Numerical optimization (i.e., function minimization) is an important aspect of many data analysis problems. Like numerical integration, the literature on optimization methods is vast. SciPy implements many of the most common and powerful techniques in the submodule `scipy.optimize`. Here we'll find the minimum of a simple function using `fmin`, which implements a Nelder–Mead simplex algorithm:

```
In [22]: from scipy import optimize

In [23]: def simple_quadratic(x):
   ...:         return x ** 2 + x

In [24]: optimize.fmin(simple_quadratic, x0=100)
Optimization terminated successfully.
         Current function value: -0.250000
         Iterations: 24
         Function evaluations: 48
Out[24]: array([-0.50003052])
```

The optimization converges to near the expected minimum value of $x = -0.5$. There are several more sophisticated optimizers available which can take into account gradients and other information: for more details use `optimize?` in IPython to see the documentation for the module.

A.7.7. Interpolation

Another common necessity in data analysis is interpolation discretely sampled data. There are many algorithms in the literature, and SciPy contains the submodule `scipy.interpolate` which implements many of these, from simple linear and polynomial interpolation to more sophisticated spline-based techniques.

Here is a brief example of using a univariate spline to interpolate between points. We'll start by enabling the interactive plotting mode in IPython:

```
In [26]: %pylab
Welcome to pylab, a matplotlib-based Python
environment [backend: TkAgg].
For more information, type 'help(pylab)'.

In [27]: from scipy import interpolate

In [28]: x = np.linspace(0, 16, 30)
         # coarse grid: 30 pts

In [29]: y = np.sin(x)
```

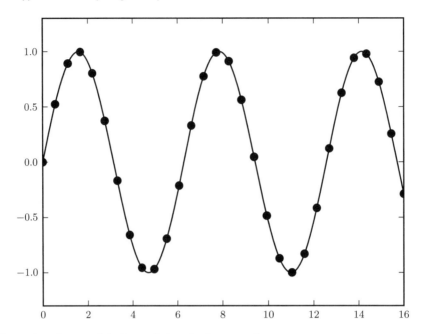

Figure A.5. Output of the interpolation plotting example.

```
In [30]: x2 = np.linspace(0, 16, 1000)
         # fine grid: 1000 pts

In [31]: spl = interpolate.UnivariateSpline(x, y,
                                            s=0)

In [32]: ax = plt.axes()

In [33]: ax.plot(x, y, 'o') # 'o' means draw points
         # as circles

In [34]: ax.plot(x2, spl(x2), '-')
         # '-' means draw a line
```

The result is shown in figure A.5. The spline fit has interpolated the sampled points to create a smooth curve. There are many additional options for spline interpolation, including the adjustment of smoothing factors, number of knots, weighting of points, spline degree, and more. Linear and polynomial interpolation is also available, using the function `interpolate.interp1d`. For more information, type `interpolate?` in IPython to view the documentation.

A.7.8. Other Submodules

There is much more functionality in SciPy, and it can be explored using the online documentation as well as the IPython help functions. Here is a partial list of remaining packages that readers might find useful:

- `scipy.spatial`: distance and spatial functions, nearest neighbor, Delaunay tessellation
- `scipy.sparse`: sparse matrix storage, sparse linear algebra, sparse solvers, sparse graph traversal
- `scipy.stats`: common statistical functions and distributions
- `scipy.special`: special functions (e.g., Airy functions, Bessel functions, orthogonal polynomials, gamma functions, and much more)
- `scipy.constants`: numerical and dimensional constants

As above, we recommend importing these modules and using IPython's ? functionality to explore what is available.

A.8. Efficient Coding with Python and NumPy

As discussed above, Python is a dynamically typed, interpreted language. This leads to several advantages: for example, the ability to write code quickly and the ability to execute code line by line in the IPython interpreter. This benefit comes at a cost, however: it means that certain code blocks may run much more slowly than equivalent code written in a compiled language like C. In data analysis settings, many of these disadvantages can be remedied by efficient use of NumPy arrays. Here we will show a few common mistakes that are made by beginners using Python and NumPy, and solutions which can lead to orders-of-magnitude improvement in computation times. Throughout this section we'll make use of IPython's `%timeit` magic command, which provides quick benchmarks for Python execution.

A.8.1. Data Structures

The Python list object is very convenient to use, but not necessarily suitable for large data sets. It functions well for small sequences, but using large lists for computation can lead to slow execution:

```
In [1]: L = range(10000000)   # a large list
In [2]: %timeit sum(L)
1 loops, best of 3: 1.37 s per loop
```

This slow execution is due both to the data layout (in this case, a linked list) and to the overhead required for dynamic typing. As in other languages, the operation is much faster if the data is stored as a contiguous array. NumPy arrays provide this sort of type-specific, contiguous data container, as well as the tools to efficiently manipulate these arrays:

```
In [3]: import numpy as np
In [4]: x = np.array(L, dtype=float)
In [5]: %timeit np.sum(x)
100 loops, best of 3: 17.1 ms per loop
```

This is a factor of nearly 100 improvement in execution time, simply by using NumPy rather than the built-in Python list. This suggests our first guideline for data processing with Python:

Guideline 1: store data in NumPy arrays, not Python lists. Any time you have a sequence or list of data larger than a few dozen items, it should be stored and manipulated as a NumPy array.[11]

A.8.2. Loops

Even when using NumPy arrays, several pitfalls remain that can lead to inefficient code. A Python user who has a C or Java background may be used to using `for` loops or `while` loops to implement certain tasks. If an algorithm seems to require loops, it is nearly always better in Python to implement it using **vectorized** operations in NumPy (recall our discussion of ufuncs in §A.5.4). For example, the following code takes much longer to execute than the equivalent C code:

```
In [6]: x = np.random.random(1000000)

In [7]: def loop_add_one(x):
   ...:       for i in range(len(x)):
   ...:           x[i] += 1

In [8]: %timeit loop_add_one(x)
1 loops, best of 3: 1.67 s per loop

In [9]: def vectorized_add_one(x):
   ...:       x += 1

In [10]: timeit vectorized_add_one(x)
100 loops, best of 3: 4.13 ms per loop
```

Using vectorized operations—enabled by NumPy's ufuncs—for repeated operations leads to much faster code. This suggests our second guideline for data processing with Python:

Guideline 2: avoid large loops in favor of vectorized operations. If you find yourself applying the same operation to a sequence of many items, vectorized methods within NumPy will likely be a better choice.

A.8.3. Slicing, Masks, and Fancy Indexing

Above we covered **slicing** to access subarrays. This is generally very fast, and is preferable to looping through the arrays. For example, we can quickly copy the first

---

[11] One exception to this rule is when building an array dynamically: because NumPy arrays are fixed length, they cannot be expanded efficiently. One common pattern is to build a list using the `append` operation, and convert it using `np.array` when the list is complete.

half of an array to the second half with a slicing operation:

```
In [11]: x[:len(x) / 2] = x[len(x) / 2:]
```

Sometimes when coding one must check the values of an array, and perform different operations depending on the value. For example, suppose we have an array and we want every value greater than 0.5 to be changed to 999. The first impulse might be to write something like the following:

```
In [12]: def check_vals(x):
    ....:     for i in range(len(x)):
    ....:         if x[i] > 0.5:
    ....:             x[i] = 999

In [13]: %timeit check_vals(x)
1 loops, best of 3: 1.03 s per loop
```

This same operation can be performed much more quickly using a boolean mask:

```
In [14]: %timeit x[(x > 0.5)] = 999
         # vectorized version
100 loops, best of 3: 3.79 ms per loop
```

In this implementation, we have used the boolean mask array generated by the expression (x > 0.5) to access only the portions of x that need to be modified. The operation in line 14 gives the same result as the operation in line 12, but is orders of magnitude faster.

Masks can be combined using the bitwise operators & for AND, | for OR, ~ for NOT, and ^ for XOR. For example, you could write the following:

```
In [15]: x[(x < 0.1) | (x > 0.5)] = 999
```

This will result in every item in the array x which is less than 0.1 or greater than 0.5 being replaced by 999. Be careful to use parentheses around boolean statements when combining operations, or operator precedence may lead to unexpected results.

Another similar concept is that of **fancy indexing**, which refers to indexing with lists. For example, imagine you have a two-dimensional array, and would like to create a new matrix of random rows (e.g., for bootstrap resampling). You may be tempted to proceed like this:

```
In [16]: X = np.random.random((1000000, 3))

In [17]: def get_random_rows(X):
    ....:     X_new = np.empty(X.shape)
    ....:     for i in range(X_new.shape[0]):
```

```
    ....:             X_new[i] = X[np.random.randint(\
    ....:                                   X.shape[0])]
    ....:         return X_new
```

This can be sped up through fancy indexing, by generating the array of indices all at once and vectorizing the operation:

```
In [19]: def get_random_rows_fast(X):
    ....:         ind = np.random.randint(0, X.shape[0],
    ....:                                   X.shape[0])
    ....:         return X[ind]
```

The list of indices returned by the call to `randint` is used directly in the indices. Note that fancy indexing is generally much slower than slicing for equivalent operations.

Slicing, masks, and fancy indexing can be used together to accomplish a wide variety of tasks. Along with ufuncs and broadcasting, discussed in §A.5.4, most common array manipulation tasks can be accomplished without writing a loop. This leads to our third guideline:

Guideline 3: use array slicing, masks, fancy indexing, and broadcasting to eliminate loops. If you find yourself looping over indices to select items on which an operation is performed, it can probably be done more efficiently with one of these techniques.

A.8.4. Summary

A common theme can be seen here: Python loops are slow, and NumPy array tricks can be used to sidestep this problem. As with most guidelines in programming, there are imaginable situations in which these suggestions can (and should) be ignored, but they are good to keep in mind for most situations. Be aware also that there are some algorithms for which loop elimination through vectorization is difficult or impossible. In this case, it can be necessary to interface Python to compiled code. This will be explored in the next section.

A.9. Wrapping Existing Code in Python

At times, you may find an algorithm which cannot benefit from the vectorization methods discussed in the previous section (a good example is in many tree-based search algorithms). Or you may desire to use within your Python script some legacy code which is written in a compiled language, or that exists in a shared library. In this case, a variety of approaches can be used to wrap compiled Fortran, C, or C++ code for use within Python. Each has its advantages and disadvantages, and can be appropriate for certain situations. Note that packages like NumPy, SciPy, and Scikit-learn use several of these tools both to implement efficient algorithms, and to make use of library packages written in Fortran, C, and C++.

Cython is a superset of the Python programming language that allows users to wrap external C and C++ packages, and also to write Python-like code which can be automatically translated to fast C code. The resulting compiled code can then be imported into Python scripts in the familiar way. Cython is very flexible, and within the scientific Python community, has gradually become the de facto standard tool for writing fast, compilable code. More information, including several helpful tutorials, is available at http://www.cython.org/.

f2py is a Fortran to Python interface generator, which began as an independent project before eventually becoming part of NumPy. f2py automates the generation of Python interfaces to Fortran code. If the Fortran code is designed to be compiled into a shared library, the interfacing process can be very smooth, and work mostly out of the box. See http://www.scipy.org/F2py for more information.

Ctypes, included in Python since version 2.5, is a built-in library that defines C data types in order to interface with compiled dynamic or shared libraries. Wrapper functions can be written in Python, which prepare arguments for library function calls, find the correct library, and execute the desired code. It is a nice tool for calling system libraries, but has the disadvantage that the resulting code is usually very platform dependent. For more information, refer to http://docs.python.org/library/ctypes.html

Python C-API Python is implemented in C, and therefore the C Application Programming Interface (API) can be used directly to wrap C libraries, or write efficient C code. Please be aware that this method is not for the faint of heart! Even writing a simple interface can take many lines of code, and it is very easy to inadvertently cause segmentation faults, memory leaks, or other nasty errors. For more information, see http://docs.python.org/c-api/, but be careful!

Two other tools to be aware of are **SWIG**, the Simplified Wrapper and Interface Generator (http://www.swig.org/), and **Weave**, a package within SciPy that allows incorporation of snippets of C or C++ code within Python scripts (http://www.scipy.org/weave). Cython has largely superseded the use of these packages in the scientific Python community.

All of these tools have use cases to which they are suited. For implementation of fast compiled algorithms and wrapping of C and C++ libraries or packages, we recommend Cython as a first approach. Cython's capabilities have greatly expanded during the last few years, and it has a very active community of developers. It has emerged as the favored approach by many in the scientific Python community, in particular the NumPy, SciPy, and Scikit-learn development teams.

A.10. Other Resources

The discussion above is a good start, but there is much that has gone unmentioned. Here we list some books and websites that are good resources to learn more.

A.10.1. Books

- *Python for Data Analysis: Agile Tools for Real World Data* by McKinney [6]. This book was newly published at the time of this writing, and offers an excellent introduction to interactive scientific computing with IPython. The book is well written, and the first few chapters are extremely useful for scientists interested in Python. Several chapters at the end are more specialized, focusing on specific tools useful in fields with labeled and date-stamped data.
- *SciPy and NumPy: An Overview for Developers* by Bressert [2]. This is a short handbook written by an astronomer for a general scientific audience. It covers many of the tools available in NumPy and SciPy, assuming basic familiarity with the Python language. It focuses on practical examples, many of which are drawn from astronomy.
- *Object Oriented Programming in Python* by Goldwasser and Letscher [3]. This is a full-length text that introduces the principles of object-oriented programming in Python. It is designed as an introductory text for computer science or engineering students, and assumes no programming background.
- *Python Scripting for Computational Science* by Langtangen [4]. This is a well-written book that introduces Python computing principles to scientists who are familiar with computing, focusing on using Python as a scripting language to tie together various tools within a scientific workflow.

A.10.2. Websites

- **Learning Python**

  - http://www.python.org/ General info on the Python language. The site includes documentation and tutorials.
  - http://www.diveintopython.net/ An online Python book which introduces the Python syntax to experienced programmers.
  - http://software-carpentry.org/ Excellent tutorials and lectures for scientists to learn computing. The tutorials are not limited to Python, but cover other areas of computing as well.
  - http://scipy-lectures.github.com/ Lectures and notes on Python for scientific computing, including NumPy, SciPy, Matplotlib, Scikit-learn, and more, covering material from basic to advanced.

- **Python packages**

  - http://pypi.python.org Python package index. An easy way to find and install new Python packages.
  - http://ipython.org/ Documentation and tutorials for the IPython project. See especially the notebook functionality: this is quickly becoming a standard tool for sharing code and data in Python.

- http://www.numpy.org Documentation and tutorials for NumPy.
- http://www.scipy.org/ Documentation and tutorials for SciPy and scientific computing with Python.
- http://scikit-learn.org/ Documentation for Scikit-learn. This site includes very well written and complete narrative documentation describing machine learning in Python on many types of data. See especially the tutorials.
- http://matplotlib.org/ Documentation and tutorials for Matplotlib. See especially the example gallery: it gives a sense of the wide variety of plot types that can be created with Matplotlib.

- **Python and astronomy**

  - http://www.astropython.org/ Python for astronomers. This is a great resource for both beginners and experienced programmers.
  - http://www.astropy.org/ Community library for astronomy-related Python tools. These are the "nuts and bolts" of doing astronomy with Python: tools for reading FITS files and VOTables, for converting between coordinate systems, computing cosmological integrals, and much more.
  - http://www.astroML.org/ Code repository associated with this text. Read more in the following section.

B. AstroML: Machine Learning for Astronomy

B.1. Introduction

AstroML is a Python module for machine learning and data mining built on NumPy, SciPy, Scikit-learn, and Matplotlib, and distributed under an unrestrictive license. It was created to enable the creation of the figures in this text, as well as to serve as a repository for Python astronomy tools, including routines for analyzing astronomical data, loaders for several open astronomical data sets, and a large suite of examples of analyzing and visualizing astronomical data sets.

In order to balance the competing desires to offer easy installation for first-time users and at the same time offer powerful routines written in compilable code, the project is split into two components. The core AstroML library is written in pure Python, and is designed to be very easy to install for any user, even one who does not have a working C or Fortran compiler. A companion library, `astroML_addons`, can be optionally installed for increased performance on certain algorithms. The algorithms in `astroML_addons` duplicate functionality existing in AstroML, but the `astroML_addons` library contains faster and more efficient implementations of the routines. In order to streamline this further, if `astroML_addons` is installed in your Python path, the core AstroML library will import and use the faster routines by default.

For access to the source code and detailed installation instructions, see http://www.astroML.org.

B.2. Dependencies

There are three levels of dependencies in AstroML. **Core** dependencies are required for the core AstroML package. **Add-on** dependencies are required for the performance `astroML_addons`. **Optional** dependencies are required to run some (but not all) of the example scripts. Individual example scripts will list their optional dependencies at the top of the file.

Core dependencies

The core AstroML package requires the following:

- Python: http://python.org version 2.6.x–2.7.x
- NumPy: http://www.numpy.org/ version 1.4+
- SciPy: http://www.scipy.org/ version 0.7+
- Matplotlib: http://matplotlib.org/ version 0.99+

- Scikit-learn: http://scikit-learn.org version 0.10+
- PyFITS: http://www.stsci.edu/institute/software_hardware/pyfits version 3.0+. PyFITS is a Python reader for Flexible Image Transport System (FITS) files, based on the C package cfitsio. Several of the data set loaders require PyFITS.

Additionally, to run unit tests, you will need the Python testing framework nose, version 0.10+. Aside from the newer Scikit-learn version, this configuration of packages was available in 2010, so most current scientific Python installations should meet these requirements (see §A.3.1 for more information).

Add-on dependencies

The performance code in `astroML_addons` requires a working C/C++ compiler.

Optional dependencies

Several of the example scripts require specialized or upgraded packages. These requirements are listed at the top of the example scripts.

- SciPy version 0.11 (released summer 2012) added a sparse graph submodule. The minimum spanning tree example (figure 6.15) requires SciPy version 0.11+.
- PyMC: http://pymc-devs.github.com/pymc/. PyMC provides a nice interface for Markov chain Monte Carlo. Many of the examples in this text use PyMC for exploration of high-dimensional spaces. The examples were written with PyMC version 2.2 (see §5.8.3 for more information).
- HealPy: https://github.com/healpy/healpy. HealPy provides an interface to the HEALPix spherical pixelization scheme, as well as fast spherical harmonic transforms. Several of the spherical visualizations in chapter 1 use HealPy.

B.3. Tools Included in AstroML v0.1

AstroML is being continually developed, so any list of available routines may quickly go out of date. Here is a partial list of submodules and routines that astronomers may find useful; for more details see the online documentation and the code examples throughout the text.

- `astroML.datasets`: Data set loaders; AstroML includes loaders for all of the data sets used throughout the book (see, e.g., §1.5).
- `astroML.stats`: Common statistics not available in SciPy, including robust point statistics such as σ_G (§3.2.2), binned statistics for creating Hess diagrams (§1.6.1), bivariate Gaussian fitting (§3.5.3), and random number generation (§3.7).
- `astroML.density_estimation`: For one-dimensional data, this includes tools for choosing optimal histogram bin widths, including Knuth's and related rules (§4.8.1), and Bayesian blocks methods (§5.7.2). For D-dimensional data, this includes tools for spatial density estimation, including kernel density estimation (§6.1.1), K-nearest-neighbor density estimation (§6.2), and extreme deconvolution (§6.3.3).

- `astroML.fourier`: Tools for approximating continuous Fourier transforms using FFT (§10.2.3, appendix E) and for working with wavelets and wavelet PSDs (§10.2.4).
- `astroML.filters`: Filtering routines, including the Wiener filter, Savitzky–Golay filter, and minimum component filter (§10.2.5).
- `astroML.time_series`: Tools for time series analysis, including generation of power-law and damped random walk light curves (§10.5), computation of Lomb–Scargle periodograms (§10.3.2), multiterm Fourier fits to light curves (§10.3.3), and autocorrelation functions (§10.5).
- `astroML.lumfunc`: Tools for computing luminosity functions, including Lynden-Bell's C^- method (§4.9.1).
- `astroML.resample`: Routines to aid in nonparametric bootstrap and jack-knife resampling (§4.5).
- `astroML.correlation`: Two-point correlation functions in D-dimensional space or on a sphere, based on the ball-tree data structure in Scikit-learn (§6.5).
- `astroML.crossmatch`: Tools for spatial cross-matching of data sets, built on fast tree-based searches.

The goal of AstroML is to be useful to students and researchers beyond those who use this text. For that reason the code, examples, and figures are made freely available online, and anybody can contribute to the package development through GitHub (see http://github.com/astroML/). We hope for it to become an active community repository for fast, well-tested astronomy code implemented in Python.

C. Astronomical Flux Measurements and Magnitudes

A stronomical magnitudes and related concepts often cause grief to the uninitiated. Here is a brief overview of astronomical flux measurements and the definition of astronomical magnitudes.

C.1. The Definition of the Specific Flux

Let us assume that the specific flux of an object *at the top* of the Earth's atmosphere, as a function of wavelength, λ, is $F_\nu(\lambda)$. The specific flux is the product of the photon flux (the number of photons of a given wavelength, or frequency, per unit frequency, passing per second through a unit area) and the energy carried by each photon ($E = h\nu$, where h is the Planck constant, and ν is the photon's frequency). Often, the specific flux per unit wavelength, F_λ, is used, with the relationship $\nu F_\nu = \lambda F_\lambda$ always valid. The SI unit for specific flux is $\mathrm{W\,Hz^{-1}\,m^{-2}}$; in astronomy, a widely adopted unit is Jansky (Jy), equal to $10^{-26}\,\mathrm{W\,Hz^{-1}\,m^{-2}}$ (named after engineer and radio astronomer Karl G. Jansky, who discovered extraterrestrial radio waves in 1933).

C.2. Wavelength Window Function for Astronomical Measurements

Strictly speaking, it is impossible to measure $F_\nu(\lambda)$ because measurements are always made over a finite wavelength window. Instead, the measured quantity is an integral of $F_\nu(\lambda)$, weighted by some window function, $W_b(\lambda)$,

$$F_b = \int_0^\infty F_\nu(\lambda) W_b(\lambda)\, d\lambda, \qquad (\text{C.1})$$

where b stands for "passband," or wavelength "bin." Here, the window function has units of λ^{-1}, and F_b has the same units as $F_\nu(\lambda)$ (the exact form of the window function depends on the nature of the measuring apparatus).

Typically, the window function can be, at least approximately, parametrized by its effective wavelength, λ_e, and some measure of its width, $\Delta\lambda$. Depending on $W_b(\lambda)$, with the most important parameter being the ratio $\Delta\lambda/\lambda_e$, as well as on the number of different wavelength bins b, astronomical measurements are classified as spectra (small $\Delta\lambda/\lambda_e$, say, less than 0.01; large number of bins, say, at least a few dozen) and photometry (large $\Delta\lambda/\lambda_e$, ranging from ~ 0.01 for narrowband photometry

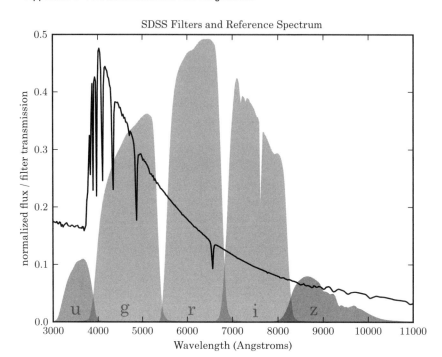

Figure C.1. The five SDSS filters, showing the total transmission taking into account atmospheric transmission and instrumental effects such as CCD efficiency. Shown for reference is the spectrum $(F_\lambda(\lambda))$ of a star similar to Vega (α Lyr), which for many years was used as a reference flux for magnitude calibration. See color plate 12.

to $\sim$ 0.1 and larger for broadband photometry, and typically fewer than a dozen different passbands).

There are many astronomical photometric systems in use, each with its own set of window functions, $W_b(\lambda)$ (which sometimes are not even known!); for more details see [1]. As an example, the SDSS passbands (the displayed total transmission is proportional to $\lambda W_b(\lambda)$) are shown in figure C.1.

C.3. The Astronomical Magnitude Systems

The above discussion of flux measurements is fairly straightforward and it applies to all wavelength regimes, from X-rays to the radio. Enter optical astronomers. For historical reasons (a legacy of Hipparchus and Ptolemy, and a consequence of the logarithmic response of the human eye to light), optical astronomers use a logarithmic scale to express flux,

$$m \equiv -2.5 \log_{10} \left(\frac{F_b}{F_b^0} \right), \tag{C.2}$$

where the normalization flux, F_b^0, depends on the passband b (at least in principle). Note also the factor of 2.5, as well as the minus sign which makes magnitudes increase as the flux decreases (a star at the limit of the human eye's sensitivity, $m = 6$, is 100 times fainter than a star with $m = 1$). Nevertheless, eq. C.2 is a simple mathematical transformation and no harm is done as long as F_b^0 is known. Traditionally, astronomical measurements are calibrated *relative* to an essentially arbitrary celestial source of radiation, the star Vega. Since Vega's flux varies with wavelength, in most photometric systems (meaning a set of $W_b(\lambda)$ and corresponding F_b^0), F_b^0 varies with wavelength as well.

As a result of progress in absolute calibration of astronomical photometric measurements, a simplification of astronomical magnitude systems was proposed by [7]. Oke and Gunn defined the so-called AB magnitudes by $F_b^0 = 3631$ Jy, irrespective of wavelength and particular $W_b(\lambda)$. The SDSS magnitudes, including data sets introduced in §1.5, are reported on AB magnitude scale. For example, the SDSS bright limit at $r = 14$ corresponds to a specific flux of 9 mJy, and its faint limit at $r = 22.5$ corresponds to $3.6\,\mu$Jy.

This query was used to download the data set discussed in §1.5.5. It can be copied verbatim into the SQL window at the SDSS CASJobs site.[1]

```
SELECT
  G.ra, G.dec, S.mjd, S.plate, S.fiberID, --- basic identifiers
  --- basic spectral data
  S.z, S.zErr, S.rChi2, S.velDisp, S.velDispErr,
  --- some useful imaging parameters
  G.extinction_r, G.petroMag_r, G.psfMag_r, G.psfMagErr_r,
  G.modelMag_u, modelMagErr_u, G.modelMag_g, modelMagErr_g,
  G.modelMag_r, modelMagErr_r, G.modelMag_i, modelMagErr_i,
  G.modelMag_z, modelMagErr_z, G.petroR50_r, G.petroR90_r,
  --- line fluxes for BPT diagram and other derived spec.
  parameters
  GSL.nii_6584_flux, GSL.nii_6584_flux_err, GSL.h_alpha_flux,
  GSL.h_alpha_flux_err, GSL.oiii_5007_flux, GSL.oiii_5007_
  flux_err,
  GSL.h_beta_flux, GSL.h_beta_flux_err, GSL.h_delta_flux,
  GSL.h_delta_flux_err, GSX.d4000, GSX.d4000_err, GSE.bptclass,
  GSE.lgm_tot_p50, GSE.sfr_tot_p50, G.objID, GSI.specObjID
INTO mydb.SDSSspecgalsDR8 FROM SpecObj S CROSS APPLY
  dbo.fGetNearestObjEQ(S.ra, S.dec, 0.06) N, Galaxy G,
  GalSpecInfo GSI, GalSpecLine GSL, GalSpecIndx GSX,
  GalSpecExtra GSE
WHERE N.objID = G.objID
  AND GSI.specObjID = S.specObjID
  AND GSL.specObjID = S.specObjID
  AND GSX.specObjID = S.specObjID
  AND GSE.specObjID = S.specObjID
  --- add some quality cuts to get rid of obviously
  bad measurements
  AND (G.petroMag_r > 10 AND G.petroMag_r < 18)
  AND (G.modelMag_u-G.modelMag_r) > 0
  AND (G.modelMag_u-G.modelMag_r) < 6
  AND (modelMag_u > 10 AND modelMag_u < 25)
```

[1] http://casjobs.sdss.org/CasJobs/

```
AND (modelMag_g > 10 AND modelMag_g < 25)
AND (modelMag_r > 10 AND modelMag_r < 25)
AND (modelMag_i > 10 AND modelMag_i < 25)
AND (modelMag_z > 10 AND modelMag_z < 25)
AND S.rChi2 < 2
AND (S.zErr > 0 AND S.zErr < 0.01)
AND S.z > 0.02
```

Note that a large number of different tables is used to get the data, for example, entries starting with GSE. come from a value-added catalog which includes galaxy model fits to observed spectra. Note also the use of CROSS APPLY, and the function fGetNearestObjEQ.

E. Approximating the Fourier Transform with the FFT

The discrete Fourier transform (DFT) is introduced in §10.2.3. The DFT is important in data analysis because it can be quickly computed via the fast Fourier transform (FFT) algorithm, and because it can be used to approximate the continuous Fourier transform. The relationship between the DFT and the continuous version is not always straightforward, however. Consider the definition of the Fourier transform,

$$H(f) = \int_{-\infty}^{\infty} h(t) \exp(-i2\pi ft)\, dt, \tag{E.1}$$

and the discrete Fourier transform,

$$H_k = \sum_{j=0}^{N-1} h_j \exp(-i2\pi jk/N). \tag{E.2}$$

Under what circumstances can $H(f)$ be approximated by H_k? Recall the discussion of the Nyquist sampling theorem in §10.2.3: if $h(t)$ is band limited, such that $H(f) = 0$ for all $|f| > f_c$, then $h(t)$ can be exactly recovered when $\Delta t \leq 1/(2 f_c)$. This is only half the story, however: because of the symmetry in the Fourier transform, there also must be a "band limit" in the time domain as well. In classical periodic analyses, this limit is the longest period of the data. For nonperiodic functions often evaluated with the continuous Fourier transform, this limit means that $h(t)$ must become constant outside the range $(t_0 - T/2) \leq t \leq (t_0 + T/2)$. If this is the case, then the Fourier transform can be written

$$
\begin{aligned}
H(f) &= \int_{t_0-T/2}^{t_0+T/2} h(t) \exp(-i2\pi ft)\, dt \\
&\approx \Delta t \sum_{j=0}^{N-1} h(t_j) \exp(-i2\pi ft_j).
\end{aligned}
\tag{E.3}
$$

In the second line, we have approximated the integral with an N-term Riemann sum, and we have defined

$$t_j \equiv t_0 + \Delta t \left(j - \frac{N}{2} \right); \; \Delta t = \frac{T}{N}. \tag{E.4}$$

If we now define

$$f_k \equiv \Delta f \left(k - \frac{N}{2} \right); \; \Delta f = \frac{1}{T}, \tag{E.5}$$

and put these into the equation, then the terms can be rearranged to show

$$\left[H(f_k)(-1)^k \frac{e^{i2\pi f_k t_0} e^{\pi i N/2}}{\Delta t} \right] = \sum_{j=0}^{N-1} \left[h(t_j)(-1)^j \right] \exp(-i2\pi jk/N). \tag{E.6}$$

Comparing this to the expression for the DFT, we see that the continuous Fourier pair $h(t)$, $H(f)$ correspond to the discrete Fourier pair

$$H_k = H(f_k)(-1)^k \frac{e^{i2\pi f_k t_0} e^{\pi i N/2}}{\Delta t},$$

$$h_j = h(t_j)(-1)^j, \tag{E.7}$$

with f_k and t_j sampled as discussed above. The expression for the inverse transform can be approximated by similar means.

Taking a closer look at the definitions of f_k and t_j, we see that f_k is defined such that $|f_k| \leq N\Delta f/2 = 1/(2\Delta t)$; that is, this sampling perfectly satisfies the Nyquist critical frequency! In this sense, the DFT is an optimal approximation of the continuous Fourier transform for uniformly sampled data.

We should note here one potentially confusing part of this: the ordering of frequencies in the FFT. Eq. E.6 can be reexpressed

$$\left[H(f_k)(-1)^k \frac{e^{i2\pi f_k t_0} e^{\pi i N/2}}{\Delta t} \right] = \sum_{j=0}^{N-1} h(t_j) \exp[-i2\pi j(k - N/2)/N], \tag{E.8}$$

where the phase terms have been absorbed into the indices j and k. Due to aliasing effects, this shifting of indices leads to a reordering of terms in the arrays h_j and H_k. Because of this, the results of a simple Fourier transform of a quantity h_j are ordered such that

$$f_k = \Delta f \, [0, 1, 2, \ldots, N/2, -N/2 + 1, -N/2 + 2, \ldots, -2, -1]. \tag{E.9}$$

Hence, in practice one can take the "raw" Fourier transform of $h(t_j)$, and obtain $H(f_k)$ by shifting the indices. Many FFT implementations even have a tool to do this: SciPy provides this in `scipy.fftpack.fftshift`, and the inverse of this operation, `scipy.fftpack.ifftshift`. Note also that H_k and $H(f_k)$ are related by a phase: if one wishes merely to compute the PSD, then it is sufficient to use

$$|H(f_k)| = \Delta t |H_k|. \tag{E.10}$$

As a quick example, we will use the FFT to approximate the Fourier transform of the sine-Gaussian wavelet discussed in §10.2.4; these functions are nice because their range is limited in both the time and frequency domain. We first define functions which compute the analytic form of the wavelet and its transform, using eqs. 10.16–10.17.

```python
import numpy as np
from scipy import fftpack

def wavelet(t, t0, f0, Q):
    return (np.exp(2j * np.pi * f0 * (t - t0))
            * np.exp(-(Q * f0 * (t - t0)) ** 2))

def wavelet_FT(f, t0, f0, Q):
    return (np.sqrt(np.pi) / Q / f0
            * np.exp(-2j * np.pi * f * t0)
            * np.exp(-(np.pi * (f - f0) / Q / f0)
            ** 2))
```

Next we choose the wavelet parameters and evaluate the wavelet at a grid of times which conforms to the notation we use above:

```python
# choose parameters for the wavelet
N = 10000
t0 = 5
f0 = 2
Q = 0.5

# set up grid of times centered on t0
Dt = 0.01
t = t0 + Dt * (np.arange(N) - N / 2)
h = wavelet(t, t0, f0, Q)
```

Now we can compute the approximation to the continuous Fourier transform via eq. E.8:

```python
# set up appropriate grid of frequencies,
# and compute the fourier transform
Df = 1. / N / Dt
f = Df * (np.arange(N) - N / 2)
H = Dt * (fftpack.fftshift(fftpack.fft(h))
          * (-1) ** np.arange(N)
          * np.exp(-2j * np.pi * t0 * f)
          * np.exp(-1j * np.pi * N / 2))
```

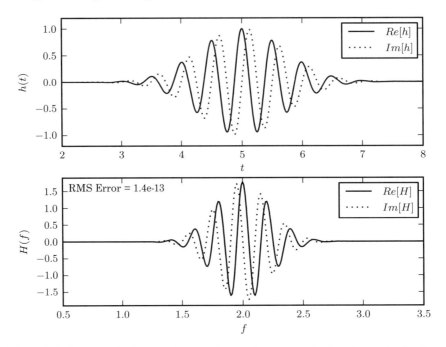

Figure E.1. An example of approximating the continuous Fourier transform of a function using the fast Fourier transform.

Finally, we check our answer against the analytic form, and visualize the results:

```
rms_err = np.sqrt(np.mean(abs(H - wavelet_FT(f, t0,
        f0, Q)) ** 2))

import matplotlib.pyplot as plt
fig = plt.figure()

ax = fig.add_subplot(211)
ax.plot(t, h.real, '-', c='black', label='$Re[h]$')
ax.plot(t, h.imag, ':', c='black', label='$Im[h]$')
ax.legend()

ax.set_xlim(2, 8)
ax.set_ylim(-1.2, 1.2)
ax.set_xlabel('$t$')
ax.set_ylabel('$h(t)$')

ax = fig.add_subplot(212)
ax.plot(f, H.real, '-', c='black', label='$Re[H]$')
ax.plot(f, H.imag, ':', c='black', label='$Im[H]$')
ax.text(0.6, 1.5, "RMS Error = %.2g" % rms_err)
ax.legend()
```

```
ax.set_xlim(0.5, 3.5)
ax.set_ylim(-2, 2)
ax.set_xlabel('$f$')
ax.set_ylabel('$H(f)$')

plt.show()
```

The resulting plots are shown in figure E.1. Note that our approximation to the FFT has an rms error of a few parts in 10^{14}. The reason this approximation is so close is due to our fine sampling in both time and frequency space, and due to the fact that both $h(t)$ and $H(f)$ are sufficiently band limited.

References

[1] Bessell, M. S., (2005). Standard photometric systems, *ARAA 43*, 293–336.

[2] Bressert, E. (2012). *SciPy NumPy: An Overview for Developers*. O'Reilly.

[3] Goldwasser, M. and D. Letscher (2008). *Object-Oriented Programming in Python*. Pearson Prentice Hall.

[4] Langtangen, H. P. (2009). *Python Scripting for Computational Science* (3rd ed.). Springer.

[5] Matsumoto, M. and T. Nishimura (1998). Mersenne twister: A 623-dimensionally equidistributed uniform pseudo-random number generator. *ACM Trans. Model. Comput. Simul. 8*(1), 3–30.

[6] McKinney, W. (2012). *Python for Data Analysis*. O'Reilly.

[7] Oke, J. B. and J. E. Gunn (1983). Secondary standard stars for absolute spectrophotometry. *ApJ 266*, 713–717.

Visual Figure Index

This is a visual listing of the figures within the book. The first number below each thumbnail gives the figure number within the text; the second (in parentheses) gives the page number on which the figure can be found.

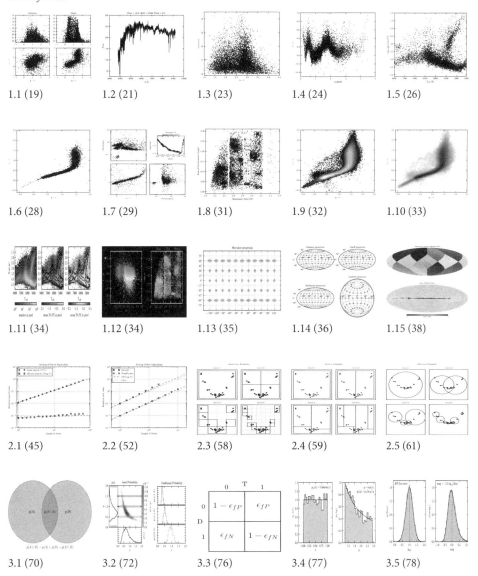

1.1 (19)   1.2 (21)   1.3 (23)   1.4 (24)   1.5 (26)

1.6 (28)   1.7 (29)   1.8 (31)   1.9 (32)   1.10 (33)

1.11 (34)   1.12 (34)   1.13 (35)   1.14 (36)   1.15 (38)

2.1 (45)   2.2 (52)   2.3 (58)   2.4 (59)   2.5 (61)

3.1 (70)   3.2 (72)   3.3 (76)   3.4 (77)   3.5 (78)

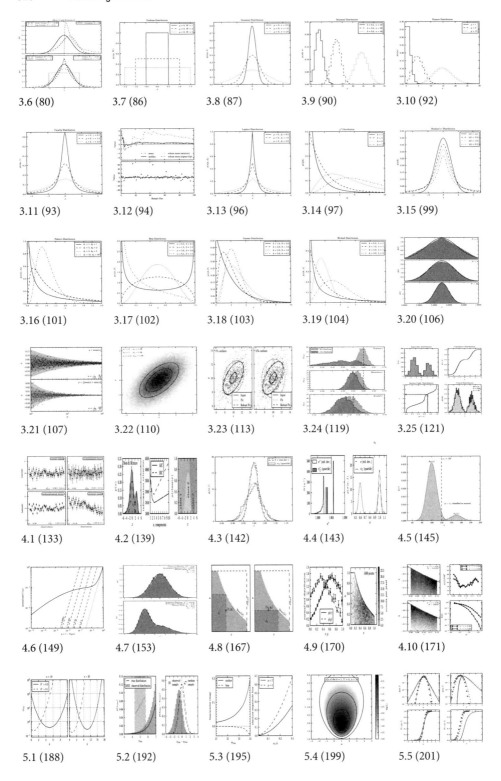

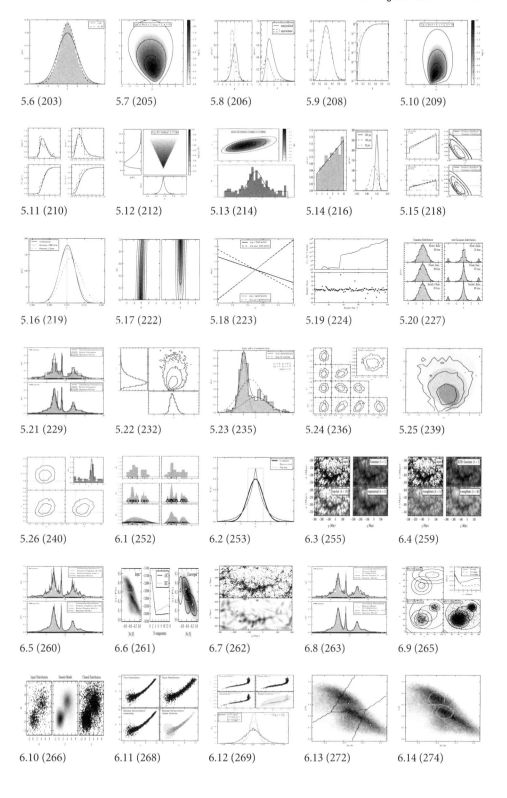

5.6 (203)   5.7 (205)   5.8 (206)   5.9 (208)   5.10 (209)

5.11 (210)   5.12 (212)   5.13 (214)   5.14 (216)   5.15 (218)

5.16 (219)   5.17 (222)   5.18 (223)   5.19 (224)   5.20 (227)

5.21 (229)   5.22 (232)   5.23 (235)   5.24 (236)   5.25 (239)

5.26 (240)   6.1 (252)   6.2 (253)   6.3 (255)   6.4 (259)

6.5 (260)   6.6 (261)   6.7 (262)   6.8 (263)   6.9 (265)

6.10 (266)   6.11 (268)   6.12 (269)   6.13 (272)   6.14 (274)

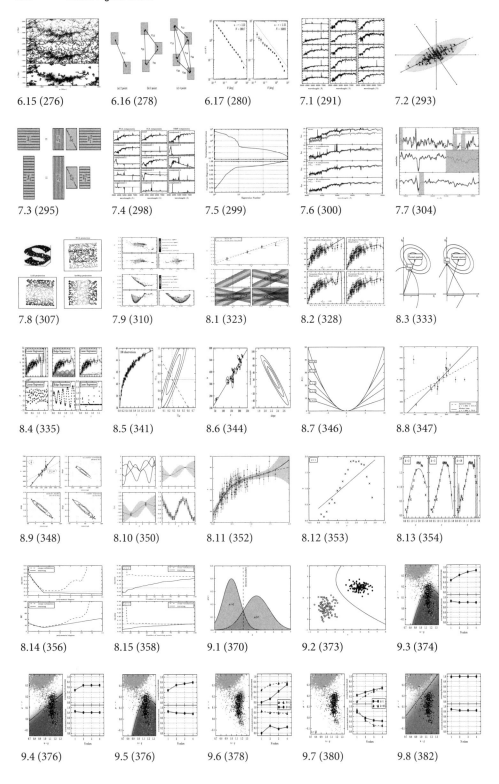

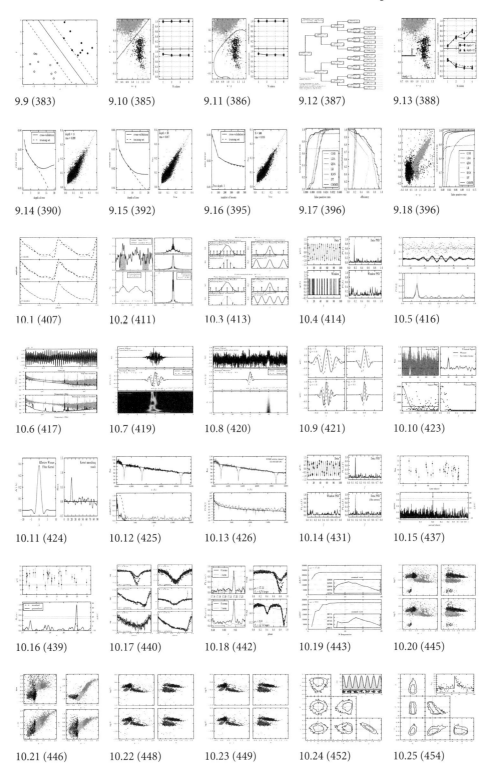

9.9 (383) 9.10 (385) 9.11 (386) 9.12 (387) 9.13 (388)

9.14 (390) 9.15 (392) 9.16 (395) 9.17 (396) 9.18 (396)

10.1 (407) 10.2 (411) 10.3 (413) 10.4 (414) 10.5 (416)

10.6 (417) 10.7 (419) 10.8 (420) 10.9 (421) 10.10 (423)

10.11 (424) 10.12 (425) 10.13 (426) 10.14 (431) 10.15 (437)

10.16 (439) 10.17 (440) 10.18 (442) 10.19 (443) 10.20 (445)

10.21 (446) 10.22 (448) 10.23 (449) 10.24 (452) 10.25 (454)

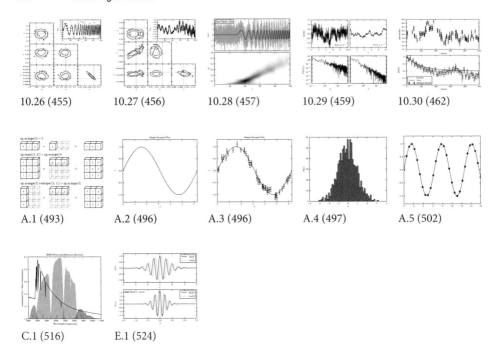

Index